Lecture Notes in Civil Engineering

Volume 339

Lecture Notes in Civil Engineering (LNCE) publishes the latest developments in Civil Engineering—quickly, informally and in top quality. Though original research reported in proceedings and post-proceedings represents the core of LNCE, edited volumes of exceptionally high quality and interest may also be considered for publication. Volumes published in LNCE embrace all aspects and subfields of, as well as new challenges in, Civil Engineering. Topics in the series include:

- Construction and Structural Mechanics
- Building Materials
- Concrete, Steel and Timber Structures
- Geotechnical Engineering
- Earthquake Engineering
- Coastal Engineering
- Ocean and Offshore Engineering; Ships and Floating Structures
- Hydraulics, Hydrology and Water Resources Engineering
- Environmental Engineering and Sustainability
- Structural Health and Monitoring
- Surveying and Geographical Information Systems
- Indoor Environments
- Transportation and Traffic
- Risk Analysis
- Safety and Security

To submit a proposal or request further information, please contact the appropriate Springer Editor:

- Pierpaolo Riva at pierpaolo.riva@springer.com (Europe and Americas);
- Swati Meherishi at swati.meherishi@springer.com (Asia—except China, Australia, and New Zealand);
- Wayne Hu at wayne.hu@springer.com (China).

All books in the series now indexed by Scopus and EI Compendex database!

P. V. Timbadiya · P. L. Patel · Vijay P. Singh · A. B. Mirajkar
Editors

Geospatial and Soft Computing Techniques

Proceedings of 26th International Conference on Hydraulics, Water Resources and Coastal Engineering (HYDRO 2021)

Springer

Editors
P. V. Timbadiya
Department of Civil Engineering
Sardar Vallabhbhai National Institute of Technology
Surat, India

P. L. Patel
Department of Civil Engineering
Sardar Vallabhbhai National Institute of Technology
Surat, India

Vijay P. Singh
Department of Biological and Agricultural Engineering, Zachry Department of Civil and Environmental Engineering
Texas A&M University
College Station, TX, USA

A. B. Mirajkar
Department of Civil Engineering
Visvesvaraya National Institute of Technology
Nagpur, India

ISSN 2366-2557 ISSN 2366-2565 (electronic)
Lecture Notes in Civil Engineering
ISBN 978-981-99-1900-0 ISBN 978-981-99-1901-7 (eBook)
https://doi.org/10.1007/978-981-99-1901-7

This Springer imprint is published by the registered company Springer Nature Singapore Pte Ltd.
The registered company address is: 152 Beach Road, #21-01/04 Gateway East, Singapore 189721, Singapore

Preface

Geospatial and Soft Computing Techniques in water resources management play a key role in creating the thematic maps and their analysis for extraction of useful data for the area of interest. Satellite images are used for identification of droughts, floods, sedimentation, land use and land cover on spatial and temporal scales. The change in land use and land cover leads to change in the characteristics of a river basin. The change detection is important for further planning and management of available resources. The Geographical Information System (GIS) and Remote Sensing (RS) are important tools and data for the assessment of reservoir sedimentation, impact assessment of water conservation planning, climate change impact while analysing various parameters and their spatiotemporal variations, waterlogging risk assessment, groundwater potential zone identification and many more.

This book covers the broader theme of the Geospatial and Soft Computing Techniques while dealing with several sub-topics such as satellite derived data for hydraulic applications, usage of Global Positioning System (GPS) in Water Resources Engineering applications, GIS and RS applications in water resources management. Several other sub-topics include hydro-informatics, rainfall and stream-flow prediction, optimization of water resources systems and data-driven and artificial intelligent-based hydrological modelling. Exploring Geospatial Technology in a basin for demarcation of groundwater potential zones and identification of recharge structures, suitable artificial intelligence techniques for multispectral image classification, application of random forest and model tree for discharge and water level estimation and prediction, comparison of spatial interpolation methods for mapping seasonal groundwater levels, water quality estimation using Remote Sensing technique through a case study, GIS and RS applications in water resources management in consumptive use with crop assessment. Comparison of a long short-term memory model with statistical-based water demand prediction models for a case study is also presented. This book also includes analysis of water distribution network for case studies of city/village by using LOOP 4.0 and/or EPANET Software, artificial intelligence-based reference evapotranspiration modelling with minimum climatic parameters and least cost path pipeline routing using spatial multi-criteria analysis for a region.

In addition, a review study on the techniques employed in prediction of northwest monsoon rainfall for peninsular region of India is also included in this book. This book will help the readers to gain an overview of the Geospatial and Soft Computing applications in water resources along with water distribution network optimization, leak detection and management through various case studies and review articles.

Surat, India — P. V. Timbadiya
Surat, India — P. L. Patel
College Station, TX, USA — Vijay P. Singh
Nagpur, India — A. B. Mirajkar

Acknowledgements

The editors are grateful for the support provided by the technical advisory committee and local organizing committee of the 26th International Conference on Hydraulics, Water Resources and Coastal Engineering (HYDRO 2021) held at Sardar Vallabhbhai National Institute of Technology (SVNIT), Surat, during December 23–25, 2021. The editors thank the Indian Society for Hydraulics (ISH) Pune, India, its office bearers and executive council members for their support in conducting the HYDRO 2021 International Conference. The editors wish to thank all the authors for their support and contribution to this book. The editors duly acknowledge the timely and sincere efforts of the reviewers in providing their valuable comments and suggestions to maintain the quality of the book. The editors would like to thank the keynote speakers, the session chairs and co-chairs, participants, and student volunteers for their contribution to the successful conduct of the conference. The editors are also thankful to the administrators of Sardar Vallabhbhai National Institute of Technology, Surat (SVNIT), India, for supporting the HYDRO 2021 International Conference. Lastly, the editors are sincerely thankful to the publishing team of Springer Nature for their support and cooperation at various steps since the beginning of the book project.

P. V. Timbadiya
P. L. Patel
Vijay P. Singh
A. B. Mirajkar

Contents

About the Editors

P. V. Timbadiya is an Associate Professor in the Water Resources Engineering section, Department of Civil Engineering, Sardar Vallabhbhai National Institute of Technology (SVNIT), Surat, India. He secured his doctoral degree and post-graduation in Water Resources Engineering from SVNIT Surat in 2012 and 2004, respectively. He did his under graduation in Civil Engineering from Sardar Vallabhbhai Regional College of Engineering Technology (now SVNIT). He has guided three Doctoral Thesis and 32 Master's Dissertations. He has more than 110 research papers to his credit, including 31 articles in peer-reviewed journals. He served as Dean (Alumni and Resources Generation), and currently serving as Sectional Head, Water Resources Engineering Section at SVNIT. He played an instrumental role in setting up infrastructure facilities in the Centre of Excellence on 'Water Resources and Flood Management' such as the Experimental Hydraulics Lab, Computational Hydraulics Lab, Water Circulation System, and others. He is appointed as 'National Consultant' for Kalpsar Project by Narmada, Water Resources, Water Supply and Kalpsar Department of Govt. of Gujarat. He is a recipient of 'Best Case Study Award—2015' award by the American Society of Civil Engineers for his publication in the Journal of Hydrologic Engineering and 'Young Engineers Award' by the Institution of Engineers (India) in the year 2015. He received 'Prof. R. J. Garde Research Award' for the year 2020 by the Indian Society for Hydraulics. He has awarded DST-SERB Core Research Grant for the project on 'Local Scouring around tandem and staggered bridge piers on Non-uniform mobile bed' in the year 2021. He is active in various professional bodies and organized numerous conferences, workshops, and short-term training programmes in his academic career.

P. L. Patel is a Professor of Hydraulics and Water Resources in the Department of Civil Engineering, Sardar Vallabhbhai National Institute of Technology (SVNIT), Surat, India. He served as the Deputy Director of SVNIT. He also worked as a Reader in the Civil Engineering Department at Delhi College of Engineering (now DTU) from 1999–2007. He served as an Assistant Executive Engineer in Border Roads Organization (BRO) from 1995–1999. He did his bachelor's in Civil Engineering from Government Engineering College, Rewa, Madhya Pradesh, India and then pursued his Master's and Doctoral degrees in Civil Engineering from the then University of Roorkee, now Indian Institute of Technology Roorkee, India. He has published more than 220 papers in peer-reviewed journals and conferences of repute. He has guided 12 Doctoral Thesis and 47 Master's Dissertation so far. He has also served in various academic positions in SVNIT Surat, such as Dean (Academics), Head of Civil Engineering Department, Dean (Research and Consultancy), Dean (PG), etc. He was also instrumental in setting up a Centre of Excellence (CoE) on 'Water Resources and Flood Management' in the Institute through a research grant from World Bank-sponsored TEQIP-II. He is a recipient of visiting International Fellowship (VIF-2017) 2017 for attending the ASCE EWRI Congress-2017 in Sacramento, California, USA. He is active in various professional bodies and organized numerous conferences, workshops, and short-term training programmes in his academic career.

Vijay P. Singh is a University Distinguished Professor, a Regents Professor, and Caroline and William N. Lehrer Distinguished Chair in Water Engineering at Texas A&M University. He received his B.S., M.S., Ph.D., and D.Sc. in engineering. He is a registered professional engineer, a registered professional hydrologist, and an Honorary Diplomate of ASCE-AAWRE. He is a Distinguished Member of ASCE, a Distinguished Fellow of AGGS, an Honorary Member of AWRA, and a Fellow of EWRI-ASCE, IAH, ISAE, IWRS, and IASWC. He has published extensively in the areas of hydrology, irrigation engineering, hydraulics, groundwater, water quality, and water resources (more than 1320 journal articles, 31 textbooks, 75 edited reference books, 110 book chapters, and 315 conference papers). He has received over 95 national and international awards, including three honorary doctorates. He is a member of 11 international science/engineering academies. He has served as President of the American Institute of Hydrology (AIH), Chair of the Watershed Council of the American Society of Civil Engineers and is currently President of the American Academy of Water Resources Engineers. He has served/serves as Editor-in-Chief of three journals and two book series and serves on editorial boards of more than 25 journals and three book series. His Google Scholar citations include 64073, h-index: 115, and I10-index: 903.

A. B. Mirajkar is presently working as an Associate Professor in the Department of Civil Engineering, Visvesvaraya National Institute of Technology (VNIT) Nagpur-Maharashtra, since January 2015. She previously worked as an Associate Professor at DIEMS Aurangabad in 2014. She also worked as a lecturer in SVNIT Surat, Gujarat from 2011–2013. She completed her Ph.D. in Civil Engineering with a specialization in Water Resources Engineering from Sardar Vallabhbhai National Institute of Technology (SVNIT) Surat in January 2014. She is a recipient of DST research Grant and also has G. M. Nawathe Award for best presentation in HYDRO 2012 and was awarded during HYDRO 2013 at IIT Chennai by Indian Society for Hydraulics. She has guided 03 doctoral thesis and 24 M.Tech. thesis. She has authored more than 33 research publications, including peer-reviewed good impact factor journals and submitted one book chapter. Her research interests are reservoir planning and operations, irrigation scheduling, watershed management, rainfall trend analysis and climate change impact assessment on water resources.

Drought Monitoring Using Satellite Soil Moisture Data Over Godavari Basin, India

Hussain Palagiri, Manali Pal, and Rajib Maity

Abstract Advancement in remote sensing (RS) technology has made the availability of large scale, fine resolution (both spatial and temporal) soil moisture (SM) dataset, which is extremely useful in agricultural drought monitoring. In this study, temporal evolution of drought status is assessed for the Godavari basin in India, through two drought indices, namely Soil Water Deficit Index (SWDI) and Soil Moisture Deficit Index (SMDI) during 2015–2020. The SWDI is derived using the daily Soil Moisture Active Passive (SMAP) enhanced Level-3 (L3) surface SM product along with the information of field capacity (FC) and wilting point (WP) derived from soil physical and chemical characteristics. The SMDI is computed using the Global Land Data Assimilation System (GLDAS) long-term SM product using a temporally incremental basis. A comparison between the SWDI and SMDI shows a good agreement for the basin as both of them are able to capture different dry and wet soil conditions. The drought indices are also able to capture the seasonal and inter-annual variability. The two SM products are validated using the fifth-generation ECMWF reanalysis (ERA5) dataset which provides a consistent view of the evolution of SM dynamics over several decades. The validation shows a reliable accuracy of both SMAP and GLDAS-SM products.

Disclaimer: The presentation of material and details in maps used in this chapter does not imply the expression of any opinion whatsoever on the part of the Publisher or Author concerning the legal status of any country, area or territory or of its authorities, or concerning the delimitation of its borders. The depiction and use of boundaries, geographic names and related data shown on maps and included in lists, tables, documents, and databases in this chapter are not warranted to be error free nor do they necessarily imply official endorsement or acceptance by the Publisher or Author.

H. Palagiri (✉) · M. Pal
Department of Civil Engineering, National Institute of Technology Warangal, Warangal 506004, India
e-mail: hussain96p@gmail.com

M. Pal
e-mail: manalipal@nitw.ac.in

R. Maity
Department of Civil Engineering, Indian Institute of Technology Kharagpur, Kharagpur 721302, India
e-mail: rajib@civil.iitkgp.ac.in

P. V. Timbadiya et al. (eds.), *Geospatial and Soft Computing Techniques*, Lecture Notes in Civil Engineering 339, https://doi.org/10.1007/978-981-99-1901-7_1

Keywords Agricultural drought · Soil moisture active passive (SMAP) · Global land data assimilation system (GLDAS) · Soil water deficit index (SWDI) · Soil moisture deficit index (SMDI)

1 Introduction

Agricultural droughts have the most significant adverse socio-economic impacts, especially in agriculture-based zones such as India. It is defined as a period of decreasing soil moisture (SM) resulting in failure of crops [1]. It is considered to begin when SM availability reaches such a low level that it negatively affects crop yield [2]. Hence, this vital role of SM in agricultural production suggests that drought indices based on SM is crucial for accurate and reliable agricultural drought monitoring. Generally, India with a diverse agro-climatic zones experiences agricultural drought almost in every season in some parts of the country [3]. Precipitation-based indices are used for monitoring agricultural drought in many parts of India due to the wide availability of meteorological databases [4]. However, the processes involved in the evolution of droughts are complex and varies for different agro-ecosystems that depend on the crop type, location and season [5]. Hence, the indices based on SM are more informative than the precipitation-based indices that typically do not consider site-specific soil properties. However, the extensive spatio-temporal variation of SM imparts difficulties in its measurements in large scale and longer time period applications. This leads to the scarcity in consistent and long-term time series of soil moisture data required to derive the drought indices and hence prevented their operational uses till date in India.

The recent advancement of several space-borne systems, especially using the Microwave (MW) sensors, providing SM data with different spatial and temporal resolutions, has made the SM data available in globally. Recently, National Aeronautics and Space Administration (NASA) launched a soil Moisture Active Passive (SMAP) satellite in January 2015 carrying sensors of an active L-band radar 3 km (1.26 GHz) and a passive L-band radiometer 36 km (1.41 GHz). The enhanced SMAP Level-3 (L3) provides a daily global surface SM at a spatial resolution of 9 km [6]. The drought indices based on SMAP soil moisture product, henceforth SMAP-SM, are established to well indicate the agricultural droughts in various parts of globe. Mishra et al. [7] developed Soil Water Deficit Index (SWDI) based on SMAP-SM product and SM information at field capacity and available water content to monitor drought conditions over the Contiguous United States. The study compared the performance of SMAP derived SWDI with the Atmospheric Water Deficit (AWD) index that is based on precipitation values. The comparison showed that SWDI provides similar drought information as AWD, making it a good agricultural drought indicator. Fang et al. [8] used SWDI and Soil Moisture Deficit Index (SMDI) to assess drought over Australia. The SWDI was developed using SMAP-SM data while the SMDI was developed by integrating GLDAS and SMAP soil moisture using a temporally incremental-based method. The derived SWDI and SMDI products can demonstrate the different patterns of dry/wet seasons in different climate zones of Australia. However, in India, the satellite soil moisture-based drought indices to study the agricultural droughts are still not used or very limited. Various studies in India made use of vegetation indices retrieved from remote sensing methods, which include Normalized Difference Vegetation Index (NDVI), Vegetation Condition Index (VCI), and Vegetation Temperature Condition Index (VTCI) to examine the trends in agricultural droughts [4, 9–11]. Chattopadhyay et al. [12] and Kulkarni et al. [13] developed combined drought indices by weighted analysis of meteorological, land-based observations and remote sensing rainfall/vegetation-based indices for agricultural drought assessment.

Thus, the successful use of these satellite SM-based drought indices over several places and the deficiency of the same in our country have formed the motivation of this study. The study attempts to evaluate the potential applicability of SMAP-SM-based drought indices over the drought-prone Godavari basin region for agricultural drought monitoring. The main objectives of this study are (1) to derive SMAP-SM-based SWDI for the Godavari basin; (2) to compare SWDI with SMDI derived from GLDAS long-term soil moisture product.

2 Study Area and Data Used

2.1 Godavari Basin

The Godavari basin (shown in Fig. 1) is considered as the study basin which lies in southern peninsular India and covering an area of 3,12,812 km^2. The major portion of basin's rainfall occurs in south-west monsoon (July to September) which is unpredictable with wide temporal and spatial variation in rainfall. The basin has an average annual rainfall of 1100 mm. South-west monsoon has a direct influence on agriculture, which is more vulnerable to extreme weather events [14].

2.2 Data Used

2.2.1 SMAP Data

The NASA launched the SMAP satellite mission in January 2015 to retrieve global soil moisture content in the top 5 cm of soil. SMAP-SM products are available in different levels with L2 products defined for half orbit, L3 products for daily composites and L4 products for model assimilation. The SMAP enhanced L3 SM product provides surface SM in m^3/m^3 at 9 km resolution with one day temporal resolution [6]. Backus-Gilbert optimal interpolation techniques are used for this enhanced L3 product for extracting information from the SMAP antenna temperatures and

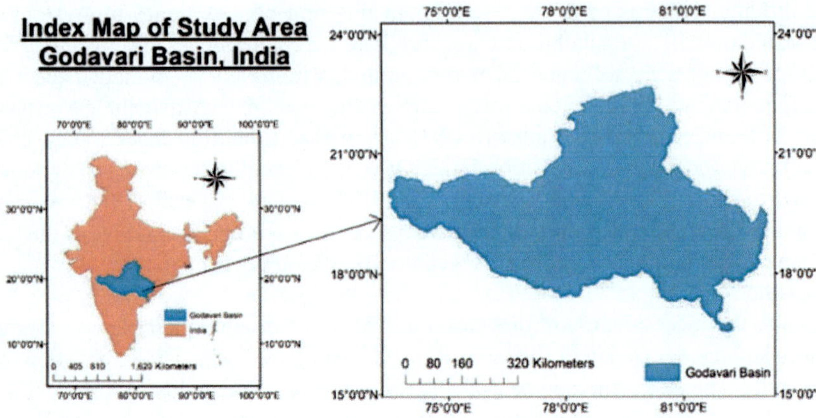

Fig. 1 Study area map

converting it to brightness temperatures. This study uses enhanced SMAP Level-3 (L3) data to retrieve soil moisture data over the Godavari basin during 2016–2020 ([15], https://lpdaacsvc.cr.usgs.gov/appeears/).

2.2.2 GLDAS Data

Global Land Data Assimilation System (GLDAS) was developed by scientists at the NASA, Goddard Space Flight Center (GSFC) and the National Oceanic and Atmospheric Administration (NOAA), National Centers for Environmental Prediction (NCEP) for generating optimal fields of land surface states and fluxes [16]. In this study, the soil moisture estimates for period 1989–2000 were obtained from 'GLDAS_NOAH025_M v2.0' and for period 2001–2020 were obtained from 'GLDAS_NOAH025_M v2.1' [17] product which delivers data with 0.25° × 0.25° spatial resolution and monthly temporal resolution at 0–10 cm layer (https://ldas.gsfc.nasa.gov/gldas/model-output).

2.2.3 HWSD Database

The Harmonized World Soil Database (HWSD) is a 30 arc-second raster with over 15,000 different soil mapping units. It was made by combining Soil and Terrain (SOTER) databases, European Soil Database (ESDB), Soil Map of China, International Soil Reference and Information Centre (ISRIC)'s World Inventory of Soil Emission Potentials (WISE), (ISRIC-WISE) with the information contained within the 1:5,000,000 scale FAO-UNESCO Soil Map of the World [18, 19]. The spatial data layers of these four databases were used as input for the GIS coverage of the HWSD.

The HWSD database provides global information of soil physical and chemical properties at 0–30 cm (topsoil) depth and at 30–100 cm (subsoil) depth. In this study, topsoil sand fraction (% weight), topsoil clay fraction (% weight), topsoil organic carbon (% weight) of the Godavari basin are extracted from the HWSD database ([20], https://data.apps.fao.org/map/catalog/).

2.2.4 ECMWF Reanalysis (ERA5) Dataset

The fifth-generation global atmospheric reanalysis data of the European Centre for Medium-Range Weather Forecasts (ECMWF), henceforth ERA5, is the latest generation created by the Copernicus Climate Change Service (C3S) [21]. The monthly ERA5-Land product includes soil moisture at four different layers at a spatial resolution of $0.1° \times 0.1°$. The first (0–7 cm) layer of soil moisture data from 2016 to 2020 (https://cds.climate.copernicus.eu/ accessed in September 2021) are used to validate the SMAP and GLDAS-SM products for this study.

3 Methodology

3.1 Soil Water Deficit Index (SWDI)

The SWDI for the Godavari Basin is computed using SMAP-SM estimates and soil properties obtained from HWSD database. The SWDI is calculated at original HWSD resolution, whereas SMAP is downscaled to HWSD resolution using nearest neighbor interpolation approach. SWDI is calculated using the following equation

$$\mathrm{SWDI} = \left(\frac{\theta - \theta_{\mathrm{FC}}}{\theta_{\mathrm{AWC}}}\right) \times 10 \tag{1}$$

where θ, θ_{FC} and θ_{AWC} denote SMAP-SM estimates, SM at field capacity, and SM at available water capacity, respectively. The available water capacity is calculated as the difference between soil moisture at field capacity (θ_{FC}) and wilting point (θ_{WP}) as following:

$$\theta_{\mathrm{AWC}} = \theta_{\mathrm{FC}} - \theta_{\mathrm{WP}} \tag{2}$$

The SM values at field capacity and wilting point are determined using Pedotransfer functions (PTFs) that express the relations between θ_{FC}, θ_{WP} and soil physical and chemical characteristics, such as soil texture (% of sand, silt, and clay), organic matter, and bulk density by the set of following equations:

$$\theta_{\mathrm{WP}} = \theta^*_{\mathrm{WP}} + \left(0.14 \times \theta^*_{\mathrm{WP}} - 0.02\right) \tag{3}$$

Table 1 Classification of SWDI for different drought classes [8]

SWDI	Drought class
≥ 0	No drought
− 2 to 0	Mild
− 5 to − 2	Moderate
− 10 to − 5	Severe
≤ − 10	Extreme

$$\theta^*_{\mathrm{WP}} = -0.024S + 0.487C + 0.006\mathrm{OM} + 0.005(S \times \mathrm{OM}) - 0.013(C \times \mathrm{OM}) + 0.068(S \times C) + 0.031$$

$$\theta_{\mathrm{FC}} = \theta^*_{\mathrm{FC}} + [1.283\left(\theta^*_{\mathrm{FC}}\right)^2 - 0.374\theta_{\mathrm{FC}} - 0.015 \tag{4}$$

$$\theta^*_{\mathrm{FC}} = -0.251S + 0.195C + 0.011\mathrm{OM} + 0.006(S \times \mathrm{OM}) - 0.027(C \times \mathrm{OM}) + 0.452(S \times C) + 0.29$$

where *S*, *C*, and OM in above set equations are the percentages of sand, clay, and organic matter derived from upgraded PTFs [22]. The OM is obtained by organic carbon (obtained from HWSD) divided by a factor of 0.58 [7, 8]. The positive values of SWDI indicate SM is higher than field capacity and the negative SWDI values indicate drought conditions. The classification of droughts based on SWDI values is provided in Table 1.

3.2 Soil Moisture Deficit Index (SMDI)

Developed by Narasimhan and Srinivasan [23], SMDI is useful for identifying and monitoring drought affecting agriculture for different crop types and seasons. The SMDI makes use of long-term SM estimates and SM estimates of the period for which drought conditions would be evaluated. Monthly GLDAS-SM data over a 32-year period from 1989 to 2020 are obtained at 0–10 cm layer to derive median, maximum, and minimum of the SM values for each month. As the median is more stable and is not influenced by few outliers, it is chosen over the mean as a measure of 'normal' [23]. An assumption is made that all the SM estimates within any GLDAS grid could be summarized by the long-term SM records of this GLDAS grid [8]. Using these long-term SM metrics and SMAP data, soil moisture deficit for each month over a period of 32 years (1989–2020) is calculated as

$$\mathrm{SD}_j = \frac{\mathrm{SW}_j - \mathrm{MSW}_j}{\mathrm{MSW}_j - \min \mathrm{SW}_j} \times 100 \text{ if } \mathrm{SW}_j \geq \mathrm{MSW}_j \tag{5}$$

where SD_j is the percentage of soil water deficit; SW_j is the SMAP soil moisture estimate of the year/season for which the drought would be evaluated; MSW_j and $\min SW_j$ are the median and minimum values of GLDAS soil moisture estimates at any given month j over the period of 1989 to 2020.

The drought index is computed on a temporally incremental basis [24] as it is calculated from the past year SMDI and current year SD_j. SMDI during any period will range from − 4 to + 4 representing dry to wet conditions and for any month j can be calculated as

$$SMDI_j = 0.5 \times SMDI_{j-1} + \frac{SD_j}{50} \tag{6}$$

where $SMDI_{j-1}$ is the SMDI from the previous year/season. However, the SMDI for the first year of the total time period considered, i.e., 2016 for our study is calculated as following:

$$SMDI_1 = \frac{SD_1}{50} \tag{7}$$

4 Results and Discussions

4.1 Comparison of SMAP, GLDAS, and ERA5 SM Product

Figure 2 compares the monthly basin-averaged SMAP-SM and GLDAS-SM time series with the ERA5 SM product for the study period, i.e., from 2016 to 2020. It is to mention that for the selected study basin the ground observation for SM is not available. Thus, the two SM products used to develop the drought indices are compared with the ERA5 reanalysis data, which provides a consistent view of the evolution of SM dynamics over several decades. The comparison of three SM data sets showed that all of them are able to explicitly show seasonal and inter-annual variations for the time period considered. The average SM values of the SMAP, GLDAS, and ERA5 SM time series are 0.26, 0.24, and 0.29 m^3/m^3, respectively that indicates little difference of the values. The difference is approximately 0.05 m^3/m^3 except for the July to September (JAS) season, when the difference goes up to 0.1 m^3/m^3. Additionally, it is found that the SMAP-SM values are slightly higher than GLDAS-SM and had higher variability for most of the months. Also, two evaluation criteria namely root mean square error (RMSE) and Nash–Sutcliffe efficiency coefficient (NSE) are computed between the ERA5 (as observed) and SMAP and GLDAS-SM data. The RMSE and NSE values of monthly basin-averaged SM for 2016 to 2020 are 0.042 and 0.819, respectively, between SMAP and ERA5 data. The same metrics are found to be 0.057 and 0.669 when GLDAS data is compared to the ERA5 data. Both the metrics show an acceptable agreement among the three soil moisture data.

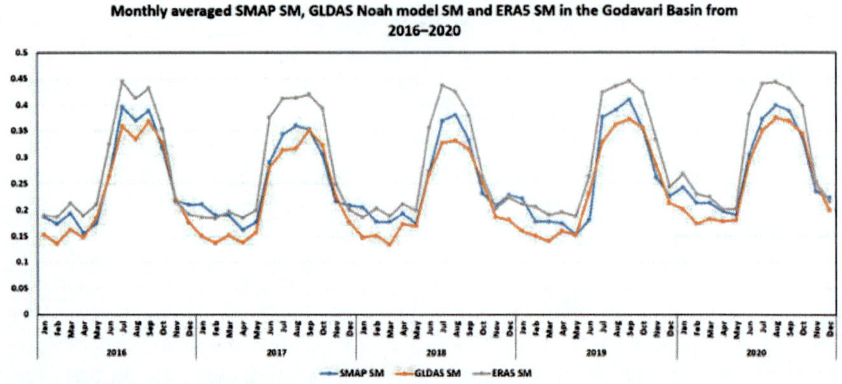

Fig. 2 Monthly basin-averaged time series plot of SMAP, GLDAS, and ERA5 SM data for the Godavari Basin during the period of 2016 to 2020

4.2 Interpretation of SWDI and SMDI Distribution

Figures 3 and 4 show the yearly averaged and seasonally averaged SWDI and SMDI maps for the study basin during the years 2016 to 2020, respectively. Figure 3 indicates the negative/positive values of SWDI over the study basin during the study period that may be interpreted as deficit/surplus of rainfall. The extreme drought conditions shown by the SWDI values (nearly − 25) from 2016 to 2020 at the western side of the study basin indicates lack of rainfall during that period. The eastern part of the basin and lower Godavari sub basins experienced moderate to extreme wet conditions for the same period (SWDI > ~ 50). The spatial distribution of SMDI values also indicate the same pattern, i.e., the dry western part of the basin indicated by negative SMDI values (< − 2) and the wet eastern zone as shown by the positive SMDI values for the period of 2016 to 2020. A pixel by pixel visual analysis shows that areas where SWDI values indicate mild to extreme drought/wet conditions, SMDI values also represent dry/wet conditions.

From Fig. 4, it can be noted that both SWDI/SMDI maps are able to capture the seasonal variability of drought condition throughout the basin. It is observed that in monsoon seasons (JAS) most of the drought stricken parts are moderate to severely dry, while winter/spring/summer seasons (October to June) are extremely dry in the western parts of the study basin for the year of 2020. The same pattern has been observed for the years of 2016 to 2019 and hence, not presented in the paper to avoid redundancy. During the monsoon seasons, the regions had the greatest wetting trend and the SWDI and SMDI values reached the highest peak of the year, i.e., SMDI of 6.03 and SWDI of 85.45. Similarly, the lowest values of these SWDI and SMDI were attained during JAS season where the minimum values of SWDI and SMDI are − 12.90 and − 1.33, respectively.

The comparison between two indices shows that in case of SWDI values, there is hardly any seasonal variation in the eastern part of the basin throughout the entire year

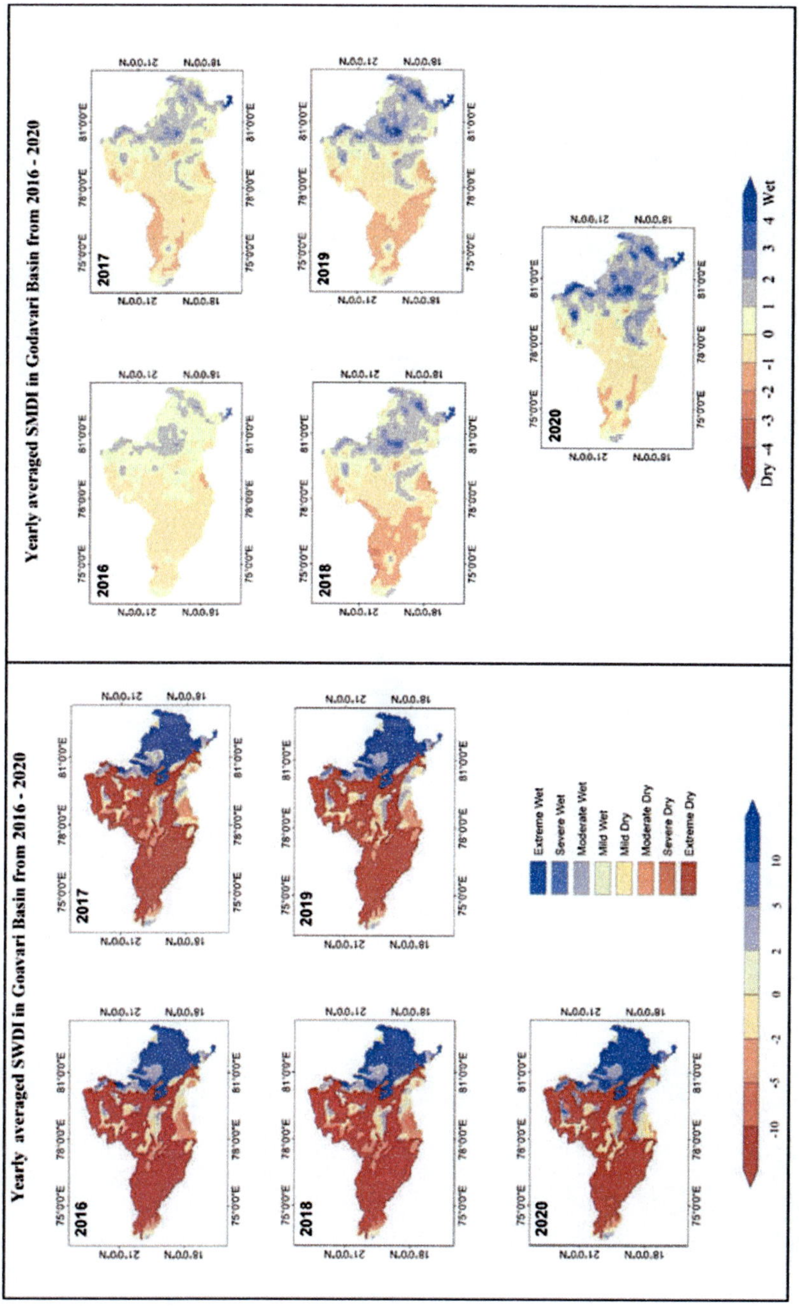

Fig. 3 Yearly averaged spatial maps of SWDI and SMDI across Godavari Basin from 2016 to 2020

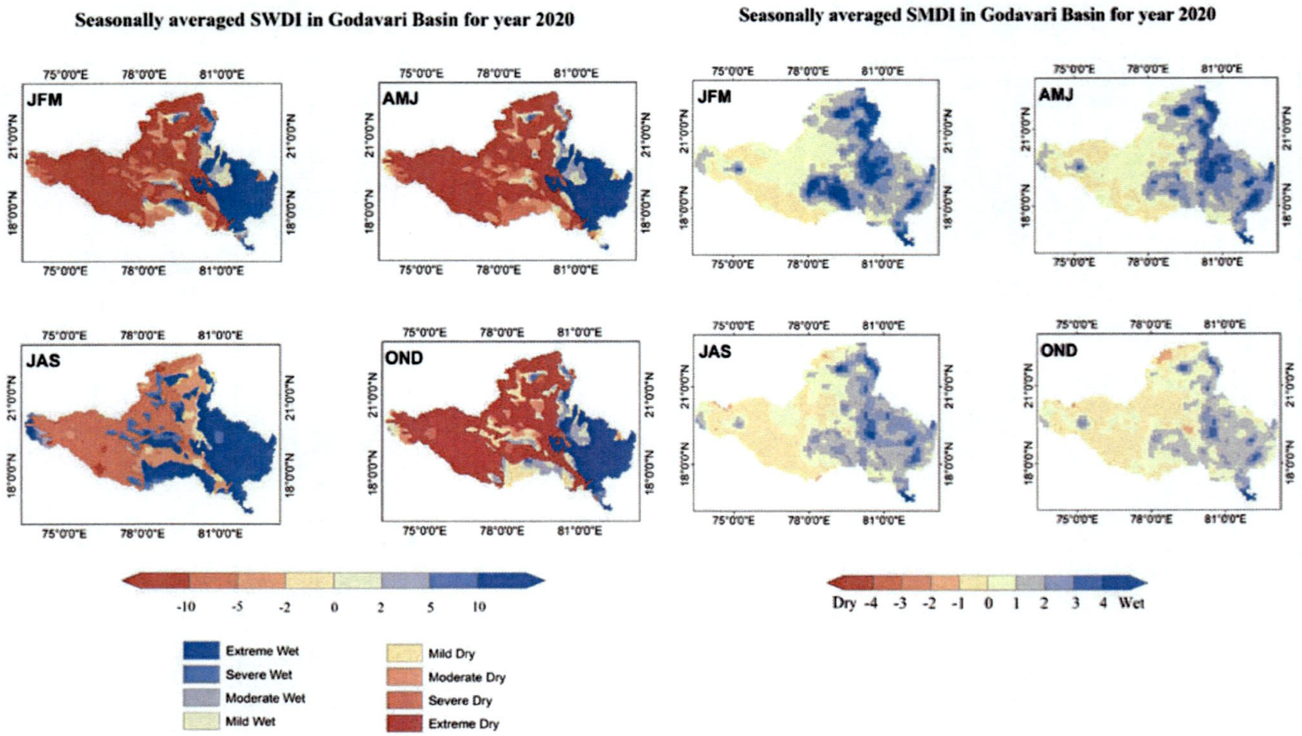

Fig. 4 Seasonally averaged spatial maps of SWDI and SMDI across Godavari Basin for year 2020

as the SWDI values are showing extreme wets for all the seasons. In contrast, a closer look suggests a slight different temporal change pattern for the case of SMDI values than the SWDI. The SMDI values showed the highest wetting trend during the peak monsoon (JAS) season and lowest wetting trends during the winter season (OND). Also, the yearly averaged SMDI values (Fig. 3) show an increasing trend of wet conditions each year for the eastern part of the study basin. This can be attributed to the fact that SMDI is a time-dependent variable and is computed based on the characteristics of long-term SM variations and SMDI of the past months. Although, the SMDI shows more inter-annual and inter-seasonal variations throughout the study basin and the study emphasizes on the use of satellite-based SM product, the short data length period of SMAP-SM products restricts the data to be used for computation of SMDI values. However, in a nutshell, the analysis suggests the acceptability of both the SWDI and SMDI indices for monitoring the agricultural droughts for the study area.

5 Conclusions

In this study, two drought indices SWDI and SMDI are used to evaluate drought conditions in Godavari basin for the period from 2016 to 2020, using SMAP and GLDAS-SM products, respectively. The two SM products are validated with ERA5 SM data for the study basin due to the unavailability of in situ soil moisture measurements which is an inherent limitation of the study. In a nutshell, the analysis derives the conclusions that both SWDI and SMDI are capable of indicating the spatial distributions of dry/wet conditions for the study basin. The inter-seasonal and inter-annual variations of the dry/wet conditions are more explicit through the SMDI values as it is time-dependent and uses the long-term SM characteristics. Yet, both the indices exhibit a correspondence in identification of the spatial distribution of drought/wet events. Finally, the outcome of the study proposes the approach to be applied for the different agro-climatic zones of the entire Indian mainland, which can give a very useful insights of the complete evolution of agricultural droughts.

Acknowledgements This study is partially supported and sponsored by the Research Seed Money (RSM) Grant, National Institute of Technology Warangal, Telengana-506004 (Ref No.: P-1094).

References

1. Mishra AK, Singh VP (2010) A review of drought concepts. J Hydrol 391(1–2):202–216
2. Panu US, Sharma TC (2002) Challenges in drought research: some perspectives and future directions. Hydrol Sci J 47(S1):S19–S30
3. Ray SS, Sesha Sai MVR, Chattopadhyay N (2015) Agricultural drought assessment: Operational approaches in India with special emphasis on 2012. In: High-impact weather events over the SAARC region. Springer, Cham, pp 349–364

4. Sandeep P, Reddy GO, Jegankumar R, Kumar KA (2021) Monitoring of agricultural drought in semi-arid ecosystem of Peninsular India through indices derived from time-series CHIRPS and MODIS datasets. Ecol Ind 121:107033
5. Salvia MM, Sanchez N, Piles M, Gonzalez-Zamora A, Martínez-Fernández J (2020) Evaluation of the soil moisture agricultural drought index (SMADI) and precipitation-based drought indices In Argentina. In: 2020 IEEE Latin American GRSS & ISPRS remote sensing conference (LAGIRS). IEEE, pp 663–668
6. Entekhabi D, Njoku EG, O'Neill PE, Kellogg KH, Crow WT, Edelstein WN, Van Zyl J (2010) The soil moisture active passive (SMAP) mission. Proc IEEE 98(5):704–716
7. Mishra A, Vu T, Veettil AV, Entekhabi D (2017) Drought monitoring with soil moisture active passive (SMAP) measurements. J Hydrol 552:620–632
8. Fang B, Kansara P, Dandridge C, Lakshmi V (2021) Drought monitoring using high spatial resolution soil moisture data over Australia in 2015–2019. J Hydrol 594:125960
9. Patel NR, Parida BR, Venus V, Saha SK, Dadhwal VK (2012) Analysis of agricultural drought using vegetation temperature condition index (VTCI) from Terra/MODIS satellite data. Environ Monit Assess 184(12):7153–7163
10. Dutta D, Kundu A, Patel NR (2013) Predicting agricultural drought in eastern Rajasthan of India using NDVI and standardized precipitation index. Geocarto Int 28(3):192–209
11. Dutta D, Kundu A, Patel NR, Saha SK, Siddiqui AR (2015) Assessment of agricultural drought in Rajasthan (India) using remote sensing derived vegetation condition index (VCI) and standardized precipitation index (SPI). Egypt J Remote Sens Space Sci 18(1):53–63
12. Chattopadhyay N, Malathi K, Tidke N, Attri SD, Ray K (2020) Monitoring agricultural drought using combined drought index in India. J Earth Syst Sci 129(1):1–16
13. Kulkarni SS, Wardlow BD, Bayissa YA, Tadesse T, Svoboda MD, Gedam SS (2020) Developing a remote sensing-based combined drought indicator approach for agricultural drought monitoring over Marathwada, India. Remote Sens 12(13):2091
14. Palanisami K, Ranganathan CR, Nagothu US, Kakumanu KR (2014) Climate change and agriculture in India: studies from selected river basins. Routledge India
15. O'Neill PE, Chan SK, Njoku EG, Jackson TJ, Bindlish R (2016) SMAP Enhanced L3 radiometer global daily 9 km EASE-grid soil moisture, Version 1. Boulder, Colorado USA. NASA National Snow and Ice Data Center Distributed Active Archive Center [Date Accessed: 04 July 2021]. In Boulder, Colorado USA. NASA National Snow and Ice Data Center Distributed Active Archive Center. https://nsidc.org/data/SPL3SMP_E/versions/1
16. Rodell M, Houser PR, Jambor UEA, Gottschalck J, Mitchell K, Meng CJ, Toll D (2004) The global land data assimilation system. Bull Am Meteor Soc 85(3):381–394
17. Beaudoing H, Rodell M (2020) NASA/GSFC/HSL (2020), GLDAS Noah Land Surface Model L4 monthly 0.25 x 0.25 degree V2.1, Greenbelt, Maryland, USA, Goddard Earth Sciences Data and Information Services Center (GES DISC), Accessed: [02-09-21]
18. FAO/UNESCO (1971–1981) The FAO-UNESCO soil map of the world. Legend and 9 volumes. UNESCO, Paris
19. Nachtergaele F, van Velthuizen H, Verelst L, Batjes NH, Dijkshoorn K, van Engelen VWP, Montanarela L (2010) The harmonized world soil database. In: Proceedings of the 19th world congress of soil science, soil solutions for a changing world, Brisbane, Australia, 1–6 Aug 2010, pp 34–37
20. FAO/IIASA/ISRIC/ISSCAS/JRC (2012) Harmonized world soil database (version 1.2). FAO, Rome, Italy and IIASA, Laxenburg, Austria. Accessed 9 Sept 2021
21. Muñoz-Sabater J, Dutra E, Agustí-Panareda A, Albergel C, Arduini G, Balsamo G, Thépaut JN (2021) ERA5-Land: a state-of-the-art global reanalysis dataset for land applications. Earth Syst Sci Data 13(9):4349–4383

22. Saxton KE, Rawls WJ (2006) Soil water characteristic estimates by texture and organic matter for hydrologic solutions. Soil Sci Soc Am J 70(5):1569–1578
23. Narasimhan B, Srinivasan R (2005) Development and evaluation of soil moisture deficit index (SMDI) and evapotranspiration deficit index (ETDI) for agricultural drought monitoring. Agric For Meteorol 133(1–4):69–88
24. Palmer WC (1965) Meteorological drought. In: U.S. Weather Bureau, Res. Pap. No. 45, p 58

An Ecohydrological and Geospatial Assessment for Urban River System: A Case Study in the Bhogdoi River, India

Anupal Baruah, Dhruba Jyoti Sarmah, and Arup Kumar Sarma

Abstract The rapid urbanization and the increasing anthropogenic activities impact the river basin health, urban flow system and endanger the aquatic ecosystem. The quantitative assessment and monitoring of urban river health are crucial in designing a sustainable ecosystem and undertaking river restoration strategies. The seasonal flow rate variations, built up and vegetation cover changes are some common indicators that can address urban river health both qualitatively and quantitatively. In this work, a GIS-based spatio-temporal analysis is carried out to quantify the Normalized Difference Vegetation Index (NDVI) and Normalized Difference Built-up Index (NDBI) in an urban reach of the Bhogdoi River, India using multi-temporal satellite remote sensing data spanning from 1990 to 2020. An ecohydrological assessment is conducted by utilizing the seasonal flow rates for 24 year periods, and flow indexes are developed to indicate the river health alterations. Results suggest that at the peak of urbanization, the NDVI reduces to 0.027 and NDBI increases to 0.160. From the quantitative assessment of the ecological status, under worst environmental scenario the minimum flow in the river is 3.031 m^3/s.

Keywords Urban river health · GIS · NDVI · NDBI · Flow indicators · Ecohydrological assessment

1 Introduction

The sustainability and biodiversity of marine ecosystem is highly influenced by the physical flow condition such as flow rate, velocity water depth, etc. The aquatic

A. Baruah (✉) · D. J. Sarmah · A. K. Sarma
Department of Civil Engineering, Indian Institute of Technology, Guwahati 781039, India
e-mail: anupalbaruah@iitg.ac.in

D. J. Sarmah
e-mail: jyoti123@iitg.ac.in

A. K. Sarma
e-mail: aks@iitg.ac.in

P. V. Timbadiya et al. (eds.), *Geospatial and Soft Computing Techniques*, Lecture Notes in Civil Engineering 339, https://doi.org/10.1007/978-981-99-1901-7_2

species are highly susceptible to the flow conditions and in absence of favorable habitat conditions may migrate to other regions as well. The increase in anthropogenic activities near the urban river banks is responsible to the river health degradation leading to the inferior flow condition. The lack of awareness about the importance of the river restoration and management among the riverine community, the aquatic health urban river system is constantly detreating. Nowadays, the GIS studies have been implemented by different investigator to study the changes in the built up index and vegetation cover of the urban river basin. For instance, Chen et al. [1], estimate the vegetation coverage assessment of ecological impact using NDVI in Hanjiang River Basin. They observed that NDVI in the HJRB increased from 2001 to 2018, and the variation rate was 0.0046 year. Moreno et al. [2], identify potentiality of the urban forest using NDVI in Temuco, Chile. The results demonstrate that, NDVI can be used as a tool to continuous monitoring of sustainable urban planning and with a better life quality for their population. Again, Zha et al. [3], using NDBI for mapping of urban land in the city of Nanjing. They observed that NBBI is a reliable method to mapping for urban land in the city of Nanjing with an accuracy of 92.6%. Li and Chen [4], studied the urbanization using NDVI and NDBI along with genetic algorithm in major Chinese cities reviles that this new method shows great potential to solve urban ecological issues. These studies suggested that with the increase in NDBI, there is a possibility of progressive degradation in the ecosystem especially in urban river system.

Depending on the system and surroundings, the environmental flow rate varies [5]. In broad sense, the tools available for the quantitative assessment of the ecological flow in a domain are characterized by hydrological, hydraulic, habitat simulation and holistic method. The probabilistic approach is mostly used in hydrological methods to analyze the recorded flow series data. There are several hydrological methods available for estimating environmental flow assessments (EFA). Out of different available methods, the flow duration curve (FDC) shifting method is a new hydrological technique [6], based on the monthly flow series.

In this work, a quantitative assessment of ecological flow is carried out in an urban stretch of the Bhogdoi River, Assam, India using flow duration curve shifting (FDCS) method. Six different environmental management classes are considered during the study to evaluate the minimum flow in the river. A GIS study is also conducted in the reach to understand the trend of anthropogenic activities in the urban reach from the last three decades.

2 Study Area and Methodology

2.1 Study Area

This study is confined in River Bhogdoi (26° 45′ N to 26° 46′ N and 94° 12′ E and 94° 15′) covering an area of about 2.33 km^2. It is located in the southern bank of the

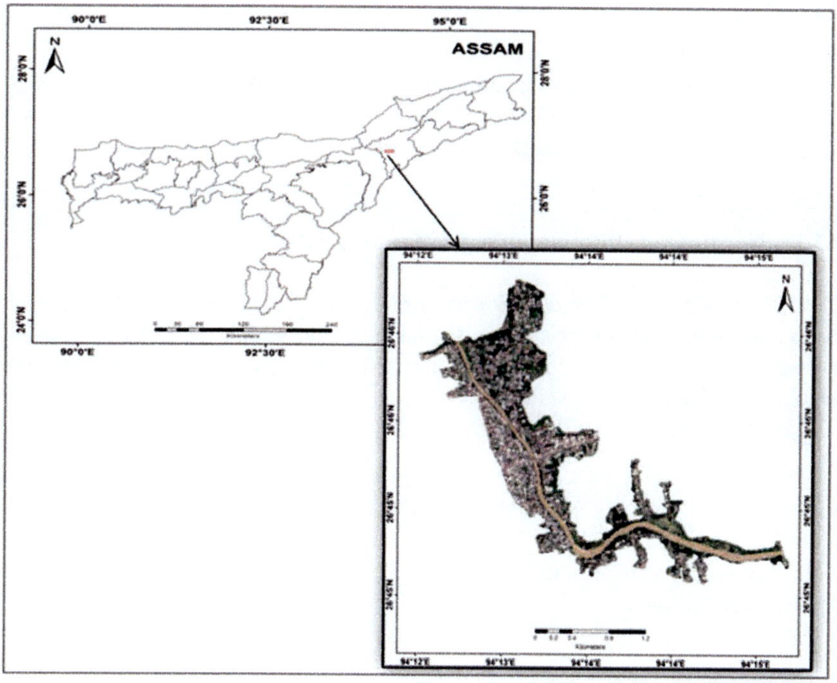

Fig. 1 Study reach

Brahmaputra River and originates from Long Samtang of Mukokchung (Naga Hills) and finally falling down in Gelabill River in North-west of Jorhat city (Fig. 1).

2.2 *Index-Based Approaches Using GIS and Remote Sensing Technique*

In the present study, to analyzing the decadal changes of a geographical environment over time by means of GIS and remote sensing technique, Normalized Difference Vegetation Index (NDVI) and Normalized Difference Built-up Index (NDBI) at different time scenes (1990 to 2020 at an interval of ten constructive years) are used.

2.2.1 Satellite Data

The present approach employs satellite images of the last 30 years from 1990 to 2020 at the interval often consecutive years. The required pre-processing was conducted before the index were calculated.

2.2.2 Normalized Differential Vegetation Index (NDVI)

The NDVI index is widely used for identifying the expansion status of vegetation cover. Vegetation absorb and re-emit solar radiation within the infrared and red regions of electromagnetic spectrum and it is very easy for satellites to identify the vegetation reflectance.

It is expressed by

$$\text{NDVI} = \frac{\text{NIR}-\text{RED}}{\text{NIR}+\text{RED}}$$

In this equation, NIR is the reflectance of the near-infrared and Red is the red bands, respectively.

2.2.3 Normalized Difference Built-Up Index (NDBI)

NDBI is basically proposed for to extract urban surfaces. According to Zha et al. [3], it utilizes short-wave infrared and near infrared bands for calculation as shown in equation

$$\text{NDBI} = \frac{\text{SWIR}-\text{NIR}}{\text{SWIR}+\text{NIR}}$$

2.3 *Hydrological Method for Ecological Flow Rate Estimation*

Hydrological alterations in the stream are governed by the south west monsoon. The Bhogdoi River is a perennial stream with high variation in seasonal flow. The present study employs the FDCS method to estimate the ecological flow requirement for the river sustainability and restoration. Monthly discharge data for 24 years is collected from the concerning department and flow duration curve is prepared.

2.3.1 Flow Duration Curve Shifting (FDCS) Method

Smakhtin [7] developed this method and estimate the EFR at some specific environmental management classes (EMC). It is a four step method starting with the construction of the FDC from the observed data corresponding to 17 fixed percentage points. Depending upon the basin modifications, six ecological management classes are identifying that replicates the river health. These classifications are primarily dependent on the anthropogenic factors and its impact over the river sustainability. At higher EMC, it is necessary to allocate high flow for the maintenance and conservation of the biodiversity in the river. Later, the environmental flow time series curves are developed by the lateral shifting to the left of the original FDC as shown in Figs. 2 and 3. A spatial interpolation technique proposed by Hughes and Smithton is used while developing these time series curves.

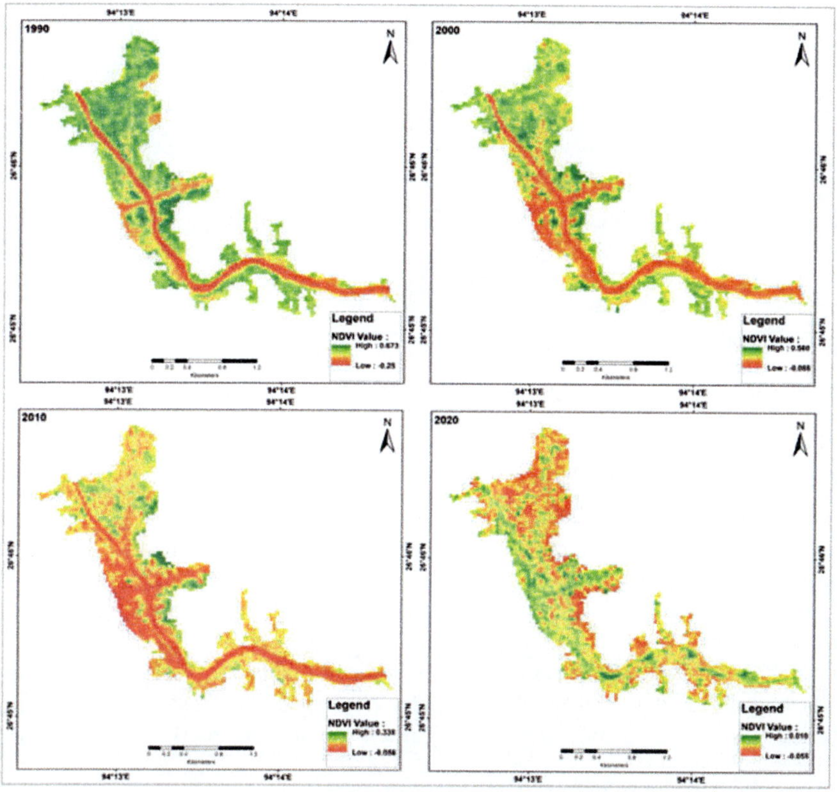

Fig. 2 Normalized difference vegetation index (NDVI) analysis of the study area

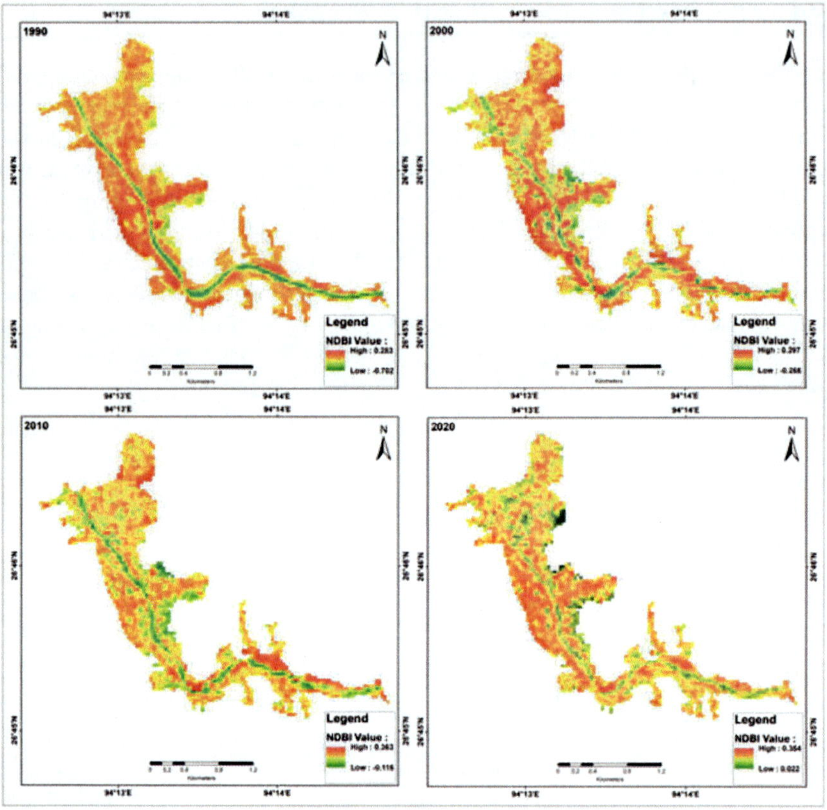

Fig. 3 Normalized difference built-up index (NDBI) analysis of the study area

3 Result and Discussion

Spatial variation of the vegetation status in the flow domain throughout the study period is as shown in Fig. 2.

The vegetation indices maps (Fig. 2) indicate the gradual decreasing trend of average greenness cover in the flow domain from, i.e., 0.297, 0.212, 0.082 and -0.027 in 1990, 2000, 2010 and 2020, respectively (Table 1). While, the built up indices maps (Fig. 3) indicate the gradual increasing trend of average built up in the study area, i.e., – 0.062, 0.071, 0.123 and 0.160 in 1990, 2000, 2010 and 2020, respectively (Table 2).

Thus from the study, it can be inferred that there is an increase in the anthropogenic activities near the flow domain. The decrease in the vegetation cover in the urban river basin leads to a degraded aquatic ecosystem which may not be favorable for the survival of the flaura and faunas. The increased built up index strongly hampers the water quality and quantity in the study area. The various degree of degradation

Table 1 List satellite imageries used in the study

Type of data	Year	Source
LANDSAT-5 TM	1990, 2000 and 2010	https://earthexplorerusgs.gov
LANDSAT-8 OLI_TIRS	2020	https://earthexplorerusgs.gov

Table 2 Average NDVI and NDBI values in the flow domain at Bhogdoi River, India

Year	NDVI mean	NDBI mean
1990	0.297	− 0.062
2000	0.212	0.071
2010	0.082	0.123
2020	− 0.027	0.16

affects the flow rate in the stream. For the quantitative assessment of the flow, six different environmental management class (EMC) are considered during the study. The flow rates at different (EMC) is estimated from the Global Environmental Flow Calculator (GEFC). GEFC uses the flow duration curve shifting (FDCS) method and quantify the flow rate at different EMC. Monthly flow data for 24 years is collected from the concerning department, maximum recorded natural flow in the domain is 90M^3/s, and an average flow rate is 24 N^3/s is observed. Flow duration curve in the reach is developed as shown in Fig. 4.

The different EMC used in the study is presented in Fig. 5. It is observed that under the critically modified condition, it is necessary to maintain the ecological flow within 12.3% of mean annual runoff (Class-F) and in an average the streamflow in the river must be more than 38.5% of MAR for river health (Class-C). Computed E-flow requirement as a percentage of MAR are shown in Table 3.

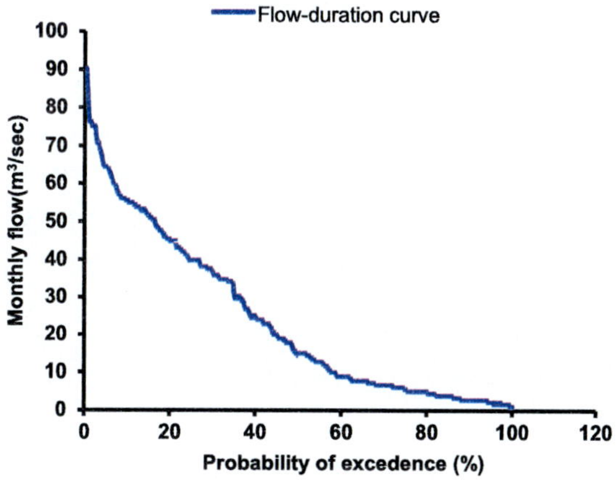

Fig. 4 Flow duration curve in the study site

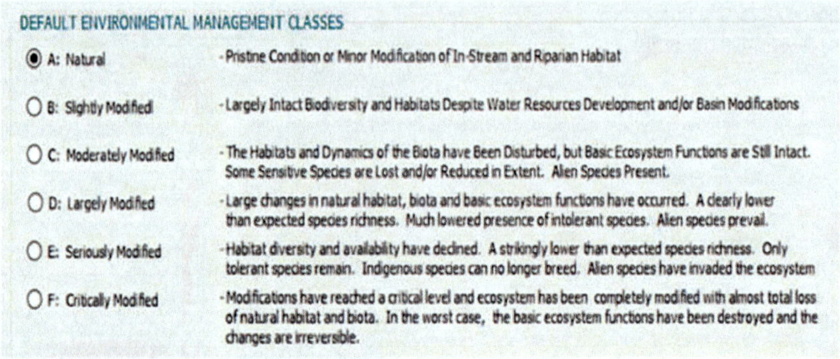

Fig. 5 Different environmental management class

Table 3 Estimated EFR at different environmental management class

Station	A.T road gauge station
Record period	1991–2015
Mean annual runoff (MAR)(N^3/s)	24.05
FDCS method (as per global environmental flow calculator)	EF (% MAR) at different EMC
Default management classes	
EMC-A: no modification condition	75.3
EMC-B: slightly modified but biodiversity prevails	54.8
EMC-C: moderate modification in the aquatic environment	38.5
EMC-D: large changes in the ecosystem	26.3
EMC-E: declination in the habitat availability and diversity	
Reduction in aquatic species	17.8
EMC-F: extreme modification in the river basin	12.3

Figure 6 Shifted FDC at different EMC. For each management class, the original FDC shifted laterally toward the left direction indicating the degradation in the ecosystem.

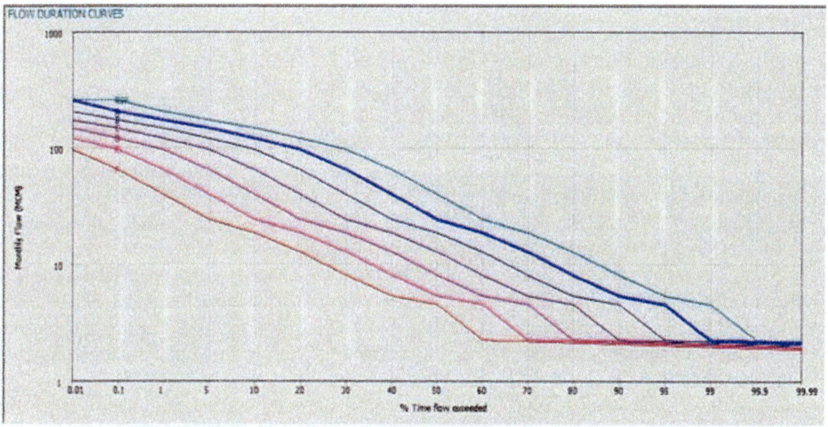

Fig. 6 Environmental flow duration curves for different EMC

Results indicate that with the degraded ecosystem, the ecological flow rate in the stream progressively decreases and reaches a critical value at EMC-F. At this level, it is difficult to maintain the biodiversity and provide habitat for the marine spices. Primarily, with the increase in anthropogenic activity as obtained from the remote sensing studies strongly suggest the degradation in the riverine ecosystem.

4 Conclusion

The sustainability of an urban aquatic ecosystem relies on the available flow rate as well as in the anthropogenic activity of the surrounding. In this work, an ecohydrological analysis is carried out along with a GIS-based study in an urban stretch. The NDVI and NDBI indices are used to evaluate the changes in the vegetation cover and the built up index in the study area. Results indicate the decrease in NDVI suggesting the reduction in the vegetation cover and the increase in NDBI. The ecological flow at six different environmental management scenario is carried out in the domain by using FDCS method. Results indicate that the ecological flow in the river must be maintained at 18.05 m^3/s. However, with the continuous degradation in the ecosystem the minimum flow in the stretch may reaches up to 3.05 m^3/s. The present study will be helpful for the stakeholders before implementing any river restoration or basin improvement project in the study area.

References

1. Chen T, Xia J, Zou L, Hong S (2020) Quantifying the influences of natural factors and human activities on NDVI Changes in the Hanjiang River Basin, China. Remote Sens 12:3780
2. Moreno R, Ojeda N, Azócar J, Venegas C, Inostroza L (2020) Application of NDVI for identify potentiality of the urban forest for the design of a green corridors system in intermediary cities of Latin America: case study, Temuco, Chile. Urban For Urban Green 55:126821
3. Zha Y, Gao J, Ni S (2003) Use of normalized difference built-up index in automatically mapping urban areas from TM imagery. Int J Remote Sens 24(3):583–594
4. Li KN, Chen YH (2018) A genetic algorithm-based urban cluster automatic threshold method by combining VIIRS DNB, NDVI, and NDBI to monitor urbanization. Remote Sens 10:277
5. Yin XA, Yang Z, Zhang E, Xu Z, Cai Y, Yang W (2018) A new method of assessing environmental flows in channelized urban rivers. Engineering 4(5):590–596. https://doi.org/10.1016/j.eng.2018.08.006
6. Smakhtin VU, Anputhas M (2006) An assessment of EF requirements of Indian river basins. Research Report 107, International Water Management Institute, Colombo, Sri Lanka
7. Smakhtin V (2008) Basin closure and environmental flow requirements. Int J Water Resour Dev 24(2):227–233

Assessment of Reservoir Sedimentation Using Geospatial Tools: A Case Study of Kadana Reservoir

Himadri Shah, Sudhanshu Dixit, and Shard Chander

Abstract Accumulation of sediments in a reservoir with time is a known phenomenon and is generally taken in consideration while estimating useful capacity of any reservoir. However, hydrographic surveys carried out for various reservoirs instills that for a significant number of reservoirs, reduction in useful storage capacity is much higher than anticipated loss. This is due to unprecedented sediment deposition in live storage zone, while lesser sediments are deposited in dead storage zone of reservoir. This indicates premature aging of such reservoirs. In this study, Kadana reservoir located on Mahi River on the border of Gujarat and Rajasthan is taken as the study area to quantify the change in storage capacity of reservoir. Remote sensing and geospatial tools are used for analyzing the scenario of sedimentation in live storage zone of Kadana reservoir. The results of this study showed that the reservoir impounded in 1979 with original live storage of 1203MCM is left with 834.84 ± 31.41 MCM live storage in 40 years, i.e., 32.11% loss in live storage against anticipated loss of 7.4% as stated in design reports.

Keywords Sediment assessment · Geospatial tools · Remote sensing · Reservoir · Kadana · Useful storage capacity

1 Introduction

India is a tropical country with vast variations in water availability in dry and wet seasons. Nearly 70% population depends directly or indirectly on agriculture. For

H. Shah (✉)
Water Resources Engineering Department, L. D. College of Engineering, Ahmedabad, Gujarat, India
e-mail: shahhimadri@iitgn.ac.in

S. Dixit
Civil Engineering Department, L. D. College of Engineering, Ahmedabad, Gujarat, India

S. Chander
Space Applications Centre, ISRO, Ahmadabad, Gujarat, India
e-mail: schander@sac.isro.gov.in

P. V. Timbadiya et al. (eds.), *Geospatial and Soft Computing Techniques*, Lecture Notes in Civil Engineering 339, https://doi.org/10.1007/978-981-99-1901-7_3

economic development of an agrarian country, it is mandatory to improve agriculture conditions, which can be achieved by proper irrigation. Hence, under the first five years plan, it was suggested to build a series of major and minor dams on the rivers across the country. The reservoirs of dammed rivers would prove as permanent storage sources of surface water. Many major dams were constructed in Independent India right from Hirakud Dam (1957), Bhakhra and Nangal Dams (1959) to Tehri Dam (2006), Sardar Sarovar Dam [1]. At present, there are 5202 large dams in India as per Central Water Commission [2]. Majority of these dams are more than 30 years old and have lost significant useful storage capacity due to sediment deposition [3]. Siltation has reduced economic benefits of many reservoirs. Sediment deposition in reservoir is one of the most technical issues faced by dam engineers, even after decades of research. On the other hand, due to increasing urban sprawl, the drastic change in land use land cover pattern has increased the rate of soil erosion significantly. Thus, it is of prime importance for dam engineers to understand the pattern of sediment deposition, sedimentation rate and its distribution at various levels of reservoir. Since the impoundment of reservoir, the sediments will settle in reservoir according to its trap efficiency. Talking about the 'sediment design life' as quoted by Army corps of Engineers, U.S. there are four stages of reservoir (Fig. 1). In the starting phase (A), there are no additional sediment loads. Gradually at stage B, coarser particles will form delta at head reach of reservoir. At stage C, the sediments start affecting the economic benefits and intake structures present in uniform. Finally, at the last stage (D), dam needs to be dismantled. These stages cannot be classified in terms of years equally for all reservoirs. Most of the dams in India are in first or second stage. For reservoirs particularly in stage B [4], it is important to keep an eye on live storage capacity variation over period of years. CWC monitors 243 large reservoirs in India, Kadana being one of them. The last sedimentation study was carried out in 2000, which showed higher rate of depletion of live storage capacity (by 20%) over dead storage capacity (by 13.34%). After 2000, there is no specific study on Kadana reservoir; hence, it is selected as case study for present analysis.

For measurement of sediment deposition, three methods are prevalent: inflow outflow method, hydrographic survey method and remote sensing method. Inflow–outflow method consists of gauging sediment concentration at every significant outflow and inflow stream. This method is very tedious and difficult to implement. Hydrographic survey method measures sediment accumulation at fixed grid points in reservoir. This method is accurate but costly. Remote sensing method analyzes change in water spread area at each reservoir elevation and thus shows sediment deposition at each level. However, remote sensing method can be applied for live storage zone and only for the reservoirs which maintain daily water level measurement records.

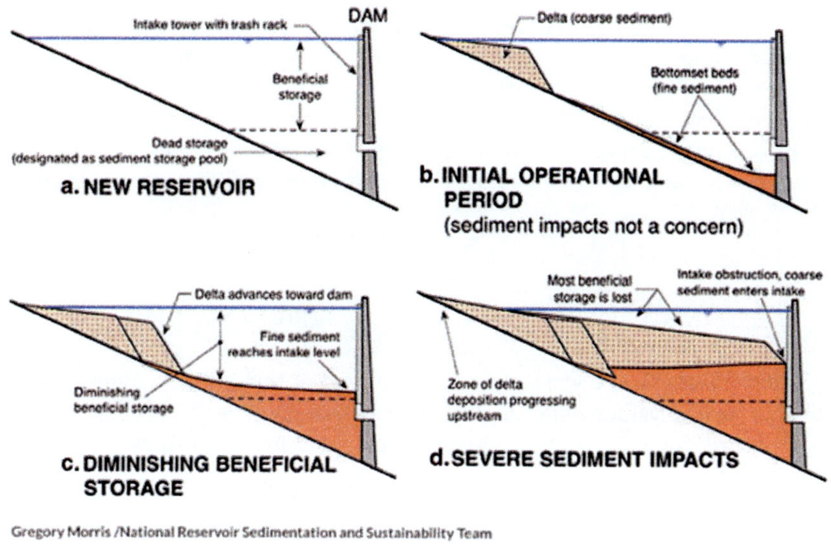

Fig. 1 Stages of reservoir sedimentation (Picture credit: Gregory Morris/National Reservoir Sedimentation and sustainability team: https://waterdesk.org/2019/11/reservoir-sedimentation-h2o-radio)

2 Materials and Methods

2.1 Remote Sensing Method for Sediment Assessment

Principle for sediment assessment is for same reservoir elevation, the water spread area decreases due to sediment deposition [6]. As the reservoir operates for various purpose, the reservoir elevation changes along the year. Corresponding water spread area is also changing. The water spread area delineation is carried out from a satellite imaginary of particular date, with known (in situ) reservoir water elevation. The water spread extent for various reservoir elevation are measured to generate area-elevation table. The volume is obtained by transforming water spread area between two elevations into volume by trapezoidal formula. The loss of storage capacity (live storage) between any two years is the amount of sediment deposition between the years. The most crucial step for sedimentation study by remote sensing method is extraction of water spread area for any given date. The classification of water pixel and non-water pixel mostly distinguish of reservoir water and wet soil pixel on reservoir periphery plays important role in extracting water spread area. The main phenomenon for differentiation is based on the fact that interaction of every earth feature with each spectral band is different. For this, normalized difference method is used in present study. In this method, selection of band is done as, one band will maximize the reflection by water features, other will minimize low reflectance of water at the same time high reflectance of other features. In present study, Normalized Difference Water Index (NDWI) developed by MCFeeters (1996) is taken, with a modification as suggested by Hanqiu Xu (2006) termed as Modified Normalized Difference Water Index (MNDWI). NDWI uses spectral reflectance of green band (GREEN: 0.53–0.59 μm) and near infrared band (NIR: 0.85–0.88 μm). MNDWI uses short-wave infrared band (SWIR 1:1.57–1.65 μm) instead of near infrared band.

$$\text{NDWI} = (\text{GREEN} - \text{NIR})/(\text{GREEN} + \text{NIR}) \quad (1)$$

$$\text{MNDWI} = (\text{GREEN} - \text{SWIR1})/(\text{GREEN} + \text{SWIR1}) \quad (2)$$

So, from Eqs. 1 and 2 when we calculate NDWI or MNDWI, the water features will give positive values while all other features will give negative values [5]. This is way to mask out non-water features, and just water features will be extracted as water spread area.

2.2 *Study Area and Data Source*

2.2.1 Kadana Reservoir and Its Catchment Area

Kadana Dam is located in the gorge cut by Mahi River through a low range of hills in Mahisagar district of Gujarat, near the border with Rajasthan. Geographic coordinates of Kadana Dam is 23° 18′ 26.12″N latitude and 73° 49′ 38.12″E longitude. Kadana Dam is an Earthen-Masonry composite and has ogee spillway with roller bucket energy dissipater (NWRWS). The construction of dam took place between 1979 and1989. Dam is 1551 m long and 66 m high from foundation level. Kadana reservoir (Fig. 2) has water spread area of 166 sq. km. and storage capacity of 1542 MCM out of which 1203 MCM is live storage. The reservoir has full reservoir level (FRL) at 127.7 m and maximum drawdown level (MDDL) at 114.3 m. The catchment area of reservoir is 25520 sq. km. Actual sedimentation rate of Kadana reservoir is higher than the estimated rate (0.501 Th.cum/sq km/yr. instead of design rate 0.13 Th.cum/sq km/yr.) [2].

2.2.2 Data Collection

Daily reservoir elevation data from 2000–2020 is obtained from IndiaWRIS web portal (https://indiawris.gov.in/wris/#/), CWC. Last hydrographic survey report (carried out in year 2000) and reservoir water depth contour are obtained from dam authority. Landsat satellite imaginaries (2000–2020) are obtained from USGS web portal. The water spread extraction analysis is carried out in QGIS software and

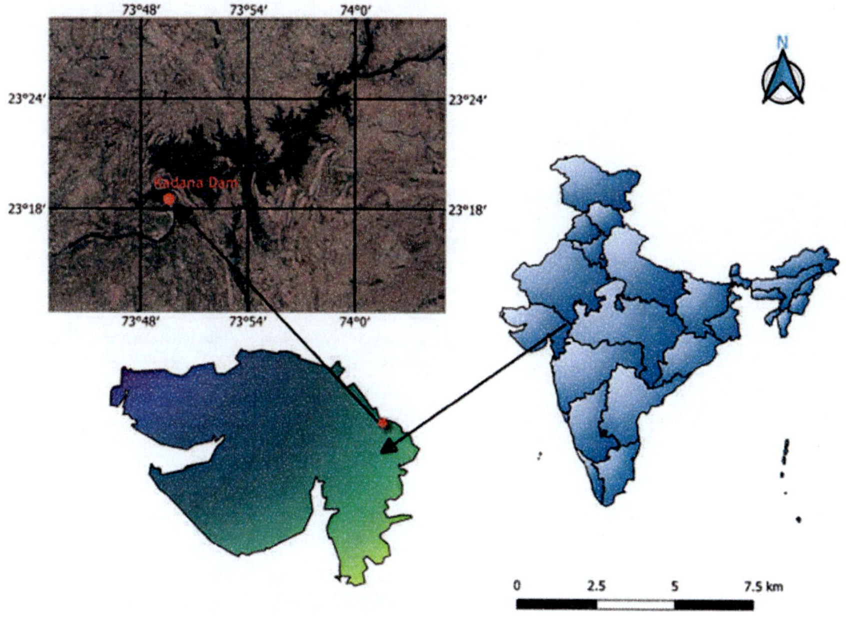

Fig. 2 Index map of study area

Google Earth Engine environment. Sentinel 2A imaginary is selected for accuracy assessment.

2.3 *Methodology*

See Fig. 3.

The satellite data is selected such that the cloud cover is less than 10% at study area. Atmospheric corrections are applied to get reflectance data. The date range selection is done so that, maximum water level fluctuation of live storage zone (i.e., from FRL to MDDL) is covered for a water year [7]. Considering all the data quality requirements, water year 2019–20 and water year 2015–16 are found suitable date ranges for analysis. The reservoir boundary is delineated using SRTM DEM (30 m) and water depth contours in QGIS software using vector analysis. Water spread area corresponding to each elevation of satellite passing in a given water year is obtained by generating NDWI and MNDWI maps using raster analysis tools in QGIS software. Individual image is processed in QGIS for water spread contour extraction. The water spread area time series is obtained at a single click in google earth engine by writing a code in Java [8]. After obtaining water spread area at various reservoir elevations, capacity between two successive elevations for same water year is calculated using trapezoidal formula. Area-elevation-capacity (live)

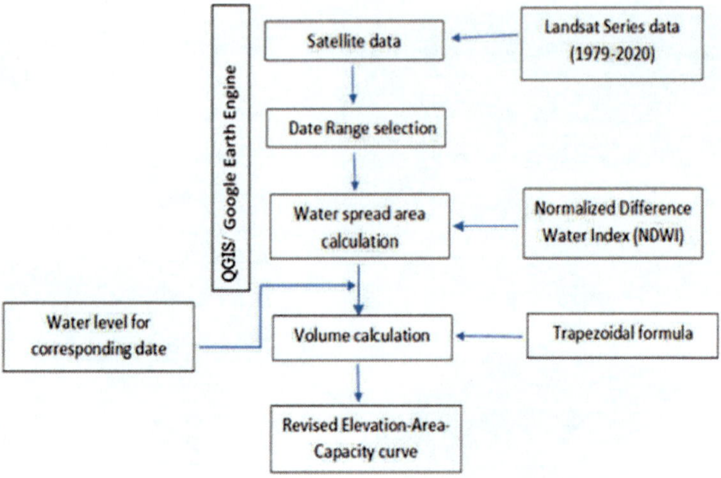

Fig. 3 Methodology flowchart

curve can be obtained by plotting elevation versus cumulative live storage at each elevation considering live storage capacity at MDDL 114.3 as zero. Comparing the elevation-capacity curve between two years give the amount of sediment deposited. For accuracy assessment and error analysis, due to lack of present ground truth data, satellite imaginary with higher spatial resolution (sentinel 2A with 10 m resolution) is considered as ground truth data. Supervised classification is carried out in QGIS software to get water spread area for same elevation, same year. Results are compared with NDWI and MNDWI maps. Statistical analysis (Kappa coefficient) is carried out for determining accuracy of analysis. For illustration, consider atmospherically corrected satellite image corresponding to February 8, 2020. The reservoir elevation on the given date is obtained as 125.17 m from IndiaWRIS portal. The image contains reflectance images of various bands. In QGIS, the reflectance of green band, NIR band and SWIR 1 band is loaded. The images are masked by reservoir boundary using vector analysis. NDWI and MNDWI images are obtained by raster analysis of green-NIR bands and green-SWIR 1 bands respectively. After setting threshold value of zero, NDWI and MNDWI images are classified in categories, water pixel and non-water pixel. The area of water pixel is obtained as 92.43 km^2 from NDWI and 95.64 km^2 from MNDWI. Similarly for analysis is carried out for January 23, 2020, corresponding to reservoir elevation of 125.83 m (0.66 m elevation difference). The water spread area as per NDWI was found to be 95.05 and 100.93 km^2. as per MNDWI. Thus, the storage capacity between these levels were calculated as per trapezoidal rule. The live storage capacity was estimated between these elevations to be 517.38MCM as per NDWI and 599.60 MCM as per MNDWI.

For accuracy assessment, Sentinel 2A image dated February 7, 2020, was selected with reservoir elevation 125.17 m (same as Landsat image). Accuracy assessment in QGIS depicted that in NDWI image, 96.08 km^2 was water spread area, with 1.66

km^2 water pixel area misinterpreted as non-water pixel area while 5.30 km^2. soil pixel area misinterpreted as water. Similarly, for MNDWI 2.86 km^2 water pixel area misinterpreted as non-water pixel area while 3.30 km^2. soil pixel area misinterpreted as water.

3 Results and Discussions

3.1 Water Spread Area Time Series

Water spread area is obtained for 2015–2016, 2019–2020 by masking non-water pixels using NDWI/MNDWI. The Raster images for each date are converted to vector, to delineate water spread area for each date (Fig. 4a, b). Tables 1, and 2 given show the dates of satellite passing with corresponding reservoir elevation. The water spread area is calculated from satellite imaginary by two normal differential indices (NDWI/MNDWI).

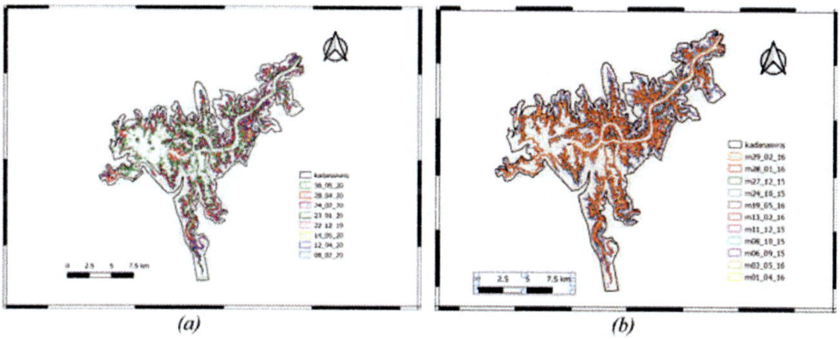

Fig. 4 Water spread area contours for Kadana reservoir **a** for year 2019–20 and **b** for year 2015–16

Table 1 Date-elevation-water spread area table 2019–2020

Date	Elevation (m)	Area (m^2) NDWI	Area (m^2) MNDWI
05-14-2020	121.18	59,341,500	62,652,600
04-28-2020	121.79	64,539,900	66,692,700
04-12-2020	122.48	71,056,800	72,924,300
02-24-2020	124.36	85,023,900	90,486,900
02-08-2020	125.17	92,435,400	95,649,300
01-23-2020	125.83	95,055,300	100,934,100
12-22-2019	126.98	110,687,555	110,659,500

Table 2 Date-elevation-water spread area table, 2015–2016

Date	Elevation (m)	Area (m^2) NDWI	Area (m^2) MNDWI
05-19-2016	116.54	33,082,200	33,867,000
05-03-2016	117.27	38,717,100	39,959,100
04-01-2016	118.54	46,454,400	46,479,600
02-29-2016	118.9	51,599,700	52,420,500
02-13-2016	119.89	52,137,900	53,973,000
01-28-2016	120.83	61,219,800	64,152,000
12-27-2015	121.87	68,179,500	71,756,100
12-11-2015	122.12	70,041,600	73,167,300
10-24-2015	123.29	78,994,800	81,990,900
10-08-2015	124.11	80,838,000	87,232,500
09-06-2015	126.42	94,320,900	103,756,500

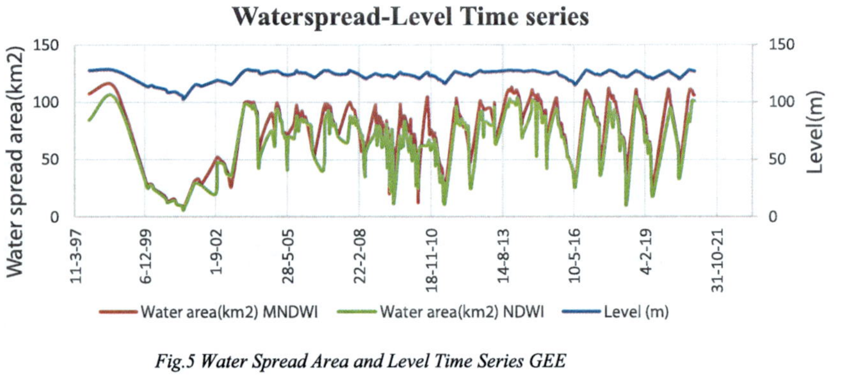

Fig. 5 Water spread area and level time series GEE

Water spread area obtained from two indices (NDWI/MNDWI) time series obtained by google earth engine is as shown in Fig. 5.

3.2 *Elevation-Area-Capacity Curve*

Capacity at each elevation is calculated using trapezoidal formula. Revised elevation-area-capacity curve (Fig. 6) is obtained from water spread area extracted in 2019–20 from MNDWI maps; as in accuracy assessment, MNDWI performed better.

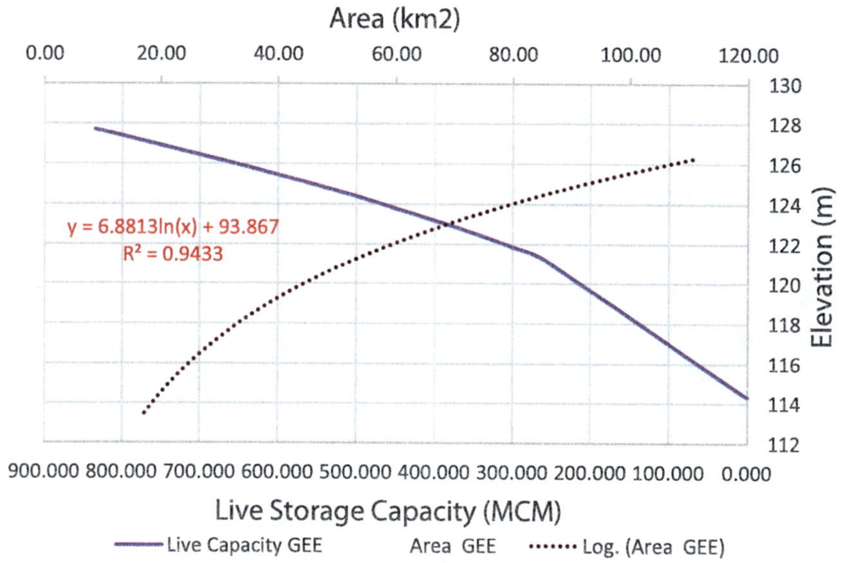

Fig. 6 Revised elevation-area-capacity curve

3.3 Spatial Distribution of Sediments

Spatial distribution of sediments can be obtained by comparing the obtained results with past hydrographic survey results. As two water years are considered in present study, spatial distribution is studied in two time-scales: from 2000 to 2015–16 (Fig. 7) and from 2015–16 to 2019–20 (Fig. 8).

For time scale 2015–16 to 2019–20, it was observed that water spread area increased for level above 124 m. This shows the position of delta formed by coarser

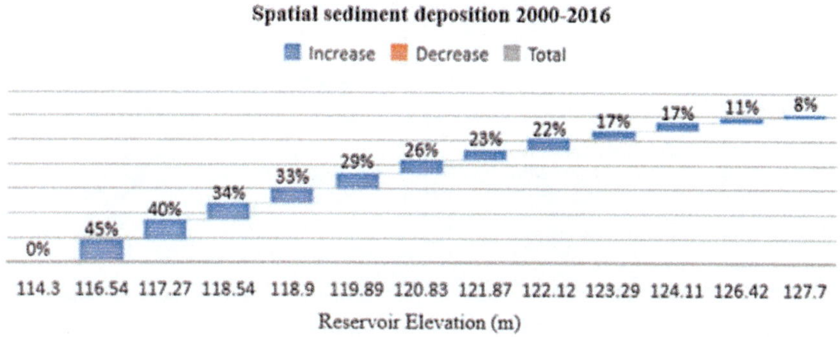

Fig. 7 Spatial distribution of sediments in 2016

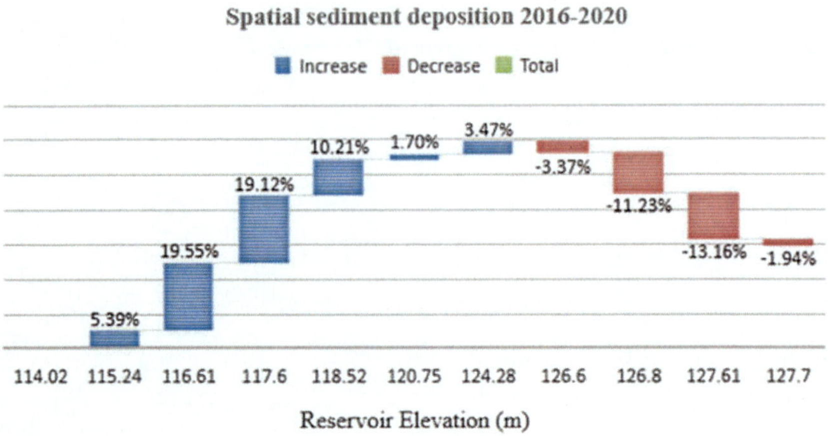

Fig. 8 Spatial distribution of sediment at each elevation

sediments. This is due to the fact that the sediment accretion and deposition take place in alternate period intervals in delta zone of sediment deposition.

3.4 Accuracy Assessment and Error Analysis

In absence of present actual reservoir data, Sentinel 2A data set is considered as ground truth data, accuracy assessment of water spread analysis is done with respect to it. The water pixels misclassified as non-water are indicated by pink color, while the non-water misclassified as water are indicated by red color. Blue color indicates reservoir, and yellow color represents non-water pixels (Fig. 9).

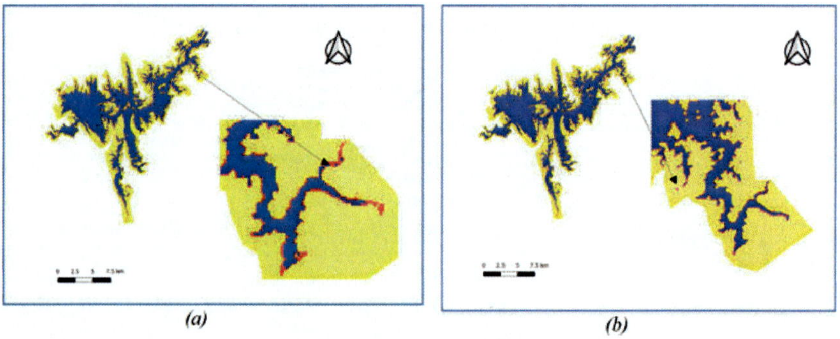

Fig. 9 Accuracy assessment for water pixel delineation **a** for NDWI classification and **b** for MNDWI classification

Overall accuracy and kappa's coefficient is found to be maximum for MNDWI analysis and is 96.75% and 0.93 respectively. Considering 96.75% accuracy in area estimation, the maximum possible error in capacity calculation is ± 34.41 MCM.

4 Conclusion and Future Scope

The revised live storage of Kadana reservoir for 2019–20 at FRL is 834.84 ± 31.41 MCM. Loss of live storage capacity since impoundment of reservoir is 386.34 MCM (32.11%) and loss of live storage capacity since last hydrographic survey (2000) is 141.38 MCM (14.75%).

The revised live storage capacity will help in revising reservoir water allocation planning, so as to get actual ratio of water demand-water supply assessment for different departments. The spatial sediment deposition analysis will help in planning desilting operation corresponding to various reservoir elevations. The temporal analysis will further help in predicting the useful life of reservoir and rate of sediment deposition.

Considering the future scope, incorporating the reservoir bathymetry profile will increase the accuracy of capacity assessment. Sedimentation issues are greater for reservoirs with less height of dam and reservoir shape like soccer. But, as most of these reservoirs are medium to small, the reservoir water level is not monitored on daily basis. With a greater number of satellites launched for earth observation, it is a new science which should be explored to monitor reservoir water depth along with extracting bathymetry using satellite data for smaller and medium size reservoirs.

References

1. Garg SK (2016) Irrigation engineering and hydraulic structures, water resource engineering, vol II, 32nd revised edition, 2016, ISBN: 81-7409-047-9
2. Central Water Commission (2015) Compendium on silting of reservoirs in India. Published by the Central Water Commission, Government of India, New Delhi, p 6
3. Froehlich DC, Narayan P, Kumar M (2017) Estimating reservoir capacity loss from sedimentation. In Proceedings of the Third National Dam Safety Conference, Roorkee, India, pp 18–19
4. Singh RD, Jain SK, Goel MK, Singh SK, Agarwal PK (2013) Mathematical representation of elevation-area-capacity curves for Indian Reservoirs. National Institute of Hydrology
5. Pandey A, Chaube UC, Mishra SK, Kumar D (2016) Assessment of reservoir sedimentation using remote sensing and recommendations for desilting Patratu Reservoir, India. Hydrol Sci J 61(4):711–718
6. Gujrati A, Jha VB (2018) Surface water dynamics of inland water bodies of india using google earth engine. ISPRS Ann Photogrammetry, Remote Sens Spatial Inf Sci 4:467–472

7. Dadoria D, Tiwari HL, Jaiswal RK (2017) Assessment of reservoir sedimentation in Chhattisgarh State using remote sensing and GIS. Int J Civ Eng Technol 8(4):526–534
8. Singh S, Dhasmana MK, Shrivastava V, Sharma V, Pokhriyal N, Thakur PK, Dhote PR (2018) Estimation of revised capacity in Gobind Sagar reservoir using Google Earth Engine and GIS. Int Arch Photogrammetry, Remote Sens Spatial Inf Sci 42:5

Water Quality Estimation Using Remote Sensing Technique: A Case Study of Bhadra Reservoir, Karnataka

Avantika Latwal, K. S. Rajan, and S. Rehana

Abstract In the twenty-first century, water quality monitoring is a growing challenge in inland water bodies. Therefore, monitoring the current state of the water bodies is highly important. The remote sensing data is important for effective monitoring that covers a large area. In recent decades, chlorophyll-a (as a proxy) has been a significant indicator of nutrient contamination and also indicates the qualitative status of the water bodies. Besides, the waterbody's temperature also plays an important role in the biological and chemical processes. Hence, the present study aims to detect the chlorophyll-a spread area and temperature of the Bhadra Reservoir, Tungabhadra River system, India. To explore the quality of the reservoir, Sentinel 2A and Landsat 8 satellite images were used for the years 2017 and 2018. The study results noticed that chlorophyll-a spread percent was higher in summer months compared with the water spread area of the reservoir, i.e., 74.5% in 2017 and 61.4% in 2018 from the total water spread area. The study also revealed that as the temperature rises, the chlorophyll-a spread also increases in the reservoir. It was also observed that the October month has a low chlorophyll-a spread, i.e., 32.04% in 2017 and 24.35% in 2018, because the one side reservoir is covered by forest area. Thus, there is a need to explore further to understand the impact of forest area on the reservoir and the temporal changes of the water quality parameters that decide the water bodies quality.

Keywords Landsat 8 · Maximum chlorophyll index · Land surface temperature · Sentinel 2

A. Latwal · S. Rehana (✉)
Hydroclimatic Research Group, Lab for Spatial Informatics, International Institute of Information Technology, Gachibowli, Hyderabad, Telangana 500032, India
e-mail: rehana.s@iiit.ac.in

A. Latwal
e-mail: avantika.latwal@research.iiit.ac.in

K. S. Rajan
Lab for Spatial Informatics, International Institute of Information Technology, Gachibowli, Hyderabad, Telangana 500032, India
e-mail: rajan@iiit.ac.in

P. V. Timbadiya et al. (eds.), *Geospatial and Soft Computing Techniques*, Lecture Notes in Civil Engineering 339, https://doi.org/10.1007/978-981-99-1901-7_4

1 Introduction

Water quality deterioration has become a severe concern worldwide due to increased water pollution (contamination of water bodies) in recent decades [19]. Poor water quality threatens water resources, especially the drinking water supply, which is linked to public health concerns and economic development. Water quality is generally described according to physical, chemical, and biological characteristics [9]. The natural resources and inland water bodies provide essential habitats for the ecosystems, including wildlife and aquatic species [2, 13], and also act as a hot spot for global carbon cycling and a key player in climate change [21]. Thus, the degradation of the water quality resources can result from pesticides, heavy metals, nutrients, microorganisms, sediments, and many others [8].

Remote sensing techniques come with various spatial, temporal, and spectral resolutions, depending on the sensor specifications and satellite orbit. It makes it possible to have a spatial and temporal view of water quality parameters [8]. The satellite data products are used to monitor water quality parameters such as chlorophyll-a efficiently, suspended sediments, turbidity, temperature, and sechhi disk depth over a large scale [5, 8]. Numerous satellite images could be used for water quality assessments and monitoring. However, Landsat and Sentinel satellite images have been used widely due to free availability, temporal coverage (Landsat 8), and spatial resolution [10, 15].

Among various water quality parameters, chlorophyll-a is one of the most commonly investigated parameters using remote sensing information [10, 17, 20]. It is directly linked to the presence of algal bloom, which directs the level of eutrophication in water bodies [5]. Hence, monitoring chlorophyll-a concentrations is essential for managing eutrophication in inland water bodies [4]. Besides, chlorophyll-a parameter temperature is also an essential parameter for any waterbody. The water bodies biological and chemical processes depend on temperature [3]. Thus, satellite information can be used to assess baseline conditions and understand the changes and variations in water quality parameters. Therefore, a remote sensing-based study was carried out to estimate water quality in an inland water body. The main focus of the present study is (i) to detect the chlorophyll-a content spread and temperature of the reservoir; (ii) to map of contamination spread area in the waterbody for the period 2017–2018.

2 Study Area and Data Source

2.1 Bhadra Reservoir

The present study was carried out in the Bhadra reservoir, a tributary of the Tungabhadra River, which lies between 13° 42′00″N latitude and 75° 38′20″E longitude in

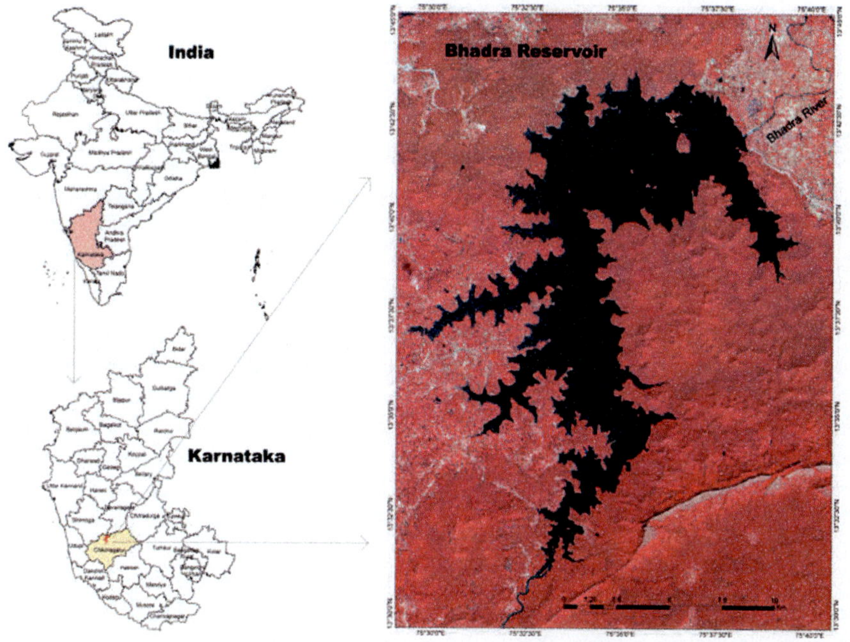

Fig. 1 Location of Bhadra Reservoir (False Color Composite Image of Sentinel 2A). *Source* Administrative areas from Survey of India and Sentinel Satellite data from European Space Agency (ESA) Copernicus

the western part of Karnataka, India (Fig. 1). The catchment area of the reservoir is 1968 km^2. It is used for irrigation, water supply, and hydropower generation.

2.2 Data Used

Two satellite data were used to detect the water quality parameter, i.e., Landsat 8 for temperature and Sentinel 2A for chlorophyll-a detection in the reservoir from European Space Agency (ESA) Copernicus (https://scihub.copernicus.eu/dhus/). The acquired dates are different for both the data source due to the unavailability of the same date satellite images; the description is in Table 1. Landsat 8 has a 30 m spatial resolution, whereas Sentinel 2A has different resolutions for different bands (10, 20, and 60 m).

Table 1 Acquired date information for Sentinel 2 and Landsat 8 satellite images

S. N.	Acquired date 2017			Acquired date 2018		
	Sentinel 2A	Landsat 8	Date difference	Sentinel 2A	Landsat 8	Date difference
1	07/01/2017	09/01/2017	2 days	22/01/2018	28/01/2018	6 days
2	17/04/2017	15/04/2017	2 days	22/04/2018	18/04/2018	4 days
3	29/10/2017	24/10/2017	5 days	24/10/2018	27/10/2018	3 days

3 Methods

To estimate the water quality parameter, two parameters were selected, i.e., chlorophyll-a and temperature. Chlorophyll-a is the proxy for nutrient contamination in the water body, and temperature plays a vital role in the reservoir to maintain the taste, color, and odor. To know the variation in chlorophyll-a and temperature in the reservoir, three-season were selected, i.e., winter (January), summer (April month was chosen for the summer season because the May image was not available for 2017), and autumn (October).

Sentinel 2A satellite data with Level-1C product were used to detect the chlorophyll-a content spread in the reservoir for 2017 and 2018. Level-1C product provides top of atmosphere (TOA) reflectance, representing the 'raw' reflectance of the Earth as measured from space. Therefore, the atmospheric correction was done through a physical model Sen2Cor [14], which converts Level 1 product into Level 2A product, providing the bottom of the atmosphere (BOA). The BOA represents the actual reflectance of the object or areas on the surface of the Earth and eliminates the effect of the atmosphere on the reflectance values.

To detect the chlorophyll-a content spread area in the reservoir, a maximum chlorophyll index (MCI) algorithm was used (Eq. 1):

$$\mathrm{MCI} = \mathrm{B5} - 1.005\left[\mathrm{B4} + \frac{(\mathrm{B6} - \mathrm{B4})(\lambda_{\mathrm{B5}} - \lambda_{\mathrm{B4}})}{\lambda_{\mathrm{B6}} - \lambda_{\mathrm{B4}}}\right] \tag{1}$$

where B4 (red band), B5, and B6 (vegetation red edge band) are the bands of satellite data and λ_{B4}, λ_{B5}, and λ_{B6} are the central wavelengths of the equivalent bands of Sentinel-2, and 1.005 is the factor, used to minimize the effect of thin clouds. The maximum chlorophyll index has been established as a valuable and multipurpose tool for monitoring intense surficial blooms.

Landsat 8 satellite data having thermal bands were used to determine the temperature of the reservoir. To evaluate the temperature, land surface temperatures were calculated. The surface temperature measures the surface soil, water, and vegetation cover on the earth's surface [18]. To determine the LST of the reservoir, the following equation was used Eqs. 2–7:

$$L_\lambda = M_L Q_{\mathrm{cal}} + A_L \tag{2}$$

where L_λ is top of atmosphere spectral radiance, M_L is radiance mult band in metadata (band 10 or 11), Q_{cal} is a thermal band of satellite data (band 10 or band 11), and A_L is radiance add band in metadata (band10 or 11)

$$T = \frac{k_2}{\ln\left(\frac{k_1}{L_\lambda} + 1\right)} \tag{3}$$

where T is brightness temperature in Kelvin (K), k_1 is band-specific thermal conversion K_1 constant from metadata (band 10 or 11), and k_2 is band-specific thermal conversion K_2 constant from metadata (band 10 or 11), and then the temperature in kelvin changes into degree Celsius.

Normalize Differentiate Vegetation Index (NDVI) is an important index for calculating, afterward, the proportional vegetation (P_V) and land surface emissivity (ε) are necessary to calculate LST, the Eqs. (4–6) used are as follows:

$$NDVI = \frac{NIR - RED}{NIR + RED} \tag{4}$$

where near infrared band (NIR) is the Band 5, and RED band is the Band 4 of Landsat 8 satellite data

$$P_v = \left[\frac{NDVI - NDVI_{min}}{NDVI_{max} - NDVI_{min}}\right]^2 \tag{5}$$

$$\varepsilon = 0.004 * P_v + 0.986 \tag{6}$$

where ε is emissivity, 0.004 is the average emissivity value of bare soil, 0.986 the average emissivity values of the vegetated areas, and P_v is the proportional vegetation calculated from NDVI [16]

$$LST = \left[T/\left(1 + \lambda * \frac{T}{\rho * ln(\varepsilon)}\right)\right] \tag{7}$$

where LST stands for land surface temperature, T is brightness temperature, λ is the wavelength of radiance emitted, ρ is Boltzmann constant (1.380649×10^{-23} J/K), and ε is emissivity.

The above calculations and all other datasets were subjected to various statistical calculations (area, percentage) using ArcGIS 10.2.2 and Excel 2016.

4 Results and Discussion

Chlorophyll-a is the most widely examined parameter for measuring water quality using remote sensing. The study detects the chlorophyll-a spread in the reservoir from

2017 to 2018. Season-wise comparisons were done through January (winter), April (summer), and October (autumn) months satellite images for both the years of 2017 and 2018 (Fig. 2). It was observed that October in both years has a high chlorophyll-a spread area, i.e., 32.18 km^2 in 2017 and 35.35 km^2 in 2018. In the winter season, the spread was 15.6 km^2 in 2017 and 14.85 km^2 in 2018, whereas, in summer, it was 31.07 km^2 in 2017 and 27.31 km^2 in 2018, respectively (Table 2). The water spread area was maximum in the month of October for both the year (2017 and 2018) due to a slight increase in inflow in the reservoir compared to January and April (Fig. 2). The inflow volume of the reservoir depends on the rainfall; it sharply increases during the flooding season. Due to the high rainfall events, the water inflow carries high nutrient input from the surroundings and also leads to inhibition of the release of nutrients from sediments, especially phosphate [12]. However, it was noticed that chlorophyll-a spread percent was higher in summer months when compared with the water spread area of the reservoir, i.e., 74.5% in 2017 and 61.4% in 2018 from the total water spread area (Table 2). April month gets highly contaminated due to very low inflow (Fig. 3) of water in the reservoir, which leads to an accumulation of contamination.

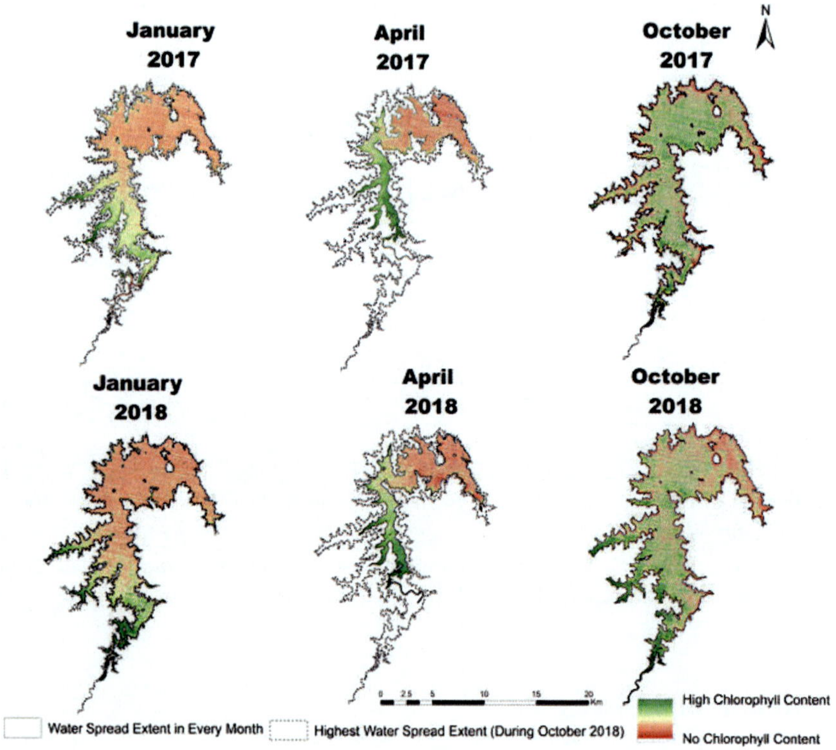

Fig. 2 Chlorophyll-a content spread derived from Sentinel 2A satellite data

Table 2 Water spread and chlorophyll-a content spread in the reservoir during the 2017 and 2018 period

Months	Water spread area (km^2)		Chlorophyll-a spread area (km^2)		Percentage of chlorophyll-a spread area from total water spread area (%)	
	2017	2018	2017	2018	2017	2018
January	72.71	93.06	15.60	14.85	21.45	15.96
April	41.69	44.49	31.07	27.31	74.54	61.38
October	100.45	112.30	32.08	35.35	32.04	24.35

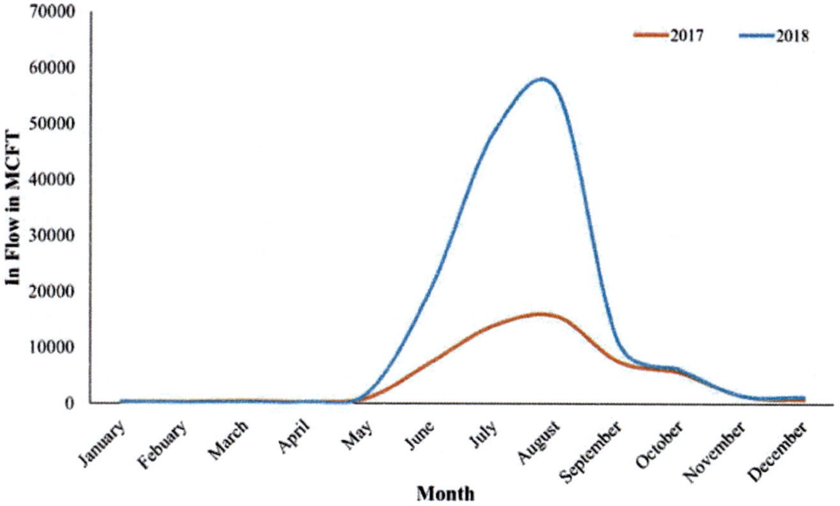

Fig. 3 Water inflow (in MCFT) of Bhadra Reservoir for 2017 and 2018 years

The temperature is also a critical water quality parameter that governs aquatic life and regulates the maximum dissolved oxygen concentration in water. The water temperature plays a significant role in ecological functioning and in monitoring the biogeochemical processes of a water body [1]. Therefore, the study calculates the temperature for the reservoir during the period 2017 and 2018, the same as the months chlorophyll-a spread was detected. It was observed that the reservoir temperature was higher in the summer of 2017 (range 23.3–32.2 °C) compared to the 2018 summer month (range 18.1–29.9 °C), as shown in Table 3. The high temperature of water increases the growth of the microorganisms and may increase problems related to taste, color, odor, and deterioration [22].

According to the many literatures, it was suggested that warming temperatures or short-term meteorological conditions (including wind, precipitation, and high temperatures) favor the growth and expansion of algal blooms in water bodies which

Table 3 Minimum, maximum, and mean temperature derived from Landsat 8 satellite image for the period 2017 and 2018

Months	Minimum temperature (°C)		Maximum temperature (°C)		Mean temperature (°C)		Standard deviation	
	2017	2018	2017	2018	2017	2018	2017	2018
January	19.8	20.2	25.1	26.3	20.6	21.4	0.5	0.7
April	23.3	18.1	32.1	29.9	24.6	21.2	0.9	1.2
October	20.0	22.2	23.9	25.9	21.5	23.3	0.6	0.3

leads to an increase in chlorophyll-a content spread [11, 23]. The studies also show that intense activity of chlorophyll-a spread was detected in the summer (April and May) and autumn (October) months [17, 20]. In the current study, it was noticed that the chlorophyll-a spread was also increasing in the reservoir as the temperature rises. In April, the chlorophyll-a spread and temperature were higher in 2017 compared to 2018, which shows that increasing temperature increases chlorophyll-a content in the reservoir. In addition, the reservoir is located near the forest area; therefore, it might be the reason for the low chlorophyll-a spread in October month (32.04% in 2017 and 24.35% in 2018) because it was observed that October month is the outbreak season for eutrophication. [7] It was observed that forest cover has a significant positive effect on water quality, which helps to maintain the quality of ideal water status. The forested area near water bodies plays a vital role in minimizing human activity impacts and maintaining water quality [6]. Thus, future research needs to be thorough study to know the impact of forest areas near the reservoir and estimate water quality.

5 Conclusions

In the present study, the two parameters were calculated for water quality monitoring, i.e., chlorophyll-a and temperature. The study noticed that the temperature and chlorophyll-a content spread parameters are comparable because as the temperature increase, the chlorophyll-a content also increases. The present study did not include in-situ measurements; the estimated values of chlorophyll-a content spread need thorough consideration. As October is an outbreak of eutrophication which leads to the spread high amount of chlorophyll in the polluted waterbody, the current study observed that in October, the reservoir has not much a high amount of chlorophyll-a spread due to the reservoir covered one side by the forest area. The forest area near the waterbody keeps water clean. Thus, there is a need to explore further to understand the impact of forest area in the reservoir and also the temporal changes in the water quality parameters.

Acknowledgements The research presented in this study was funded by the Ministry of Science & Technology, Department of Science & Technology, TMD (Energy, Water & Others), Water Technology Initiative, Project no. DST/TMD-EWO/WTI/2K19/EWFH/2019/306. The authors sincerely thank Dr. P. Somaeskhar Rao, Technical Director at the Advanced Centre for Integrated Water Resources Management (ACIWRM), Bengaluru, Karnataka, India, for providing Bhadra reservoir data.

References

1. Alcântara EH, Stech JL, Lorenzzetti JA, Bonnet MP, Casamitjana X, Assireu AT, de Moraes Novo EML (2010) Remote sensing of water surface temperature and heat flux over a tropical hydroelectric reservoir. Remote Sens Environ 114(11):2651–2665
2. Beeton AM (2002) Large freshwater lakes: present state, trends, and future. Environ Conserv 21–38
3. Bhateria R, Jain D (2016) Water quality assessment of lake water: a review. Sustain Water Resourc Manag 2(2):161–173
4. Carlson RE (1977) A trophic state index for lakes. Limnol Oceanogr 22:361–369
5. Chawla I, Karthikeyan L, Mishra AK (2020) A review of remote sensing applications for water security: quantity, quality, and extremes. J Hydrol 585:124826
6. de Mello K, Valente RA, Randhir TO, Vettorazzi CA (2018) Impacts of tropical forest cover on water quality in agricultural watersheds in southeastern Brazil. Ecol Ind 93:1293–1301
7. Duffy C, O'Donoghue C, Ryan M, Kilcline K, Upton V, Spillane C (2020) The impact of forestry as a land use on water quality outcomes: an integrated analysis. Forest Policy Econ 116:102185
8. Gholizadeh MH, Melesse AM, Reddi L (2016) A comprehensive review on water quality parameters estimation using remote sensing techniques. Sensors 16(8):1298
9. Gorde SP, Jadhav MV (2013) Assessment of water quality parameters: a review. J Eng Res Appl 3(6):2029–2035
10. Ha NTT, Thao NTP, Koike K, Nhuan MT (2017) Selecting the best band ratio to estimate chlorophyll-a concentration in a tropical freshwater lake using sentinel 2A images from a case study of Lake Ba Be (Northern Vietnam). ISPRS Int J Geo Inf 6(9):290
11. Hansen CH, Steven JB, Philip ED, Gustavious PW (2020) Evaluating historical trends and influences of meteorological and seasonal climate conditions on lake chlorophyll a using remote sensing. Lake Reservoir Manage 36(1):45–63
12. Huang T, Li X, Rijnaarts H, Grotenhuis T, Ma W, Sun X, Xu J (2014) Effects of storm runoff on the thermal regime and water quality of a deep, stratified reservoir in a temperate monsoon zone, in Northwest China. Sci Total Environ 485:820–827
13. Johnson N, Revenga C, Echeverria J (2001) Managing water for people and nature. Science 292:1071–1072
14. Main-Knorn M, Pflug B, Louis J, Debaecker V, Müller-Wilm U, Gascon F (2017) Sen2Cor for sentinel-2. In: Image and signal processing for remote sensing XXIII. Int Soc Opt Photonics 10427:1042704
15. Malahlela OE, Oliphant T, Tsoeleng LT, Mhangara P (2018) Mapping chlorophyll-a concentrations in a cyanobacteria-and algae-impacted Vaal Dam using Landsat 8 OLI data. S Afr J Sci 114(9–10):1–9
16. Palafox-Juárez EB, López-Martínez JO, Hernández-Stefanoni JL, Hernández-Nuñez H (2021) Impact of urban land-cover changes on the spatial-temporal land surface temperature in a tropical city of Mexico. ISPRS Int J Geo Inf 10(2):76
17. Peppa M, Vasilakos C, Kavroudakis D (2020) Eutrophication monitoring for lake Pamvotis, Greece, using sentinel-2 data. ISPRS Int J Geo-Information 9(3):143

18. Prasad R, Mani K (2015) Estimation of spatial variability of land surface temperature using Landsat 8 Imagery. Int J Eng Sci (IJES) 4(11):19–23
19. Scanlon BR, Jolly I, Sophocleous M, Zhang L (2007) Global impacts of conversions from natural to agricultural ecosystems on water resources: quantity versus quality. Water Resour Res 43(3)
20. Teja KT, Rajan KS (2016) Understanding the behaviour of contamination spread in Nagarjuna Sagar Reservoir using temporal Landsat Data. Int Arch Photogrammetry, Remote Sens Spatial Inf Sci 41
21. Tranvik LJ, Downing JA, Cotner JB, Loiselle SA, Striegl RG, Ballatore TJ, Dillon P, Finlay K, Fortino K, Knoll LB (2009) Lakes and reservoirs as regulators of carbon cycling and climate. Limnol Oceanogr 54:2298–2314
22. World Health Organisation (2017) Guidelines for drinking-water quality: Fourth Edition Incorporating the First Addendum; World Health Organisation: Geneva, Switzerland, ISBN 9241546964
23. Zong JM, Wang XX, Zhong QY, Xiao XM, Ma J, Zhao B (2019) Increasing outbreak of cyanobacterial blooms in large lakes and reservoirs under pressures from climate change and anthropogenic interferences in the Middle-Lower Yangtze River Basin. Remote Sens 11(15):1754

Impact Assessment of Water Conservation Planning Using RS and GIS Techniques—A Case of "Buldhana Project"

M. H. Rana and D. P. Patel

Abstract Today more than 56% of India is in the water-stressed zone. The groundwater levels around the country have drastically lowered, which has become a matter of concern for the government. There is an immediate requirement to stop lowering the groundwater table and work on recharging for its rising. Rain water conservation is one of the methods which can be effectively used for the purpose. Rain water conservation can be a by-product of infrastructure projects. For a recent example, an infrastructure project in the highways sector is acknowledged for setting up an example of model convergences of National Highway improvement/construction with water conservation and recharging groundwater. Khamgaon–Chikhli Highway, located in the water scarcity area of central India, is a trendsetter and example to set its unique way of groundwater conservation. This highway project which inter alia involves the consumption of a large quantum of earthwork. After a detailed study of the topography of the area, soil required for highway construction was extracted in such a way to create new ponds at potential location or enhancement of the capacity of existing ponds, deepening the nallas and river bed which were dry and contributed into small–medium irrigation projects and helped in recharging of the groundwater table. The present case describes the utilization of RS and GIS techniques for the impact assessment of water conservation planning. Recently, the 75 number of ponds which were constructed under the conservation and recharging planning of Khamgaon–Chikhli highway. This unique idea was applied to fully fill the earth materials requirements for the highway by consorting the ponds nearby. Here, RS and GIS

Disclaimer: The presentation of material and details in maps used in this chapter does not imply the expression of any opinion whatsoever on the part of the Publisher or Author concerning the legal status of any country, area or territory or of its authorities, or concerning the delimitation of its borders. The depiction and use of boundaries, geographic names and related data shown on maps and included in lists, tables, documents, and databases in this chapter are not warranted to be error free nor do they necessarily imply official endorsement or acceptance by the Publisher or Author.

M. H. Rana (✉) · D. P. Patel
Department of Civil Engineering, Pandit Deendayal Energy University, Gandhinagar 382426, India
e-mail: mrunalini.rphd20@sot.pdpu.ac.in; ranamrunalini@gmail.com

D. P. Patel
e-mail: dhruvesh.patel@sot.pdpu.ac.in

P. V. Timbadiya et al. (eds.), *Geospatial and Soft Computing Techniques*, Lecture Notes in Civil Engineering 339, https://doi.org/10.1007/978-981-99-1901-7_5

techniques were utilized to identify their impact on water conservation and recharge. Pre- and post-images are analyzed of this region using Google Engine. The significant changes in terms of water storage, vegetation cover and land use land cover are identified. Suitable maps are prepared to identify the pre- and post-impacts of water conservation planning.

Keywords Water conservation · Water recharge · Groundwater table · Tank

1 Introduction: Water Challenges Faced Globally

70% of the human body is made of water, and we are aware of the fact that water is the prime requirement for our lives. 70% of the earth is covered with water, yet we are running out of water. Such a contradict! Man is so greedy that he can use that 70% of saline water as well. Due to our high living standards in cities, we have already exceeded per capita consumption set forth by Central Public Health & Environmental Engineering Organization (CPHEEO). Further, rapid urbanization created concrete jungles across the world. Man has encroached and paved the earth with concretization and left no scope for water penetration. This has undoubtedly affected on the ecosystem. Rivers are dying, because the hydrological cycle is not getting completed due to imbalance in the environment and crazy man-made haphazard construction. There are countless examples of cities and urban settlements facing water availability challenges. In summer 2019, Chennai after soaring heat was facing a shortage of water. Similarly, the scarcity of water in Karnataka had become so severe that the state government had to pass an order to ban construction activities in Bengaluru city for five years. Today not only India but other countries around the globe are also facing water shortages and related challenges. In 2017, Cape-Town was declared to be water scarcity area and Zero-day. How miserable is that! "Zero-day" means no water for usage. In the history of Australia 1997–2007, Australia was also running out of water. Singapore is also one of the water scarce counties. This small county was buying 60% of water for its use, but they have utilized its topography in the best way to conserve water and control floods. Marina Barrage has solved its water crisis problem and boosted the water supply by 10%. The main objective of this paper is to showcase how highway construction activity can solve the purpose of recharging the ground considering the natural topography of the terrain. Hence, the twofold activity, that is using soil for road construction and creating ponds/tanks for recharging the ground, will be helpful for conserving the rainwater and in turn, will be supportive for social economic welfare and paybacks. Along with this, it assists to accomplish the restoration of ponds and tanks of the area and complete the ecological imbalance. To do so RS and GIS techniques are utilized. The water storage system is a key component for groundwater recharge, particularly in environments with poor rainfall. The identification of suitable water harvesting sites is a vital step in increasing land productivity by ensuring that water resources in the area are efficiently utilized and managed [1].

Water Conservation There are many methods for conserving water like rooftop rainwater harvesting, recharging ground water, etc. If we look at the water-deficient countries, the solution is right with us. It's just how we actually implement to the conditions with the correct solutions. First and foremost is the development of consciousness among people. People are utilizing water carelessly; therefore development of consciousness for utilization of water is as must as education. For example, Australia has made people aware by displaying the available water level of reservoir on the billboards. This concept made people consciously use water and come up with innovative methods to control the use of water. Many arid and semi-arid countries came up with an innovative technique for growing agriculture with limited use of water, with the idea of more crops per drop. Corporate offices set up company-level challenges to save water in UAE, the matter of the fact that the employee stated saving water up to 35% of water consumption. In a nutshell, unless and until a proper vigilant management system is not set, we will never be able to pay back to the earth. It is simple what we take from earth needs to be reverted back to our planet. Secondly, saving water through intelligent ways of water conservation is also important. Many Indian cities are located on the sea coast, like Mumbai, Bhuvneshwar, etc. wherein rainwater is allowed to disperse in to the sea due to insufficient water conservation resources and fewer methods for water conservation. Suppose smart methods and means can be developed for water conservation which would save significant rainwater which is getting wasted due to the merger with the sea. Similarly, smart utilization of available water is also the most important aspect of water conservation. For example, in most parts of India, there is no separation between drinking water and water being used in the industries. Due to growing industrialization, there is huge demand of water for industries. There has to be a difference between drinking water and industry water. Water demand of many industries can be served with treated water claimed from wastewater treatment plants. This will help in lowering freshwater consumption. Likewise, water conservation can also be a by-product of infrastructure projects. For example, during the development of infrastructure projects, lakhs of cubic meters of soil are required to be extracted from the ground from various locations. Instead of random extraction, if it is appropriately planned, these soil extraction points can be converted into potential sources of water conservation. This will not only serve the demand of soil for building infrastructure but also to create storage to conserve rainwater. For regions affected with water scarcity, by application of a scientific approach and the creation of multiple conservation storage can prove to be an effective measure for the improvement of the groundwater table [2, 3]. Several Landsat photos from various time periods (1972, 1982, 1987, 2000, 2003, and 2008) were used to analyze changes in the shoreline of Lake Nasser, Egypt, as well as the river volume [4]. Similarly, mapping of the ponds of Buldhana is done to see the pre- and post-monsoon effect after the execution of the highway project. Furthermore, studying this will aid in determining land use land cover change, which is necessary for understanding present changes and forecasting future highway development modifications. A similar concept was adopted in the development of National Highway Projects in the Buldhana district, which is severely affected with a scarcity of water.

2 Materials and Methods

2.1 Contribution Toward Water Conservation and Groundwater Recharge in Highway Projects

Ministry of Road Transport & Highway, Government of India (GoI) is implementing the development of National Highways, which inter alia involves considerable quantum of Earthwork/murrum. Ministry had decided to link this requirement with the deepening/de-silting of nallas, rivers, village tanks, minor and medium irrigation projects, etc., in the vicinity of the National Highway. To achieve this, ministry has issued a guideline in 2017 to all state governments to implement this, especially in drought regions. Maharashtra state government has promptly responded by allowing royalty-free material for the purpose. A detailed topographic survey was conducted to identify potential locations to create water storage or to utilize existing storage, enhancing its capacity by way of deepening/de-silting. Approximately 52.10 lakh cubic meter of material had been extracted from various such sources, i.e., nalla, river, old percolation tank, and small irrigation projects and thereby creating 5510 thousand cubic meter of additional storage capacity in the district during May 2016—June 2019, Photograph 1.

Photograph 1 Progressive improvement

3 Study Area and Construction Method Followed in the Project

3.1 Buldhana

This district is located in the Vidarbha area of Maharashtra State, Fig. 1. Buldhana has an area of about 9670 km^2. It is situated between 19° 51′ and 21° 17′ north latitudes and 75° 57′ to 76° 59′ east longitudes and has a population of 2,588,039 according to 2011 census data. The people's primary occupation is agriculture. The district is located in the Godavari and Tapi basins and the rivers Purna and Penganga run through the territory. Lonar in Buldhana is the 2nd caterer of Basaltic rock in the world. Except during the southwest monsoon season, which runs from June to September, the district's climate is typified by a scorching summer and general dryness throughout the year. 13 °C is the average minimum temperature, and 42.3 °C is the average maximum temperature [5]. The district's average yearly rainfall is between 711 and 911 mm. The area falls under scanty rainfall, which is the main reason for water scarcity in the district. Artificial recharging and rainwater harvesting were almost NIL up to the 2013 record [6]. The government has taken an initiative to recharge the groundwater through the development of new potential locations for rainwater storage and enhancement of existing storage by way of capacity building of existing ponds, deepening the nallas and river bed during the development of National Highway Construction Projects in the District.

3.2 The Method Followed in Buldhana District for the Collection of Construction

The murrum for road construction was located and utilized from the pond, tanks, river, and nallas. The topography survey was done by which the distance from the pond to the construction work site was known. The soil specimen was tested from these locations to check the suitability for its utilization in the construction activity of the highway. The ponds and tanks were deepened and frequently overflowing due to lesser storage capacity. The contour was deepened to enhance storage capacity considering FTL, Photograph 2. The small nalla flowing through the fields were encroached by the farmers on either side, making it narrower which resulted in damaging the crops due to over flooding. Hence, restoration of nallas helped in generating the smooth flow of rainwater toward storage. Similarly, for rivers, instead of continuous excavation, it was proposed to deepen in the stretch of 150 m with a gentle slope so that rainwater will continuously store in the ground, Photograph 3.

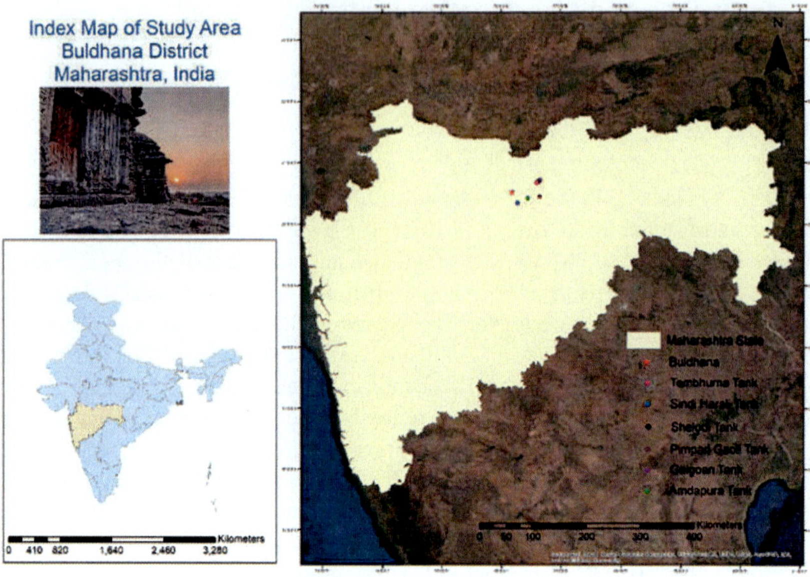

Fig. 1 Index map of the study area, Buldhana

Photograph 2 Amdapur tank

Photograph 3 Bandhara @ 88 + 050

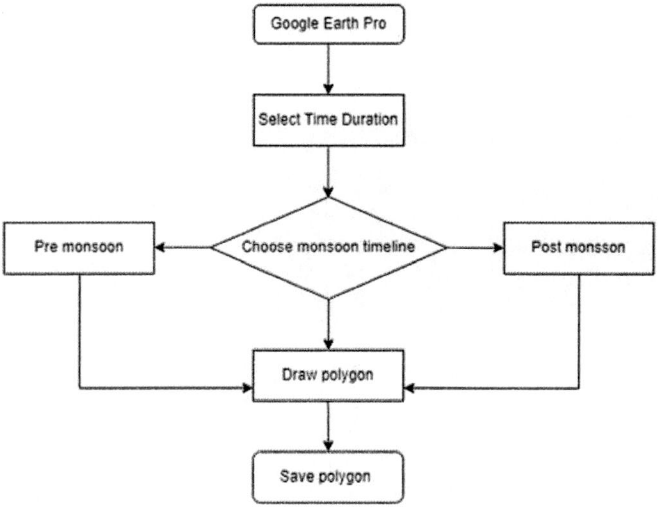

Fig. 2 Flow chart

3.3 The Method Followed in Buldhana District to Find the Pre- and Post-monsoon Effect of the Implementation of This Project

Google Earth Pro software is used to track the area of pre- and post-monsoon images. The time duration was selected for the years 2016, 2019, and 2021. The pre- and post-images were chosen for 2019 to show the impact of the highway project before and after the execution of the project. Polygon tool was used to map the areas of these timelines. As shown in flow chart, Fig. 2, the radical effect was observed in the pond area after the construction activity for the highway, as shown in Table 1.

3.4 Actual Field Implementation of Tanks in Buldhana District

Case 1 Shelodi Irrigation Tank, Near Jagruti Ashram, Taluka: Khamgaon, District: Buldhana.

Shelodi is an irrigation tank having an original storage capacity of 400 TCM. This tank was developed under a water supply scheme for Tembhurna and Shelodi villages. Due to water scarcity, during the summer season, both villages are highly dependent on water tanker. By implementing this twofold activity, the NHAI contractor has extracted about 2,500,000 cum soil from the tank and utilized it for constructing the embankment layer for the construction of National Highway No.6. It developed an

Table 1 Images of tanks in different years, before and after the project implementation

Name/year	2016	2019 (Pre-monsoon)	2019 (Post-monsoon)	2021 (Pre-monsoon)
Shelodi	 Area: Nil	 Area: Nil	 Area: 81,565 m^2	 Area: 41,313 m^2
Temburna	 Area: Nil	 Area: Nil	 Area: 369,721 m^2	 Area: 219,092 m^2

(continued)

Table 1 (continued)

Name/year	2016	2019 (Pre-monsoon)	2019 (Post-monsoon)	2021 (Pre-monsoon)
Sindi Harali	 Area: 3824 m^2	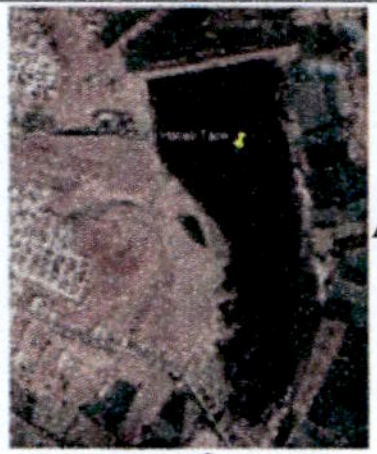 Area: 157,753 m^2	 Area: 313,123 m^2	 Area: 225,316 m^2
Amdapur	 Area: Nil	 Area: Nil	 Area: 96,425 m^2	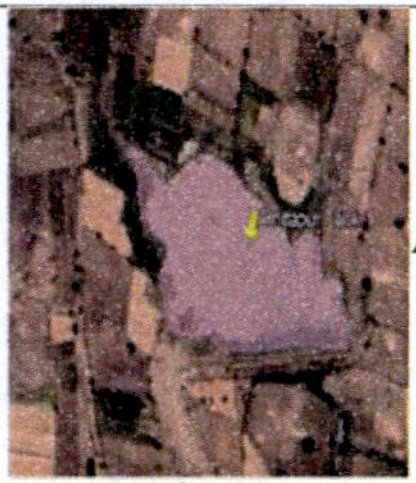 Area: 91,425 m^2
Pimpari Gaoli	 Area: 1967,941 m^2	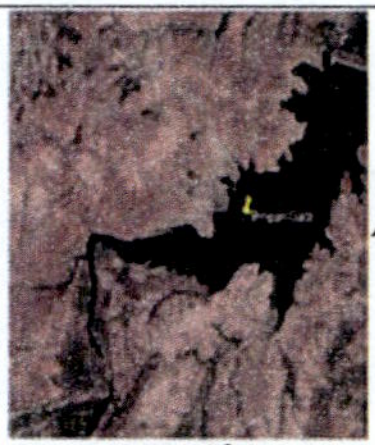 Area: 2087,987 m^2	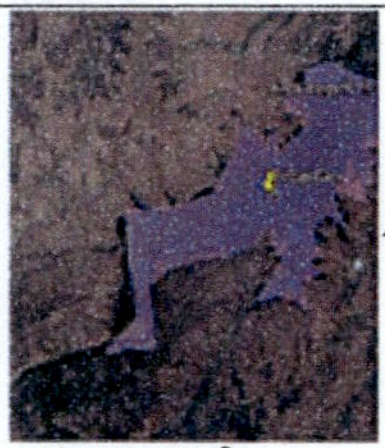 Area: 4165,885 m^2	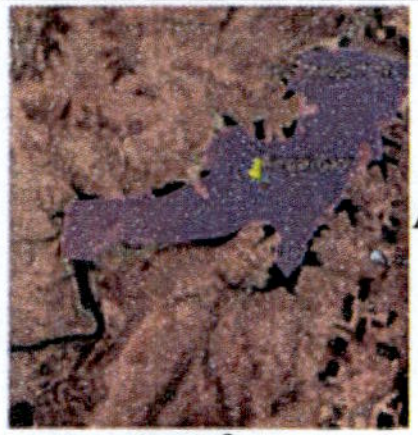 Area: 3171,301 m^2

(continued)

Table 1 (continued)

Name/year	2016	2019 (Pre-monsoon)	2019 (Post-monsoon)	2021 (Pre-monsoon)
Gaigaon	 Area: Nil	 Area: Nil	 Area: 150,818 m^2	 Area: 84,318 m^2

additional storage capacity of 250 TCM. Now the storage capacity of this tank is 650 TCM (i.e., 62.50% increase). During monsoon 2019, the tank got overflowed.

Benefits:

- Creation of additional storage of 250 TCM
- The expected increase in the irrigated area through recharged wells-33 ha
- Recharged wells—500 Nos
- Tanker free village—2 Nos
- Removal of encroachments in tank basin 2 ha
- Strengthen the source of water supply scheme for 2 villages.

Case 2 Tembhurna Percolation Tank, Taluka: Khamgaon, District: Buldhana.

Tembhurna is an earthen percolation tank that was initially developed by Govt. of Maharashtra, having a capacity of 340 TCM. The topsoil of the tank was fertile soil; therefore, at the time of deepening initially, the fertile soil at the top of the tank basin was extracted by the machines provided by an NGO named Bhartiya Jain Sanghthana. This fertile soil was taken away by farmers to their fields. About 45,000 cum of such soil was extracted and used in fields by farmers and as a result of that around 50 acres of land is made more fertile. Further for improvement to NH-6 and Khamgaon—Mehkar NH-548C, about 325,000 cum murrum which was available below fertile soil was extracted by the contractor, creating additional storage of 325 TCM. As mentioned above, the original capacity of the tank was 340 TCM which increased to 625 TCM. During monsoon 2019, the tank was overflowing.

Benefits:

- Improvement of farm fertility—50 Acres
- Creation of additional storage with no cost—325 TCM (95% extra)
- No of wells recharged—490 No
- Irrigation area increased through recharged wells—45 ha.

Case 3 Sindi Harali Percolation Tank, Taluka: Chikhli, District: Buldhana.

Sindi Harali tank is located just adjacent to Sindi Harali and Shelud villages having a population of 4132 nos. The original storage capacity of this tank was 131.20 TCM. About 50,000 cum soil from the tank basin was extracted and utilized for the construction of the embankment/subgrade layer of Chikhli–Khamgaon Highway NH 548 CC. It helped in creating additional storage of 50 TCM. The increased total storage capacity is 181.20 TCM as against the original 131.20 TCM. During this monsoon 2019, tank overflowed. Nearly 330 nos. of wells in Sindi Harali and Shelud village got recharged.

Case 4 Amdapur Irrigation Tank, Taluka: Chikhli, District: Buldhana.

Amdapur irrigation tank is located at km. 64/000 on Chikhli–Khamgaon Highway NH 548 CC. Amdapur village is having population 12,740 nos. The original storage capacity of the tank was 322.28 TCM, and during deepening, about 114,382 cum soil was extracted by the contractor engaged for the construction of Chikhli–Khamgaon Highway NH 548 CC which created additional storage of 114.38 TCM with 35.45%

additional. During monsoon 2019, the tank was overflowed. About 300 wells around the tank got recharged.

Case 5 Lanjud (Chikhali) Irrigation Tank (State Irrigation), Taluka: Khamgaon, District: Buldhana.

This state irrigation tank is located at km. 308 + 100 (off 0.50 km) of Balapur—Khamgaon—Malkapur—Chikhli NH 6. The original storage capacity of this tank was 1737 TCM with an expected irrigation area of 272 ha. This tank is a primary source for the water supply schemes of about 12 villages, viz. Chikhli, Lanjud, Saujatpur, Amboda, Amsari, Sutala, Pahurjira, Morgaon, Parkhed, Pimpri, H-Ghatouri and Wadi. The total population of the above villages is 42,877 nos. During summer, all these villages are tanker feed villages. Approximately 600,000 cum murrum was extracted from the tank basin and utilized for the construction of embankment/subgrade by the contractor appointed by NHAI (National Highway Authority of India). The storage capacity increased by 600 TCM (35% increase). During monsoon 2019, the tank got overflowed. About 700 wells got recharged, through which irrigated area increased by 95 ha.

Benefits:

- Creation of additional storage—600 TCM (35%)
- Expected Increase in the irrigated area due to recharged wells—95 ha
- Tanker free Village—12 Nos.
- Removal of encroachments in tank basin—5 ha
- Strengthening of water source for water supply scheme for 12 Nos. of villages.

Case 6 Tintrao Percolation Tank, Taluka: Shegaon, District: Buldhana.

The original storage capacity of the tank was 255.02 TCM. Approximately 200,000 cum soil was extracted from the tank and utilized for the construction of the embankment/subgrade of Khamgaon—Shegaon—Deori Highway NH 548 C. Additional storage of 200 TCM was created. Increased storage available is 455.02 TCM, i.e., 78% more than the original capacity.

Benefits:

- Creation of additional storage—200 TCM
- Wells recharged—205 No.
- Encroachment removal in basin area—2 ha
- The expected increase in irrigation area through recharged wells—19 ha
- Benefitted villages—Tintrao & Gaigaon

4 Results and Discussions

The year-wise mapping has been carried out for the tanks: (1) Shelodi, (2) Temburna, (3) Sindi Harali, (4) Amdapur, (5) Pimpari Gaoli and (6) Gaigoan using Google Earth Pro, and the results are described as below. For each tank, the area was mapped using the Google Earth Pro; the image captured in the tiles clearly shows the impact in the storage capacity of the tanks before and after the project implementation. The area of the tanks was measured pre- and post-monsoon of execution of the highways project for 2016, pre-monsoon 2019, post-monsoon 2019 and for the year 2021. Findings show that all tanks were dry in the year 2016, with the area covered by water being NIL in all tanks except in Pimpari Gaoli, with an area covered 1,967,941 km^2. in 2016. Later, post-monsoon in 2019 during the execution of the project, we could find the images were deepening of the tanks taken place with the area covered for each tanks recorded post-monsoon as mentioned and shown in Table 1.

(1) Shelodi-Area: 81,565 m^2, (2) Temburna-Area: 369,721 m^2, (3) Sindi Harali-Area: 313,123 m^2, (4) Amdapur- Area: 96,425 m^2, (5) Pimpari Gaoli- Area: 4,165,885 m^2 and (6) Gaigoan- Area: 150,818 m^2.

5 Conclusions

The outcome of the convergence pilot project in the Buldhana district in the state of Maharashtra has been very encouraging. After the project completion-post-monsoon 2019, the additional water storage has turned to 12,605 thousand cubic meters. Recharging of water table thereby increases the water level in wells. It also benefits agricultural kharif and rabi crop patterns through the augmented surface and groundwater storage. Socio-economic benefits via this project to the society—By deepening of nalla / river tank, there is an increase in rainwater storage capacity. Also, there is a creation of surface water storage without any additional expenditures, and henceforth, there is a good contribution in the groundwater table recharging wells, Photograph 4.

"Save Environment" was very well achieved else for construction of highway, there is a necessity of huge quantum of earthwork which is extracted from potential water storage locations. This is also supportive to soil conservation. The lands that were getting waterlogged were not just saved but also contributed to well-irrigated lands. There goes good support to Kharif and Rabi cropping due to the augmented surface and groundwater storage. A drastic change in the images is seen in pre- and post-monsoon images of Shelodi Tank, Tembhurna Tank, Sindi Harali Tank,

Photograph 4 Recharge well

Amdapur Tank, Pimpari Gaoli Tank and Gaigaon Tank, shown in Table 1. There was a creation of additional storage of 250 TCM in Shelodi Irrigation Tank observed to the original capacity of 400 TCM. As mentioned above, original capacity of tank was 340 TCM which increased to 625 TCM in Tembhurna Tank. During monsoon 2019, the tank was overflowing. The original storage capacity of Sindi Harali Tank was 131.20 TCM. About 50,000 cum soil from tank basin was extracted and utilized for construction of embankment/subgrade layer of highway construction, which helped in creating additional storage of 50 TCM. Increased total storage capacity is 181.20 TCM as against original 131.20 TCM. The original storage capacity of Amdapur Tank was 322.28 TCM, and by deepening, about 114,382 cum soil was extracted which created additional storage of 114.38 TCM with 35.45% additional. The original storage capacity of Pimpari Gaoli Tank was 1737 TCM. Approximately 600,000 cum murrum was extracted from the tank basin and utilized for the construction of the embankment/subgrade by the contractor appointed by National Highway Authority of India (NHAI). The storage capacity increased by 600 TCM up to 35% increase. The original storage capacity of Tintrao Tank was 255.02 TCM. Approximately 200,000 cum soil was extracted from the tank and utilized for the construction of embankment/subgrade for highway. Additional storage of 200 TCM created. Increased storage available is 455.02 TCM, i.e., 78% more than original capacity.

Acknowledgements The men behind the success of the project are Mr. Balasahen Theng, Chief Engineer (former), PWD Maharashtra and Mr. Vijay Patil, Executive Engineer (former), who were the pioneer in the implementation of the Model Project right from planning, investigation and identification the existing water bodies and coordinating and monitoring of the project. The authors are also thankful to Iron Triangle Limited who provided the site information and required data whatever and whenever required.

References

1. Patel D, Samal DR, Prieto C, Eslamian S (2021) Application of RS and GIS for locating rainwater harvesting structure systems
2. Mukherjee D (2016) A review on artificial groundwater recharge in India
3. Hashemi H, Berndtsson R, Kompani-Zare M, Persson M (2013) Natural versus artificial groundwater recharge, quantification through inverse modeling. Hydrol Earth Syst Sci 17:637–650. https://doi.org/10.5194/hess-17-637-2013
4. El Gammal EA, Salem SM, El Gammal AEA (2010) Change detection studies on the world's biggest artificial lake (Lake Nasser, Egypt). Egypt J Remote Sens Space Sci 13:89–99. https://doi.org/10.1016/j.ejrs.2010.08.001
5. Deshmukh DT, Lunge HS (2013) A Study Of Temperature And Rainfall Trends In Buldana District Of Vidarbha, India. Int J Sci Technol Res
6. Rafiuddin MK Govt of India, Ministry of Water Resources, Central Ground Water Board, Ground Water Information, Buldhana District-Maharashtra

Application of GIS and RS for Morphometric and Hypsometric Analysis of Pargaon Watershed: A Case Study

S. G. Wagh and V. L. Manekar

Abstract This study for the Pargaon watershed, a study area in the western portion of Maharashtra state with significant socio-economic implications, reports a morphometric analysis to comprehend the hydrological process and a hypsometric analysis for the catchment to expose the stages of geomorphic evolution. The study's approach included the use of digitized toposheets, georeferencing, and ArcGIS software. Knowing the type of the watershed yields morphometric parameters such as basin, areal, linear, shape, and landscapes. The highest stream order obtained was eighth order, the watershed has a dendritic drainage pattern, and the mean bifurcation ratio is 4.39, according to the analysis. The drainage density of 2.03 km^2/ km^2 indicates a low drainage density, categorizing the research area as coarse drainage texture. The hypsometric integral is 0.42, and the hypsometric curve is concave and convex, indicating that the Paragon watershed is fully developed.

Keywords Geographical information system (GIS) · Remote sensing (RS) · Morphometric analysis · Hypsometric analysis · Watershed

S. G. Wagh (✉) · V. L. Manekar
Department of Civil Engineering, Sardar Vallabhbhai National Institute of Technology Surat, Surat 395007, India
e-mail: ds16ce009@ced.svnit.ac.in

V. L. Manekar
e-mail: vlm@ced.svnit.ac.in

P. V. Timbadiya et al. (eds.), *Geospatial and Soft Computing Techniques*, Lecture Notes in Civil Engineering 339, https://doi.org/10.1007/978-981-99-1901-7_6

1 Introduction

The development of land and water resources on a sustainable basis without deterioration and with a steady growth in productivity is the lifeblood of humanity [1]. Soil is one of the most important natural resources on the planet's surface for supporting life. The state of the country's economy is also influenced by the soil cover [2]. The process of detachment and transfer of surface soil components is known as soil erosion. The entire land area affected by human-caused soil degradation is estimated to be 2 billion hectares worldwide. The land area impacted by erosion due to water and wind is estimated to be 1100 Mha for water erosion and 550 Mha for wind erosion [3]. Out of a total area of 3,280,000 km^2, 1,750,000 km^2 in India is vulnerable to soil erosion. As a result, over 53% of India's entire land area is vulnerable to erosion [1]. Soil erosion is a problem in every part of the planet. Because their farming practises are unable to replenish the loss of soil and nutrients, developing countries are experiencing increased issues with soil erosion [4]. Climate, hydrology, ground cover, and land use all have an impact on sediment output, as can temporarily invariable elements like lithology and watershed area [5]. Soil erosion and sediment yield are often higher in agricultural, degraded forest regions than in uncultivated areas.

A watershed is a natural hydrological feature where runoff is collected and directed into collecting channels, streams, or rivers. The size of the watershed is determined by the magnitude of the stream or river's interception. For watershed delineation, drainage density and distribution are also critical. Morphometric analysis is required to comprehend the watershed's hydrological behaviour in order to develop and manage natural resources [6]. As a result, determining the basic properties of a watershed in order to comprehend its geomorphology is necessary [7]. Lithologic, geomorphic, morphometric, land use land cover (LULC), and soil factors all play a role in long-term watershed management [8, 9].

Basin hydrology is vital for correctly managing water resources and flood threats in a basin, as the hydrological response of the upstream basin region influences the downstream basin area's hydrological response [10]. Soil erosion quantification could not provide an indication of management studies for the impacted watershed [11]. As a result, the morphological research will be useful in watershed management to prevent soil erosion. Morphometric analysis is an important part of a landform's geomorphology [12]. Morphometric analysis of drainage basins provides important information for basin characterization as well as topographical and geological data, as well as demonstrating the tendency for runoff and soil loss. The stream network can be created in one of two ways: by digitizing the streamlines from toposheets or by outlining the stream network in GIS using the Digital Elevation Model (DEM). The hypsometric curve is the relationship between horizontal cross-sectional areas and watershed altitudes, as determined through hypsometric analysis. The hypsometric curves and integrals are important indications of the state of a watershed. The hypsometric analysis, which gives data to quantitatively investigate the structural forms in a watershed, is regarded a useful method for understanding the geomorphic evolution of watersheds. The hypsometric analysis is used to identify the status of soil erosional proneness in the watershed and to establish its geomorphological stage, which includes monadnock (ancient), equilibrium (mature), and in-equilibrium (young) [13].

For planners and decision-makers to make effective and right decisions and efficient designs, remote sensing (RS) and Geographic Information System (GIS) techniques have become increasingly significant [14]. Remote sensing (RS), Geographic Information Systems (GIS), and Multi-Criteria Decision Making (MCDC) models have improved quantitative drainage network assessment, morphometric and hypsometric parameter analysis, and thematic mapping of soil erosion-prone locations [7]. Using the GIS technology, this study proposes watershed management options. To identify locations appropriate for implementing soil and water conservation measures, thematic maps are combined with morphometric and hypsometric criteria. When compared to customarily and manually collected data, sophisticated RS and GIS technology proved to be powerful, fast, reliable, and cost-effective for better data management, data processing, and mapping generation [7]. The goal of this study is to employ DEM to investigate morphometric parameters. The toposheets and DEM are used to demarcate the watershed border. This case study uses remote sensing and GIS to describe the Pargaon watershed's morphometric and hypsometric analysis results in order to better understand the watershed's behaviour and geomorphological stages.

2 Study Area and Data Source

2.1 Pargaon Watershed

The Paragon watershed is used as a research area in this study. The study area spans 18° 18′ N to 19° 05′ N latitude and 73° 19′ E to 74° 25′ E longitude in Maharashtra's western region. A geographical area of 6283.45 km^2 is covered by the watershed. The Survey of India (SoI), the country's National Mapping Agency (NMA), has toposheet numbers 47E12, 47F5, 47F6, 47F7, 47F9, 47F10, 47F11, 47F13, 47F14, 47F15, 47J1, 47J2, 47J3, 47J6, and 47J7 that cover the area. The geology of the research area is characterized by granite gneiss. 1134 m and 318 m, respectively, are the highest and lowest reliefs obtained. The research area contains a wide range of soil types. Clayey loam, loam, and sandy clay loam are the soil types. [15]. The mountain ranges in the Western Ghats reach heights of around 1134 m, resulting in a steep slope. The slopes are moderate in the northern section of the catchment, but they range from moderately steep to steep in the southern part [16]. Figure 1 shows the index map of the Pargaon watershed.

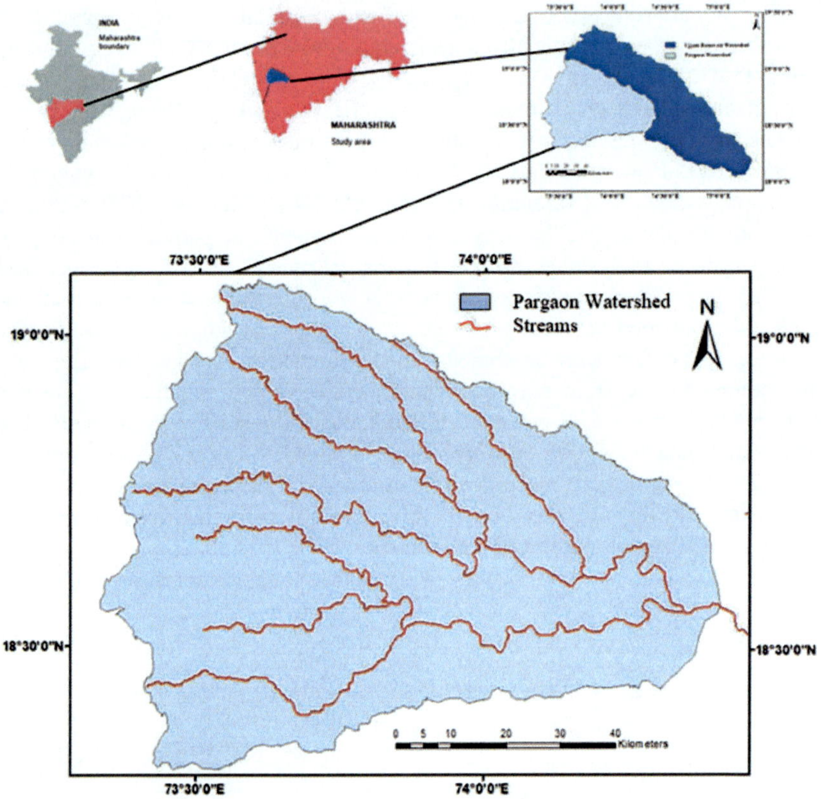

Fig. 1 Index map of the study area

2.2 Data Used

The Pargaon watershed has an area of 6283.45 Km^2. The study region is delineated using toposheets produced by the Survey of India at a scale of 1:50,000. Toposheets are used to not only demarcate watersheds and micro-watersheds, but also to create a base map with information about drainage, contours, and other features. A total of 15 toposheet data sets are employed in this method. The scanning of toposheets is part of the approach. Georeferencing of masked, mosaicked toposheets and resampling into Universal Transverse Mercator (UTM) projection WGS 1984, zone 43 North in TIFF format, importing the scanned toposheets in.img format, georeferencing of masked, mosaicked toposheets and resampling into Universal Transverse Mercator (UTM) projection WGS 1984, zone 43 North. The toposheets are used to demarcate the study area in the ERDAS IMAGINE programme 2014. Contours are scanned at 20 m intervals using a mosaic of resampled toposheets. ArcGIS software version 10.3 is used to create a DEM with a spatial resolution of 20 m.

Morphometric parameters

To comprehensively represent watershed geometry, morphometric parameters such as basic, linear, areal, slope, and landscape characteristics are required. The morphometric parameters reflect the causative elements affecting surface runoff and sediment output from the watershed, either directly or indirectly [6]. The numerous mathematical relationships for quantifying morphometric characteristics are shown in Table 1.

Hypsometric integral

It is used to determine the geomorphic stages of watershed development and to represent how mass is distributed from bottom to top within a watershed [14]. The elevation–relief ratio method proposed by Pike (1971) is used to calculate the hypsometric integral [17]. $(H_{\text{mean}} - H_{\text{min}})/(H_{\text{max}} - H_{\text{min}})$. Table 2 shows the hypsometrical integral (HI) ranges and stages.

Table 1 Global climate models used in the present analysis

S. No.	Morphometric parameter	Quantifying relation
1	Stream order (N)	Hierarchical order
2	Stream length (L_s)	Length of stream (Km)
3	Number of streams (N_s)	Total number streams of all order
4	Bifurcation ratio (R_b)	$R_b = (N_u/N_u + 1)$ Where, N_u and $N_u + 1$ are total number of stream segments of order u and stream segments of next higher order
5	Drainage density (D_d)	$D_d = L_s/A$ Where L_s is the total steam length of all orders
6	Stream frequency (F_s)	$F_s = N_s/A$
7	Texture ratio (R_t)	$R_t = N_s/P$
8	Length of overland flow (L_{of})	$L_{\text{of}} = (1/2 * D_d)$
9	Constant of channel maintenance (C_{cm})	$C_{\text{cm}} = 1/D_d$
10	Form factor (F_f)	$F_f = (A/Lb\text{^}2)$
11	Elongation ratio (R_e)	$\text{Re} = \left(\frac{2}{Lb}\right) \times \left(\frac{A}{\P}\right)^{0.5}$ ^0.5
12	Compactness coefficient (C_c)	$C_c = 0.2821 * (P)/(A\text{^}0.5)$
13	Circulatory ratio (R_c)	$R_c = 4 \P A/P^2$
14	Basin relief (R)	$R = H_{\text{max}} - H_{\text{min}}$ Where H_{max} = maximum elevation H_{min} = minimum elevation
15	Relief ratio (R_r)	$R_r = R/L_b$
16	Relative relief (R_{rf})	$R_{\text{rf}} = R/P$
17	Relief peakedness (R_{pk})	$R_{\text{pk}} = H_{\text{mean}}/H_{\text{max}}$

Table 2 Hypsometric integral ranges and corresponding geomorphic stages

S. No.	Hypsometric integral range	Remarks
1	Less than 0.3	Monadnock (old stage)
2	0.3–0.6	Mature stage
3	More than 0.6	Young stage

Table 3 Basic watershed parameters

Parameter	Value
Area (A)	6283.45 Km^2
Perimeter (P)	418.00 km
Basin length (L_b)	113.00 km
Maximum elevation (H_{max})	1134 m
Minimum elevation (H_{mean})	318 m
Mean elevation (H_{min})	661 m

3 Results and Discussions

The study's findings are divided into two sections: the first covers the morphometric investigation, and the second covers the hypsometric study. Basic, linear, areal, slope, and landscape morphometric parameters are investigated, while hypsometric analysis in the form of the hypsometric curve and hypsometric integral is performed and discussed.

3.1 *Basic Morphometric Parameters*

For evaluating linear, areal, slope, and landscape indices, the basic parameters are critical [18]. Table 3 shows the basic watershed parameters.

3.2 *Linear Morphometric Parameters*

The GIS environment is used to quantify linear metrics such as stream order (N), number of streams (N_s), stream length (L_s), bifurcation ratio (R_b), and mean bifurcation ratio (R_{bm}) of watersheds. Table 4 shows the findings of the parameters that were acquired.

Table 4 Linear watershed parameters

Stream order (N)	No. of Streams (N_s)	Total stream length (L_s) (Km)	Bifurcation ratio (R_b)	Mean bifurcation ratio (R_{bm})
1	26,973	6617.88	–	4.39
2	11,556	3128.94	2.33	
3	6548	1552.97	1.76	
4	3547	689.72	1.85	
5	2309	371.17	1.54	
6	1863	233.61	1.24	
7	829	138.27	2.25	
8	42	7.59	19.74	
	53,667	12,740.14		

3.3 Areal Morphometric Parameters

The GIS environment measures drainage density (D_d), stream frequency (F_s), length of overland flow (L_{of}), and constant of channel maintenance (C_{cm}). In Table 5, the areal morphometric characteristics are listed.

Drainage density is determined by the permeability of subsurface materials, vegetation type, and terrain relief. Lower drainage density suggests a permeable subsurface, adequate flora, and low roughness, whereas high drainage density indicates the reverse. [7, 19]. The low drainage density indicates a very porous subsoil with a coarse drainage texture in the Pargaon watershed. The Pargaon watershed's drainage density value is calculated to be 2.03 km/km^2. The frequency of a stream is determined by its recurrence rate; the stream frequency is calculated to be 8.54. The flow of precipitated water that flows over the ground surface and into the stream channel is referred to as overland flow. The length of the overland flow is estimated to be 0.24 km, and channel maintenance is continual. The overland flow length is predicted to be 0.24 km, and the channel maintenance constant is 0.49 km^2/km, indicating that the Pargaon watershed is less prone to erosion.

Table 5 Areal watershed parameters

Parameter	Value
Drainage density (D_d)	2.03
Stream frequency (F_s)	8.54
Length of overland flow (L_{of})	0.24
Constant of channel maintenance (C_{cm})	0.49

Table 6 Shape watershed parameters

Parameter	Value
Form factor (F_f)	0.49
Elongation ratio (R_e)	0.79
Compactness coefficient (C_c)	1.49
Circulatory ratio (R_c)	0.45

3.4 Shape Morphometric Parameters

For the Pargaon watershed, the shape parameters form factor (F_f), elongation ratio (R_e), compactness coefficient (C_c), and circulation ratio (R_c) are investigated. In Table 6, the shape morphometric parameters are listed.

The dimensionless ratio of basin area to square of basin length is known as the form factor. The form factor in the study region is assessed to be 0.49. A lower form factor number implies that the watershed is extended, with a flatter flow for a longer period of time. The watershed's elongation ratio is 0.79, indicating considerable relief with steep slopes. Infiltration is related to the compactness coefficient. The compactness coefficient is calculated to be 1.49, indicating that it has an elongated shape. The circulatory ratio is a number that ranges from 0 to 1. The circulatory ratio of the Pargaon watershed is 0.45, indicating that the topography of the watershed is mature.

3.5 Landscape Morphometric Parameters

To examine terrain characteristics, landscape factors of the watershed are linked to elevation features. For the Pargaon watershed, the metrics basin relief (R), relief ratio (R_r), relative relief (R_{rf}), and relief peakedness (R_{pk}) are investigated. In Table 7, the landscape morphometric characteristics are listed.

The Pargaon watershed's basin relief is estimated to be 816 m, implying little infiltration and high runoff. The relief ratio with a larger value suggests a mountainous terrain, whereas the relief ratio with a lower value indicates a plain area. The watershed's estimated relief ratio of 7.22 indicates low-to-moderate relief. Erosion

Table 7 Landscape watershed parameters

Parameter	Value
Basin relief (R)	816
Relief ratio (R_r)	7.22
Relative relief (R_{rf})	1.95
Relief peakedness (R_{pk})	0.58

Table 8 Estimation of hypsometric curve

Elevation (m)	Area (km^2)	Altitude (m)	Elevation difference (m)	e/E	Cumulative area (km^2)	a/A
318	0	1134	816	1	0	0.00
318–400	0.76	1100	782	0.96	0.76	0.00
400–500	1.52	1000	682	0.84	2.28	0.00
500–600	2088.02	900	582	0.71	2090.3	0.33
600–700	2338.57	800	482	0.59	4428.87	0.70
700–800	1144.76	700	382	0.47	5573.63	0.89
800–900	497.67	600	282	0.35	6071.3	0.97
900–1000	191.24	500	182	0.22	6262.54	1.00
1000–1100	19.77	400	82	0.10	6282.31	1.00
1100–1134	1.14	318	0	0.00	6283.45	1.00

is directly related to relative relief and relief peakedness. Its lower value represents an old stage topography in the study area.

3.6 Hypsometric Analysis

It is used to represent how mass is distributed within a watershed from bottom to top and to establish the geomorphic stages of development of watersheds [14]. Table 8 shows the hypsometric curve estimation for the Pargaon watershed.

By plotting the hypsometric curve of the Pargaon watershed on the x-axis and elevation on the y-axis, the curve for the Pargaon watershed is obtained as shown in Fig. 2. As elevation rises, the convex shape represents a younger, undissected landscapes stage, whereas the concave shape implies an older, severely dissected, and pediplain landscapes stage [13].

The hypsometric qualities of the Pargaon watershed were investigated using the hypsometric integral (HI) and the hypsometric curve obtained, and the HI value was found to be 0.42. Table 2 shows that the Pargaon watershed is at an advance stage of development.

4 Conclusions

The following conclusions are derived from the foregoing study:

- In comparison to traditional methods, advanced RS and GIS technologies are powerful, fast, dependable, and cost-effective for better data management, data

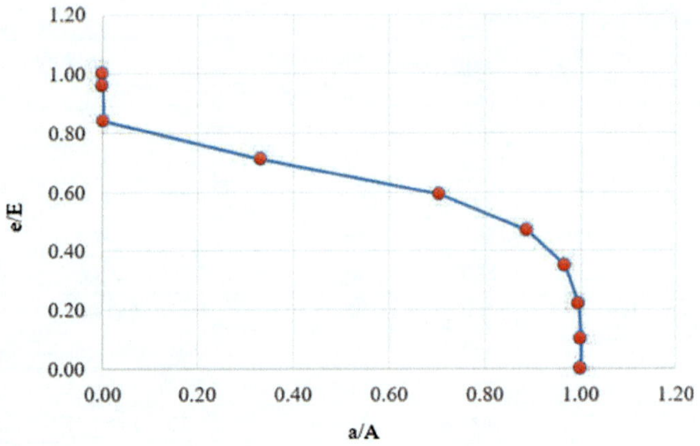

Fig. 2 Hypsometric curve of the pargaon watershed

processing, and map creation. As a result, the current research aimed to use GIS and RS for morphometric and hypsometric analysis.

- The systematic description of watershed geometry and the geomorphic stages of development of watersheds was revealed through quantitative morphometric and hypsometric analysis.
- The research area's drainage pattern is dendritic, showing homogeneous erosion resistance, and the greatest stream order obtained is the eighth order. The bifurcation ratio implies that the geological features have disrupted the watershed. The watershed's shape parameters imply that the watershed is more extended. The landscape parameters, on the other hand, imply that the watershed has low to medium relief.
- The shape of the hypsometric curve and the hypsometric integral (0.42) indicate that the Pargaon watershed has reached mature equilibrium, and soil and water conservation should be given high attention for its long-term growth.
- This study recommends soil conservation measures such as installing catchment area treatments such as check dams and plant cover to reduce sediment production reaching the study area's reservoirs. Extreme indices during the historical period at annual scale.
- The digitization of toposheets for the creation of DEMs is recommended in this study to reduce the uncertainty produced by other DEMs or data sources and to improve the accuracy and quality of the work.

5 Disclaimer

The presentation of material and details in maps used in this chapter does not imply the expression of any opinion whatsoever on the part of the Publisher or Author concerning the legal status of any country, area or territory or of its authorities, or concerning the delimitation of its borders. The depiction and use of boundaries, geographic names and related data shown on maps and included in lists, tables, documents, and databases in this chapter are not warranted to be error free nor do they necessarily imply official endorsement or acceptance by the Publisher or Author.

References

1. NIH (2000) Watershed prioritization of Ukai dam catchment using Remote sensing and GIS. National Institute of Hydrology, Roorkee
2. Naqvi HR, Athick AMA, Ganaie HA, Siddiqui MA (2015) Soil erosion planning using sediment yield index method in the Nun Nadi watershed, India. Int Soil Water Conserv Res 3(2):86–96
3. Ganasri BP, Ramesh H (2016) Assessment of soil erosion by RUSLE model using remote sensing and GIS-A case study of Nethravathi Basin. Geosci Front 7(6):953–961
4. Erenstein OC (1999) The economics of soil conservation in developing countries: the case of crop residue mulching
5. Noori H, Karami H, Farzin S, Siadatmousavi SM, Mojaradi B, Kisi O (2018) Investigation of RS and GIS techniques on MPSAIC model to estimate soil erosion. Natl Hazards 91:221–238
6. Malik A, Kumar A, Kandpal H (2019) Morphometric analysis and prioritization of sub-watersheds in a hilly watershed using weighted sum approach. Arab J Geosci 12(4):118
7. Arabameri A, Tiefenbacher JP, Blaschke T, Pradhan B, Tien BD (2020) Morphometric analysis for soil erosion susceptibility mapping using novel gis-based ensemble model. Remote Sens 12(5):874
8. Bhattacharya RK, Chatterjee ND, Das K (2020) Sub-basin prioritization for assessment of soil erosion susceptibility in Kangsabati, a plateau basin: A comparison between MCDM and SWAT models. Sci Total Environ 734:139474
9. Balasubramanian A, Duraisamy K, Thirumalaisamy S, Krishnaraj S, Yatheendradasan RK (2017) Prioritization of subwatersheds based on quantitative morphometric analysis in lower Bhavani basin, Tamil Nadu, India using DEM and GIS techniques. Arab J Geosci 10(24):552
10. Pawar AD, Sarup J, Mittal SK (2014) Application of GIS and RS for morphometric analysis of upper Bhima Basin: a case study. J Inst Eng (India): Ser A 95(4):249–257
11. Snehit B, Sahoo SN (2019) Prioritization of sub watersheds for soil conservation management using morphometric characteristics. In: Proceedings of the 24th hydro international conference Hyderabad. India, pp 1399–1409
12. Sahoo S, Meher J (2019) Estimation of sub watershed wise morphometric parameters of Baitarani basin using Remote Sensing and Geoinformation System. In: Proceedings of the 24th hydro international conference Hyderabad. India, pp 2559–2568
13. Rajitha E, Ravikumar AS (2019) Morphometric and hypsometric analysis using remote sensing and GIS technique. In: Proceedings of the 24th hydro international conference Hyderabad. India, pp 2846–2854
14. Meshram SG, Sharma SK (2015) Prioritization of watershed through morphometric parameters: a PCA-based approach. Appl Water Sci 7(3):1505–1519
15. Shendge R, Chockalingam M, Saritha B, Ambica A (2018) A. Swat modelling for sediment yield: a case study of Ujjani reservoir in Maharashtra India. Int J Civ Eng Technol 9(1):245–252
16. Wagh S, Manekar V (2021) Assessment of reservoir sedimentation using satellite remote sensing technique (SRS). J Inst Eng (India): Ser A 102(3):851–860
17. Pike RJ, Wilson SE (1971) Elevation relief ratio, hypsometric integral and geomorphic area altitude analysis. Geol Soc Am Bull 82:1079–1084

18. Hembram TK, Saha S (2018) Prioritization of sub-watersheds for soil erosion based on morphometric attributes using fuzzy AHP and compound factor in Jainti River basin, Jharkhand, Eastern India. Environ Dev Sustain 22(2):1241–1268
19. Kadam AK, Jaweed TH, Kale SS, Umrikar BN, Sankhua RN (2019) Identification of erosion-prone areas using modified morphometric prioritization method and sediment production rate: A remote sensing and GIS approach. Geomat Nat Haz Risk 10(1):986–1006

Hypsometric Analysis of Brahmani–Baitarani Basin Using ArcGIS

R. N. Sankhua and K. P. Samal

Abstract Surface and subsurface water runoff in a river basin plays a very crucial role in geomorphic stages of a drainage basin. It is also an essential tool to measure and represent the form of a watershed and its evolution. Hypsometric analysis is useful for understanding the stages of geomorphic development of the watershed under consideration. Elevation, relief ratio method was also used for calculation of hypsometric integral values. In this study, hypsometric integral values of sub-basins of Brahmani–Baitarani basin were evaluated. In the study area, two stages of erosional cycle development namely youthful to mature stage are identified. The development of stream segment is affected by slope and local relief. These factors produce differences in values of drainage density among the sub-basins. Hypsometric integral quantifies the geologic stages of development and erosional proneness of the watershed. The representation of horizontal cross-sectional area with respect to elevation (area-altitude analysis) are found out by the morphological analysis. Brahmani–Baitarani basin was divided in to eight watersheds using 30 m SRTM DEM. The stream ordering of the watershed also has been done. Hypsometric curve was derived and analyzed for sub-basins of the divided Brahmani–Baitarani basin. The high-medium hypsometric integrals/elevation relief ratios indicate a youthful to mature stage landscape, medium to complex geological processes. The analysis of linear morphological changes, hypsometric integrals, and curves of this river basin are attempted using spatial tools. This paper emphasizes the hypsometric integral and corresponding geological age of the eight sub-basins of Brahmani–Baitarani basin.

Keywords Hypsometric curve; Brahmani · Baitarani river basin

R. N. Sankhua
NWDA, Ministry of Jal Shakti, Hyderabad, India

K. P. Samal (✉)
KIIT Deemed University, Bhubaneswar, India
e-mail: kshyanaprava.samalfce@kiit.ac.in

P. V. Timbadiya et al. (eds.), *Geospatial and Soft Computing Techniques*, Lecture Notes in Civil Engineering 339, https://doi.org/10.1007/978-981-99-1901-7_7

1 Introduction

Area-altitude analysis reveals the distribution of horizontal cross-sectional area of a drainage area with respect to elevation. Erosional landforms at different stages during their evolution can be differentiated through hypsometric analysis. In this paper, the statistical parameters of the hypsometric analysis are tabulated, which comprises the hypsometric integral (I), hypsometric curve, hypsometric skewness, indicating the watershed conditions. The hypsometric integral indicates the area beneath the curve, which gives the interrelationship between drainage area and altitude of basin, the percentage of total relief to cumulative percent of area and shape of the area-elevation curve, thereby indicating age of the drainage area.

Relative landform age and the extent of drainage area dissection can be interpreted using the hypsometric curves. High integrals depict younger geomorphic stages and un-anatomized landscapes, S-shaped curves represents equilibrium or mature stage, and deeply dissected landscapes are showed by concave-up with low integrals as found by Strahler [1]. The inverse correlation of the hypsometric integral can be seen for total relief, slope steepness, drainage density, and channel gradients too. This gives erosion status of the drainage area hinting about the soil conservation and water measures. The value of hypsometric integral of the drainage area explains about the erosion that had taken place. No hypsometry study was undertaken for the Brahmani–Baitarani basin to know about the watershed health so far because of availability of data for the study. Now, the study has facilitated to have plenty of idea about the landform characteristics for throwing light on the nature of erosion, run off pattern, etc.

2 The Study Area

The combined Brahmani–Baitarani basin extends over a geographical area of 53,902 km^2, and the basin is bounded on the north by the Chhota Nagpur Plateau, on the west and south by the ridge separating it from Mahanadi basin and on the east by the Bay of Bengal. Through intersection of state administrative boundaries and basin boundaries (derived for the present study), state-wise drainage areas are computed. The drainage area of the basin lies in the states of Odisha (33,923 km^2), Jharkhand (15,479 km^2), and Chhattisgarh (1367 km^2). Out of the total basin area, major part of 66.82% is covered in Odisha, 30.49% in Jharkhand, and 2.69% in Chhattisgarh. The basin is bounded by 20° 29′ 00″ to 23° 37′ 47″ North latitude and 83° 53′ 49″ to 87° 1′ 27″ East longitude.

Fig. 1 Study area

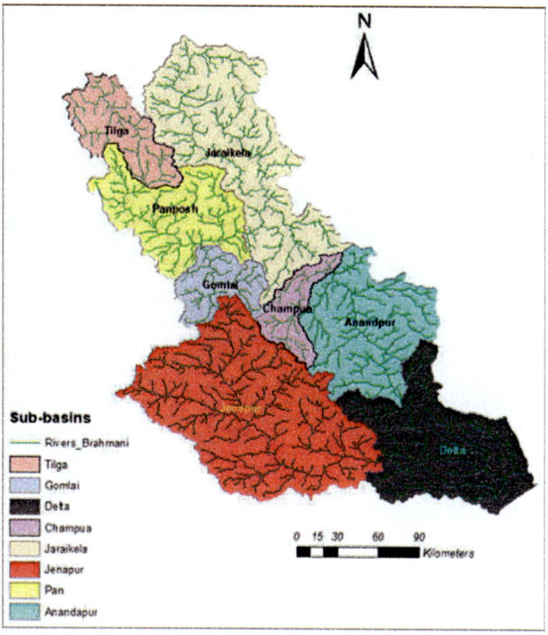

3 Data and Methodology

Figure 1 shows the study area. The DEM reveals that the elevation ranges from 0 m to a maximum of 1174 m in the basin, whereas the mean elevation is about 341 m. The freely available Digital Elevation Model (DEM) map of 30 m spatial resolution data of Shuttle Radar Topographic Mission (SRTM) were obtained from because of its better Vertical Dilution of Precision (VDOP). The tiles were mosaicked together and the sub-basin boundaries were delineated following the same pattern as has been done on water resources availability study by Central Water Commission in 2019. Figures 2 and 3 show the clipping of SRTM DEMs and topography maps with reference to basin boundary. As suggested by Kokkas [2], the area-elevation references were drawn from the DEM for Brahmani–Baitarani basin landscape characterization.

Hypsometric integral Calculation

As proposed by Pike and Wilson [3], the integration of the hypsometric curve was evaluated which gives the hypsometric integral (HI), which is equivalent to the elevation-relief ratio (E). Mathematically, it is defined as

$$E \approx HI = \frac{(\text{Mean Elevation} - \text{Minimum elevation})}{(\text{Max Elevation} - \text{Minimum elevation})} \tag{1}$$

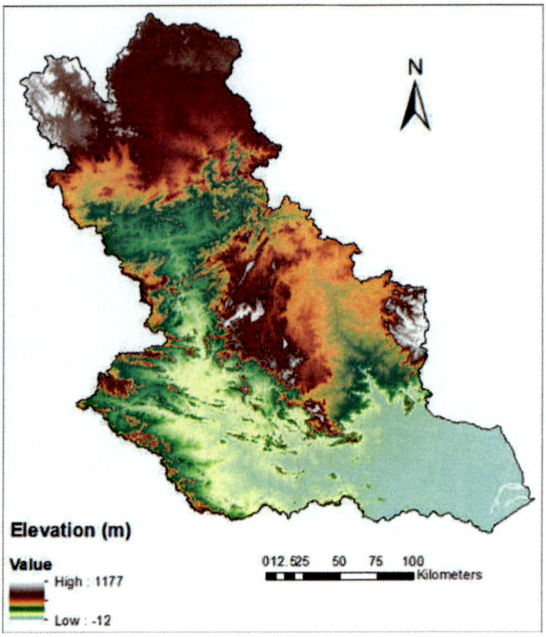

Fig. 2 DEM map of study area

Fig. 3 Topography map of the study area

As is evident, the degradation and topological changes in the basin are generally weathered by surface runoff, sediment transportation, and stream erosion. The complex, long-term geomorphic processes in nature are many; thus, it becomes very tedious to really interpret the changes that take place over period of time. Thus, the vital indicators like hypsometric curve (HC) and hypsometric integral (HI) are analyzed to arrive at the degree of disequilibria in the balance of erosive and tectonic forces [4]. As discussed earlier, the characteristics of the curve thus derived could be utilized to get the classification of the eight sub-basins as younger (convex upward curves), young matured (S-shaped hypsometric curves which is concave upward at high elevation and convex downward at low elevation) and peneplain distorted (concave upward curves) as narrated by Strahler [1]. The basin shape, relief, and drainage area control the numerical hypsometric integral value as suggested by Lifton and Chase [5], Masek et al. [6], Hurtrez et al. [7, 8], Chen et al. [9]. The steepness, slope, drainage density, total relief, and gradient correlate the hypsometric integral inversely. This in turn quantifies the landform development and status of erosion in the basin. For example, the higher value of HI advocates a younger geomorphic stage and less eroded area, and the comparatively lower values tend toward matured stages and old or monadnock stages, which in turn indicates remnant of the present volume as compared to the original volume and erosion cycle. Erosion cycle of the basin can be classified in to three distinctive stages of the geomorphic cycle.

(i) If the HI $\geq$ 0.60, then young stage (basin is highly susceptible to erosion and is under development) [1].
(ii) If 0.35 $\leq$ HI $\leq$ 0.60, then equilibrium or mature stage (the basin development has attained steady state condition).
(iii) If HI $\leq$ 0.35, then old stage of geomorphic, in which the basin is fully stabilized.
(iv) Equilibrium or mature stage, in which integration of the hypsometric curve gives the hypsometric integral (I). As suggested and proved mathematically by Pike and Wilson (1971) that the elevation-relief ratio (E), i.e., integration of the hypsometric curve gives the hypsometric integral (I). Table 1 shows the hypsometric integral values of Brahmani–Baitarani basin and its eight sub-basins along elevation, relative area, elevation (max, min and mean), and HI values information.

Relief

The amplitude of the topography, or relief governing the erosional and tectonic processes is used to describe the vertical dimension. The study under consideration groups into three categories, viz. high, moderate, and low. The area has been divided into three relief classes (Fig. 2).

(i) Relief value < 60 m, low relief zone which occupies about 33,937 km^2 of the basin area
(ii) Relief value < 60 m to 120 m, medium relief zone which occupies about 6203 km^2 of the basin area
(iii) Relief value > 120 m reckoned as high relief zone which occupies about 13,767 km^2 of the total basin area.

Table 1 Hypsometric integral values of Brahmani–Baitarani basin and its sub-basins

Basin/Sub-basins	Area (km^2)	Elevation (m)			Hypsometric Integral (HI)	Geomorphic stage
		Max	Min	Mean		
Brahmani-Baitarani basin	53902	1174	− 12	350.97	0.305	Old stage
Anandapur	6825	1117	31	442.35	0.379	Late maturity
Champua	1802	1116	375	591.65	0.292	Old stage
Delta	9362	1041	− 12	71.16	0.079	Old stage
Gomlai	2317	1068	136	349.14	0.229	Old stage
Jaraikela	10605	1084	198	541.25	0.613	Late maturity
Jenapur	14272	1174	9	203	0.167	Old stage
Panposh	5537	911	161	360.32	0.266	Old stage
Tilga	3182	1111	376	732.9	0.486	Middle maturity

It is evident from the study that major portion of the basin area falls under low relief zone area (63%) dominates other classes. The second relief zone is high relief (25.50%) and only 11.5% of the total area characterize as moderate relief zone.

Aspect

Describing the aspect map, this basin is bounded by the Chhota Nagpur Plateau in the north, and in the west and south by the ridge separating it from Mahanadi basin, Bay of Bengal on the east. Physiographically, there are four regions: the northern plateau, the Eastern Ghats (both hilly and forested), the coastal plains, and the central table land. The aspect plays an important role for determining the soil moisture regime.

Land Use and Land Cover Classification

Land use and land cover as well as soil maps are depicted in Figs. 4 and 5. The Landsat 8 image corresponds to the 2004–05 year processed under supervised classification using ERDAS Imagine Professional Software 2016 and consists of 17 different classes. For image interpretation, ERDAS Imagine 16 software was used to prepare land use category map of the study area considering field data using the base map to identify different categories of land uses. A field data was conducted to find out latitude and longitude of specific land use category and recorded. From the knowledge of field data, 17 land use classes were made. The collected land use data was used to find out the color tone of 2004–2005 Landsat image while training the dataset. Cross checking of same land use category was verified through pixel color tone. After validation of the signature tone for specific land use class, forest cover forms as the major constituent (31.9%), followed by crop area (30%) and current fallow (28.25%). The remaining 10.7% of basin area is covered by built up land, plantation, littoral swamp, grassland, gullied land, scrubland, and other waste land and water bodies. The crop area is further categorized as Kharif only (23.64%), Rabi only (0.9%), Zaid only (0.26%), and Double/Triple (4.35%) classes.

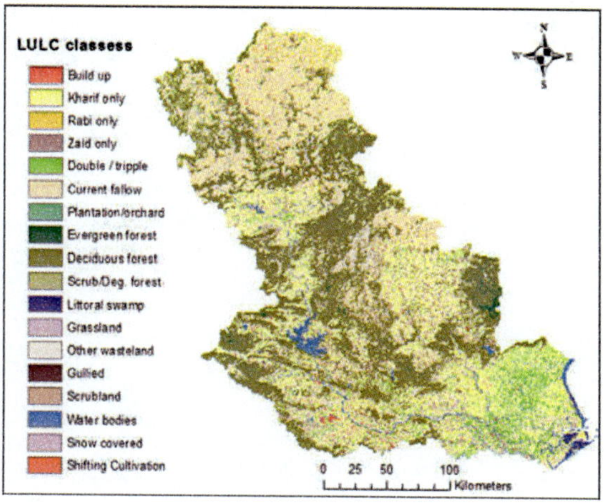

Fig. 4 LULC map of Brahmani–Baitarani basin (2014–15)

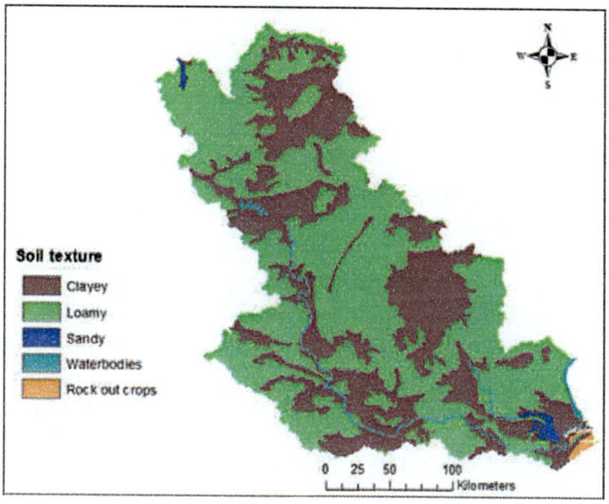

Fig. 5 Soil texture map of Brahmani–Baitarani basin

The hypsometric integral (HI) values have been derived for the study area along with its 8 sub-basins as illustrated in Fig. 5 and presented in Table 1. The HI value of the overall basin is computed to be 0.305, which reveals that only 30% of the land masses remain in the basin to be eroded and entire basin reflects decline of in-equilibrium stage. The calculated HI values for all the sub-basins of the basin ranged from 0.079 to 0.613 (Table 1). Out of the eight sub-basins, only five sub-basins, namely Champua, Delta, Gomlai, Jenapur and Panposh, fall under older stage, one sub-basin, namely Jaraikela belongs to late youthful stage state of its development. The remaining two sub-basins Anandapur and Tilga belong to mature stage of landscape evolution. Only Anandapur sub-basin represents late mature stage of landforms and reaching toward monadnock stage. Figures 6 and 7 depict the relief map and aspect map of the study area.

Fig. 6 Relief map of the study area

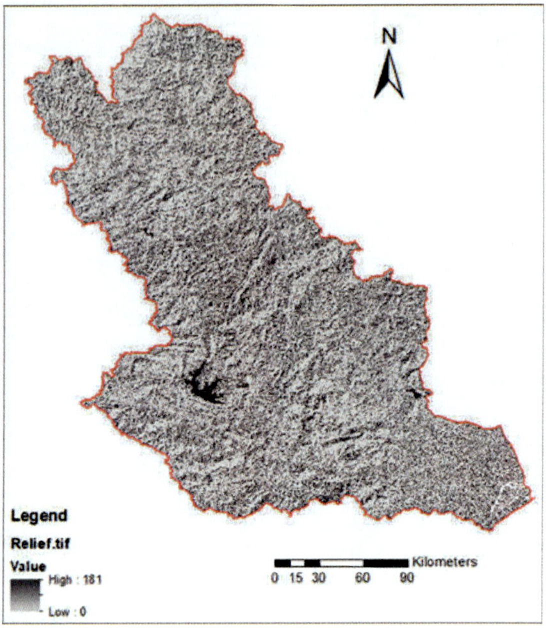

Fig. 7 Aspect map of the study area

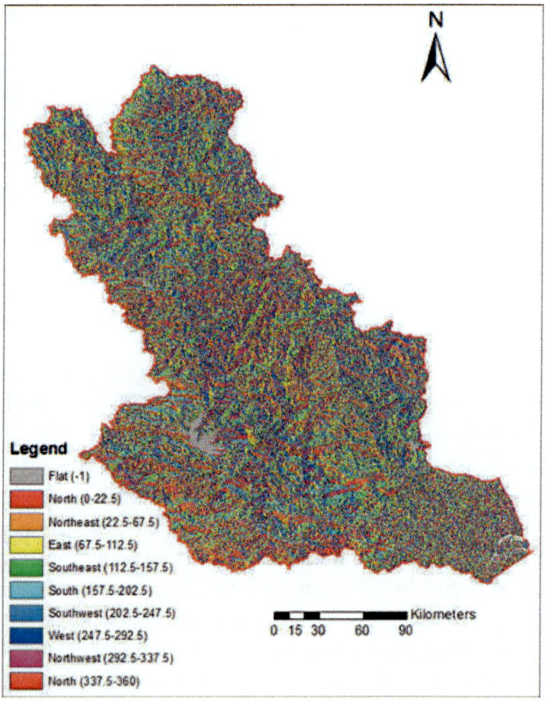

Hypsometric Curve for Brahmani–Baitarani Basin

Hypsometric curve for the basin is shown in Fig. 8. The plotted hypsometric curve described the distribution of elevations across the study area, which has been used to evaluate the geomorphic and tectonic evolution of landforms. In Fig. 8, the ordinate represents the ratio of relative elevation (*h*/*H*), and the abscissa represents the ratio of relative area (*a*/*A*). The relative elevation is calculated as the ratio of the height of a given elevation contour (*h*) from the minimum base plane to the maximum basin elevation (*H*). The relative area has been taken as the ratio of the area above a certain contour (*a*) to the total area of the basin above the outlet. The value of relative area (*a*/*A*) varies from one to zero. Thus, the lowest point in the basin ($h/H = 0$) and at the highest point the ($h/H = 1$).

In the basin relief graph of the considered basin, the upper concave shaped hypsometric curve revealed the proportion of land area that exists at various elevations by plotting relative area against relative height. For example, if a value for one of these variables is given, and we will determine the value of the other. For example, at 0.6 in *X*-axis would read 0.2 in cumulative altitude. When a percentage is read on the horizontal axis that is cumulative area, this cuts the elevation in the curve, if vice versa, then the elevation above which the given percent of the basin surface can be found.

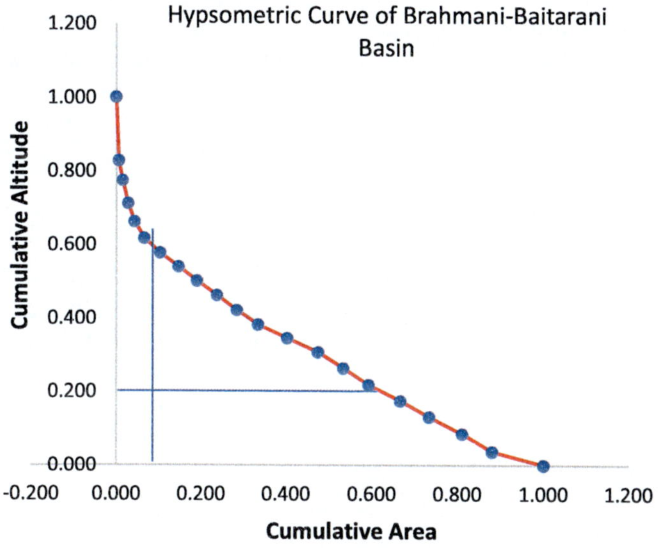

Fig. 8 Hypsometric curve for Brahmani–Baitarani basin

4 Results and Discussions

Figure 8 depicts the results of hypsometric analysis for the basin. Due to heavy rainfall and corresponding runoff, more kinetic energy because of hills which cut away the landmass faster, so the curve elevation -area curve fell off more quickly. In the planes of Brahmani–Baitarani basin, sapping is present in the eastern part. This has lower energy process and so its curve appeared flatter. In the study area, approximately more than 80% of area (or volume) lay at elevations than mean elevation. It was also observed that there was a combination of moderate convex-concave and slightly **S** shape of the hypsometric curves for the basin reasons out for the soil erosion from the basin and down slope movement of topsoil and bedrock material.

Residual and transported soils are the two broadly classes of soil in Brahmani sub-basin based on formation. The upper basin which is part of Chhota Nagpur Plateau is predominantly red gravel soil, red earth. Mixed red and black, red loams dominate the Central Table land of Odisha while the lower basin characterizes red loam lateritic and laterite soils. The hydrologic response of the sub-basins having younger geomorphic stages will have high to moderate rate of erosion during peak runoff, which needs appropriate catchment area treatment for soil and water conservation measures. The sub-basins like Anandpur and Tilga with mature geomorphic stages express medium to complex denudational processes. Five sub-basins, namely Champua, Delta, Gomlai, Jenapur and Panposh fall undehaving older stage of geomorphic process, will have lower erosion. Further, CalHypso analysis in GIS can reveal the details of the statistics related to the hypsometric curve by applying polynomial fits.

5 Conclusions

Some parts of this basin like Bhadrak and Jajpur are prone to severe soil erosion because of recurring floods. About 35% of the basin being in hills, there is also moderate erosion. For control of further erosion from the basin reducing the runoff, enhancement of ground water potential, hypsometric analysis could suffice the information on various stages of landform processes in the Brahmani–Baitarani basin. In the basin, considerable land remains rainfed; thus, watershed management in the upper regions may be resorted to as revealed by the study.

References

1. Strahler AN (1952) Hypsometric (area-altitude) analysis of erosional topography. Geol Soc Am Bull 63:1117–1141
2. Kokkas NA, Miliaresis G (2008) Geomorphometric mapping of grand canyon from the 1-Degree USGS DEMs, ISPRS, proceedings, XXXV. https://www.isprs.org/proceedings/XXXV/congress/comm4/papers/460.pdf. Accessed on 20 Feb 2011

3. Pike RJ, Wilson SE (1971) Elevation-relief ratio, hypsometric integral and geomorphic area-altitude analysis. Geol Soc Am Bull 82:1079–1084
4. Ritter DF, Kochel RC, Miller JR (2002) Process geomorphology. McGraw Hill, Boston
5. Lifton NA, Chase CG (1992) Tectonic, climatic, and lithologic influences on landscape fractal dimension and hypsometry: implications for landscape evolution in the San Gabriel Mountains California. Geomorphology 5(1–2):77–114
6. Masek JG, Isacks BL, Gubbels TL, Fielding EJ (1994) Erosion and tectonics at the margins of continental plateaus. J Geophys Res 99(B7):13941–13956
7. Hurtrez JE, Lucazeau F (1999) Lithological control on relief and hypsometry in the Herault drainage basin (France), Comptes Rendues Acade mie des Sciences de la terre et des planets. Earth Planet Sci 328(10):687–694
8. Hurtrez JE, Sol C, Lucazeau F (1999) Effect of drainage area on hypsometry from an analysis of small-scale drainage basins in the Siwalik hills (Central Nepal). Earth Surf Proc Land 24:799–808
9. Chen YC, Sung Q, Cheng K (2003) Along-strike variations of morphotectonics features in the western foothills of Taiwan: tectonic implications based on stream-gradient and hypsometric analysis. Geomorphology 56(1–2):109–137

Climate Change Impact and Adaptive Measures for Green Cover Assessment at District Level

Amol Dhokchaule, Anita Morkar, Santosh Wagh, and Makarand Kulkarni

Abstract Green cover maps are critical for a variety of water resource and environmental applications. Green space distribution is critical in urban planning because it significantly improves the ecological quality of metropolitan areas. It improves air quality, urban health, biodiversity conservation, noise reduction and so on. One of the most negative effects of urbanization is the loss of vegetation cover. As a result, proper distribution of green spaces in urban environments is becoming increasingly important for sustainable development and healthy living. The goal of this research is to identify and map green cover using remote sensing technology. This paper proposes a supervised classification methodology for district-level green cover mapping. This study made use of the Resourcesat-II satellite's LISS-III sensor (23.5 m) resolution. Nashik district has received post-monsoon and pre-monsoon statistics for the water year 2019–2020. About 580 ground truthing sets in the form of point, line and polygon data collected with handheld GPS instruments from various tahasils in Nashik district for analysis and validation. The method's accuracy is evaluated using ground truthing data. The proposed method yields accurate results. Nashik district has a land area of 1,551,489 Ha. The estimated post-monsoon green cover area is 739876 ha, or around 47.68 percent of the geographical area, and the estimated pre-monsoon green cover area is 537909 ha, or around 34.67 percent of the geographical area. The findings discussed in this paper will provide critical information to state and local governments for the protection and restoration of green covers and the conservation of natural ecosystems in the Nashik district. The study also finds its utility for biogeochemical models where the type and extent of cover is the major input to monitor and sustain the ecological balance.

Keywords Green cover · Supervise classification · Pre-monsoon · Post-monsoon · Ground truthing

A. Dhokchaule (✉) · A. Morkar · S. Wagh · M. Kulkarni
Resources Engineering Centre, Maharashtra Engineering Research Institute, Nashik 422004, India
e-mail: dhokchauleamol14@gmail.com

P. V. Timbadiya et al. (eds.), *Geospatial and Soft Computing Techniques*, Lecture Notes in Civil Engineering 339, https://doi.org/10.1007/978-981-99-1901-7_8

1 Introduction

Nature has been bountiful to us providing the substance which has been the basis of all our material and cultural progress. Green space distribution plays an imperative role in climate change [1]. The importance of forests and forest resources in the economy of the state/country is really very important as they perform productive, protective and aesthetic functions and confer other advantages to the community. The forests conserve and enrich soil, helping in maintaining geographical, geological and climatic conditions. The natural resources are now depleting very fast causing serious environmental problem. Natural and human-induced environmental changes in an urban environment are of concern today due to deterioration of the environment and human health [2]. Monitoring the world's remaining forests is now an important component of a number of global initiatives relating to sustainability, the environment and climate change [3]. The study of changes in green cover is critical for proper planning, utilization and management of natural resources. Traditional methods for gathering demographic data, censuses and environmental sample analysis are insufficient for multi-complex environmental studies, as many problems are frequently presented in environmental issues. Several studies have been conducted to improve classification accuracy using various remote sensing and/or GIS-based ancillary data at various stages of classification.

Due to unevenness of rainfall, development activities, rapid expansion of population, urbanization, industries, climate change, etc. in various areas, green cover mapping at district level is becoming most important [4]. Remote sensing technique proves to be cost-effective and accurate for doing spatially large area. The remote sensing technique for assessing green cover is primarily based on mapping of vegetation-spread areas during satellite overpasses. The primary source of spatial information about the Earth's surface cover and contribution is remote sensing [5]. Different sensors collect a wealth of information about the Earth, which can be used by scientists who want to monitor spatial data in real time. The following are the goals of this paper: (1) to investigate the impact of climate change on green cover at the district level, (2) to generate Taluka-wise green cover statistics for the Nashik district for the post- and pre-monsoon seasons in 2019 and (3) conduct spatial and temporal analysis of green cover data from Indian remote sensing satellites.

2 Study Area and Data Source

2.1 Study Area

Nashik district is included in the study area. Nashik is located in the north-west corner of Maharashtra, between 19° 35′ and 20° 52′ North latitude and 73° 15′ to 74° 56′ East longitude. It is located 565 m above mean sea level. Several rivers drain the western slope of the Ghats, including the Daman-Ganga River, which flows west to the Arabian Sea.

The majority of the Nashik district's eastern portion, which is located on the Deccan Plateau, is open, fertile and well cultivated. The Godavari River rises in the district and flows eastwards to the Bay of Bengal. Godavari tributaries include the Kadwa and Darna. The Girna River and its tributaries flow eastwards into the Tapi River through fertile valleys. The index map of Nashik district is shown in Fig. 1.

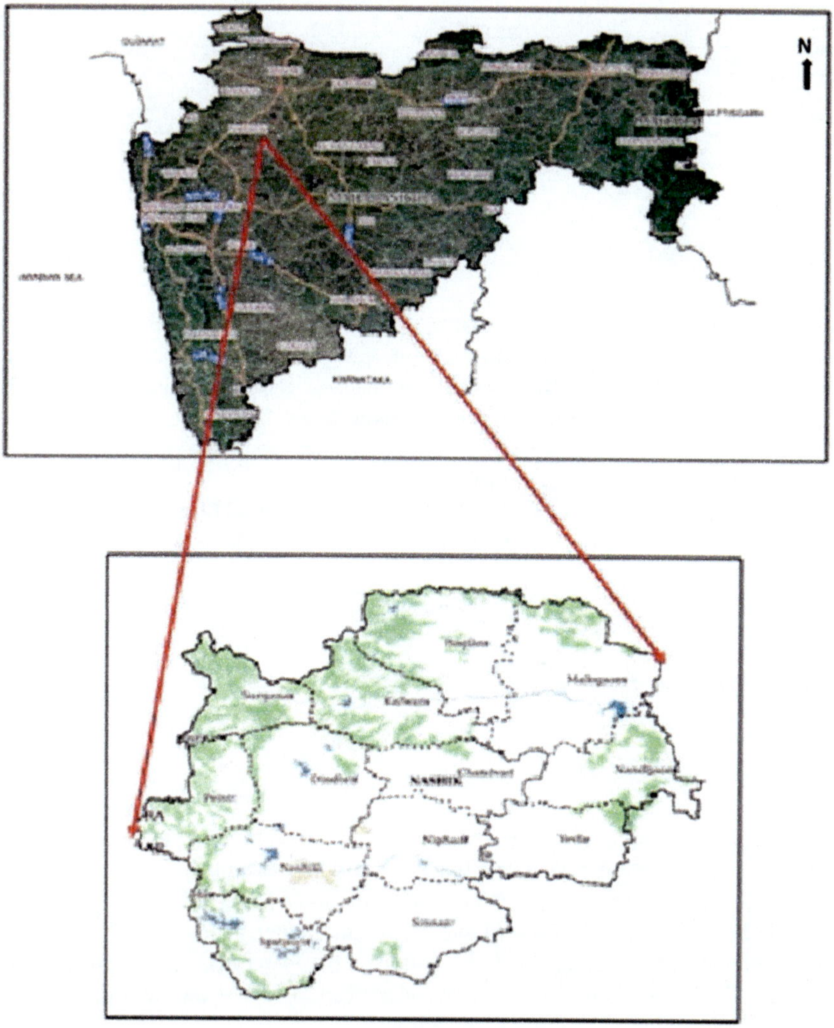

Fig. 1 Index map of Nashik district, Maharashtra [5]

Table 1 Dates of pass for satellite data

Satellite	Sensor	Path	Row	Date of pass for pre-monsoon	Date of pass for post-monsoon
Resourcesat-II	LISS-III	94	58	06/05/2019	08/12/2019
Resourcesat-II	LISS-III	95	58	11/05/2019	13/12/2019
Resourcesat-II	LISS-III	95	59	11/05/2019	13/12/2019
Resourcesat-II	LISS-III	96	58	16/05/2019	24/11/2019

Table 2 Details of SOI toposheet used

S. No.	Toposheets	Number
1	46H5 to 46H15	11
2	46L1 to 46L15	11
3	46I10 to 46I13	3

2.2 Data Used

Satellite Data National Remote Sensing Centre, Hyderabad, website has been browsed to prepare a list of dates of satellite pass over the Nashik district for the year 2019. The list of cloud-free satellite images have been prepared for the both pre-monsoon and post-monsoon period. Resource sat II satellite data have been finalized and used for the analysis. These images cover the total area of interest. The details of satellite pass are given in Table 1.

SOI toposheet. Survey of India (SOI) toposheets of 1:50,000 scales have been used for georeferencing of satellite images for study area. These details are as given in Table 2.

3 Results and Discussions

3.1 Ground Truthing

Remote sensing techniques require certain amount of field observation called "ground truth" in order to convert pixel data into meaningful information. Such work involves visiting number of sites. The green cover data dictionary was created for recording the field information such as barren land, fallow land, water, vegetation, forest, scrub, crop, urban and crop waste, etc. The location of the features is recorded using the handheld GPS. The standard pro forma was used to record the field data. Field photographs are also taken during the field visit.

GPS readings were taken with the help of Trimble Juno Handheld GPS device, and simultaneously, latitude/longitude values have been recorded during the visit. In above all field visits, almost 315 point features, 162 line features and 103

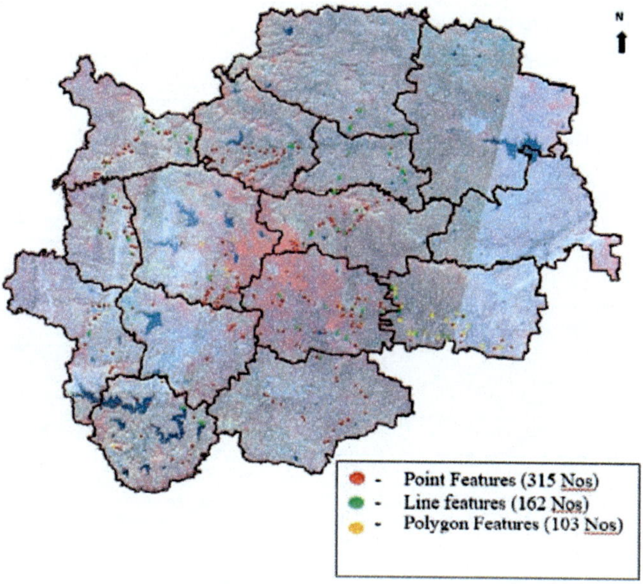

Fig. 2 Ground truth features of Nashik district

polygon features, total 580 features from 13 talukas, i.e., Nashik, Dindori, Niphad, Yeola, Sinnar, Igatpuri, Chadwad, Peth, Surgana, Kalwan, Trimbakeshwar, Deola and Baglan of Nashik district have been collected. The collected ground truth features of Nashik district are shown in Fig. 2.

3.2 *Taluka-Wise Green Cover Statistics and Maps*

Green cover statistics followed by green cover map for each taluka in pre-monsoon and post-monsoon is given to have a fairly good idea about the distribution pattern and density of green cover in the taluka. Following Tables 3 and 4 give the statistics of green cover in pre-monsoon and post-monsoon, respectively. Similarly, Figs. 3 and 4 show the green cover maps in pre-monsoon and post-monsoon of Nashik district for the year 2019. A bar chart is prepared to show the comparison of green cover area in pre-monsoon and post-monsoon in the year 2019 for Nashik district. Figure 5 shows the comparative bar chart of pre- and post-monsoon green cover area in Nashik district for the year 2019.

Table 3 Taluka-wise green cover area (pre-monsoon)

Taluka	Geographic area (Ha)	Green cover area (Ha)	% of total green cover Area	% of taluka geographic area
Baglan	145,910	59,567	11.07	40.82
Chandwad	95,558	14,599	4.38	24.66
Deola	57,112	13,597	2.53	23.81
Dindori	132,049	59,107	10.99	44.76
Igatpuri	86,180	39,986	5.77	36.00
Kalwan	85,464	48,440	8.50	53.53
Malegaon	179,408	40,421	7.51	22.53
Nandgaon	109,813	25,628	4.76	23.34
Nashik	89,295	31,634	5.53	33.31
Niphad	104,979	55,142	10.25	52.53
Peth	55,675	26,227	4.88	47.11
Sinnar	133,819	26,786	4.98	20.02
Surgana	81,368	34,716	6.45	42.67
Trimbakeshwar	89,287	48,185	8.96	53.97
Yeola	105,570	13,875	3.43	17.49
Total	1,551,489	537,909	100.00	

Table 4 Taluka-wise green cover area (post-monsoon)

Taluka	Geographic area (Ha)	Green cover area (Ha)	% of total green cover area	% of taluka geographic area
Baglan	145,910	74,023	10.00	50.73
Chandwad	95,558	47,867	5.26	40.71
Deola	57,112	24,876	3.36	43.56
Dindori	132,049	63,221	8.54	47.88
Igatpuri	86,180	31,024	5.40	46.40
Kalwan	85,464	45,747	6.55	56.68
Malegaon	179,408	87,101	11.77	48.55
Nandgaon	109,813	46,664	6.31	42.49
Nashik	89,295	29,741	4.28	35.43
Niphad	104,979	60,753	8.21	57.87
Peth	55,675	27,479	3.71	49.36
Sinnar	133,819	56,533	7.64	42.25
Surgana	81,368	50,171	6.78	61.66
Trimbakeshwar	89,287	50,988	6.89	57.11
Yeola	105,570	43,687	5.28	37.04
Total	1,551,489	739,876	100.00	

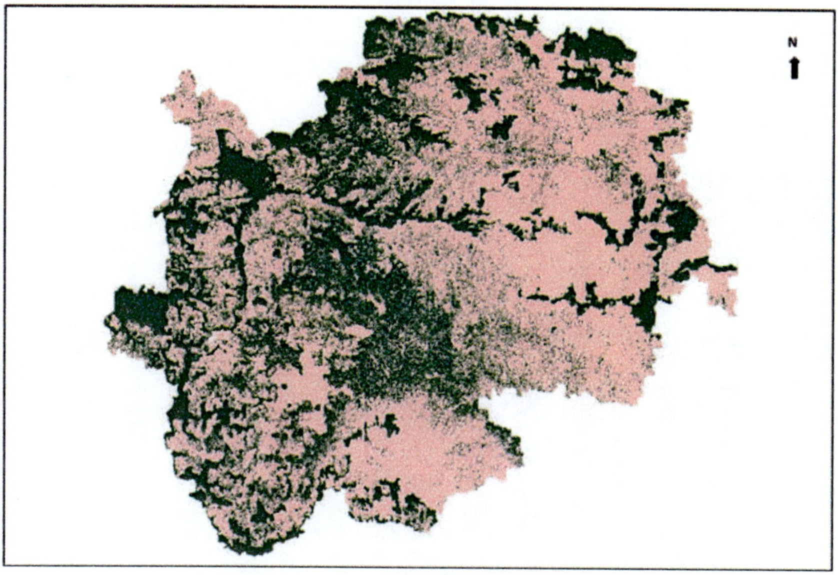

Fig. 3 Green cover maps of Nashik district (pre-monsoon) year 2019

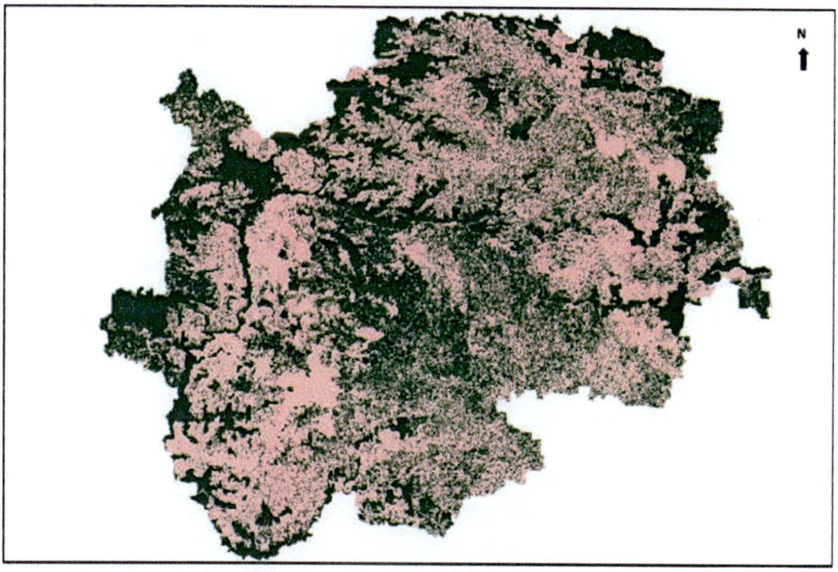

Fig. 4 Green cover maps of Nashik district (post-monsoon) year 2019

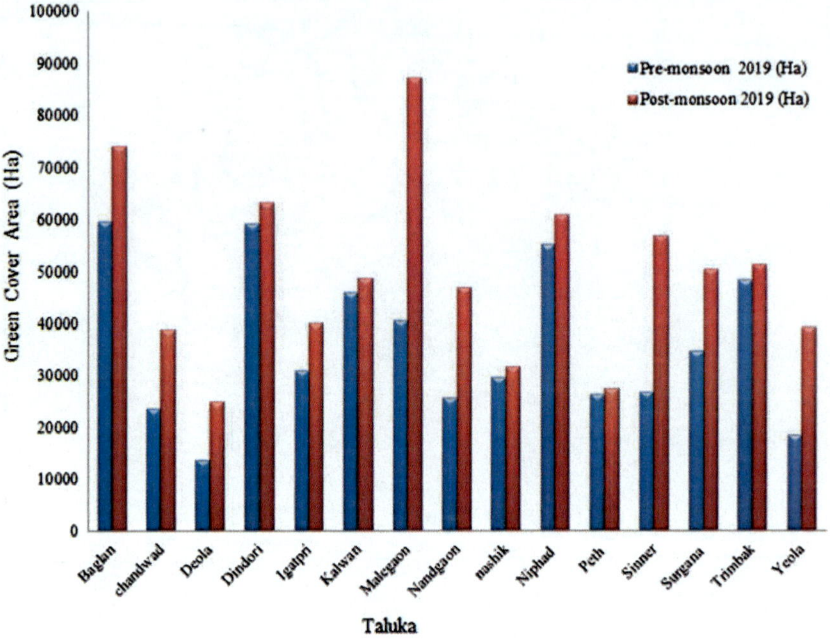

Fig. 5 Pre-monsoon and post-monsoon taluka-wise distribution of green cover in Nashik district, year 2019

4 Conclusions

The following conclusions are derived from the foregoing study:

- Green cover area in Nashik district as per geographical area in post-monsoon is 47.69% and in pre-monsoon it is 34.67%.
- In pre-monsoon, the green cover in each taluka as per percent of total district Green cover area ranged from as 2.53% to 11.07%.
- In post-monsoon the green cover in each taluka as per percent of total district green cover area ranged from as 3.36% to 11.77%.
- Nine talukas in Nashik district are rich talukas contributing green cover more than 5 percent of total green cover area in pre-monsoon. Similarly, 12 talukas in Nashik district are rich talukas contributing green cover more than 5 percent of total green cover area in post-monsoon.
- Two talukas in Nashik district contributing less than 5 percent green cover area in both pre-monsoon and post-monsoon.
- The result of this study will provide fundamental information to state as well as local authorities for the protection and restoration of green covers and conservation of natural ecosystem in Nashik district.

Acknowledgements The authors would like to acknowledge the Resources Engineering Centre, Maharashtra Engineering Research Institute (MERI) Nashik Water Resources Department, Government of Maharashtra, for providing funding for this research. The authors would also appreciate the infrastructural support provided by Resources Engineering Centre Nashik on "Climate Change Impact and Adaptive Measures for Green Cover Assessment", Govt. Of Maharashtra.

References

1. Gandhi MG, Thummala N, Christy A (2015) Urban green cover assessment and site analysis in Chennai, Tamil Nadu—a remote sensing and GIS approach. ARPN J Appl Sci 10(5):2239–2243
2. Chandramathy I, Kitchley J (2018) Study and analysis of efficient green cover types for mitigating the air temperature and urban heat island effect. Int J Glob Warning 10(25):1–22
3. Estoque RC, Johnson BA, Gao Y, DasGupta R, Ooba M, Togwa T, Hijioka Y, Murayama Y, Gavina LD, Lasco RD, Nakamura S (2021) Remotely sensed tree canopy cover-based indicators for monitoring global sustainability and environmental initiatives. Environ Res Lett 16(2021):044047
4. Maharashtra Engineering Research Institute (MERI) (2020) A research report on land use land cover assessment of Nashik District, Maharashtra, through satellite Remote Sensing
5. Space Application Center (SAC) Ahmadabad (2010) Report on National Wetland Atlas, Maharashtra

Analysis of Land Use Land Cover Changes in the Netravati Basin, Karnataka, India

N. Nayana, Dinu Maria Jose, and G. S. Dwarakish

Abstract Analysis and mapping of land use land cover (LULC) are essential to improve our understanding of the human-nature interactions and their effects on land use changes. In this study, LULC maps of the Netravati river basin for the years 2000, 2010, and 2020 were obtained using maximum likelihood classifier on Landsat images. The classifier produced LULC maps of 2000, 2010, and 2020 with overall accuracy of 87.34%, 85.74%, and 86.3%, respectively. The results of this study showed that there is an increase in the spatial extends of the urban area (3.54–9.21%) and agriculture (18.2–21.09%) during the period 2000 to 2020. In contrast, forest (55.48–51.02%), bare soil (6.61–5.91%), water bodies (1.64–1.23%), and vegetation (14.53–11.54%) cover have decreased from the year 2000 to 2020. The results of this study can be used for proper LULC management in the basin. This study is a prerequisite for the prediction and management of future urban growth.

Keywords Land use and land cover · Maximum likelihood classifier · Netravati basin · Watershed management

Disclaimer The presentation of material and details in maps used in this chapter does not imply the expression of any opinion whatsoever on the part of the Publisher or Author concerning the legal status of any country, area or territory or of its authorities, or concerning the delimitation of its borders. The depiction and use of boundaries, geographic names and related data shown on maps and included in lists, tables, documents, and databases in this chapter are not warranted to be error free nor do they necessarily imply official endorsement or acceptance by the Publisher or Author.

N. Nayana · D. M. Jose (✉) · G. S. Dwarakish
Department of Water Resources and Ocean Engineering, National Institute of Technology Karnataka, Surathkal 575025, India
e-mail: dinumariajose@gmail.com

G. S. Dwarakish
e-mail: dwaraki@nitk.edu.in

P. V. Timbadiya et al. (eds.), *Geospatial and Soft Computing Techniques*, Lecture Notes in Civil Engineering 339, https://doi.org/10.1007/978-981-99-1901-7_9

1 Introduction

LULC has two separate terms that are often used interchangeably. Land cover refers to the biophysical characteristics of earth's surface, including the distribution of vegetation, water, soil, and other physical features of the land, while land use refers to the way in which land has been used by humans and their habitat, usually with an emphasis on the functional role of land for economic activities [1]. Evaluation of LULC changes is essential for effective river basin management. It can alter precipitation process and thus change hydrological processes like surface runoff, percolation, lateral flow, and evapotranspiration [2]. Land use land cover changes are unceasing process and must be understood using more dynamic information [3].

2 Materials and Method

2.1 Study Area

The Netravati river basin (Fig. 1) has an area of approximately 3416 km^2 and lies between latitude of 12° 30′ N–13° 10′ N and longitudes of 74° 50′ E–75° 50′ E. From the Mean Sea Level, the elevation of basin varies from 0 to 1884 m. The upper portion of river is mountainous and covered with dense forest, whereas the lower portion is an undulated area where agriculture and urban lands are predominant. Kumaradhara, Kallaji hole, Belthangadi hole, Gowri hole, Netravati hole, shisla hole and Neriya hole are the sub basins of Netravati river basin. The longest path of river in this basin is 107 km. The climate is characterized by heavy rainfall, high humidity, and harsh weather in summer season [4]. The heavy rainfall supports in better growth of vegetation in the region [5]. The basin attains an average annual precipitation of 3076 mm and has an average air temperature of 20 to 26 °C. Monsoons (June–September) contribute 70 to 80% of the annual precipitation, and maximum humidity is experienced in June and July.

2.2 Data Source and Methodology

Landsat imageries (Level 2) for the time periods of 2000, 2005, 2010, and 2020 are obtained from the United States Geological Survey (https://earthexplorer.usgs.gov/). Table 1 shows the details of Landsat data used. After downloading, the study area is delineated, layer stacked using ArcGIS, and ERDAS IMAGINE, respectively, for creating the LULC maps. The LULC datasets are classified into 6 classes, namely water bodies, forest, urban, agriculture, bare soil, and vegetation for the years 2000, 2010, and 2020 using maximum likelihood classifier (MLC).

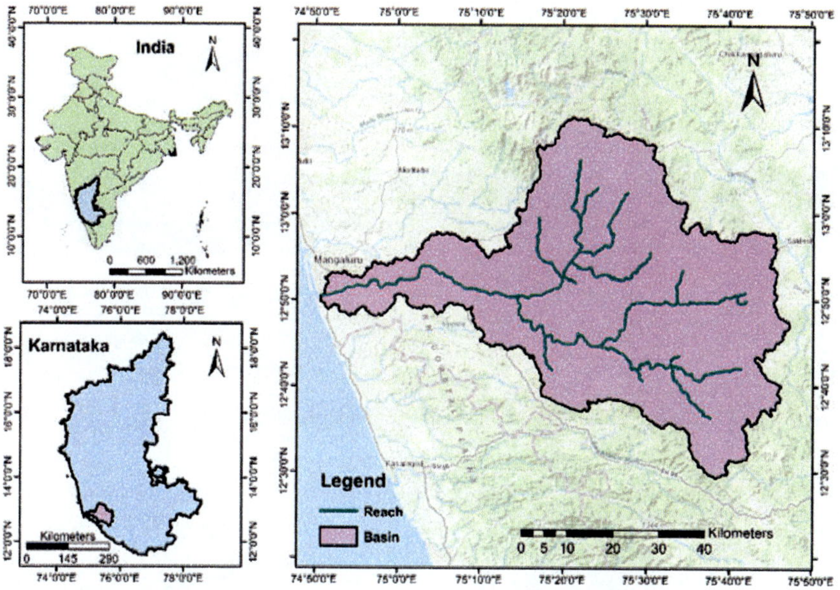

Fig. 1 Index map of study area

Table 1 Landsat data for LULC classification

Year	Satellite	Acquisition date	Sensor	Resolution	Path/ row
2000	Landsat 7	20/12/2000	Enhanced thematic mapper plus	30	145/51
2010	Landsat 7	30/01/2010	Enhanced thematic mapper plus	30	145/51
2020	Landsat 8	03/02/2020	Operational land imager and thermal infra-red scanner	30	145/51

3 Results and Discussion

Landsat imageries are classified in to LULC datasets by performing supervised classification using MLC algorithm in Erdas Imagine software. Landsat dataset for the time period of 2000, 2010, and 2020 is classified into 6 classes. According to the classification results, the land use area (in percentage) for the different time period is shown in Figs. 2, 3, and 4. Based on the results, it is observed that the Netravati river basin is mainly influenced by forest (55.48, 54.55, and 51.02%) in all the three years.

All the three LULC datasets are validated and overall accuracy of 87.34, 85.74, and 86.3% with kappa coefficient values of 0.792, 0.773, and 0.784 for the year 2000, 2010, and 2020, respectively. Table 2 shows the distribution of 6 classes over

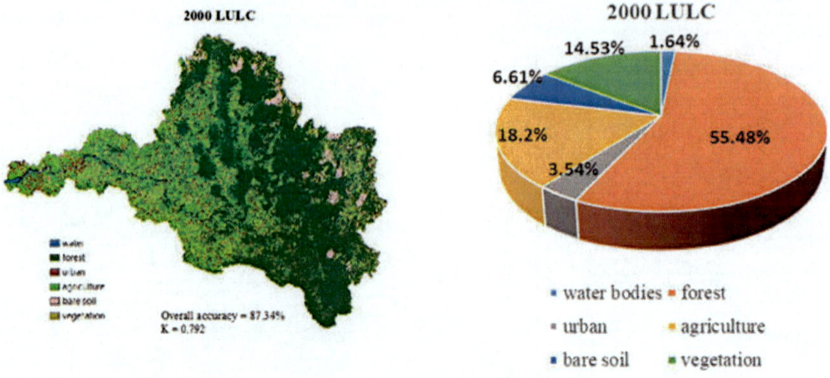

Fig. 2 LULC map and percentage area distribution for the year 2000

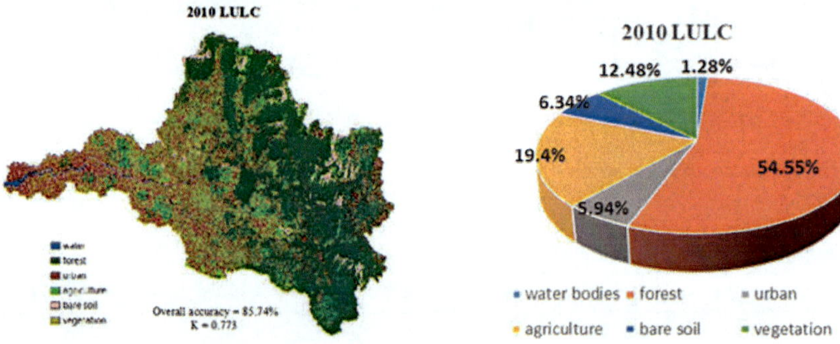

Fig. 3 LULC map and percentage area distribution for the year 2010

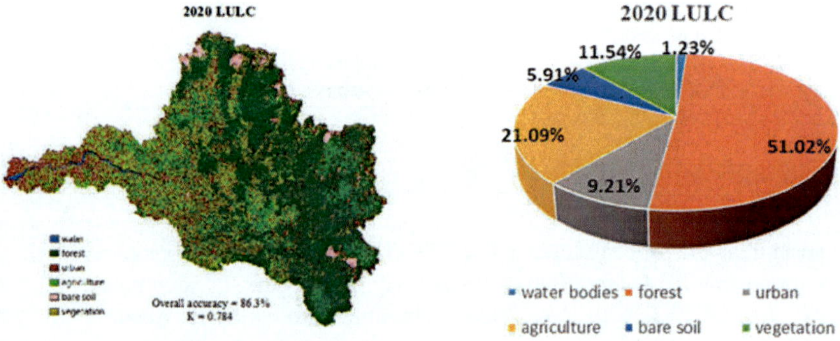

Fig. 4 LULC map and percentage area distribution for the year 2020

Table 2 LULC distribution in 2000, 2010, and 2020

LULC classes	2000		2010		2020	
	Area (km^2)	Percentage area	Area (km^2)	Percentage area	Area (km^2)	Percentage area
Water bodies	56.02	1.64	43.72	1.28	42.02	1.23
Forest	1895.20	55.48	1863.43	54.55	1742.84	51.02
Urban	120.93	3.54	202.91	5.94	314.61	9.21
Agriculture	621.71	18.2	662.70	19.4	720.43	21.09
Bare soil	225.80	6.61	216.57	6.34	201.89	5.91
Vegetation	496.34	14.53	426.32	12.48	394.21	11.54

the study area. From 2000 to 2020, there is net decrease of 14 km^2 of water bodies, 152.36 km^2 of forest, 23.91 km^2 of bare soil, and 102.13 km^2 of vegetation over the Netravati river basin. Whereas, agriculture and urban area showed an increasing trend from 2000 to 2020 with increase in area of 98.72 km^2 and 193.68 km^2, respectively.

4 Conclusions

In this study, Landsat imageries for the year 2000, 2010, and 2020 are used to generate the LULC maps of Netravati river basin for the different time periods. LULC maps were developed using MLC and showed overall classification accuracy more than 75%. Based on the results, it is observed that the study area is mainly dominated by forest and agricultural land. The results indicated an increasing trend in agricultural (18.2–21.09%) and urban areas (3.54–9.21%). In contrast, water bodies (1.64–1.24%), forest (55.48–51.02%), bare soil (6.61–5.91%), and vegetation (14.53–11.54%) showed a decreasing trend from 2000 to 2020. Proper environmental urban management can be done using these LULC maps. These maps can be further used for prediction of LULC changes.

Acknowledgements The authors would like to acknowledge National Institute of Technology for providing the necessary support for this research work.

References

1. Abburu S, Golla SB (2015) Satellite image classification methods and techniques: a review. Int J Comput Appl 119:20–25. https://doi.org/10.5120/21088-3779
2. Ganasri BP, Dwarakish GS (2015) Study of land use/land cover dynamics through classification algorithms for Harangi catchment area, Karnataka State, INDIA. Aquatic Procedia:1413–1420

3. Mondal MS, Sharma N, Garg PK, Kappas M (2016) Statistical independence test and validation of CA Markov land use land cover (LULC) prediction results. Egypt J Remote Sens Sp Sci 19:259–272. https://doi.org/10.1016/j.ejrs.2016.08.001
4. Sajikumar N, Remya RS (2015) Impact of land cover and land use change on runoff characteristics. J Environ Manage 161:460–468. https://doi.org/10.1016/j.jenvman.2014.12.041
5. Sinha RK, Eldho TI (2018) Effects of historical and projected land use/cover change on runoff and sediment yield in the Netravati river basin, Western Ghats. India. Environ Earth Sci 77:111. https://doi.org/10.1007/s12665-018-7317-6

Spatiotemporal Land Use Land Cover Change Impacts on Groundwater Table in Surat District, India

Prajakta Jadhav, V. L. Manekar, and J. N. Patel

Abstract For maintaining life, groundwater resource is the important factor. In this study, the impact of Land Use Land Cover (LULC) alterations on groundwater table in Surat district (India) which is semi-arid to dry sub-humid region was determined. Influence of LULC on groundwater table variation is quantified based on time-series using Landsat imagery (2000, 2010, 2015, and 2020). Supervised classification is used in this study for LULC feature classification. In this study, groundwater data (2000–2019) was used for groundwater table variation analysis The result shows that Groundwater Level (GWL) in the study area was declined 5 m below ground level (mbgl) during pre-monsoon and 1.5 mbgl during post-monsoon season. LULC maps developed in this study showed an overall accuracy of 82% and the kappa coefficient as 0.73. It is revealed from the analysis that a significant relation is in exists between LULC change of each class and pre-monsoon and post-monsoon GWL. The LULC results showed a 4.22% rise in built-up land, 8.93% decline in barren land, 7.63% rise in agricultural land, 3.64% decline in forest, and 0.3% decline in surface water bodies. The results of the monitoring of LULC changes as well as the GWL fluctuations of the study area provide a scientific data for decision-making, water resource management, governance, and protection.

Disclaimer: The presentation of material and details in maps used in this chapter does not imply the expression of any opinion whatsoever on the part of the Publisher or Author concerning the legal status of any country, area or territory or of its authorities, or concerning the delimitation of its borders. The depiction and use of boundaries, geographic names, and related data shown on maps and included in lists, tables, documents, and databases in this chapter are not warranted to be error free nor do they necessarily imply official endorsement or acceptance by the Publisher or Author.

P. Jadhav (✉) · V. L. Manekar · J. N. Patel
Department of Civil Engineering, Sardar Vallabhbhai National Institute of Technology, Surat 395007, India
e-mail: ds19ce004@ced.svnit.ac.in

V. L. Manekar
e-mail: vlm@ced.svnit.ac.in

J. N. Patel
e-mail: jnp@ced.svnit.ac.in

P. V. Timbadiya et al. (eds.), *Geospatial and Soft Computing Techniques*, Lecture Notes in Civil Engineering 339, https://doi.org/10.1007/978-981-99-1901-7_10

Keywords Land use land cover · Groundwater level · Kappa coefficient · Supervised classification · LANDSAT imagery

1 Introduction

Change in land use, increase in population, and increasing demand and consumption patterns all have a significant impact on water security, as per the United Nations Educational, Scientific, and Cultural Organization (UNESCO) World Water Assessment Programme (WWAP 2015) [4].

Alteration in land use is one of measure human operations which affect system of groundwater flow, and it will pursue to have an impact on recharge dynamics and the vadose zone around the world in future. Land use change is a complicated, dynamic process with significant impact for soil, atmosphere, and the groundwater. As a result, sustainable natural resource management requires a knowledge of the effect of changes in Land Use and Land Cover (LULC) on the hydrologic cycle [5]. The creation of LULC and groundwater maps was necessary for evaluating schematic growth and monitoring land use patterns [4].

Due to increases in the alarming rate of the population, the extraction rate from groundwater is far more than the recharge rate, so that Groundwater Level (GWL) has depleted in recent decades in most of the region of India [3]. Over the previous few decades, variations in LULC as well as the amount of surface and groundwater resources have come from the fast development of anthropogenic activities such as urbanization, industrialization, and irrigation. The decreasing scenario of GWL has created adverse effects on environment flow and social development [1].

Surat district is witnessing tremendous growth in terms of accommodating and growing population and is also witnessing changes in terms of built-up area expansion. GIS-based land use transformation studies are effective and appropriate for land use planning. Planning can be performed based on changes in land use trends. Urbanization and LULC patterns significantly affect the GWL of the Surat district.

2 Materials and Method

2.1 Classification of LULC

LULC classification image pre-processing classification system and study of the different LULC classes was conducted using two Landsat 8 OLI / TIRS for the year (2015 and 2020) and one LANDSAT 5 TM for the year (2000 and 2010). The images were analyzed in the software ArcGIS 10.5. Supervised classification with maximum likelihood classification used to prepare the land use in ArcGIS 10.5. For individual classification, accuracy assessment is critical. The accuracy of LULC is determined by comparing two datasets, one is resulting from remote sensing data analysis and the second is reference information, mentioned to as "ground truth." To determine the degree of classification accuracy, the non-parametric Kappa test was used.

3 Study Area and Data Source

3.1 Surat District

Surat district has a total geographical area of 4414 km^2. Surat district is semi-arid to dry sub-humid region of south Gujarat. (http://cgwb.gov.in/District_Profile/Gujarat/Surat.pdf). The administrative divisions of the Surat district are as shown below in Fig. 1.

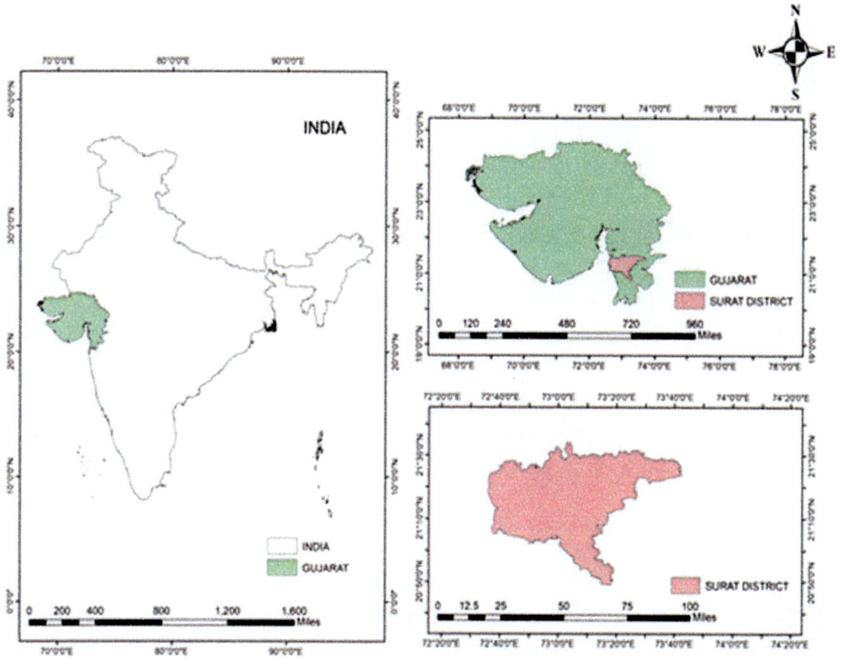

Fig. 1 Index map showing the study area

3.2 Data Used

Pre-monsoon and post-monsoon GWL data were downloaded from Centre Groundwater Board (CGWB) (India WRIS: Website http://india-wris.nrsc.gov.in/wris.html). For long-term trend analysis, there are 37 observation wells which are considered which have availability of data in pre-monsoon and post-monsoon for the period of 2000–2019. Landsat 5 satellite images for year 2000 and 2010 of the date 16/1/2000 and 27/11/2010 and Landsat 8 images for year 2015 and 2020 of the date 27/12/2015, and 2/12/2020 were procured from USGS Earth Explorer (from Website: http://earthexplorer.usgs.gov). The images having 30 m resolution. These images are primarily used for LULC classification.

4 Results and Discussion

The spatial and temporal variation of groundwater-level analysis results was compared with Central Ground Water Board records. According to the data, throughout the pre-monsoon and post-monsoon seasons, about GWL in 68 percent of wells are more than 4 meter below ground level (mbgl).

4.1 Annual GWL Scenario of Surat District (2000–2019)

The average annual depth of GWL (mbgl) for pre-monsoon and post-monsoon was plotted for 20 years (2000–2019) to analyze the future groundwater variation and shown in Figs. 2 and 3. The plot shows that the average GWL has declined below 6 mbgl during 2013, 2014, 2015, and 2016 in pre-monsoon and 4 mbgl during 2013, 2014, 2015, and 2016 in post-monsoon.

4.2 Rainfall Data Analysis

The gridded yearly rainfall data for the period 2000–2019 is generated for the variation of 26 years. A trend line is drawn, and the difference in the value given the variation in the rainfall data during the whole period of study as shown in Fig. 4.

Results indicated that the annual precipitation for the years 2000–2019; rainfall trend is decreasing. At the same time, analysis of historical rainfall data indicated that rainfall decreased by 500 mm, during the years 2000–2019. This concludes that as the rainfall is decreasing, GWL is also decreasing with time.

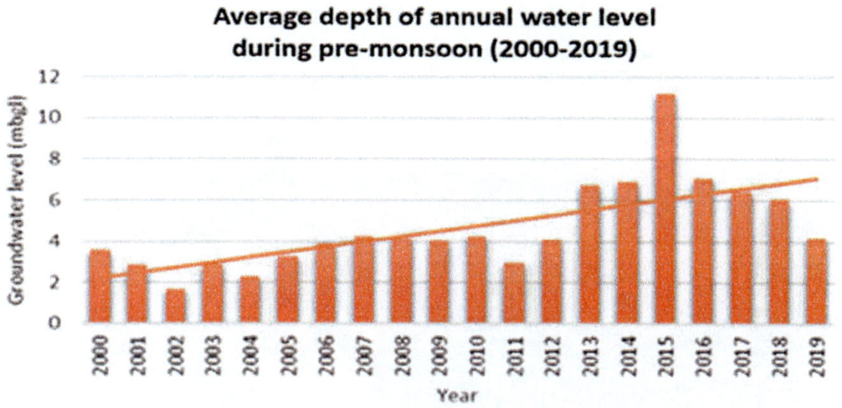

Fig. 2 Annual water level average depth for pre-monsoon period of years from 2000 to 2019

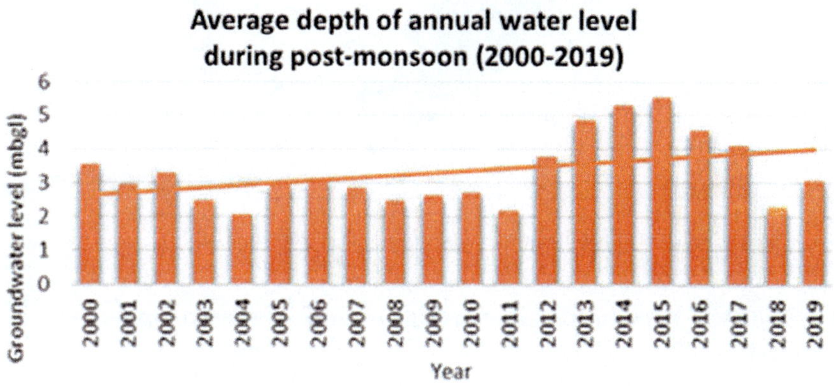

Fig. 3 Annual water level average depth for post-monsoon period of years from 2000 to 2019

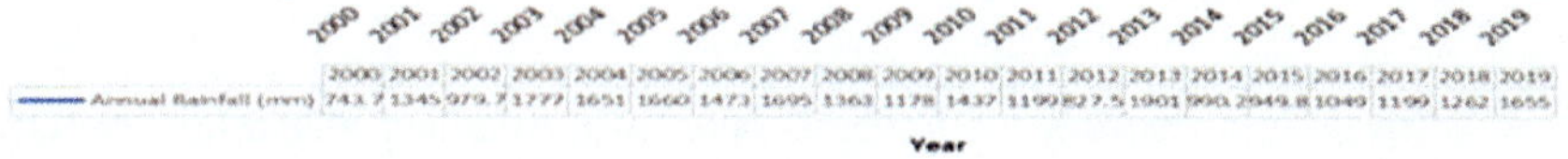

Fig. 4 Rainfall variation during whole period of study

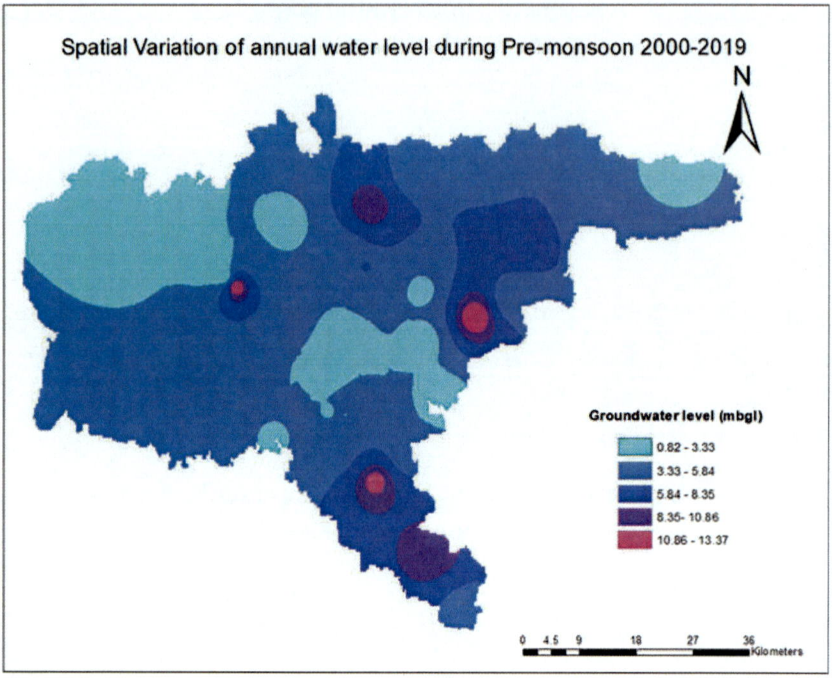

Fig. 5 Spatial variation of pre-monsoon period of years from 2000 to 2019

4.3 Spatial Variation of Pre-monsoon and Post-monsoon GWL

Spatiotemporal analysis of GWL was attempted for the Surat district using spatial interpolation methods (IDW). The water level data of 37 observation wells (dug well and bore well) which are well distributed in whole basin has been download for period of 20 years (2000–2019). Figures 5 and 6 show areal distribution of the basin with various range of GWL.

4.4 LULC of the District

LULC map has been prepared for the years 2000, 2010, 2015, and 2020 using Landsat 5 and Landsat 8 images of 30 m resolution. The whole area is classified into 5 major classes as showed in Fig. 7a–d. The results from LULC 2000, 2010, 2015, and 2020 indicate that there is an increase in agriculture, water body, and built-up area whereas the decrease in the forest and barren land area. Area-wise comparison of LULC classes of 2000, 2010, 2015, and 2020 is shown in Table 1.

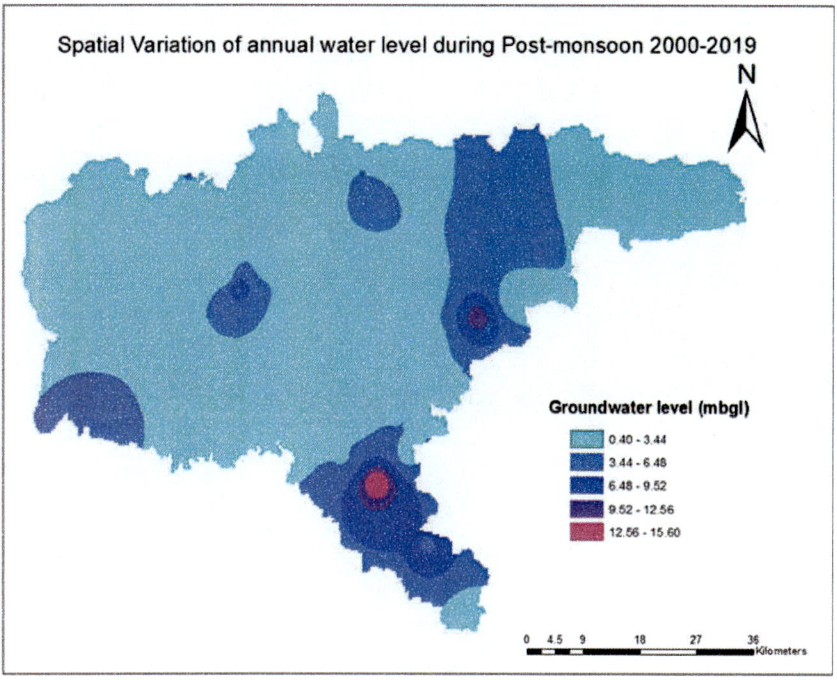

Fig. 6 Spatial variation of post-monsoon period of years from 2000 to 2019

The total area of the Surat district is 4414 km^2. The results show a 4.22% rise in built-up land, 8.93% decrease in barren land, 7.63% increase in agricultural land, 3.64% decrease in forest, and 0.3% reduction in surface water bodies during year 2000 to 2020. The outcomes were observed through LULC change monitoring.

Google Earth is a powerful and appealing source of positional data that may be utilized for research and preliminary studies with sufficient accuracy and at a low cost. Since high-resolution images from Google Earth are freely available to the public, they can be utilized directly in land use and land cover accuracy assessments. After the image was classified, ArcMap was used to generate a set of 100 random points, and the values of each random point were compared to the Google Earth image.

Accuracy assessment of classifications of years 2000, 2010, 2015, and 2020 is shown that kappa values are 0.73, 0.71, 0.75, and 0.75, respectively. Figure 8 shows the variation in the LULC classes in the years 2000, 2010, 2015, and 2020. The analysis indicates a rapid increase in agriculture, water body, and built-up area whereas the decrease in the forest and barren land area.

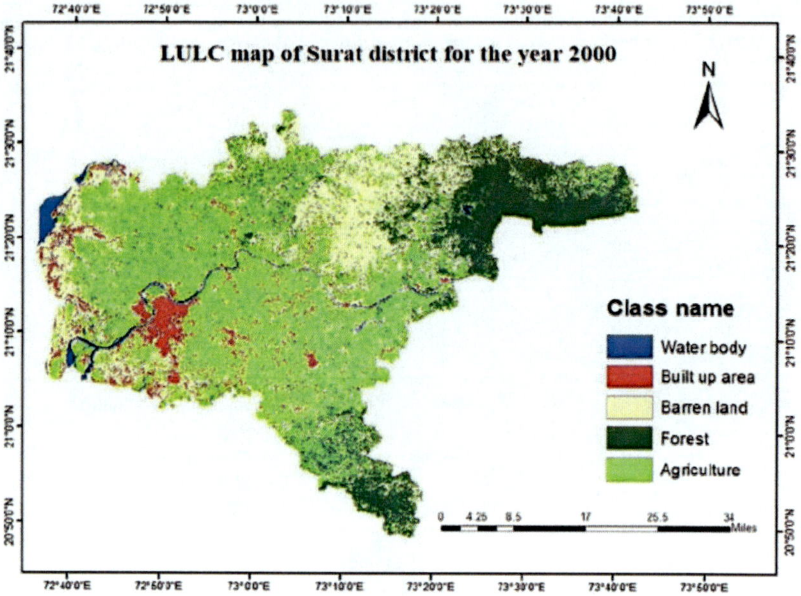

(a)

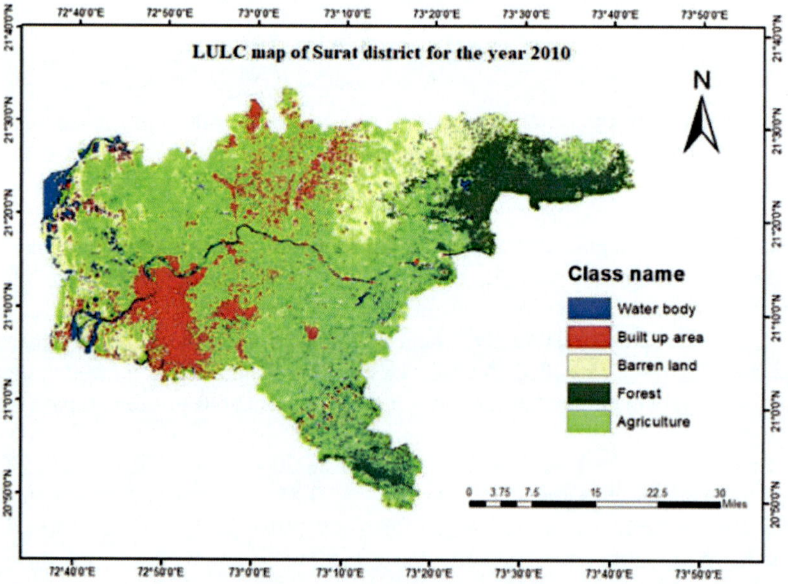

(b)

Fig. 7 **a** LULC map of Surat district for the year 2000, **b** LULC map of Surat district for the year 2010, **c** LULC map of Surat district for the year 2015, **d** LULC map of Surat district for the year 2020

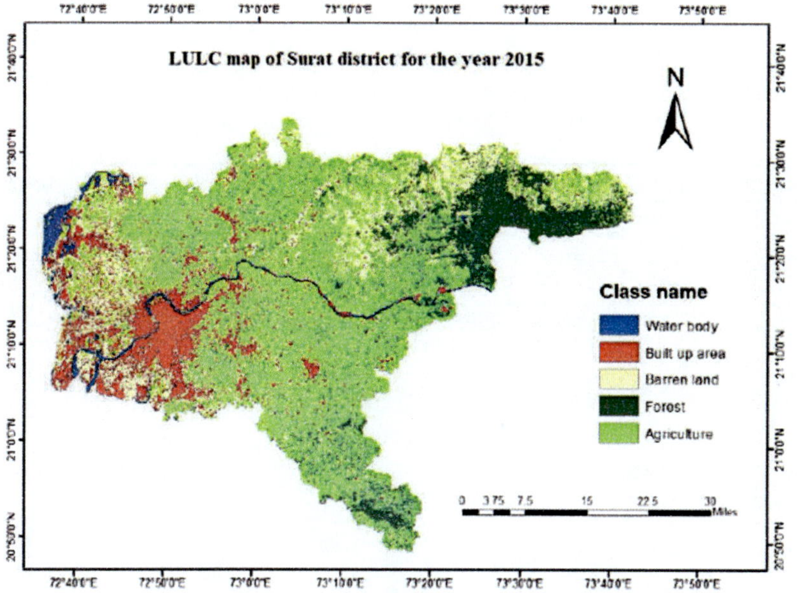

(c)

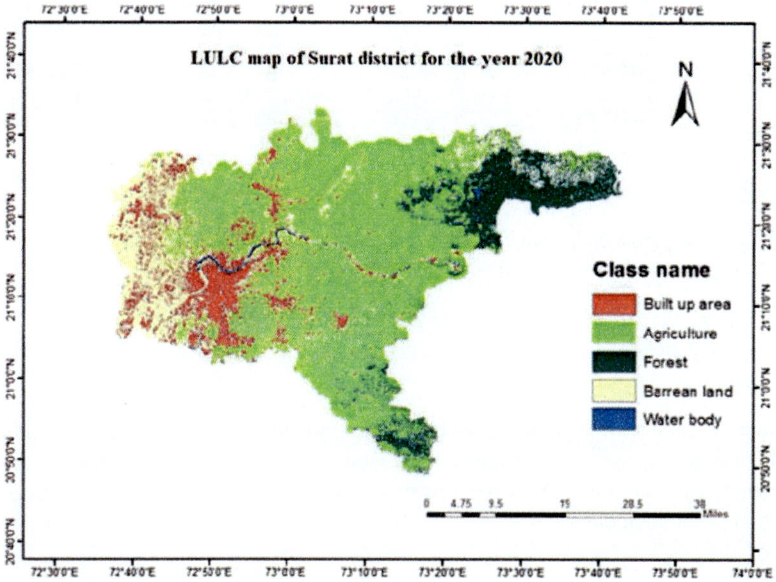

(d)

Fig. 7 (continued)

Table 1 Area-wise comparison of LULC classes of 2000, 2010, 2015, and 2020

	2000		2010		2015		2020	
Class Name	Area (km^2)	% of total area	Area (km^2)	% of total area	Area (km^2)	% of total area	Area (km^2)	% of total area
Water body	52	1.18	49	1.11	42	0.95	39	0.88
Built up	208	4.71	286	6.48	338	7.66	394	8.93
Barren land	1219	27.62	1147	25.99	958	21.70	825	18.69
Forest	698	15.81	638	14.45	622	14.09	537	12.17
Agriculture	2251	51.00	2291	51.90	2424	54.92	2588	58.63

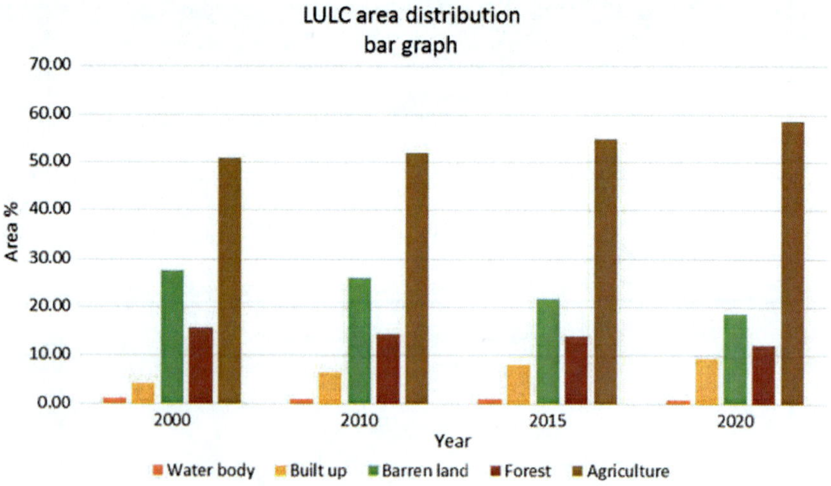

Fig. 8 LULC area distribution graph

4.5 *Correlation Between Groundwater Change Values and LULC*

The coefficients of determination between groundwater change values and LULC are determined using a simple linear regression model. As shown in Table 2, there is a significant relationship between pre-monsoon and post-monsoon GWL and land use change for each class.

Table 2 Coefficient of determination between LULC of each class and GWL

Sr. No.	Land use and land cover	Pre-monsoon GWL	Post-monsoon GWL
		Coefficients of determinations (R^2)	
1	Water body	0.64	0.92
2	Built-up area	0.96	0.76
3	Barren land	0.95	0.91
4	Forest	0.97	0.64
5	Agriculture	0.97	0.86

5 Conclusions

The following conclusions are derived from the foregoing study:

Groundwater data from 2001 to 2012 was used to analyze groundwater table change in the study area.

- The result shows that GWL declined by 5 mbgl during the pre-monsoon period and by 1.5 mbgl during the post-monsoon period in the study area. The results show a 4.22% rise in built-up land, 8.93% decrease in barren land, 7.63% increase in agricultural land, 3.64% decrease in forest, and 0.3% reduction in surface water bodies during year 2000 to 2020. The analysis indicates a rapid increase in agriculture, water body, and built-up area whereas the decrease in the forest and barren land area.
- The classification results show that the supervised maximum likelihood classification overall accuracy of 82% and the kappa coefficient of 0.73.
- Analysis shows that there is a significant correlation between LULC change and (pre-monsoon and post-monsoon) GWL.

Acknowledgements The authors are also thankful to Central Ground Water Board (CGWB) for providing necessary data to conduct the present study.

References

1. Anand B, Karunanidhi D, Subramani T, Srinivasamoorthy K, Suresh M (2020) Long-term trend detection and spatiotemporal analysis of groundwater levels using GIS techniques in Lower Bhavani River basin, Tamil Nadu, India. Environ Developm Sustain 22(4):(2779–2800)
2. Kayet N, Chakrabarty A, Pathak K, Sahoo S, Mandal SP, Fatema S, Das T (2019) Spatiotemporal LULC change impacts on groundwater table in Jhargram, West Bengal, India. Sustain Water Resour Managem 5(3):(1189–1200)

3. Patil VBB, Pinto SM, Govindaraju T, Hebbalu VS, Bhat V, Kannanur LN (2020) Multivariate statistics and water quality index (WQI) approach for geochemical assessment of groundwater quality—a case study of Kanavi Halla Sub-Basin, Belagavi, India. Environ Geochem Health 42(1–18):(2667–2684)
4. Verma P, Singh P, Srivastava SK (2020) Impact of land use change dynamics on sustainability of groundwater resources using earth observation data. Environ Developm Sustain 22(6):(5185–5198)
5. Zomlot Z, Verbeiren B, Huysmans M, Batelaan O (2017) Trajectory analysis of land use and land cover maps to improve spatial–temporal patterns, and impact assessment on groundwater recharge. J Hydrol 554:(558–569)

Critical Appraisal of Satellite Data for Land Use/Land Cover Classification and Change Detection: A Review

Zeenat Ara, Ramakar Jha, and Abdur Rahman Quaff

Abstract Satellite images are used in various fields, for example, climatology, agriculture, biodiversity landscape, geology, and forestry. Land use/land cover (LULC) classification of satellite images is one of the essential applications established from earth surveillance satellites. This paper analyzes the progresses in LULC classification techniques using satellite images. The initial techniques of land cover classification using satellite images were done in 1970s, monitored by supervised and unsupervised pixel-based classification techniques using K-means, iterative self-organizing data analysis technique (ISODAT), and maximum likelihood classifiers. Later 1980, further methods such as knowledge-based, sub-pixel, object-based image analysis (OBIA), contextual-based, and hybrid methods became prevalent in land cover classification. Satellite image classification needs selection of suitable method based on the necessities. Exact and dependable information about LULC is important for change detection and checking of the identified area. Lots of progresses in LULC classification techniques of satellite images have happened in recent past.

Keywords Satellite image · Classification · Change detection · Remote sensing (RS) · Geographical information systems (GISs) · LULC

1 Introduction

Satellite and remote sensing images provide qualitative and measurable information. Satellite image classification provides details of remote sensing images, spatial data

Z. Ara (✉)
MANUU, Gachibowli, Hyderabad, Telangana 500032, India
e-mail: zeenatiitd@rediffmail.com

R. Jha · A. R. Quaff
NIT, Patna, Bihar 800005, India
e-mail: rj@nitp.ac.in

A. R. Quaff
e-mail: arquaff@nitp.ac.in

P. V. Timbadiya et al. (eds.), *Geospatial and Soft Computing Techniques*, Lecture Notes in Civil Engineering 339, https://doi.org/10.1007/978-981-99-1901-7_11

mining, and study of plants types such as agriculture and foresters, learning urban land uses, and its regulation in an area [1]. Interfacing of GIS technology with remote sensing provides information and ability to do analysis, and hence, it is useful for land use developers. The availability of high spatial resolution satellite imaging has created chances for a better understanding of issues related to land use, source evaluation, and environmental observation [8].

The term land use communicates to human act or economic purpose related with detailed portion of land. Residential, urban, and agricultural land use are all examples of land use. Because the runoff reaction from the catchment region is dependent on its land use feature, land use classification is an important aspect in water resource engineering [4].

Because of fast growth of urbanization, population, industrialization, the farming land is decreasing day by day. LULC is playing a vital role in global environmental change. For provincial and enormous level management and planning, LULC plays an vital role by using GIS and RS [23]. It is likely to make the planning and pronouncement-making process more practical and efficient. The LULC is nil, yet it is the physical substance on the Earth's surface that is encircled by various sections of the landscape or man-made components on the Earth's surface, such as soil, rocks, water bodies, and flora. The LULC changes, in which human deeds play a significant role, interrelate with the environment, have dominant impacts on the ecosystem at the local, regional, and mesoscale, and hence have a large impact on global climate change, either directly or indirectly [6]. It has long been understood that human-caused land use changes, such as deforestation and agricultural practises, have an impact on the climate. The human activity impact on the landscape has increased tremendously both in intensity and on scale over the past centuries, mainly through the growth of agriculture, which has been the most important ancient modification in LULC [24]. In-appropriate land use is causing numerous kinds of ecological degradation. The LULC changes are related with composite relations between the climate fluctuations, availability of resources, and a very wide set of socio-economic aspects. The significances of LULC changes are reflective. Environment loss by changing the unusual forest cover is the most significant factor in the worldwide biodiversity calamity.

LULC contains coverage from ice, crop type, and snow to biomes which contains tundra, boreal or tropical rain forest, and unfertile land. Land use dynamics should be monitored in both space and time to deal with the changing demands of an expanding population and to better understand of the link between natural events and human activities [5]. Change detection for GIS is a task that shows how the feature of a particular area has changed between periods. Change detection usually involves relates satellite imagery or aerial photographs of the area taken at different time periods. Change detection has been mostly used to estimate an agricultural scheme in which a person usages a part of land, abandons or alters the initial use a short time later, urban growth, deforestation, effect of natural disasters like earthquakes, tsunamis and LULC. Change detection is an effective method for operating, managing, and monitoring a current geographical distribution of land resources. The rapidly changing terrain is the primary driver of global environmental change and a major concern for global sustainability. LULC changes are foremost problems of universal environment change. The satellite remote sensing data with their tedious nature have verified to be relatively useful in mapping LULC shapes and variations with time. GIS abilities can be used to quantify such changes, even if the produced spatial datasets are of different scales/determinations [3]. These types of studies have facilitated in considerate the changing aspects of human activities in spatio-temporal. Land use mentions to human's deeds.

Virtuous change detection exploration should be responsible for the following information:

- Change rate and area
- Change types spatial distribution
- Change trajectories of land cover types and
- Evaluation of the outcomes of change detection's accuracy.

The accurateness of change detection results depends on many aspects [11], for example,

- Accessibility of superiority ground truth data
- Accurate multi-temporal images between geometric registration
- Normalization or graduation between multi-temporal images
- The complication of landscape and environs of the study area
- Classification and change detection systems
- Change detection techniques or procedures used and
- Familiarity and information of the study area.

2 Evolution of Satellites Data

The CORONA programme was a series of American strategic exploration satellites developed and operated by the Directorate of Science and Technology of the Central Intelligence Agency (CIA), with considerable support from the United States Air Force. Beginning in June 1959 and concluding in May 1972, the CORONA satellites were used to photograph the Soviet Union (USSR), China, and other places. The Landsat database is a sequence of Earth-detecting satellite operations managed conjointly by NASA and the U.S. Geological Survey. On July 23, 1972 in collaboration with NASA, the Earth Resources Technology Satellite (ERTS-1) was launched. It was later retitled Landsat 1 [9]. Extra Landsat satellites tracked in the 1970s and 1980s. Landsat 7 was launched in 1999 followed by Landsat 8, hurled on February 11, 2013. Together Landsat 7 and Landsat 8 are presently in collecting data and orbit. Landsat 9 was launched on September 27, 2021. The widely successful Landsat programme was created to collect satellite images of Earth. Landsat is the first continuous Earth observing satellite imaging programme. Visual Landsat imagery has been composed at 30 m resolution since the initial 1980s. Launch with Landsat 5 thermal infrared imagery was also composed (at coarser spatial resolution than the optical data). Eight spectral bands with a spatial resolution of 15 to 60 m (49 to 197 ft) and a

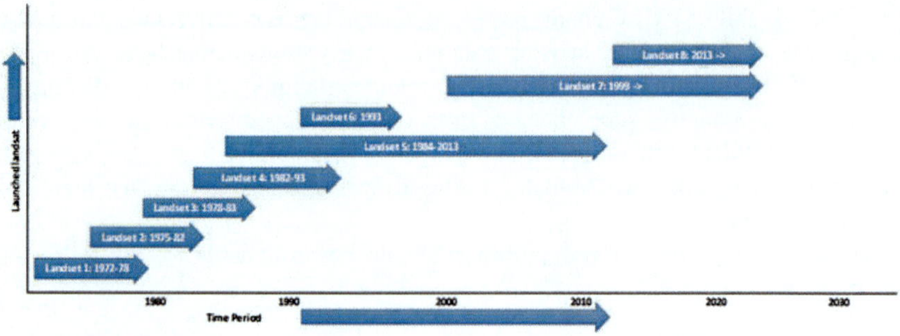

Fig. 1 Landsat satellites launched since 1972

temporal resolution of 16 days make up the Landsat 7 data. For ease of downloading, Landsat images are typically grouped into scenes. The length and width of each Landsat picture are around 115 miles (or 100 nautical miles long and 100 nautical miles wide or 185 km long and 185 km wide). Figure 1 provides Landsat satellites launched since 1972.

The IKONOS satellite instantaneously assembles 1-m panchromatic and 4-m multispectral images, given that moneymaking and technical community with a theatrical development in spatial resolution over before prevailing satellite imagery [7].

NASA deployed the Moderate Resolution Image Spectroradiometer (MODIS), an imaging payload sensor built by Santa Barbara Remote Sensing, into Earth orbit with the Terra (EOS AM) satellite in 1999 and the Aqua (EOS PM) satellite in 2002. Terra MODIS and Aqua MODIS are watching the whole Earth's surface each one to two days. The devices capture data in 36 spectral bands ranging in wavelength from 0.4 μm to 14.4 μm and at changeable spatial resolutions (29 bands at 1 km 5 bands at 500 m and 2 bands at 250 m).

The European Space Agency (ESA) is presently evolving the Sentinel collection of satellites. Presently, 7 missions are scheduled, each for a different application.

Sentinel-1 is a polar-orbiting, all-climate, and day-and-night radar imaging task for land and ocean amenities. Sentinel-1A and Sentinel-1B were launched in orbit on 3 April 2014 and 25 April 2016, respectively.

Sentinel-2 is a polar-orbiting, multispectral, and high-resolution imaging mission for land observation, allowing for the observation of things like plant, soil, water cover, and inland and coastal waterways. Sentinel-2 can provide information to emergency services as well. Launch dates for Sentinel-2A and Sentinel-2B are June 23, 2015, and March 7, 2017, respectively.

Sentinel-3's measure the topography of the sea's surface, the temperature of the land and sea, the colour of the sea, and the colour of the land with extreme precision and consistency. Along with supporting environmental and climate monitoring, the mission also assists ocean forecasting plans. On 16 February, 2016, Sentinel-3A was launched, and on 25 April, 2018, Sentinel-3B joined its twin in orbit. Sentinel-4 is a European Earth opinion mission established to maintenance the European Union Copernicus Programme for observing the Earth. In order to support the Copernicus Atmosphere Monitoring Service (CAMS), the primary goal of the Sentinel-4 mission is to monitor important air quality, trace gases, and aerosols over Europe with a high degree of three-dimensional resolution and a short re-evaluation interval.

Sentinel-5 continuously monitors the composition of the Earth's atmosphere and does air quality monitoring. Additionally, it offers data with broad worldwide coverage to track the quality of the air everywhere. Sentinel-5P's primary goal is to conduct high spatiotemporal resolution atmospheric observations of air quality, climate forcing, ozone, and UV radiation. On October 13, 2017, the satellite was successfully launched from the Plesetsk Cosmodrome in Russia.

Sentinel-6 Michael Freilich is an Earth Observation satellite mission developed to provide enhanced continuity to the very stable time series of mean sea level measurements and ocean sea state that started in 1992, with the TOPEX/Poseidon mission. The first satellite was launched into orbit on November 21, 2020 on a SpaceX Falcon 9 rocket from the Vandenberg Air Force Base in California, US.

ASTER data are used to produce complete maps of land surface temperature, altitude, and reflectance. The coordinated system of Earth Observation Satellites (EOS), containing Terra, is a main constituent of NASA's Science Mission Directorate and the Earth Science Division. NASA Earth Science's mission is to advance science's understanding of the Earth as a cohesive system, how it responds to difference, and how to effectively anticipate weather, climate change, and natural disasters.

The Meteosat-2 geostationary meteorology satellite started operationally to supply imagery data on August 16, 1981. Eumetsat has worked the Meteosats since 1987.

3 Satellite Data for LULC Classification Techniques

3.1 Necessity of Satellite Image Classification

The extraction and understanding of relevant information from vast satellite images are aided by satellite image categorization for (a) spatial data mining, (b) thematic map development, (c) interpretation of visual and digital satellite image, (d) extract information for an application, (e) field surveys, (f) effective decision making, and (g) disaster management, and satellite image categorization is necessary.

3.2 *Satellite Image Technique*

There are a number of techniques and procedures for satellite image classification. Figure 2 shows pyramid of satellite image classification techniques, and Fig. 3 shows corresponding progresses of LULC classification techniques.

Satellite image classification method can be generally classified into three types as follows:

- Computerized
- Handbook
- Hybrid.

3.2.1 Computerized

Computerized satellite image classification techniques use procedures that applied analytically the whole satellite image to cluster pixels into meaningful groupings.

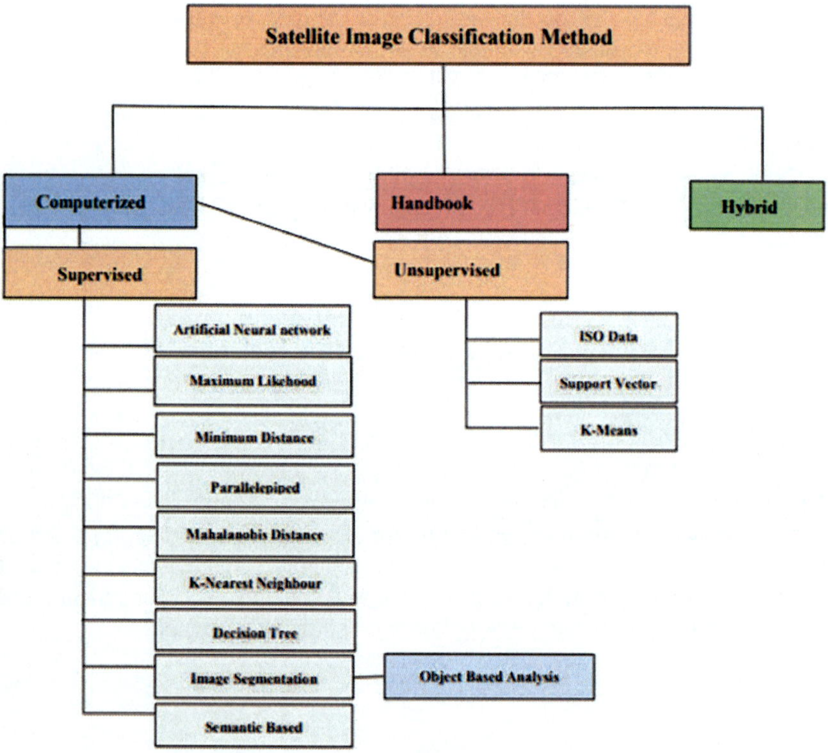

Fig. 2 Satellite image classifications techniques pyramid

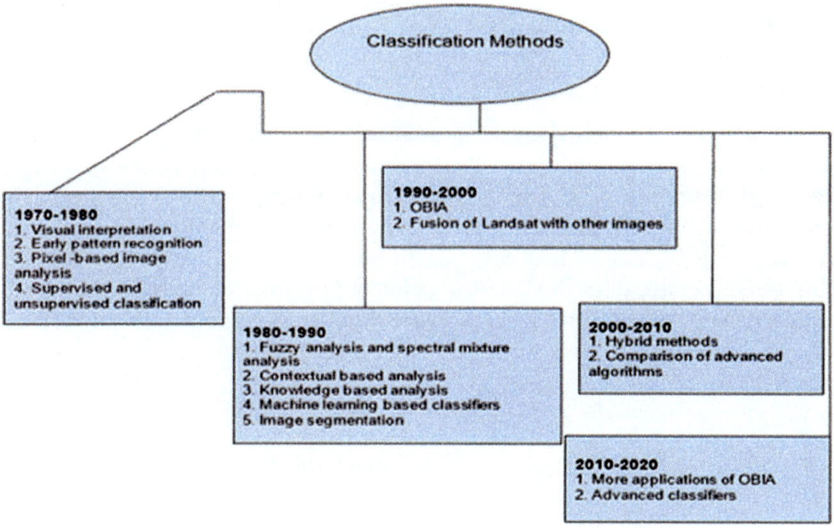

Fig. 3 Progresses of LULC classification techniques

Computerized satellite image classification techniques additionally classified into two groups, i.e., supervised and unsupervised classification methods.

3.2.2 Handbook

Handbook satellite image classification techniques are robust, operative, and efficient; however, these consume more time. In handbook techniques, the predictor must be aware with the area covered by the satellite image. Predictor knowledge and acquaintance with the subject field are key factors in productivity and accuracy of the classification.

3.2.3 Hybrid

Hybrid satellite image classification techniques combine the advantages of computerized and handbook techniques. Hybrid method uses computerized satellite image classification techniques to perform initial classification; further, handbook techniques are used to improve classification and correct faults.

3.3 Early Satellite Data for LULC Classification: Visual Approach

Initially, satellite data for LULC classification techniques were related to those used in predictable aerial photo analyzes in the 1950s and 1960 [17]. Usually, satellite images were used in the identical way as aerial photographs, which were ironic sources of info for spatially describing landscapes on cartographic maps with dissimilar scales. In the initially 1970s, Landsat land cover classification was visual and labour-intensive. This was completed through the investigation of printed aerial images [9] declared that the images were in print layout and were acquired as black and white combined or individual bands.

3.4 Satellite Data for LULC Classification: Digital Approach

In nowadays, world of unconventional skill where most remote sensing data are noted in digital format, nearly all image clarification and investigation, contains some part of digital processing. Digital image dispensation may involve various trials including configuring and modifying of the data, digital enrichment to enable improved visual explanation, or even automatic classification of marks and geographies completely by computer. In command to process remote sensing imagery digitally, the data must be logged and accessible in a digital form appropriate for storing on a computer tape or disk. Clearly, the other obligation for digital image dispensation is a computer system, sometimes mentioned to as an image analysis system, with the suitable hardware and software to procedure the data. Numerous commercially obtainable software systems have been advanced specifically for remote sensing image dispensation and investigation.

3.4.1 Digital Numbers

The numerical method of digital number (DN) or brightness values (BV) of remote sensing pictures has led to advancements in digital land cover classification (Fig. 4). As a result, digital images are made up of a collection of picture parts, or pixels, that connect each column and row of an image. The lower the DN value the lesser the radiance in that particular pixel [17]. The deviation of the land cover surfaces is represented by differences in radiance values in the pixels. Satellite photos are multi-spectral photographs that record in multiple bands of the electromagnetic spectrum at the same time.

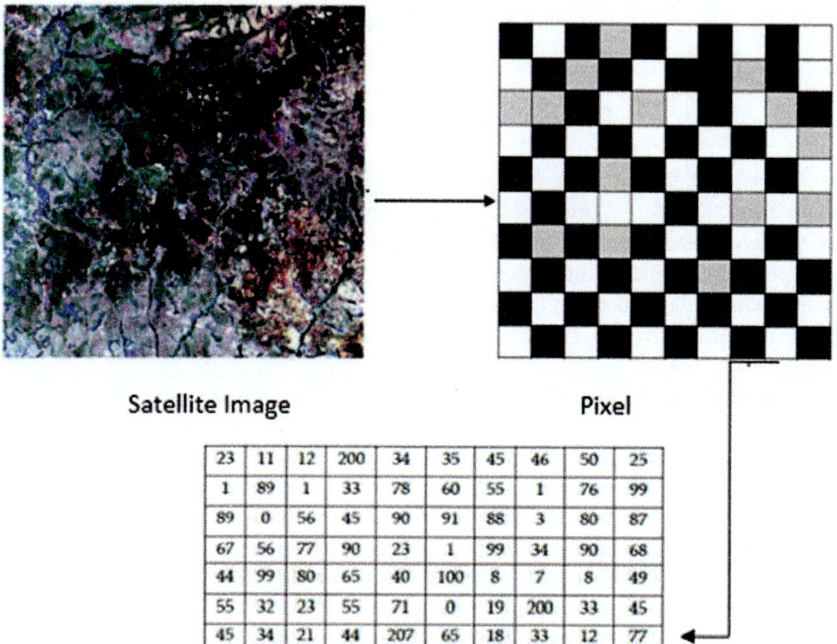

23	11	12	200	34	35	45	46	50	25
1	89	1	33	78	60	55	1	76	99
89	0	56	45	90	91	88	3	80	87
67	56	77	90	23	1	99	34	90	68
44	99	80	65	40	100	8	7	8	49
55	32	23	55	71	0	19	200	33	45
45	34	21	44	207	65	18	33	12	77
88	45	22	33	45	66	78	89	0	77
9	10	13	57	89	88	90	200	208	100
11	23	12	7	209	56	78	45	88	78

Digital Number

Fig. 4 Example of a remote sensing image; the arrow indicates the level of detail of information which can be extracted from the images [17]

3.4.2 Digital Image Processing

Digital image processing includes the mathematical alteration of digital values to form useful data involving to LULC types. Image processing generally consists of four main steps as follows:

- Pre-processing technique
- Image enhancement technique
- Image transformation technique
- Image classification and analysis technique.

Pre-processing functions contain those procedures that are normally essential preceding to the main data analysis and abstraction of information and are usually assembled as radiometric or geometric corrections.

Image enhancement is the process of modifying digital photographs to improve their quality and aesthetic appeal so that the outputs are more suited for display or additional image analysis. You could, for instance, remove noise, sharpen, or brighten an image to make it easier to spot important details. Most often, data from many spectral bands are processed in tandem during image modifications. The creative bands are combined and transformed into 'new' images using arithmetic operations (such as addition, subtraction, multiplication, and division) in order to highlight or better depict certain characteristics in the visual field. Pixels in the data are digitally categorized and classified using image analysis and classification algorithms. Usually, multi-channel datasets (A) are used for classification, and this process assigns each pixel in an image to a certain class or theme (B) based on statistical characteristics of the pixel intensity values. There is a variability of methodologies taken to implement digital classification. Classification processes can be shattered down into two wide sections created on the method used: supervised classification and unsupervised classification.

3.5 LULC Classification Ideologies for Satellite Images

Fractional filtering techniques or mathematical classification methods are initial automatic methods of image processing classification. Sharpening, smoothing, and feature abstraction are amongst the techniques used in spatial filtering approaches to transform images into more useable forms. Modern land cover classification methods are built on the numerical classification method. In most cases, pattern recognition is used to classify land cover types by looking for similarities between items. Machine learning and artificial intelligence were used to develop new classification methods derived from the early techniques on computer-automated applications [19].

3.6 Pixel-Based Classification

The first pixel-based LULC classification algorithms were created in the early 1970s on Landsat MSS. Pixel-based categorization is a method of assigning a class to each pixel by treating each pixel as a separate unit. Pixels in the same class are more spectrally similar to each other than pixels in different classes. The satellite images were classified using a pixel-based spectrum angle mapper (SAM) classifier for pixel-based classification. The sign file was created, which included the class preparation. The land cover classes of water body, vegetation, bare-soil, and built-up were trained using areas of interest (AOI). For each class, arbitrary samples were taken across the research region based on pixel bands. The SAM algorithm, a supervised methodology, became useful at that time. The spectral angle mapper (SAM) process is based on the hypothesis that a single pixel of remote sensing images symbolizes one certain ground cover substantial, which can be exclusively dispersed to only one ground cover class. This process is based on the capacity of the spectral resemblance between two spectra. The spectral resemblance can be obtained by considering each spectrum as a vector in q-dimensional space, where q is the number of bands.

Unsupervised and Supervised Classification

Supervised (human-led) and unsupervised (software-based) classification are the two basic types of image classification techniques. These classifications are pixel-based. In other words, it creates square pixels with a class for each one. In supervised classification, the images are trained, but in unsupervised classification, no training is done.

The outputs (groups of pixels with similar appearances) of unsupervised classification are based on a software study of a picture without the usage of sample classes provided by the user. The computer utilizes algorithms to figure out which pixels are connected and groups them into groups. The user can choose which algorithm the software will use and how many output classes they want; however, it does not help with categorization. When the computer creates clusters of pixels with common physical properties, the user must have information about the area being classed. This includes information about developed areas, marshes, and coniferous woods, amongst other things. ISODATA, support vector machine (SVM), and K-means are the most common unsupervised satellite image classification algorithms.

A predictor is required for supervised classification procedures. The training set is the input. In supervised satellite image classification algorithms, the training sample is the most important aspect. The precision of the procedures is strongly dependent on the training samples. There are two sorts of training samples: one for sorting and another for checking classification accuracy. The following statistical methods are used in the main supervised classification techniques:

- Binary decision tree (BDT)
- Image segmentation
- Artificial neural network (ANN).

Several classification methods are based upon dissimilar kinds of resemblance matching techniques. Additional features of supervised classification include assessing input data, creating training samples and signature files, and specifying the standard of the training samples and signature files. Binary decision tree (BDT) satellite image classification processes are based on machine learning methods. Decision tree method includes a set of binary rules that describe meaningful classes to be linked to separate pixel. Different decision tree software is accessible to create binary rules. The software takes training set and additional data to describe actual rules.

In satellite image processing, investigation, and pattern recognition, segmentation plays an important role. Image categorization is not directly related to satellite image segmentation methods/algorithms. Image segmentation divides pixels into segments that are comparatively homogeneous. Image segmentation algorithms provide factors that help predictors guess the size and shape of segments. As an alternative to pixel level classification, segmented images can be classified at the segmentation level. The categorization of satellite images at the segmentation level is substantially faster than classification at the pixel level.

Algorithms fall under artificial neural network (ANN), and it simulates human knowledge procedure. ANN-based satellite image classification procedures are easy to include additional data in the classification method and develop accurateness.

3.7 Non-parametric and Parametric Classifiers

When the probability density function is unknown, non-parametric classifiers are employed to estimate it. Parametric classifier is based on the statistical probability distribution of each class.

3.8 Sub-pixel Image Classification

Because most lands are made up of distinct land cover categories that are difficult to separate during standard pixel-based classification, sub-pixel-based classification was created to address this. A pixel is assumed to be made up of one related land cover type in pixel-based classification; nevertheless, numerous pixels may record multiple land cover categories [12]. Because Landsat's ground resolution varies between 30 and 60 m, a variety of land cover classes can combine to form a single pixel. Landsat pictures frequently contain numerous land cover categories in one pixel, which can be reduced using sub-pixel approaches. In the 1980s, Dempster-Shafter theory, fuzzy set theory, and certainty factor theory were used to develop sub-pixel categorization. Fuzzy set methods and spectral mixture analysis are the best collective sub-pixel classification methods (SMA).

3.9 Fuzzy Approach

Fuzzy classification methods for maximum likelihood categorization were created to increase classification accuracy based on a fuzzy set technique. In this technique, each pixel obtains a fractional relationship of all possible classes. Hence, the level of each class inside each pixel can be predictable. In this method, each land cover is allocated a fuzzy membership dependent on its quantity in each pixel. The quantities are in form of ratios, percentages, or chances which are changed to actual areas on the ground. The idea of the fuzzy set is vital for a fuzzy classification. A pixel can have fractional and numerous memberships to the contender classes. The comparative strengths of class membership are termed as fuzzy membership values (FMVs) [25]. FMVs typically vary from 0.0 to 1.0, and they are directly tied to a pixel's class membership in the competing classes. A pixel's FMVs typically add up to 1.0 across all potential classes.

3.10 Spectral Mixture Analysis (SMA)

SMA was developed as one of the best techniques to work with the sub-pixel methods, mainly for middle resolution imageries like Landsat. This technique was established in the early 1980s and has been useful widely on Landsat land cover classification [17]. SMA is a convenient image processing technique to classify the basic parts and proportion of miscellaneous endmembers from the spectral features of each pixel.

Generally, SMA includes three steps as follows:

- Assessment of data dimensionality to select suitable spectral bands.
- Determination of the type and number of endmembers and ensuring endmembers purity.
- Pixel-wise evaluation of endmembers fractions using a spectral likeness measure.

3.11 Object-Based Image Analysis (OBIA)

OBIA sections an image by group pixels. It does not make single pixels. In its place, it produces objects with diverse geometries. If you have the correct image, objects can be so eloquent that it does the digitizing for you. In Fig. 5 example, the separation results below highlight buildings.

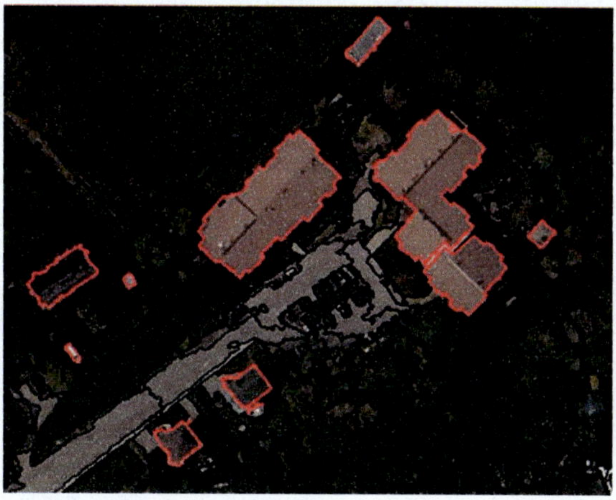

Fig. 5 Separation results highlighting buildings

3.12 Comparative Evaluation of Satellite Image Classification Techniques

Many researchers have compared the classification correctness and kappa coefficient of supervised and unsupervised satellite image classification systems, as well as the combination of the two. This section compares and contrasts various scholars. Table 1 compares and contrasts the conclusions of various scholars. Scientists' opinions on how to improve satellite image classification techniques are divided. Furthermore, there is a requirement to investigate climatic satellite image categorization algorithms in order to demonstrate reliance on the test dataset.

Table 1 Critical appraisal of various scholar's satellite image classification

Scholars	Classification technique	Check data	Best technique from the scholar study
Rozenstein and Karnieli [18]	ISODATA Maximum likelihood Hybrid method	Desert outlay datasets	Hybrid method
Akgün et al. [2]	Minimum distance parellelpiped, maximum likelihood	Landsat 7 ETM + images	Maximum likelihood
Tamouk et al. [22]	Minimum distance parallelepiped, chain method	Landsat 5TM images	Chain method
Shila and Ali [20]	Unsupervised supervised hybrid method	Landsat7 ETM + data	Hybrid method
Niknejad et al. [16]	Support vector machine maximum likelihood Mahalanobis distance Minimum distance, spectral information divergence binary codes parallelepiped	Landsat7 ETM + data	Support vector machine
Manoj et al. [14]	K-means ISODATA minimum distance maximum likelihood parallelepiped seeded region growing enhanced seeded region growing	Landsat, SPOT, and IRS datasets	Enhanced seeded region growing
Malgorzata et al. [13]	Pixel-based classification object-oriented classification	Multi-spectral satellite images	Object-oriented classification

4 Change Detection Techniques

Bi-temporal change investigation is one of the best mutual methods of change detection for the direct assessment of pairs of images or categorizations [10]. Change is computed via spectral (image) or thematic (categorization) contrast.

4.1 Discussions Before Applying Change Detection

Lu et al. [11] defined four substantial features of change detection for observing natural resources; identifying if a change has happened, classifying nature of the change, calculating aerial range of the change, and evaluating the three-dimensional arrangement of the change and listed five classes of sources that predisposed land cover change; long-term regular changes in geomorphological, climate conditions, and biological processes such as vegetation succession, soil erosion, and human-induced modifications of vegetation cover and sceneries such as land degradation, deforestation, inter-annual climate changeability, and the greenhouse effect produced by human actions. The time-based, spectral, spatial, and radiometric determinations of remotely sensed data have a substantial effect on the achievement of a remote sensing change detection project.

Change detection is a technique for identifying changes in an entity's or an event's status by sensing it at various intervals. The fundamental idea behind employing remote sensing data for change detection is that variations in land use and land cover must lead to changes in vivacity values, and variations in radiance owing to these changes must be much larger than variations in vivacity brought on by other factors [21]. These 'other' reasons consist of (a) variances in atmospheric circumstances, (b) changes in Sun angle, and (c) changes in soil moisture. The effect of these features may be moderately reduced by choosing the suitable data. For example, Landsat data fitting to the same time of the year may decrease problems from Sun angle variances and changes in vegetation.

For detecting and analyzing the change on the earth's surface, various methods are active. Earlier learning about several change detection methods, it is essential to know about the technique of change detection. To detect the changes of the surface of the earth, main steps are important as mentioned by Mishra et al. [15] which are as follows.

- Detecting the type of change
- Data selection from remote sensing
- Image processing
- Image processing and classification
- Selection change detection algorithm
- Evolution of change detection result.

4.2 *Various Detection Techniques Best Suited for*

Post classification—Land use land cover classification and change, urban sprawl evaluating, change detection by unsupervised classification.

Image rationing—LANDSAT is used to track changes in the environment.

Image differencing—Change identification in forest ecosystems using SPOT HRV imaging of urban land cover changes.

Principal component analysis—Vegetation regrowth, destruction caused by brush-fires, identification and analysis of changes in land use and land cover.

Direct multi-date classification—Land cover change detection.

5 Conclusions

This paper summarizes analyzes of classification techniques for satellite images and is focussed on the progress of use of satellite data for land cover classification techniques and defining the optimal ways of using satellite images in land cover classification. Pixel-based, sub-pixel-based, and object-based techniques to satellite imaging can all be generically classified. Although land cover classification methods have improved over the previous decades. The pixel-based classification method which was developed in the 1970s is still the most often used methodology on satellite pictures. The majority of satellite data land cover categorization studies have demonstrated OBIA's superior performance in a variety of landscapes, including agricultural areas, urban areas, wetlands, and forests. The fundamental benefit of OBIA is that it represents categorization units as real-world items on the ground, and it has shown to be very useful in both industry and academic research.

The use of change detection method in remote sensing can be used to plot land use land cover change which has followed over time. These changes on the earth are affected due to human actions. The change detection has got its several uses; few of them are crop monitoring, deforestation, moisture content of soil, urban planning, and water quality.

Many image processing techniques depend on the accuracy with which they recognize meaningful edges. It is one of the methods for identifying digital image intensity discontinuities. The analysis summarized in this paper may help scholars to select suitable satellite image classification technique based on the requirements. Change detection analysis is being performed using Landsat 8 images of 30 m resolution to assess impact of deforestation on climate changes by author in current on-going research work.

References

1. Abburu S, Golla SB (2015) Satellite image classification methods and techniques: a review. Int J Comput Appl 119(8)
2. Akgün A, Eronat AH, Türk N (2004) Comparing different satellite image classification methods: an application in Ayvalik District, Western Turkey. In: The 4th international congress for photogrammetry and remote sensing, Istanbul, Turkey
3. Anil NC, Sankar GJ, Rao MJ, Prasad IVRKV, Sailaja U (2011) Studies on land use/land cover and change detection from parts of South West Godavari District, AP–using remote sensing and GIS techniques. J Ind Geophys Union 15(4):187–194
4. Ara Z (2021) Land use classification using remotely sensed images: a case study of Eastern Sone Canal Bihar. In: Water management and water governance: hydrological modeling, pp 47–60
5. Das S, Angadi DP (2021) Land use land cover change detection and monitoring of urban growth using remote sensing and GIS techniques: a micro-level study. GeoJournal, pp 1–23
6. Deng X, Zhao C, Yan H (2013) Systematic modeling of impacts of land use and land cover changes on regional climate: a review. Adv Meteorol
7. Dial G, Bowen H, Gerlach F, Grodecki J, Oleszczuk R (2003) IKONOS satellite, imagery, and products. Remote Sens Environ 88(1–2):23–36
8. Gaur MK, Moharana PC, Pandey CB, Chouhan JS, Goyal P (2015) High resolution satellite data for land use/land cover mapping-a case study of Bilara Tehsil, Jodhpur district. Ann Arid Zone 54(3&4):125–132
9. Haack BN (1982) Landsat: a tool for development. World Dev 10(10):899–909
10. Hansen MC, Loveland TR (2012) A review of large area monitoring of land cover change using Landsat data. Remote Sens Environ 122:66–74
11. Lu D, Mausel P, Brondizio E, Moran E (2004) Change detection techniques. Int J Remote Sens 25(12):2365–2401
12. Makinde EO, Salami AT, Olaleye JB, Okewusi OC (2016) Object based and pixel based classification using rapideye satellite imager of ETI-OSA, Lagos, Nigeria. Geoinf FCE CTU 15(2):59–70
13. Malgorzata VW, Aniko K, Erzsebet V (2012) Comparison of different image classification methods in urban environment. In: International scientific conference on sustainable development & ecological footprint
14. Manoj P, Astha B, Potdar MB, Kalubarme MH, Bijendra A (2013) Comparison of various classification techniques for satellite data. Int J Sci Eng Res 4(2):1–6
15. Mishra S, Shrivastava P, Dhurvey P (2017) Change detection techniques in remote sensing: a review. Int J Wirel Mobile Commun Ind Syst 4(1):1–8
16. Niknejad M, Zadeh VM, Heydari M (2014) Comparing different classifications of satellite imagery in forest mapping (case study: Zagros forests in Iran). Int Res J Appl Basic Sci 8(9):1407–1415
17. Phiri D, Morgenroth J (2017) Developments in Landsat land cover classification methods: A review. Remote Sens 9(9):967
18. Rozenstein O, Karnieli A (2011) Comparison of methods for land-use classification incorporating remote sensing and GIS inputs. Appl Geogr 31(2):533–544
19. Shi W, Zhang M, Zhang R, Chen S, Zhan Z (2020) Change detection based on artificial intelligence: state-of-the-art and challenges. Remote Sens 12(10):1688
20. Shila HN, Ali RS (2010) Comparison of land covers classification methods in Etm+ satellite images (Case Study: Ghamishloo Wildlife Refuge). J Environ Res Develop 5(2):279–293
21. Singh A (1990) Digital change detection techniques using remotely-sensed data 1989. Int J Remote Sens 10(6):989–1000
22. Tamouk J, Lotfi N, Farmanbar M (2013) Satellite image classification methods and Landsat 5TM Bands. arXiv preprint arXiv:1308.1801

23. Vibhute AD, Gawali BW (2013) Analysis and modeling of agricultural land use using remote sensing and geographic information system: a review. Int J Eng Res Appl 3(3):081–091
24. Yang Y, Zhang S, Yang J, Chang L, Bu K, Xing X (2014) A review of historical reconstruction methods of land use/land cover. J Geogr Sci 24(4):746–766
25. Zhang J, Foody GM (1998) A fuzzy classification of sub-urban land cover from remotely sensed imagery. Int J Remote Sens 19(14):2721–2738

Land Use/Land Cover Monitoring and Change Detection of Sabarmati River Basin Using GIS and Remote Sensing

Rekha Verma, Mohammed Sharif, and Azhar Husain

Abstract India has been experiencing a significant change in the land use/land cover (LULC) pattern over several decades. So, in order to understand long-term patterns of the usage of natural resources and achieve sustainable development, effective observation and mapping of LULC maps are necessary using the spatial data. In this paper, LULC changes in the Sabarmati River basin were analyzed using Decadal land use map of 1985, 1995, and 2005. The land use map was processed in ArcGIS 10.1 software for all the three years, and land use map of Sabarmati River basin was extracted. The LULC map thus obtained showed eleven different land use classes which were then reclassified into five classes agricultural, forest, built-up land, barren land, and waterbodies. A comparative analysis was made between two decades LULC pattern, i.e., 1985–2005 to detect the LULC changes. The results depicted that there was a decrease in the agricultural land by 0.54% (−165.27 km^2) and forest area by 0.40% (−121.70 km^2) and an increase in the barren land by 0.01% (4.07 km^2) and the built-up area by 0.55% (166.71 km^2). It was also depicted that the water bodies area also increased by 0.38% (116.18 km^2). The increase in the area of the barren land and built-up land results as an important parameter leading to increase in chances of floods every time as monsoon hits Gujarat. The study shows that GIS and remote sensing techniques serve as an important tool for the temporal analysis of

Disclaimer: The presentation of material and details in maps used in this chapter does not imply the expression of any opinion whatsoever on the part of the Publisher or Author concerning the legal status of any country, area or territory or of its authorities, or concerning the delimitation of its borders. The depiction and use of boundaries, geographic names, and related data shown on maps and included in lists, tables, documents, and databases in this chapter are not warranted to be error free nor do they necessarily imply official endorsement or acceptance by the Publisher or Author.

R. Verma (✉) · M. Sharif · A. Husain
Department of Civil Engineering, Jamia Millia Islamia, New Delhi 110025, India
e-mail: rkhvrm87@gmail.com

M. Sharif
e-mail: msharif@jmi.ac.in

A. Husain
e-mail: ahusain3@jmi.ac.in

P. V. Timbadiya et al. (eds.), *Geospatial and Soft Computing Techniques*, Lecture Notes in Civil Engineering 339, https://doi.org/10.1007/978-981-99-1901-7_12

spatial data and detecting changes in LULC consuming less time and cost effective with better precision.

Keywords Decadal land use map · ArcGIS · LULC changes · Remote sensing · Spatial data

1 Introduction

Land use/land cover are two separate terms which are used interchangeably. Land cover refers to the physical characteristics of the earth's surface, in terms of vegetation, soil, water, and other physical features of the land, while land use refers to the way that the land has been used by humans and their habitat, usually in accordance with the functional role of land for economic activities. The LULC pattern of any region and its utilization by man depends upon the natural, social, and economic factors. For optimal use of land to fulfill the needs of the increasing population, proper LULC information is necessary. To assess change globally in different spatio-temporal scales, the LULC changes have been observed as an efficient tool [7]. As the population is increasing day by day, more pressure is been exercised on the use of limited available natural resources of any country and account to land cover changes. LULC changes are a continuous and dynamic process, and so research on the pattern of LULC should be done extensively with its implications in both environmental and social aspect using different spatio-temporal scales [8].

The vegetation cover is an important factor which helps in partitioning the rainfall into various hydrologic components like surface runoff, ground water flow, base flow, evapotranspiration, etc. Therefore, the land use change pattern studies play an important role in watershed management and hydrological modeling. GIS and remote sensing act as an efficient tool in collecting proper information about the land use/land cover pattern. These result in saving time as collecting information physically from the site is time consuming. Remote sensing technology provides proper information about the vegetation and the morphological and terrain features variations. Due to the existing alterations in the spatial resolution of the remote sensing satellite data, it has been a challenging task to generate the LULC change information over several decades [2, 12, 16]. Several studies have been made across the world to analyze the LULC changes of the watershed using different techniques/methods. Kumaraswamy et al. [6] revealed a significant decrease in the area covered by grassland and forest resulting due to urbanization and expansion of agricultural land in Cauvery River basin. Gadrani et al. [3] analyzed the land use changes from 1987–2016 in Tbilisi, Georgia using GIS and remote sensing. Their results revealed an increase of 13.9% in the built-up area. Islam et al. [5] assessed LULC changes in Chunati Wildlife Sanctuary over a period of 10 years, i.e., 2005–2015. The results revealed that there was an increase in forest area showing satisfactory Kappa coefficient value. Haque and Basak [4] studied the LULC change pattern over 30 years of Tanguar Haor and revealed that about 40% of the LC in the entire watershed has been converted into other classes, and forest and vegetation land have been disappearing.

Rawat and Kumar [9] exemplified the spatio-temporal LULC dynamics of Hawabagh block in Almora by using TM of the year 1990 and 2010. The results showed growth in the vegetation area and urban land while a decline in barren land, agricultural land, and water bodies. Butt et al. [1] made use of maximum likelihood-supervised classification algorithm to analyze LULC changes in Simly catchment, Pakistan. Tian et al. [14] identified the LULC change pattern of India for the time period 1980–2010 showing a decrease in forest area as a result of increase in built-up land and cropland. Zhu and Li [17] assessed the LULC change of Little River, Tennessee to identify its long-term impact on the stream flow from 1984–2012. Tahir et al. [13] using remote sensing data studied the LULC changes of over 25 years in Mekelle city, Ethiopia. The results indicated a gain in urban structures and grasslands and a loss in bare land and farm land. Tiwari and Saxena [15] signified the capability of GIS and remote sensing in collecting spatial and temporal data and its use for assessing the LULC changes. LULC changes were studied for a watershed in Egypt using Landsat images of 1987 and 2001 by Shalaby and Tateishi [11], and the results showed reduction in vegetation and increase in urban settlements.

Considering the efficiency of remote sensing and GIS in determining the LULC changes, this technique was used in this study. The main aim of this study was to make use of GIS and remote sensing technique to assess the LULC map of Sabarmati River basin for two decades, i.e., 1985, 1995, and 2005 using the Decadal land use map of India and making a relative analysis of the different LULC map.

2 Study Area and Data Acquisition

Sabarmati River is a major river of India in the west. The Sabarmati River basin stretches over Rajasthan and Gujarat covering an area of about 30,700 km^2 with maximum length and width of 300 km and 150 km, respectively. Its coordinates are 70°58’ to 73°51’ east and 22°15’ to 24°47’ north. The river basin has been divided into 2 sub-basins—upper sub-basin covering about 64.58% of the total area and lower sub-basin covering 35.42% of the total geographical area of the basin. The tributaries of the river are Hathmati, Wakal, Harnav, and Watrak River toward the left and Sei River toward the right. The major dam project of the river basin is at Dharoi, and some other minor dam projects are Harnav, Guhai, and Hathmati Dam. The location map of river basin is represented in Fig. 1.

Three years Decadal LULC map for the years 1985, 1995, and 2005 were downloaded from the Decadal land use map of India site in the form of raster file (https://daac.ornl.gov/VEGETATION/guides/Decadal_LULC_India.html). The source of the Decadal LULC map been prepared of India is shown in Table 1. The accuracy assessment of the map was done using stratified random samples using the ground truth data. Confusion matrix was created using the map and ground reference points to find out the user’s and Cohen’s Kappa accuracy. Kappa accuracy achieved

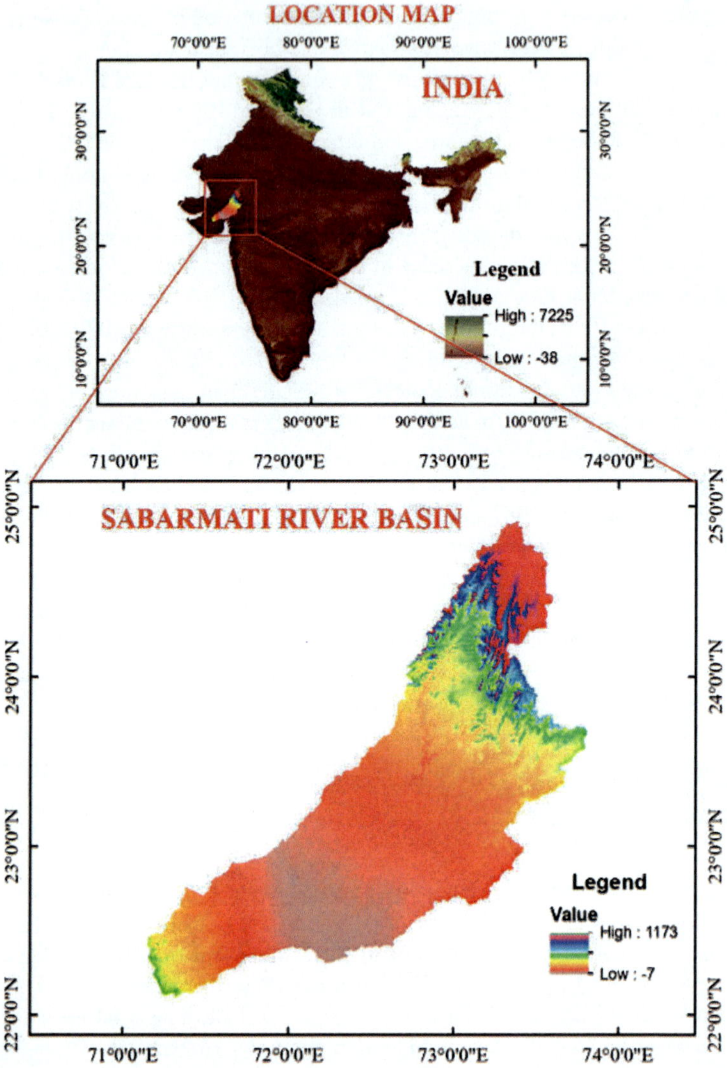

Fig. 1 Location map of Sabarmati River basin

for 2005 data was 0.9445, and overall mapping accuracy of 94.46% was obtained. The Kappa and mapping accuracies for the year 1985 and 1985 maps can be assumed to be similar with that of 2005 [10].

Table 1 Remote sensing satellite data used for Decadal LULC map

Year	Satellite	Sensor	Spatial resolution
1984–85	Landsat 4	MSS	60 m
1994–95	IRS 1B and Landsat 5	TM, ETM, LISS-I	30 m and 72 m
2004–05	Landsat 5 and Resourcesat	ETM +, LISS III	30 m and 23.5 m, respectively

3 Methodology

The methodology applied in this study is explained in detail. The first and the foremost step was to download the LULC map for the years 1985, 1995, and 2005 from the Decadal land use map of India site. These LULC maps were then uploaded in ArcGIS 10.1 software along with the shapefile of the study area, i.e., Sabarmati River basin. From the LULC map of India, the land use/land cover map of Sabarmati River basin was extracted using "clip" command in ArcGIS software. Eleven land use classes of the watershed were classified, namely deciduous broadleaf forest, built-up land, cropland, mixed forest, barren land, shrubland, fallow land, water bodies, waste land, grassland, and plantations. Then, these 11 classes were reclassified into 5 LULC classes, namely agricultural land, barren land, built-up area, forest, and water bodies. Anderson LULC classification method was used to reclassify the LULC classes. The same procedure was used to generate the LULC map for all the three years. Then, the land use map generated for the three years was compared and analyzed to identify the changes in the LULC pattern of the watershed.

4 Results and Analysis

The land use map of Sabarmati River basin for the three years was extracted from the Decadal LULC map of India downloaded through online source. The LULC map thus obtained shows eleven land use classes in the watershed. These classes were reclassified into five classes, namely agricultural land, barren land, built-up area, forest, and water bodies as represented in Fig. 2a and b for all the three years, i.e., 1985, 1995, and 2005.

The area occupied by each class of LULC for the two decades is shown in Table 2.

The percentage area coverage by each LULC class of land use for the two decades is represented in Fig. 3. The land use map exhibits that the maximum area of the watershed falls under agricultural land covering 79% of the area. The second most class of land use in the area is forest covering an area of approximately 15%. The

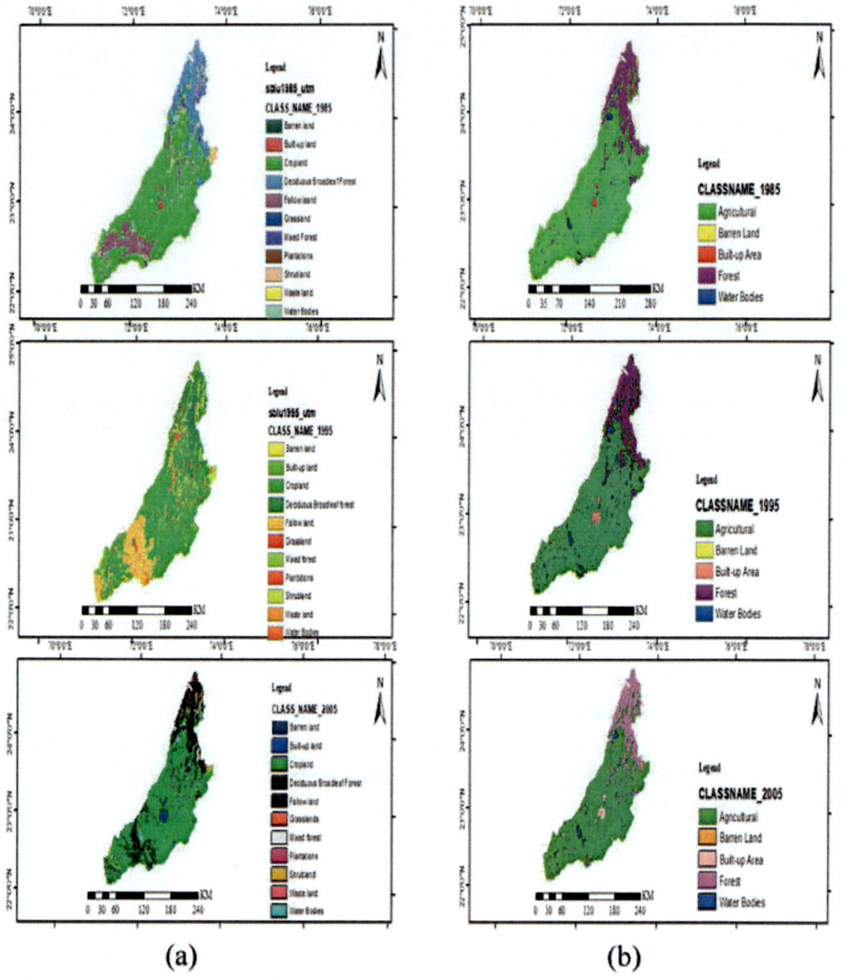

Fig. 2 **a** Land use map of Sabarmati River basin extracted from Decadal land use map of India for 1985, 1995, and 2005, **b** reclassified land use map of Sabarmati River basin for 1985, 1995, and 2005

built-up area in the watershed covers an area of about 0.7–1.3%. The barren land covers a very minimum area of about 0.13–0.15%.

The LULC changes taken place during the two decades, i.e., from 1985–2005 are shown in Table 3. A comparative analysis between LULC 1985, LULC 1995, and LULC 2005 was made. The results specify changes in the LULC map over the two decades. There was a decrease in the agricultural land by 0.54% (-165.27 km^2) and forest area by 0.40% (-121.70 km^2). The results also exhibited an increase in the barren land by 0.01% (4.07 km^2) and the built-up area by 0.55% (166.71 km^2). It

Table 2 Area coverage of LULC for the three years

Sabarmati River Basin						
Land use	1985		1995		2005	
	Area km^2	%	Area km^2	%	Area km^2	%
Agricultural	24,237.64	79.61	24,005.54	78.85	24,072.37	79.07
Barren land	42.78	0.14	38.78	0.13	46.85	0.15
Built-up area	225.12	0.74	375.29	1.23	391.83	1.29
Forest	4678.65	15.37	4604.19	15.12	4556.95	14.97
Water bodies	1259.52	4.14	1419.90	4.66	1375.70	4.52
Total	30,443.70	100	30,443.70	100.00	30,443.70	100

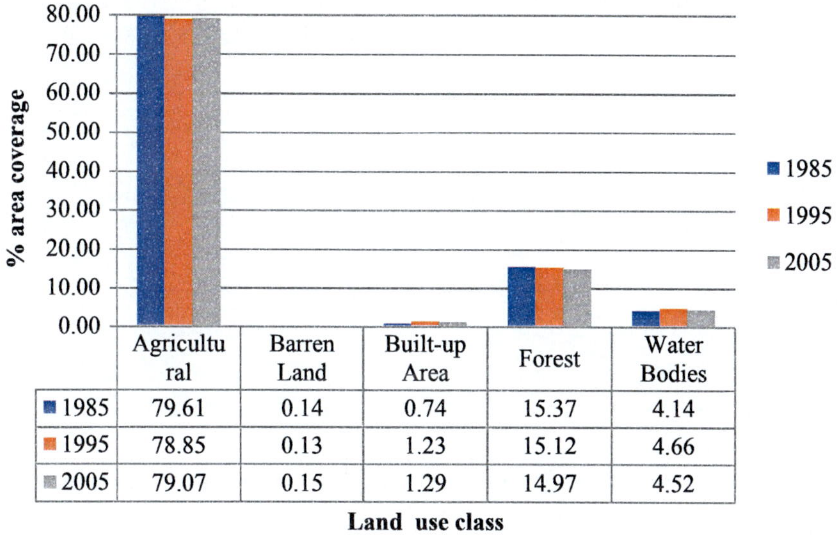

Fig. 3 Percentage of total area covered by different LULC classes for the two decades

was also depicted that the water bodies area also increased by 0.38% (116.18 km^2) that means the river has meandered a little bit.

The graphical representation of the changes in the LULC pattern in terms of the percentage changes in area occupied by different LULC classes is shown in Fig. 4. The increase in the barren land as well as the built-up land results as an important parameter leading to increase chances of floods. Rapid urbanization results due to the increase in population in the study area, thereby causing a decrease in the forest land in the watershed. This also results in the process of converting the fertile land into built-up area, thus forming an impervious layer on the top of the land, thereby causing escalation in the flood risk scenario.

Table 3 LULC changes in the two decades 1985–2005

LULC class	Area in km^2 1985	Area in km^2 2005	Change in area km^2	% change in area
Agricultural	24,237.63709	24,072.36787	−165.27	−0.54
Barren land	42.78268	46.85255	4.07	0.01
Built-up area	225.11598	391.826014	166.71	0.55
Forest	4678.649081	4556.952957	−121.70	−0.40
Water bodies	1259.521095	1375.697575	116.18	0.38

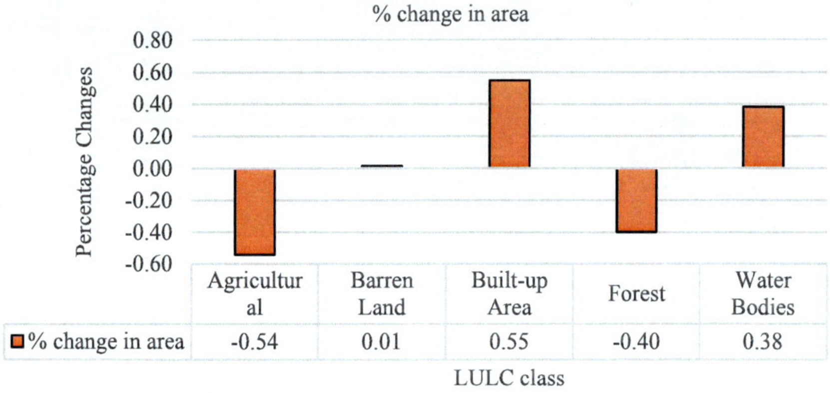

Fig. 4 Percentage changes in area covered by LULC classes from 1985–2005

5 Conclusions

LULC can be considered as the top most layer of the earth's surface which changes due to anthropogenic and natural factors. Remote sensing satellites are effectively used to analyze the changes in LULC pattern using different spatial resolutions. In this study, two decades LULC map of India for the years 1985, 1995, and 2005 were downloaded from the Decadal land use site. The LULC map was processed, and the land use map of Sabarmati River was extracted for all the three years in ArcGIS software. Then, a comparative analysis was made between the different LULC map obtained for two decades. The results obtained show a decrease in the agricultural land by 0.54% (−165.27 km^2) and forest area by 0.40% (−121.70 km^2) while an increase in the barren land by 0.01% (4.07 km^2) and the built-up area by 0.55% (166.71 km^2). It was also depicted that the area covered by water bodies also increased by 0.38% (116.18 km^2) that means the river has meandered a little bit. The increase in the barren land as well as the built-up land results as an important parameter leading to increase chances of floods. The study shows that GIS and remote sensing techniques serve as an important tool for the temporal analysis of spatial data and detecting changes in LULC consuming less time and cost effective with better precision.

To ascertain sustainable development, emphasize should be made on the use of green construction technologies/methods such as ground water recharge, roof top farming, vertical gardens, and farms as well as vertical displacement of the population. It can be suggested from this study that the results obtained will help in assessing and comparing the LULC changes over decades and also help our policy makers to plan out an eco-friendly and effective land use policy so as to deal with future water crisis.

Acknowledgements The authors would like to acknowledge the Indian National Committee on Climate Change, Ministry of Jal Shakti, Department of Water Resources, River Development & Ganga Rejuvenation, Government of India (GoI) for providing funding for this research.

References

1. Butt A, Shabbir R, Ahmad SS, Aziz N (2015) Land use change mapping and analysis using remote sensing and GIS: a case study of Simly watershed, Islamabad, Pakistan. The Egyptian J Remote Sens Space Sci 18(2):251–259
2. Fonji SF, Taff GN (2014) Using satellite data to monitor land-use land-cover change in North-eastern Latvia. Springerplus 3(1):1–15
3. Gadrani L, Lominadze G, Tsitsagi M (2018) F assessment of landuse/landcover (LULC) change of Tbilisi and surrounding area using remote sensing (RS) and GIS. Annals of Agrarian Sci 16(2):163–169
4. Haque MI, Basak R (2017) Land cover change detection using GIS and remote sensing techniques: a spatio-temporal study on Tanguar Haor, Sunamganj, Bangladesh. The Egyptian J Remote Sens Space Sci 20(2):251–263
5. Islam K, Jashimuddin M, Nath B, Nath TK (2018) Land use classification and change detection by using multi-temporal remotely sensed imagery: the case of Chunati wildlife sanctuary, Bangladesh. The Egyptian J Remote Sens Space Sci 21(1):37–47
6. Kumaraswamy TR, Ravishankar SS, & Nagaraja BC (2021) Three decadal land use and land cover changes in the Cauvery River Basin, India, pp S138-S143
7. Lambin EF (1997) Modelling and monitoring land-cover change processes in tropical regions. Prog Phys Geogr 21(3):375–393
8. López E, Bocco G, Mendoza M, Duhau E (2001) Predicting land-cover and land-use change in the urban fringe: a case in Morelia city Mexico. Landscape and Urban Plann 55(4):271–285
9. Rawat JS, Kumar M (2015) Monitoring land use/cover change using remote sensing and GIS techniques: a case study of Hawalbagh block, district Almora, Uttarakhand, India. The Egyptian J Remote Sens Space Sci 18(1):77–84
10. Roy PS, Roy A, Joshi PK, Kale MP, Srivastava VK, Srivastava SK, Kushwaha D (2015) Development of decadal (1985–1995–2005) land use and land cover database for India. Remote Sens 7(3):2401–2430
11. Shalaby A, Tateishi R (2007) Remote sensing and GIS for mapping and monitoring land cover and land-use changes in the Northwestern coastal zone of Egypt. Appl Geogr 27(1):28–41
12. Soulard CE, Wilson TS (2015) Recent land-use/land-cover change in the Central California Valley. J Land Use Sci 10(1):59–80
13. Tahir M, Imam E, Hussain T (2013) Evaluation of land use/land cover changes in Mekelle City, Ethiopia using remote sensing and GIS. Comput Ecol Softw 3(1):9
14. Tian Y, Yin K, Lu D, Hua L, Zhao Q, Wen M (2014) Examining land use and land cover spatiotemporal change and driving forces in Beijing from 1978 to 2010. Remote Sens 6(11):10593–10611

15. Tiwari MK, Saxena A (2011) Change detection of land use/landcover pattern in an around Mandideep and Obedullaganj area, using remote sensing and GIS. Int J Technol Eng Syst 2(3):398–402
16. Zhao Y, Zhang K, Fu Y, Zhang H (2012) Examining land-use/land-cover change in the Lake Dianchi watershed of the Yunnan-Guizhou Plateau of Southwest China with remote sensing and GIS techniques: 1974–2008. Int J Environ Res Public Health 9(11):3843–3865
17. Zhu C, Li Y (2014) Long-term hydrological impacts of land use/land cover change from 1984 to 2010 in the Little River Watershed, Tennessee. Int Soil and Water Conserv Res 2(2):11–21

Shoreline Changes and Sediment Distribution Studies for India's West Coast

Kavitha Natarajan, P. K. Suresh, and R. Sundaravadivelu

Abstract The study investigates the rate of erosion and accretion on shorelines along the West Coast of Gulf of Combat, Gujarat, India. GoC is selected for the study falls within the latitudes and longitudes of 20° 56′ 31″ to 21° 17′ 25″ and 72° 2′ 18″ to 72° 49′ 55″ with a coastal length of 276 km. The objective was accomplished through the use of remote sensing and GIS techniques, and 45 years of shoreline data were used for the analysis between 1973 and 2017. Landsat multispectral images-enhanced thematic mapper LISS III & LISS IV data covering the period of 12 years between 2005, 2011, 2014, and 2017 and was used for the Thapi coast. Net shoreline movement (NSM) and end point rate (EPR) were calculated using ArcGIS-DSAS computing statistical methods. DSAS calculates the change metrics by measuring the distance between the baseline and each shoreline intersection along a transect and combining date (year) information and positional uncertainty for each shoreline. Transport of sediment: According to the findings of the NSM study, the accretion rate is greater than the erosion rate. Between 1973 and 2005, erosion and accretion were approximately 0.6 km and 53.9 km, respectively. From 2005 to 2011, the erosion rate was 0.2, with an accretion of 38.7 km. For the years 2011 to 2014, erosion and accretion had no notable values; both rates were 2 km. For the years 2014–2017, erosion and accretion are approximately 0.6 km and 2.8 km. According to EPR findings, the erosion rate (-1.30 km/year) and accretion rate (3.60 km/year) between 2005 and 2011 are higher than in preceding years. The rate of erosion is slightly higher for the years 2005 to 2011 (− 1.30 km/year), and then, it falls to − 0.7 km/year (2011–2014) and − 10.2 km/year after that (2014–2017). Results of accumulation show that the rate was high, with a value of 3.6 km/year, for the

K. Natarajan
GIS Specialist, TNIAMP, WRD, Chepauk, Chennai 600005, India

P. K. Suresh (✉)
Meenakshi Sundararajan Engineering College, Kodambakkam, Chennai 600024, India
e-mail: sureshpk2000@gmail.com

R. Sundaravadivelu (✉)
Department of Ocean Engineering, Indian Institute of Technology, Madras, Chennai 600036, India
e-mail: rsun@iitm.ac.in

P. V. Timbadiya et al. (eds.), *Geospatial and Soft Computing Techniques*, Lecture Notes in Civil Engineering 339, https://doi.org/10.1007/978-981-99-1901-7_13

years 2005–2011. Then, it displays a declining tendency for the years 2011–2014 and 2014–2017, with a value of 0.6 km/year.

Keywords Sediment · Shoreline changes · RS · GIS-DSAS

1 Introduction

Coastal landforms are dynamic systems that function over a range of temporal and spatial scales. Dominant physical factors such as wave height, wave energy, tidal range, and littoral drift are responsible for shaping coastal landforms [1]. Multi-dated satellite images can be used to monitor shoreline changes by measuring sedimentation, erosion, and accretion [2]. Seasonal variations in wave energy, wave height, and wave direction have resulted in asymmetric shoreline erosion by altering sediment movement trends in the near shore area [3–6]. In the assessment of shoreline change, Landsat MSS, TM, ETM, and spot imagery have produced consistent results [7]. For the purposes of coastline extraction, assessment of erosion and accretion, and coastal morphology change detection at local and regional scale, numerous authors have exploited multi-temporal satellite imagery [8, 9].

The shorelines extracted from multi-temporal Landsat TM and ETM images were analyzed with DSAS software to determine the rate of erosion and retreating along the coastal area [10]. United States Geological Survey (USGS) developed Digital Shoreline Analysis System (DSAS) soft- ware, an add-on tool to ArcGIS used for the statistical analysis to compute the shoreline rate of change. The linear regression rate (LRR) and end point rate (EPR) statistics were used to identify the eroding, accreting, and stable shoreline for the study area [11].

According to current information by the Ministry of Earth Sciences (MoES), of the 6907 km of Indian coastline, about 34% is eroding, 26% is accreting, and the remaining 40% is constant. National Centre for Coastal Research (NCCR), which is part of the MoES, has been keeping an eye on coastline erosion since 1990 and it is continuously monitored using GIS and RS from 1990 to 2018. Rate of erosion of Gujarat (1946 km)—27%, Tamil Nadu (991 km)—43%, Kerala (593 km)—46%, West Bengal (534 km)—60%, and Puducherry (42 km)—23% make up the major coastal status for the years 1990–2018 [12]. The West Coast of India stretches from north to south and is comprised of the I Konkan, (ii) Karnataka, and (iii) Kerala coasts. It stretches from the Gulf of Cambay (Gulf of Khambhat) in the north to Cape Comorin (Kanniyakumari); on this, the two main estuaries in this strip are the Narmada and Tapi. In Gulf of Cambay (GoC), a large tidal range during high and low tides gives rise to strong tidal currents and develops a mechanism of sediment transportation. Interestingly, the inverted funnel shape of GoC has largely contribute to the sediment deposition in this region. During high tide, the tide currents move into the Gulf and encroaches the river mouth, whereas during low tide, they move out. This regular phenomena since long period on geological time scale has modified the geomorphological features in this region [13].

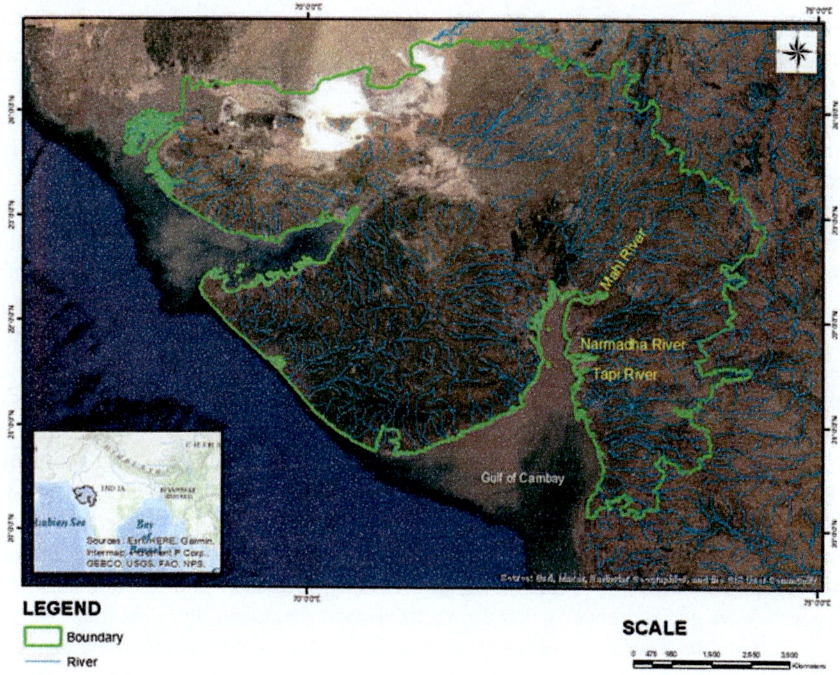

Fig. 1 Location map of the study area

2 Study Area

Due to its significance and highly influenced by the tidal currents other than geological and structural setup, the GoC is selected for the study within the latitudes and longitudes of 20° 56′ 31″ to 21° 17′ 25″ and 72° 2′ 18″ to 72° 49′ 55″ (Fig. 1) with a coastal length of 276 km. The South Gujarat coast trending almost north to south which is uniform and broken for few indentations. The major rivers traversing are Tapi and Mindhol. The following are the primary goals of the remote sensing and GIS research:

- To identify changes in the Tapi River estuary region
- To quantify the rate of erosion and identify the sediment distribution pattern

3 Materials and Method

The above goal was accomplished through the use of remote sensing and GIS techniques, 45 years of shoreline data were used for the analysis, base data was taken from the SOI toposheet, and the other's remote sensing data was used. For calculating the rate of sediment transport analysis, digital shoreline analysis system (DSAS) method was adopted. The US Geological Survey created the DSAS arc map extension (USGS). DSAS works better in analyzing shoreline change and observing specific damaged sites in smaller areas, whereas the latest remote sensing techniques with geographic information system (GIS) have proven to be very useful in monitoring coastline changes and more effective in terms of both cost and time than conventional techniques. To obtain data and results, the DSAS system requires some shoreline from various dates. Some researchers are employing the device, which incorporates several shoreline DSAS with both long- and short-time scales. The primary goal of this research is to compare the changes in shoreline that occur within the study area. The two coastlines and five shorelines are combined using DSAS to identify and measure erosion and accretion. The GIS layers of multi-date shorelines were used as input for the DSAS model to calculate the rate of change over a 45-year period from 1972 to 2017 [14].

The Survey of India (SOI) Topographical maps were used as base map and was georeferenced using its latitude and longitudinal values of the 4 corners and projected as geographic WGS 84 and for the area calculation purpose further projected as UTM Zone 43° North. Using the visual interpretation techniques [15], HTL and LTL maps were prepared on 1:50,000 scale for the year 1972. In similar way, the remote sensing products were georeferenced using this rectified topo sheet, and shoreline maps were prepared for the year for various years (Table 1). Landsat multispectral images-enhanced thematic mapper LISS III & LISS IV data covering the period of 12 years between 2005, 2011, 2014, and 2017 and was used for the Thapi coast (Fig. 2). Erdas imagine software was used for Raster Analysis and ArcGIS 10.1 software used for vector analysis.

The US Geological Survey's (USGS) arc map extension compatibility DSAS was used to analyze shoreline changes. Net shoreline movement (NSM) and end point rate (EPR) were calculated using ArcGIS-DSAS computing statistical methods. DSAS calculates the change metrics by measuring the distance between the baseline and each shoreline intersection along a transect and combining date (year) information and positional uncertainty for each shoreline. NSM was used for distance measurement, and EPR was used for statistical analysis:

NSM: Net shoreline change measured by distance rather than mean value. NSM refers to the date, and only two shorelines are required, i.e., the total distance between the earliest and latest shoreline in each transect.

Table 1 .

S. No.	Data type	Sensor	Year	Resolution (m)	Bands
1	Landsat 7	ETM +	2005	30	8
2	IRS 1C	LISS III	2011	23.5	4
3	Resourcesat 2	LISS III	2014	23.5	4
4	Resourcesat 2	LISS IV	2017	5.8	3

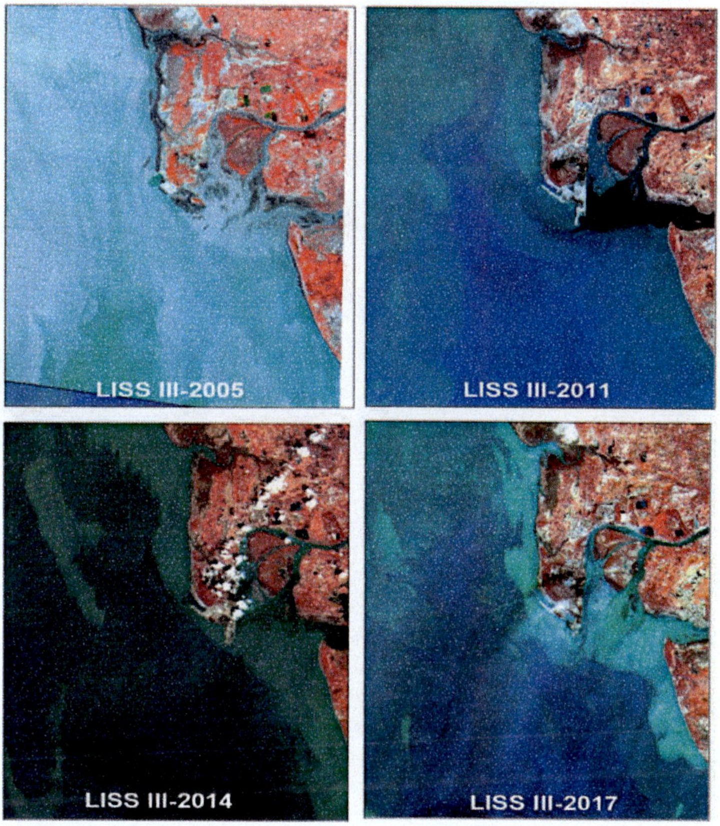

Fig. 2 Remote sensing data for the year 2005, 2011, 2014 & 2017

EPR: It is calculated by dividing the distance between the oldest and youngest shorelines by the time elapsed between them [14].

The SOI 1972 data was used to establish the baseline. A buffer of 300 m was built to the baseline's left and right (toward land and sea directions). Baselines were chosen based on the dominant change direction and in parallel with the general shoreline orientation, and 138 transect lines were established at 20 m intervals along the entire coastline. The detailed methodology was shown in (Fig. 3). The seaward shift of the shoreline along the transect is considered a positive value (accretion) based on the position with reference to the baseline at each transect, while the landward shift is considered a negative value (decrement) (erosion). Two main statistical modules, NSM and EPR, are used among the various computational functions to measure the rate of change for a time series of shoreline layers [16]. The NSM is used to calculate the rate of change between 1972 and 2017 shorelines at each transects intersecting point, and the EPR is used to estimate the per year rate of change between the same periods. This is accomplished by dividing the distance of NSM at specific transects

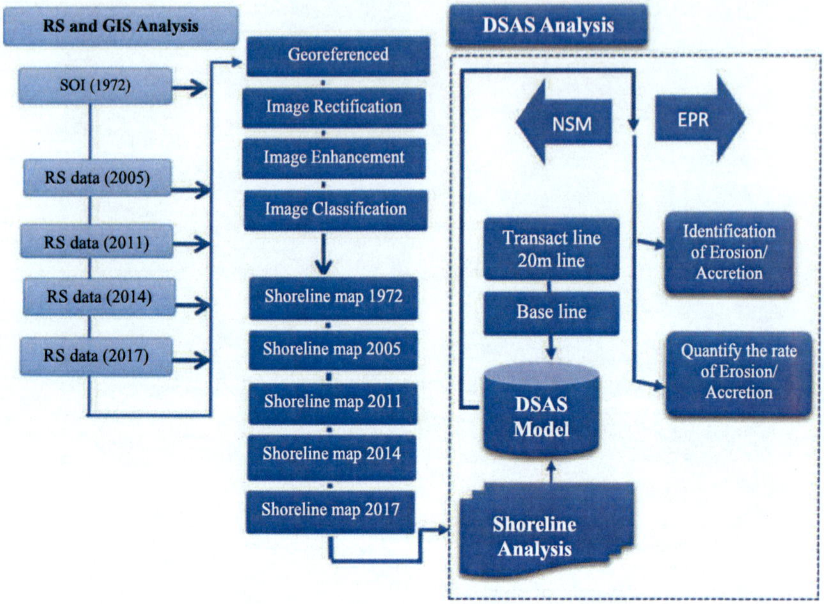

Fig. 3 Framework of the study

Table 2 Net shoreline movement rate 1972–2017

Year	Length in sq km			
	1973–2005	2005–2011	2011–2014	2014–2017
Erosion	− 6.0	− 0.2	− 1.7	− 0.6
Accretion	53.9	38.7	1.2	2.8

by the time elapsed between the two. Table 2 represents the shoreline classification of the study (EPR).

4 Results and Discussion

As Complex morphodynamic process operating at various spatial and temporal scales influence the coastal processess like erosion and accresion [17, 18]. Additional factors such as sediment availability and human interventions can influence the exact response of a coast [19, 20]. Calculation of shoreline change and sediment distribution are the methodology used for the study. Fig. 4 shows the shoreline maps derived from the SOI and remote sensing data for the years 1972–2017, and for the same NSM and EPR, measurements were estimated to calculate the rate of erosion and accretion from 1972 to 2017 (Fig. 4).

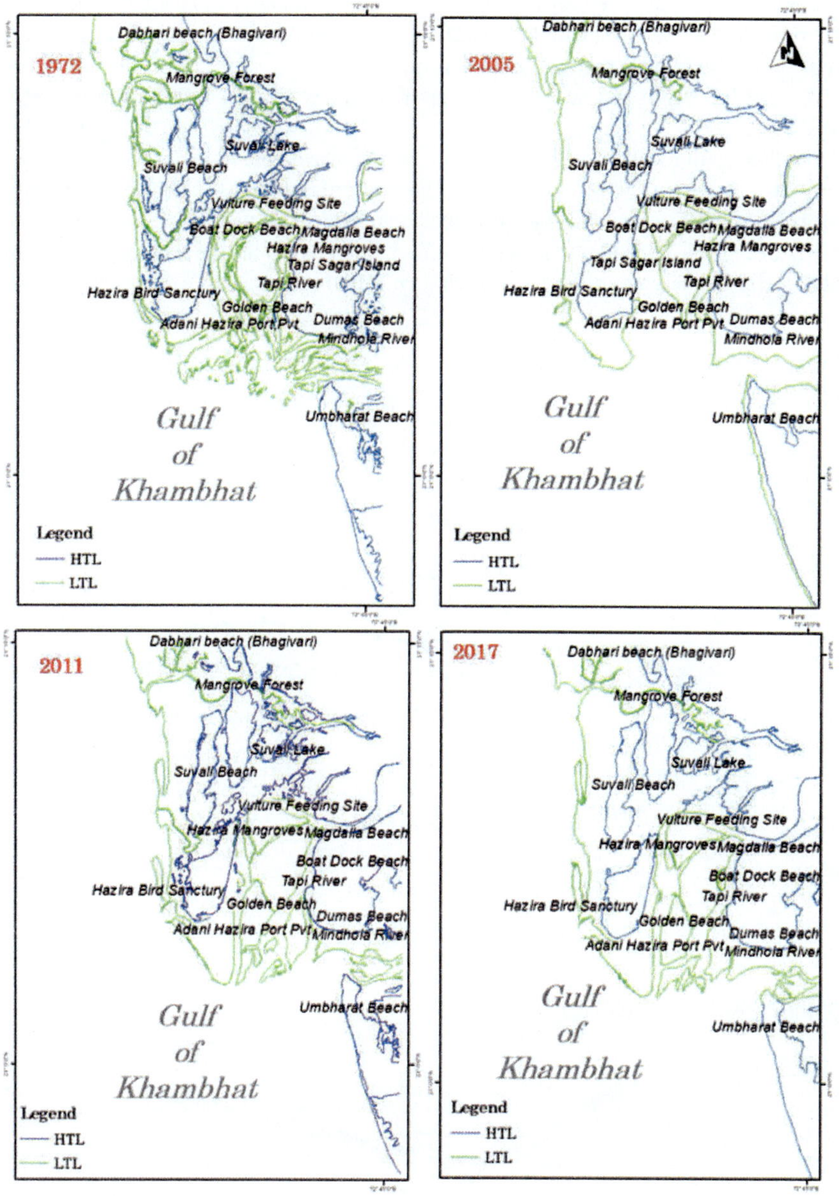

Fig. 4 HTL and LTL lines for the year 1972, 2005, 2011 & 2017

NSM studies on shoreline changes for the Tapi River Coast are shown in (Table 2 and Chart 1). Transport of sediment: According to the findings of the NSM study, the accretion rate is greater than the erosion rate. Between 1973 and 2005, erosion and accretion were approximately 0.6 km and 53.9 km, respectively. From 2005 to 2011, the erosion rate was 0.2, with an accretion of 38.7 km. For the years 2011 to 2014, erosion and accretion had no notable values; both rates were 2 km. For the years 2014–2017, erosion and accretion are approximately 0.6 km and 2.8 km (Chart 2 and Table 3).

Hydrodynamic processes like wave energy, wave direction, tidal fluctuation, and littoral currents have an impact on coastal erosion and the accretion of landform features like beaches, beach ridges, sand dunes, and estuaries. In addition, bathymetry, coastal slope, sea-level variation, and coastal artificial constructions all play a significant role in the pace of change. The research area's primary portions

Chart 1 NSM for the year 1973-2017

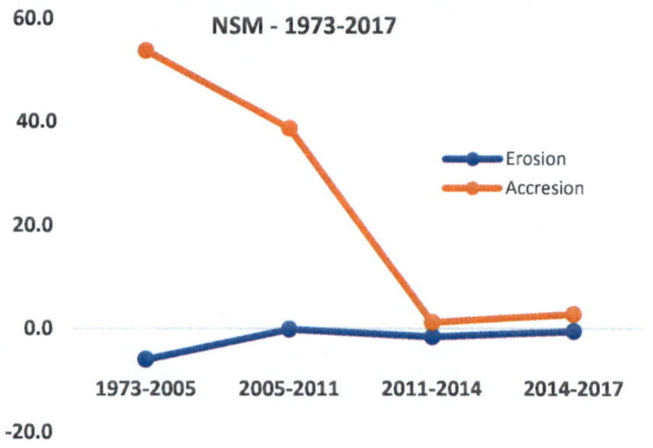

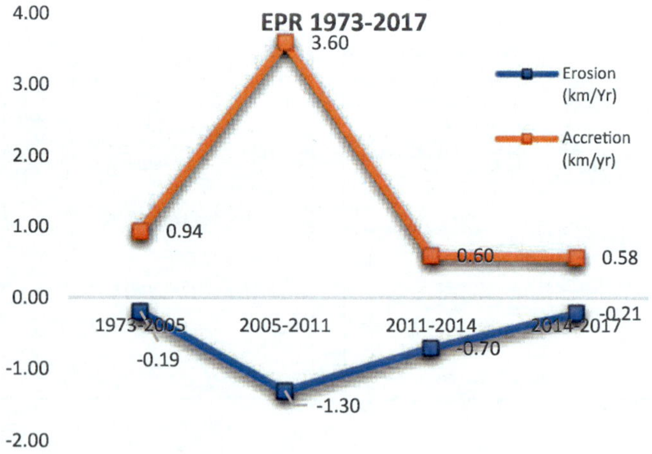

Chart 2 EPR for the year 1973 -2017

Table 3 Represents the EPR rates for the year 1972-2017

Year	Erosion (km/Yr)	Accretion (km/yr)	Maximum (km/yr)	Minimum (km/yr)
1973–2005	– 0.19	0.94	0.01	– 0.01
2005–2011	– 1.30	3.60	0.06	– 0.03
2011–2014	– 0.70	0.60	0.05	– 0.07
2014–2017	– 0.21	0.58	0.04	– 0.03

are seen to have extremely variable erosion and accretion characteristics. It displays changes in coastal dynamics on both a long-term and short-term scale [16]. Parallel to the Tapi Sagar Island (Fig. 5a) near Duma and golden Beach areas Kadifalla (Fig. 5b), erosion is high for the year 2011 compared to other years. While moving toward north near Magdalla Beach, Boat Doak beach (Fig. 5d) erosion is high for the year 2014. In 2017, there was a lot of erosion around the Tapi River mouth (Fig. 5e). In the year 2011, there was significant erosion up to 500 m from the Kandifella coastal settlement area, and 400 m beyond that there was accretion. It is noticeable between the years of 2011 and 2015, when up to 300 m of erosion at the southern end of Boat Dock Beach occurred. After that, 500 m of erosion occurred again (Fig. 5d). In comparison with the previous years, erosion is seen practically across a stretch of about 400 m along the northern and southern parts of Dumas Beach (Fig. 5c). Between 2005 and 2011, there was a rate of accretion at a lower level. Accretion was slow in process between 2005 and 2011 at a rate of 9.5 km, and high accretion was observed between 1973 and 2005 and 2005 to 2011 at 15.6 km and 8.6 km, respectively. Additionally, during the year 2017, accretion was discovered for around 400 m in a deeper fashion in the area north of Boat Duck Beach (Fig. 5d). The erosion effect of sea-level rise was expounded. The main advantage of Bruun Rule is that it provides a mechanism for obtaining quantitative estimates for erosion induced by past, present, and future sea-level rise. For the practical application of Bruun Rule, determination of the appropriate limit of exchange depth and its offshore extent is one of the most perplexing problems, suggested that a typical for limiting depth for active transport of the eroded material offshore by wave action would be between 13 and 18 m. Further, he recommended that it is possible to evaluate the outer limit of exchange of beach material from sedimentological investigation, indicating the decrease in sediment size toward offshore side [21].

According to EPR findings, the erosion rate (– 1.30 km/year) and accretion rate (3.60 km/year) between 2005 and 2011 are higher than in preceding years. The rate of erosion is slightly higher for the years 2005 to 2011 (– 1.30 km/year), and then, it falls to – 0.7 km/year (2011–2014) and – 10.2 km/year after that (2014–2017). Results of accumulation show that the rate was high, with a value of 3.6 km/year, for the years 2005–2011. Then, it displays a declining tendency for the years 2011–2014 and 2014–2017, with a value of 0.6 km/year.

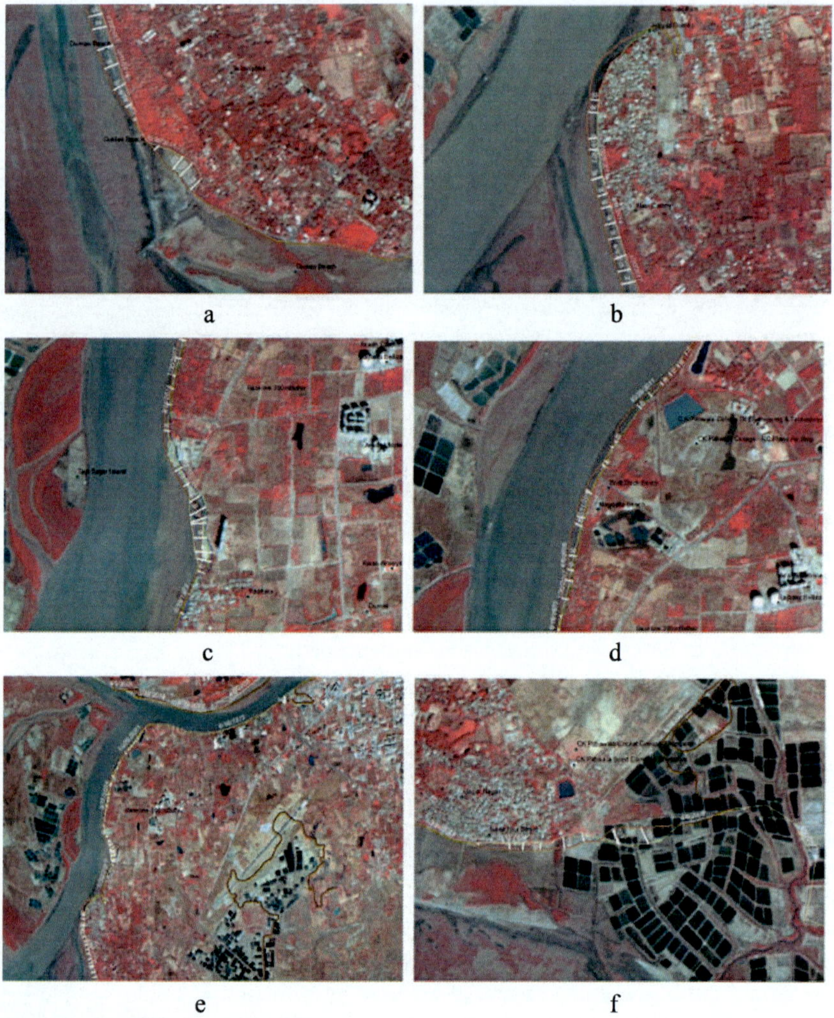

Fig. 5 Zoomed portion of the erosion and accresion area

5 Conclusions

According to the research of shoreline change evaluation utilizing remote sensing techniques, the study area between 1973 and 2017 saw high rates of accretion relative to erosion. The sediment distribution pattern along the Tapi River is primarily due to natural and anthropogenic factors. For evaluating shoreline change and observing specific damaged locations in smaller areas, DSAS performs well. However, the most recent methods of remote sensing with geographic information systems (GIS) have shown to be much more efficient than traditional methods in terms of both cost and time in monitoring changes to coastlines.

References

1. (2022). Retrieved from EOS: https://eos.com/find-satellite/landsat-7/
2. Benumof BT, SC (2000) The relationship between incident wave energy and seacliff erosion rates: San Diego County, rates: San Diego Country. J Coast Res 16(4)
3. Bruun P (1962) Sea level rise as a cause of shore erosion. J. Waterways and Harbors Div. Proc Amer Soc Civil Eng 88:117–130
4. Chauhan P, NS (1996) Remote sensing of suspended sediments along the Tamil. J Ind Soc Remote Sens 24(3):105–114
5. De Vriend H (1991) Mathematical modeling and largescale coastal behavior. Part 1: physical processes. J Hydralic Res 29:727–740
6. De Vriend H (1991) Mathematical modeling and largescale coastal behavior. Part 2—predictive models. J Hydralic Res 29:741–753
7. Dean R, Houston J (2016) Determining shoreline response to sea level rise. Coast Eng (114):1–8
8. Dewidar KM, FO (2010) Automated techniques for quantification of beach change rates using Landsat series along the North-eastern Nile delta, Egypt. J Oceanogr Mar Sci 2:28–39
9. El Asmar HM, HM (2011) Change detection of the coastal zone east of the Nile Delta using remote sensing. Environ Earth Sci 62:769–777
10. Himmelstoss EA, RE (2018) Digital shoreline analysis system (DSAS) version 5.0 user guide. USGS Publications Warehouse. https://doi.org/10.3133/ofr20181179
11. Georgiou IY, SJ (2009) Wave forecasting and longshore sediment transport gradients along a transgressive barrier island:Chandeleur Islands, Louisiana. Geo-Mar Lett 29:467–476
12. https://eos.com/find-satellite/landsat-7/ (2022) Retrieved from EOS: https://eos.com/find-satellite/landsat-7/
13. Kaliraj S, SA (2013) Impacts of wave energy and littoral currents on shoreline erosion/accretion along the south-west coast of Kanyakumari, Tamil Nadu using DSAS and geospatial technology. Environ Earth Sci 71. https://doi.org/10.1007/s12665-013-2845-6
14. Nassar K, E-A A (2022) Quantitative appraisal of naturalistic/anthropic shoreline shifts for Hurghada: Egypt. Mar Georesour Geotechnol 40(5):573–588. https://doi.org/10.1080/1064119X.2021.1918807
15. Kiran P (2022) Paper II. In: prakashan K (ed) UPSC, p 56. www.kiranbooks.com. Retrieved from http://currenthunt.com/2022/04/indian-coastline/
16. Le Cozannet G, Bulteau T, Castelle B, Ranasinghe R, Woppelmann G, Rohmer J, Bernon N, Idier D, Louisor J, Salas-y-Mélia D (2019) Quantify uncertainties of sandy shoreline change projections as sea level rises. Sci Rep (9):42
17. ME H (2011) Mapping coastal erosion at the Nile Delta western promontory using Landsat imagery. Environ Earth Sci 64:1117–1125
18. Sheik M, CN (2011) A shoreline change analysis along the coast between Kanyakumari and Tuticorin, India, using digital shoreline analysis system. Geo-spatial Inf Sci 14(4):282–293
19. Nayak RA (2003) Tides in the Gulf of Khambhat, west coast of India. Estuar Coast Shelf Sci 57:249–254. https://doi.org/10.1016/S0272-7714(02)00349-9
20. Ryabchuk D, SM (2012) Long term and short term coastal line changes of the Eastern Gulf of Finland. J Coast Conserv 16:233–242
21. Saravanan S, CN (2011) An overview of beach morphodynamic classification along the beaches between Ovari and Kanyakumari, Southern Tamil Nadu coast. India. Phys Oceanogr 21(2):130–141

22. Kaaliraj S, DA (2012) Geo-processing model on Coastal vulnerability index to explore risk zone along the South West coast of Tamilnadu, India. Int J Earth Sci Eng 5:1138–1147. Retrieved from https://www.researchgate.net/publication/312446890_Geo-processing_model_on_Coastal_vulnerability_index_to_explore_risk_zone_along_the_South_West_coast_of_Tamilnadu_India/citation/download
23. Seenipandi, DC (2012) Geo-processing model on Coastal vulnerability index to explore risk zone along the South West coast of Tamilnadu, India. Int J Earth Sci Eng 5:1138–1147. Retrieved from https://www.researchgate.net/publication/312446890_Geo-processing_model_on_Coastal_vulnerability_index_to_explore_risk_zone_along_the_South_West_coast_of_Tamilnadu_India/citation/download
24. Thomas M. Lillesand RW (2017) Remote sensing and image interpretation. Wiley, Incorporated
25. White K, EA (1999) Monitoring changing position of. Geomorphol 29:93–105

Assessment of Reservoir Sedimentation Using Remote Sensing and GIS Techniques

Bikram Prasad and H. L. Tiwari

Abstract Sedimentation occurs naturally in all water bodies; however, it is more prevalent in reservoirs as they hold a significant volume relative to the size of their watershed. Deposition of eroded silt particles, carried into the reservoir by the passage of water behind the dam, causes it. The impact of sedimentation on the gross and live storage capacity of the Harsi reservoir, Gwalior, India was assessed using satellite imagery. By developing false color composites, the water spread at different heights was calculated using the normalized difference water index (NDWI) produced by combining both green and NIR bands photographs (FCC). ILWIS software had been used to analyze seven days of IRS-(R2/P6) L-3 satellite pictures encompassing live reservoir storage in elevation as per field reports acquired from the reservoir location. In the last 84 years, sedimentation in the reservoir lost its gross storage capacity by 47.441MCM which is 23.04% and live storage capacity by 35.779 MCM which is 18%. The rate of silting was found to be 0.426 Mm^3/year considering sedimentation to be constant during an 84 year timeframe. Reservoir silting rate has been compared with Central Water Commission (CWC) sedimentation evaluated in 2007 and empirical formulas of Khosla's and Joglekar's.

Keywords ILWIS · Harsi reservoir · Normalized difference water index (NDWI) · Satellite images · Sedimentation

1 Introduction

Soil erosion is caused by heavy rains and high winds, and eroded soil particles are mainly transported during floods that must be taken into account while evaluating sedimentation [1, 2]. Topography, meteorology, land use, land cover, and drainage

B. Prasad (✉)
Civil Engineering Department, Bansal Institute of Science and Technology, Bhopal, India
e-mail: bikram2010@gmail.com

H. L. Tiwari
Civil Engineering Department, MANIT, Bhopal, India

P. V. Timbadiya et al. (eds.), *Geospatial and Soft Computing Techniques*, Lecture Notes in Civil Engineering 339, https://doi.org/10.1007/978-981-99-1901-7_14

features of the basin all have a role in soil erosion and sedimentation [3]. The presence of sediment in the reservoir hinders the project's functioning and mobility. Severe bed scour in the middle of river may put flood control systems and ecological safety in danger [4]. Over 3000 river basin projects have been created by the Indian government to address irrigation and hydropower needs, household and industrial water supply, flood control, and other needs across the country. It is extremely important to maintain the productive lifespan and live storage of a reservoir since there are topographical restrictions on getting large storage sites, which are quite high connected with planning and building reservoirs [5]. Natural streams annually transport a significant amount of sediment to reservoirs, lakes, and seas. Approximately, 1–2% of the live storage capacity in various reservoirs is lost each year owing to sedimentation [6]. The operation of the reservoir and the availability of water for achieving the fundamental aims of its creation are both impacted by sedimentation in the reservoir and the subsequent decrease in capacity. India has a soil deposition rate of roughly 0.16 tons per square kilometer per year, of which 10% are taken directly from reservoirs and 29% are eventually disposed of in the ocean and sea [7]. The important lifespan and storage capacity of the reservoirs usually decrease fast in comparison with their expected rates due to severe siltation and rapid erosion [8–11]. According to the sedimentation analysis results of 43 reservoirs around the country, the sedimentation rate ranges from 30 to 2785 cubic meters per kilometer per year, according to the 2004 survey report. Furthermore, soil protection is directly or indirectly linked to agriculture and industrialization, while agricultural productivity is inversely proportional to soil erosion [12]. Therefore, the scientific community is confronted with challenges from soil erosion and increased sedimentation in reservoirs and lakes [13]. Hydrographic surveys need people with the right skills, a high price, and a lot of time. The majority of hydrographic surveys in India are carried out using an echo sounder. Although this necessitates a significant quantity of data, some empirical methods are also employed to anticipate an approximate sedimentation profile for a reservoir. The same factors have led to the development of mathematical models like HEC-6, GSTARS, FLUVIAL, TABS, and others [14]. In the present era, remote sensing and geographic information systems have improved the convenience, speed, and affordability of this assessment [9, 15–18]. Harsi dam was constructed to irrigate 7085 hectares of Kharif and 20,243 hectares of Rabi and 3036 hectares of Sugarcane. However, it was discovered that not enough research had been done to evaluate the reduction in storage capacity caused by sedimentation. Therefore, it is crucial to evaluate the sedimentation in the reservoir in order to determine how much storage capacity has been lost and to determine what additional steps can be taken to make up for that loss. In this work, storage loss due to sedimentation in the reservoir is evaluated using elevation-area-capacity curves updated using satellite data in the reservoir's live storage zone.

2 Description of Study Area

The Harsi projects are situated in the Gwalior district at a latitude and longitude of 25°45' N and 77°58' E in Madhya Pradesh (M.P.), respectively, as illustrated in Fig. 1. The Gwalior State built it between 1928 and 1935. It is situated along the Parvati River, a branch of the Sindh River and a tributary of the Yamuna River. It offers irrigation amenities in the approximately 91,057 ha gross command area (GCA) and 62,675 ha cultural command area (CCA) in the Gwalior District. Table 1 gives an overview of the dam's key characteristics.

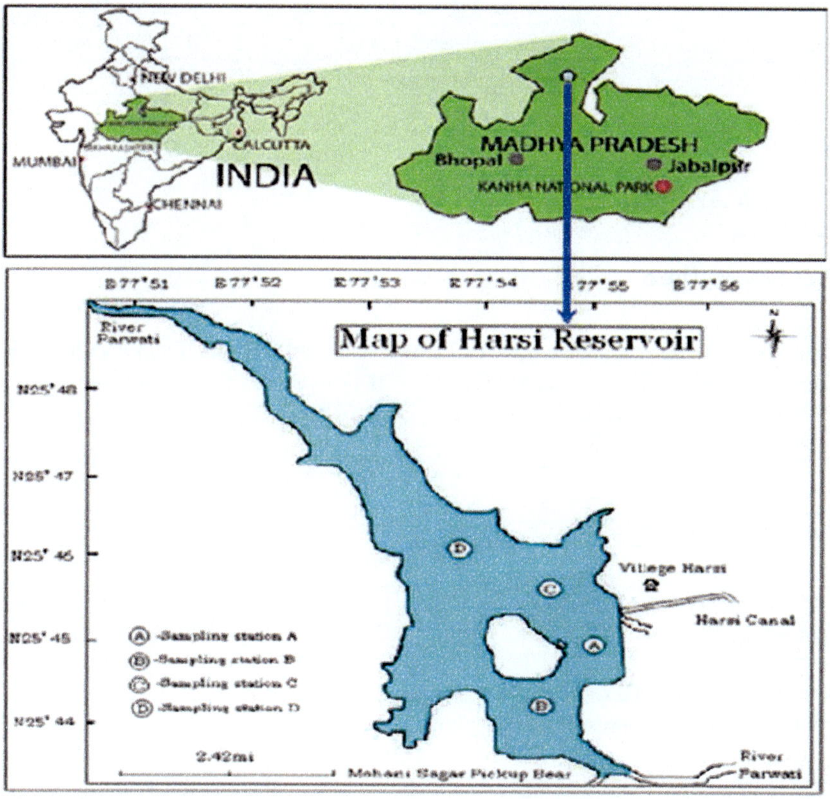

Fig. 1 Study area Harsi reservoir

Table 1 Sailent features of Harsi reservoir

Reservoir data	
Catchment area (km^2)	1960 (km^2)
Mean monsoon yield	214.07 MCM
75% dependable yield	206.02 MCM
Gross storage capacity	206.30 MCM
Dead storage capacity	13.64 MCM
Live storage capacity	192.66 MCM
Full tank level (F.T.L.)	R.L. 264.93 m
Maximum water level (M.W.L)	R.L. 267.31 m
Top bund (T.B.)	R.L.270.36 m
Lowest sill level (L.S.L.)	R.L. 252.07 m
Water spread area at F.T.L	25.05 km^2
Water spread area at M.W.L	28.43 km^2

3 Data Availability

3.1 Field Data

Data about the reservoir that was needed was acquired from the Dam's authorities. The Harsi reservoir has a riverbed level of 250.18 m and a full supply level (F.S.L.) of 264.93 m, respectively.

3.2 Satellite Data

The National Remote Sensing Centre (NRSC), located in Hyderabad, India, provided the satellite data that was utilized. For the study and delineation of the revised water distribution region, the LISS-III data of IRS-1D (Path 53, Row 96) of seven different dates for the years 2017–2019 were selected, as shown in Table 2. It uses a LISS-III satellite sensor with a resolution of 23.5.

4 Methodology

We calculated the water surface area in relation to the relevant date of the satellite pass at the known height thanks to the study based on remote sensing. The area's perimeter is defined using digital and visual methods for deciphering remote sensing data. This technology is fully dependent on expert and qualified interpretation. At the reservoir's edge, it is impossible to clearly discern between the water's top and

Table 2 Satellite data sensor information

Date of pass satellite	Reservoir level (meter)	Satellite sensor	Path/Row
11–03–2017	254.20	LISS-III	097/053
04–04–2017	256.37	LISS-III	097/053
23–05–2019	258.60	LISS-III	097/053
05–04–2019	260.05	LISS-III	097/053
12–03–2019	262.24	LISS-III	097/053
25–12–2019	264.79	LISS-III	097/053
27–08–2019	265.12	LISS-III	097/053

submerged bottom [19]. Water pixels may be mistaken for land and vice versa, and the different bands of the corresponding satellite imagery have been fully analyzed using digital technology [20]. The detailed methodology has been shown in Fig. 2.

4.1 Georeferencing, Satellite Band Import, and Band Stacking

The National Remote Sensing Centre (NRSC), in India, provided the IRS-1D satellite data and LISS-III sensor data for different dates, which were imported into the ILWIS system. The basic pixel size of the LISS-III sensor is 23.5 m, whereas the processed pixel size is 24 m. Red, green, blue, and near infrared (NIR) were the four available bands for each specific date, and in ILWIS 3.0, they were stacked together to create Maplist. False color composites (FCCs) were produced using Maplist and a combination of various bands. Except for the soil–water interface and reservoir's periphery, the water spread region of the reservoir area was extremely obvious and distinct in the false color composite (FCC) [15, 16]. The top surface of reservoir was free from cloud and noise in all of seven satellite images.

4.2 Interpreting Water Pixels

In contrast to its low absorption and reflection, water has a significant transmission in the visible spectrum (400–700 nm). While reflectance and transmittance are low in the near-infrared band, water absorption rate rises quickly. Near-infrared wavelengths gave the appearance of a black body for water. A composite image of all four bands—green, blue, red, and near infrared—clearly showed the spectral impression of water in relation to other land uses like urban areas, vegetation, and soil surfaces. In order to prevent confusion between water and soil, especially near the water body's edge, water at shallow depths was carefully researched. Deep water bodies provided a very

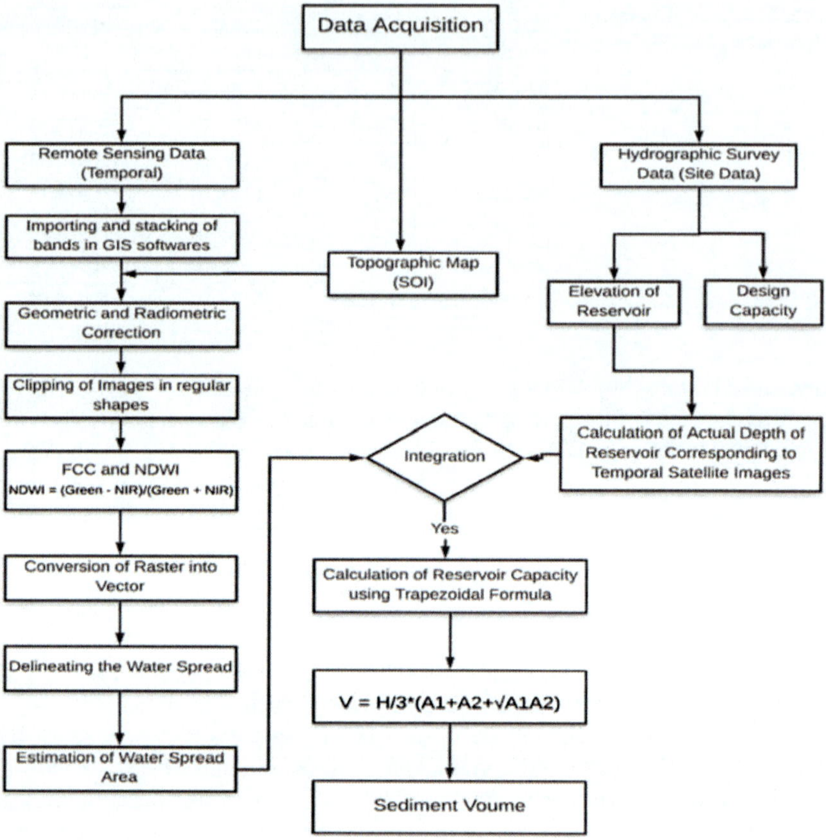

Fig. 2 Flowchart of adopted methodology

clear and distinct depiction when compared to shallow water bodies. So, multiple bands had been studied based on digital numbers in order to clearly distinguish pixels along the water-soil boundary. The combination of two or more bands in a particular association provides a definite indication based on the reflection and absorption properties of the water surface [9]. The combination of numerous bands for greater digital number values helped to delineate water bodies because these raster pictures were represented by digital number values on each spatial grid.

4.3 NDWI Approach

The normalized difference water index (NDWI), created by McFEETERS in 1996, was employed after spectral behavior (reflectance) of water pixels was examined in many imageries. This technique compares a pixel's digital number (DN) value utilizing data from various bands in order to clearly detect water pixels. The definition of the index, NDWI, is

$$\text{NDWI (Green} - \text{NIR)(Green NIR)} \tag{1}$$

NIR stands for reflected near-infrared light, while green band denotes reflected green light. To increase the reflectance of water features, the two bands (green and near-infrared) were used. In multispectral aerial photographs, water features were given positive values, whereas soil and vegetation features were given negative values since they reflected more NIR than green light. Each image's NDWI was calculated independently in ILWIS using the spatial tool's raster calculator. While identifying water pixels, the algorithm checks each pixel for the following factors [22, 23].

The pixels were classified as being water for NDWI > 0 and as being non-water for NDWI $\leq$ 0.

4.4 *Water Spread Delineation After Eliminating Pixel Gaps, Tails, and Channels*

From the NDWI image, the reservoir catchment region in question has been cropped as a regular form that carries pixel gaps along the water body's perimeter [15]. To avoid overestimating the water spread area of the reservoir at a specific date and water level, some discontinuous water pixels at the periphery had been removed.

The main stream at the reservoir's end was visible in the NDWI image, along with a large number of other minor contributing streams that met the water body along the perimeter from the surrounding sides. The long tail and channels were taken out of the final cropped raster images. This approach of deleting discontinuous pixels, tails, and channels was manually carried out using an NDWI image. The NDWI raster imagery was converted into a polygon using the ILWIS 3.0 command (raster to polygon), and the final reservoir water spread for satellite photos taken at various times was digitized to estimate the water spread area. The enhanced continuous contours of the water spread were digitalized by remote sensing analysis.

4.5 *Estimation of Modified Capacities*

The area of the polygon (reservoir water spread) had been calculated by computing the geometry in the original projected coordinate system as carried by the satellite pictures received after the raster images had been converted into polygons. The following equation was used to calculate the reservoir capacity between two subsequent reservoir water levels:

$$V = \Delta H * \left(A1 + A2 + \sqrt{/}A1A2\right)/3 \quad (2)$$

where *V* is the reservoir's volume between subsequent water levels 1 and 2, A1 and A2 are the water spread contour's areas at the corresponding altitudes, and *H* is the change in both water spread levels [9, 10, 15–17]. The initial elevation-capacity table used in the dam's design was given by the dam authority. The original elevation-capacity chart was used to compute the planned capacity at the intermediate water levels using linear interpolation. When the revised capacity was compared to the original capacity at the same water level, the difference between the two capacities resulted in a loss in storage due to sedimentation. At the lowest level observed, cumulative original and revised capacity were taken to be equal or zero (250.18 m). To get the final cumulative revised capacity at the highest observed water level, the following intermediate capacities were added together above this elevation (264.93 m).

5 Results and Discussion

False color composite (FCC) of each imagery was made, and extracted water spread area of each image was constructed and shown in Fig. 3.

5.1 Calculation of Volume of Sediments Deposition

Each image was given a histogram in the ILWIS GIS software system that displayed the value of the updated water spread area at the proper height. The elevation of the reservoir at the time of the satellite pass was reported by the reservoir authorities. The volume at the elevation of the satellite's pass date was calculated using the prismoidal formula. It is necessary to determine the revised area at the river-bed level and the revised river-bed level, and their values were quantified using the best fit curve in order to calculate the sedimentation loss. Their value was also employed in the analysis to calculate the sedimentation loss and reservoir capacity loss. Sedimentation has reduced reservoir storage capacity, as seen by the shift in the curve from the initial capacity elevation to the revised capacity elevation. Utilizing linear interpolation, the original reservoir value was used to determine the original capacity at the intermediate altitudes (reservoir heights on the day of satellite pass). Using the known values of the original and revised areas at different elevations, the matching original and revised capacities were determined. These initial data were used to build the original elevation-area-capacity curves, which are depicted in Fig. 4.

The capacity loss in the research zone is represented by the discrepancy between the original and revised cumulative capacities. Table 3 provides the calculation of sediment deposition.

From the analysis, it was observed that Harsi reservoir lost its gross storage capacity in between the maximum and minimum levels (264.93 and 250.18) from

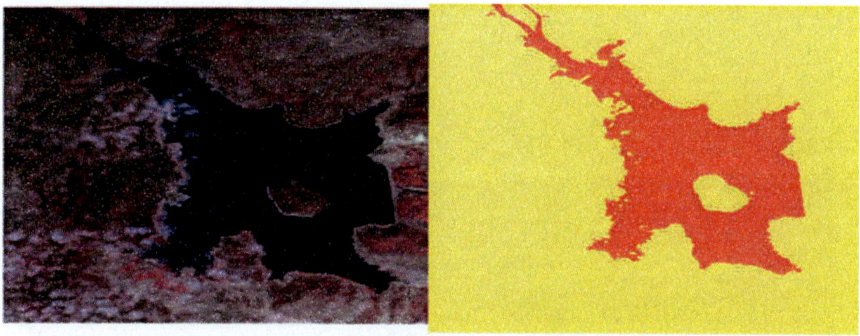

(a) FCC and extracted water spread area on 11- 03-2017 at an elevation 254.20 M

(b) FCC and extracted water spread area at elevation 256.37 on 04-04-2017

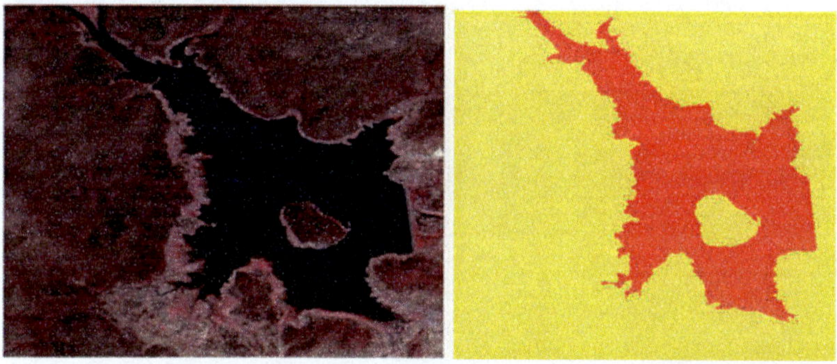

(c) FCC and extracted water spread area at an elevation of 258.60 m on 23-05-2019

Fig. 3 FCC and extracted water spread area at different elevation and on different dates

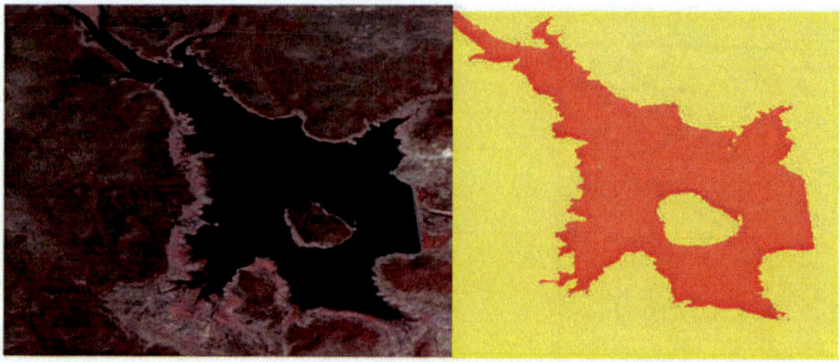

(d) FCC and extracted water spread area at an elevation of 260.05 m on 05-04-2019

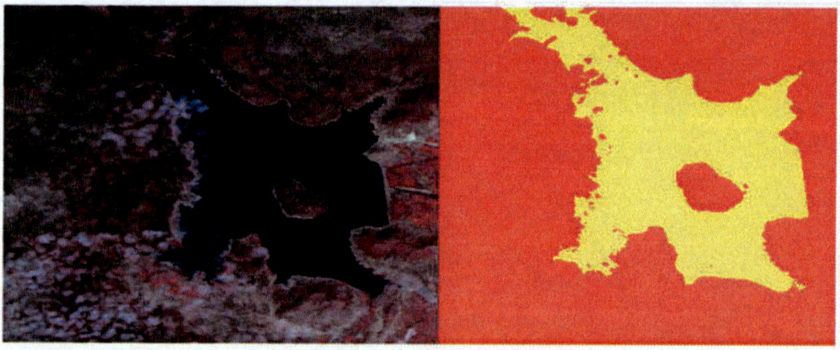

(e) FCC and extracted water spread area at an elevation of 262.24 m on 12-03-2019

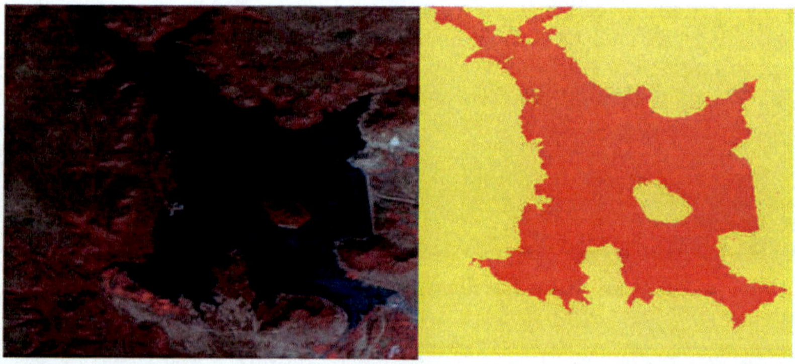

(f) FCC and extracted water spread area at an elevation of 264.79 m on 26-12-2019

Fig. 3 (continued)

(g) FCC and extracted water spread area at an elevation of 265.05 m on 28-08-2019

Fig. 3 (continued)

Fig. 4 Comparison of the reservoir's total original and upgraded capacity

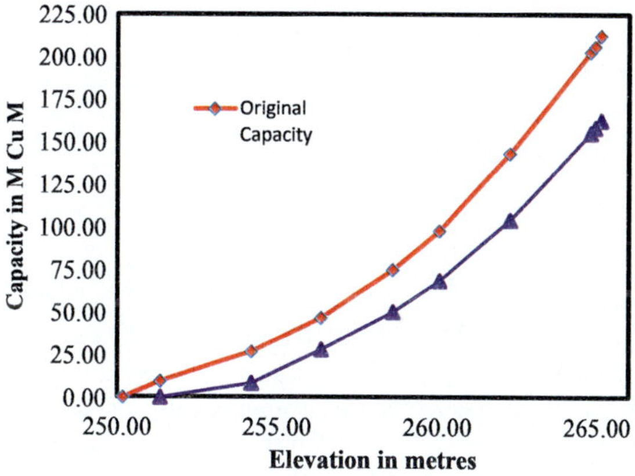

205.950 to 158.509, i.e., 47.441 MCM which is 23.04%, live storage capacity from 192.21 to 156.431 MCM, i.e., 35.779 MCM which is 18%, and the dead storage loss from 13.74 to 2.078 MCM. If the uniform rate of sedimentation is assumed in 84 years of occurrence of reservoir, then the sedimentation rate in this zone is 0.565 MCM/year which is 0.27%, of gross storage and 0.426 MCM/year which is 0.22%.

This average sediment deposition rate was compared with Khosla's and Joglekar's formula equation [24]. The Khosla's and Joglekar's equation written as follows:

Khosla's equation

$$\mathrm{Qs} = 0.323/A^{0.28} \tag{3}$$

Table 3 Estimation of reservoir capacity loss

Date of pass satellite	Reservoir elevation (meter)	Original capacity (MCM)		Revised capacity (MCM)		Loss in cum. capacity (MCM)	% Loss in cumulative capacity
		Volume	Cumulative capacity	Volume	Cumulative capacity		
River bed	250.18	0.000	0.000		0.000		
Revised river bed	251.35	9.387	9.387	0.000	0.000	9.387	100.00
11-Mar-17	254.20	17.333	26.720	8.226	8.226	18.494	69.21
04-Apr-17	256.37	19.910	46.630	19.776	28.001	18.629	39.95
23-May-19	258.60	28.210	74.840	22.239	50.240	24.600	32.87
05-Apr-19	260.05	22.790	97.630	18.199	68.440	29.190	29.90
12-Mar-19	262.24	45.603	143.233	35.615	104.055	39.178	27.35
25-Dec-19	264.79	59.497	202.730	51.286	155.341	47.389	23.38
FSL *	264.93	3.220	205.950	3.167	158.509	47.441	23.04
27–08-2019	265.12	6.390	212.340	4.300	162.809	49.531	23.33

Joglekar's equation

$$\text{Qs} = 0.597/A^{0.24} \tag{4}$$

where A is the catchment area in km^2. and Qs is the annual silting rate from a 100 km^2. watershed area (Mm^3/100 km^2/year). The sedimentation rate determined using Khosla's and Joglekar's equations is 0.038 Mm^3/100 km^2/year and 0.097 Mm^3/100 km^2/year, respectively. The catchment area of the Harsi dam is 1960 km^2. The calculated results of 0.029 Mm^3/100 km^2/year showed that the estimated sedimentation using remote sensing technology is rather less than that calculated by Khosla and Joglekar.

According to a research conducted by the Central Water Commission (CWC, 2020), the live capacity loss for the years 1935 to 2007 was found to be 21.64 MCM or 0.015 Mm^3/100 km^2/year. After analyzing the reservoir sedimentation using remote sensing and GIS techniques, it was concluded that 35.779 MCM of live storage of Harsi reservoir had been lost in last 84 years (1935 to 2019) which comes out to be 0.0217 Mm^3/100 km^2/year. From 2007 to 2019, i.e., 13 years, the loss in sedimentation live capacity comes out to be 14.139 MCM which comes out to be 0.055 Mm^3/100 km^2/year. So now, the rate of sedimentation is increasing, and calculated rate is on the higher side as per the Khosla's rate of sedimentation.

Because the updated capacity cannot be determined below the lowest observed reservoir water levels or above the highest observed reservoir water levels, the remote sensing-based technique has a substantial drawback. Sedimentation rate computation is only possible within reservoir's water level fluctuation zone. A hydrographic study inside the water spread area corresponding to the lowest recorded elevation, along

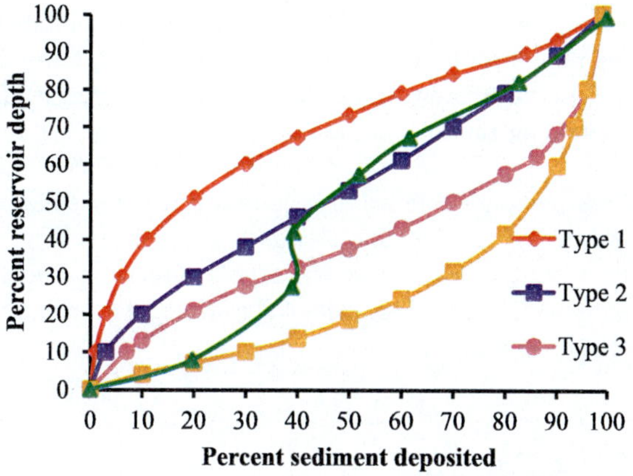

Fig. 5 Sediment deposition pattern of Harsi reservoir as per remote sensing results superimposed on Borland and Miller curves

with the analysis of remote sensing data, may be carried out if sedimentation in the entire reservoir is to be determined. As a result, less effort will be required to complete the hydrographic study.

5.2 Classification of Reservoir as Per Borland and Miller Curve

Based on an analysis of reservoir sediment data inside the United States [25], classified reservoirs into four different types. The results show a direct correlation between reservoir morphology and the amount of silt deposited at various depths across the reservoir. The type of reservoir had been identified using the reciprocal value (M) of the slope of the line produced by plotting reservoir depth as ordinate and reservoir capacity as abscissa on a log–log scale. For the bulk of reservoirs in India, standard curves of this type have been shown to be reliable [26]. From the results of remote sensing analysis, sediment deposition pattern was close to Type-II as shown in Fig. 5.

5.3 Conservation Measures

To decrease the intake of silt and sediment into the Harsi reservoir, it is required to implement the proper soil conservation measures in the watershed. The preventive measures can either involve some land reformation work upstream of the reservoir or spilling away the excess sediments that have been deposited at the reservoir site. In the vicinity of the reservoir, there are plantations, farms, human settlements, and sediments, so a buffer zone must be established. In order to stop soil erosion, planting should be done on a regular basis in the catchment region. Farmers need to be motivated to plant fruit-bearing trees. After building earthen coffer dams to allow the region to drain, silt should be removed via dry excavation. The expansion of aquatic vegetation inside the reservoir as a result of the influx of natural waste-containing runoff and the development of agriculture on the provincial crest was another improvement to the reservoir. It is crucial to regularly remove weeds from the reservoir's flooded area, prohibit illegal fishing in the reservoir, and halt the transfer of considerable amounts of pollutants from agricultural areas in order to maintain the lake's capacity and cleanliness. It can also be managed via dredging, empty flushing, storing clear water while releasing turbid water, and turbidity current release techniques.

6 Conclusions

After assessing reservoir sedimentation using remote sensing and GIS techniques, it was determined that the Harsi reservoir had lost 47.441 MCM of gross storage and 35.779 MCM of live storage during the last 84 years (1935 to 2019).

If the uniform rate of sedimentation is assumed in 84 years of occurrence of reservoir, then the reservoir capacity lost per year 0.27% of gross storage and 0.22% of the live storage.

From 2007 to 2019, i.e., 13 years, the loss in sedimentation live capacity comes out to be 14.139 MCM which is 0.055 $Mm^3/100\ km^2$/year. From the analysis, it can be concluded that in recent years, the rate of sedimentation is increasing, and the calculated rate is on higher side as per the Khosla's rate of sedimentation.

According to original elevation-capacity relation, Harsi reservoir lies in Type–II, i.e., flood plain-foothill type as per four standard type deposition patterns by Borland and Miller.

Hydrographic surveys of reservoirs are normally conducted every 5–15 years, but the recommended frequency is every 5. Traditional methods, like these, involve a lot of time and money, and they also demand manual skills. To assess capacity loss, remote sensing techniques can be employed quickly and efficiently.

Additionally, it was discovered that the accuracy of the calculated water spread area, water level data, and the original elevation-area-capacity table was very sensitive to the estimation of sedimentation by remote sensing. If the water level information is correct, and the interpretation of the water spread area is correct, it is possible to locate the revised elevation-area-capacity curves rather precisely.

Acknowledgements The National Institute of Hydrology, Bhopal, and the Civil Engineering Department of the Maulana Azad National Institute of Technology, Bhopal, both offered support to the author in the form of research help and essential institutional support, respectively.

References

1. Guo Q, Zheng Z, Huang L, Deng A (2020) Regularity of sediment transport and sedimentation during floods in the lower Yellow River, China. Int J Sediment Res 35:97–104
2. Tadesse A, Dai W (2019) Prediction of sedimentation in reservoirs by combining catchment based model and stream based model with limited data. Int J Sediment Res 34(1):27–37
3. Prasad B, Tiwari HL (2019) Sedimentation analysis and remedial measures for upper Lake Bhopal using remote sensing and GIS. Int J Innov Technol Explor Eng 8:1008–1013
4. Huang Y, Wang J, Yang M (2019) Unexpected sedimentation patterns upstream and downstream of the three gorges reservoir: future risks. Int J Sediment Res 34(2):108–117
5. Singh S, Prasad B, Tiwari HL (2021) Sedimentation analysis for a reservoir using remote sensing and GIS techniques. ISH J Hydraul Eng 1–9
6. Dhruvanarayana VV, Ram B (1983) Estimation of soil erosion in India. J Irrig Drain Eng 109(4):419–434
7. Goel MK, Jain SK, Agarwal PK (2002) Assessment of sediment deposition rate in Bargi Reservoir using digital image processing. Hydrol Sci J 47(sup1):S81–S92
8. Dutta S (2016) Soil erosion, sediment yield and sedimentation of reservoir: a review. Model Earth Syst Environ 2(3):123
9. Foteh R, Garg V, Nikam BR, Khadatare MY, Aggarwal SP, Kumar AS (2018) Reservoir sedimentation assessment through remote sensing and hydrological modelling. J Indian Soc Remote Sens 46(11):1893–1905
10. Pandey A, Chaube UC, Mishra S, Kumar D (2014) Assessment of reservoir sedimentation using remote sensing and recommendations for desilting Patratu Reservoir, India. Hydrol Sci J/J Des Sci Hydrol 61
11. Yang X, Lu XX (2014) Estimate of cumulative sediment trapping by multiple reservoirs in large river basins: an example of the Yangtze River basin. Geomorphology 227:49–59
12. Prasad B, Tiwari HL (2019) Assessment of soil erosion in the watershed of upper lake. Bhopal using remote sensing and GIS 6:456–462
13. Prasad B (2016) Gis based soil erosion modelling. vol 7(6). pp 166–171
14. Morris J, Gregory L, Fan (1998) Reservoir sedimentation handbook: design and management of dams, reservoirs and watersheds for sustainable use. McGraw-Hill Book Co., New York
15. Jain SK, Singh P, Seth SM (2002) Assessment of sedimentation in Bhakra Reservoir in the western Himalayan region using remotely sensed data, October, vol 47. pp 203–212
16. Jaiswal R, Thomas T, Singh SK, Colony M, Nagar V (2009) Assessment of sedimentation in Ravishankar Sagar reservoir using digital image processing techniques
17. Rathore DS, Choudhary A, Agarwal PK (2006) Assessment of sedimentation in harakud reservoir using digital remote sensing technique. J Indian Soc Remote Sens 34(4):377–383
18. Prasad B, Tiwari HL (2022) A comparative study of soil erosion models based on GIS and remote sensing. ISH J Hydraul Eng 28(1):98–102
19. Karim M, Maanan M, Maanan M, Rhinane H, Rueff H, Baidder L (2018) Assessment of water body change and sedimentation rate in Moulay Bousselham wetland, Morocco, using geospatial technologies. Int J Sediment Res
20. Pandey A, Chaube UC, Mishra SK, Kumar D (2016) Assessment of reservoir sedimentation using remote sensing and recommendations for desilting Patratu Reservoir, India. Hydrol Sci J 61(4):711–718
21. McFEETERS SK (1996) The use of the normalized difference water index (NDWI) in the delineation of open water features. Int J Remote Sens 17(7):1425–1432
22. Dadoria D, Tiwari H, Jaiswal R (2017) Assessment of reservoir sedimentation in Chhattisgarh state using remote sensing and GIS. Int J Civ Eng Technol 8:526–534

23. Kumar P et al (2013) Remote sensing study on geomorphological degradation of Sarda Sagar reservoir. J Environ Biol 34:1065–1068
24. Subramanya K (2008) In: Engineering hydrology. 3rd edn. Tata McGraw-Hill
25. Borland WM, Miller CR (1960) Distribution of sediment in large reservoirs, Trans ASCE, 125(I):166–180
26. Murthy BN (1977) Life of reservoir, Technical Report No. 19. Delhi, India

Assessment of Vertical Accuracy of Freely Available Global Digital Elevation Models for Heterogeneous Terrains in India

V. Nandam and P. L. Patel

Abstract Digital elevation model (DEM) is one of the key inputs widely used in the field of water resources engineering. The need for hydrologic and hydraulic modeling using advanced remote sensing and GIS technology led to the development of global DEMs. The vertical accuracies of such DEMs vary spatially for different terrain conditions. In the present study, the global DEMs–SRTM DEM, ALOS PALSAR's AW3D, CartoDEM, MERIT DEM, and TanDEM-X 90 are checked for their vertical accuracies on two heterogeneous terrain areas in lower Tapi basin (LTB) (flat terrain) and upper Tapi basin (UTB) (hilly terrain). The ice, cloud, and land elevation satellite (ICESat) altimetry is used as ground truth data for the study. The descriptive statistics such as RMSE, mean absolute error (MAE), linear error (LE) at 90% and 95% significance levels are computed. The results for both LTB and UTB indicated that the TanDEM-90 had been found to be superior DEM than SRTM based on vertical accuracies. The RMSE, MAE, LE90, and LE95 are 1.23 m, 0.74 m, 2.01 m, 2.41 m, and 0.89 m, 0.6 m, 1.45 m, and 1.73 m for LTB and UTB, respectively, for TanDEM-X 90. Similarly, for SRTM DEM, above statistics are 1.9 m, 1.51 m, 3.12 m, and 3.73 m and 2.11 m, 1.67 m, 3.47 m, 4.15 m for LTB and UTB, respectively. The order of ranking based on RMSE of the DEMs is TanDEM-X 90, ALOS PALSAR's AW3D, MERIT DEM, CartoDEM, and SRTM DEM. From estimated statistics, it can be inferred that chosen DEMs give better results for hilly terrain vis-à-vis flat terrain.

Disclaimer: The presentation of material and details in maps used in this chapter does not imply the expression of any opinion whatsoever on the part of the Publisher or Author concerning the legal status of any country, area or territory or of its authorities, or concerning the delimitation of its borders. The depiction and use of boundaries, geographic names and related data shown on maps and included in lists, tables, documents, and databases in this chapter are not warranted to be error free nor do they necessarily imply official endorsement or acceptance by the Publisher or Author.

V. Nandam (✉)
Department of Civil Engineering, SVNIT, Surat 395007, India
e-mail: d18ce004@ced.svnit.ac.in

P. L. Patel
Sardar Vallabhbhai National Institute of Technology, Surat 395007, India
e-mail: plpatel@ced.svnit.ac.in

P. V. Timbadiya et al. (eds.), *Geospatial and Soft Computing Techniques*, Lecture Notes in Civil Engineering 339, https://doi.org/10.1007/978-981-99-1901-7_15

Keywords Digital elevation model · TanDEM · Altimetry · ICESat · Lower Tapi basin · Upper Tapi basin

1 Introduction

With the rapid advancement in remote sensing (RS) and geographic information system (GIS) technologies, an element of improvement in accuracies to the space-borne derived data caught momentum. This led to various researches on the available data products. One such fundamental data is digital elevation models (DEMs), having a wide range of applications in Earth's geography and hydrology [1–6]. DEM is a generalized term used interchangeably for both digital surface models (DSMs) and digital terrain models (DTMs). The first global DEM made publicly available is Shuttle Radar Topography Mission (SRTM) DEM, followed by ASTER, many more region-specific DEMs, derived DEMs, and the latest being TanDEM-X. Comparing the DEMs for their accuracies is challenging as these DEMs generated are with reference to any standard datums. Even, the datums are updated with respect to time. The vertical accuracy of DEM plays a key role when modeling a drainage basin [5]. The physiographic landforms of the area impact the accuracies. The hilly terrain shows a significant deviation in accuracy compared to flat terrains [7]. Attempts have been made to evaluate accuracies of the DEMs ever since SRTM and ASTER GDEM are open to the public [8, 9].

Ground truth elevations recorded serve as a measure for accuracy assessment of a DEM. Not all the land surface has ground surveyed data. Therefore, reliability on the most accurate RS and GIS altimetry cater to the need. One such data is ICESat altimetry [10]. The ICESat satellite carries geoscience laser altimeter system (GLAS) instrument which was primarily designed to extract information of ice sheets [11, 12]. The elevations of GLAH14 provide global land surface altimeter elevations with reference ellipsoid as Topex/Poseidon. The geodetic corrections to water levels of an Amazon River were worked out using ICESat altimetry [12], comparison of DEMs using GLAH14 is performed by [1, 8, 13, 14,] performance assessment of TanDEM-X by [15, 16]. Therefore, the data may be considered near-ground-truth elevation data.

An extensive literature on the Tapi basin quantifying various aspects of geomorphology, sediment yield, flooding, hazard and vulnerability, dam break analysis, and hydrologic modeling is available in the open domain of the research community [17–19]. Most of the literatures relied on SRTM DEM or ground surveyed spatial data wherever available to accomplish the task. An attempt is made in this study to check the efficacy of the open altimetry data of ICESat satellite which is applied to many foreign watersheds to serve as a near-ground-truth dataset and the RS and GIS DEM products to replicate the ground conditions for a heterogeneous basin in India.

2 Study Area

The two heterogeneous regions considered for the present investigation are part of the Tapi basin, located in the western Indian sub-continent. The Tapi River originates at Multai in the Betul district of Madhya Pradesh and flows toward the west to meet the Arabian Sea in the Gulf of Khambhat. The basin is subdivided into upper Tapi basin (UTB), middle Tapi basin (MTB), and lower Tapi basin (LTB). The Tapi River from origin to Hathnur Dam is called as upper Tapi river, having a drainage area of 29,430 km^2, and is known as UTB, from Hathnur dam to the Ukai dam is called as middle Tapi river, with a drainage area of 28,970 km^2 known to be MTB, and the lower Tapi river with the drainage area of 6745 km^2 is LTB. The present work is undertaken to understand the accuracies of global DEMs for UTB and LTB regions (Fig. 1). The geographic location of UTB lies in between 75° 55′ and 78° 17′ East longitudes and 20° 05′–22° 03′ North latitudes; the LTB lies in between 72° 33′ and 73° 39′ East longitudes and 20° 3′–21° 39′ North latitudes.

The terrain is hilly in the UTB and is relatively flat downstream, varying from 1563 to 211 m above MSL. The majority of the UTB is covered with forest, and hence, elevation profiles are more sensitive with respect to space. Sediment yield is high in the UTB due to steep and hilly topography [17]. The LTB has a flat terrain with the maximum altitude being 405 m with respect to MSL, and a minimum (below MSL) is observed at the confluence of the river and the Arabian sea. The LTB region has dominating agricultural fields, fallow lands, and a major urban growth center, the Surat city, which has the nation's economic and historical importance, is situated just

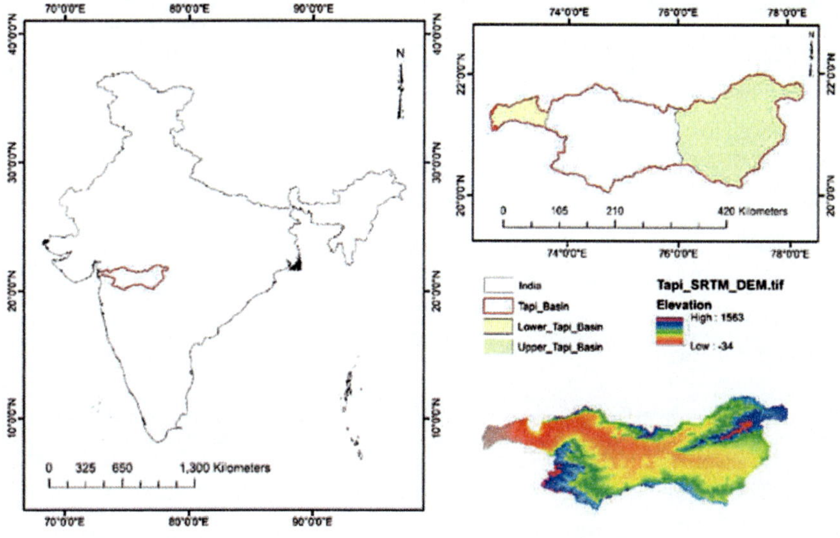

Fig. 1 Study area showing the geographic location (to the left), lower Tapi basin and upper Tapi basin, and elevational spatial variation (toward the right)

before the confluence of the river with the sea. The city is prone to frequent riverine flooding. The flood events that occurred in the years 1727, 1776, 1782, 1829, 1837, 1872, 1944, 1959, 1968, 1970, 1994, 1998, 2006, and 2013 caused massive damage, including loss of lives and properties [17, 18, 20].

3 Data Sources

The study is primarily intended to check the efficacy of satellite datasets in the accurate representation of ground features. ICESat altimetry datasets are reliable datasets to extract land elevations along the laser orbits at a specified interval. The freely available DEMs considered for the study are TanDEM-X 90, SRTM DEM, MERIT DEM, ALOS PALSAR's AW3D 30, and CartoDEM (Table 1).

Table 1 Description of DEMs

DEM	Year	Bands	LE90 (m)	Reference system, resolution
SRTM	11–22 Feb 2000	X and C-band system	16	WGS84 ellipsoid in the horizontal direction EGM96 geoid in the vertical direction 3 arc-second
MERIT DEM	11–22 Feb 2000	C-band radar interferometry	12	WGS84 ellipsoid in the horizontal direction EGM96 geoid in the vertical direction 3 arc-second
TanDEM-X 90 DEM	Dec 2010–Jan 2015	X-band signal	10	WGS84 ellipsoid in the horizontal direction WGS84 ellipsoid in the vertical direction The heights of the TanDEM-X DEM products are ellipsoidal, 3 arc-second
CartoDEM	May 2005	0.5–0.85 μm	8	WGS84 ellipsoid in the horizontal direction WGS84 ellipsoid in the vertical direction, 30 m
AW3D30	Derived from SRTM and NED datasets			WGS84 ellipsoid in the horizontal direction EGM96 geoid in the vertical direction, 30 m

3.1 ICESat Altimetry

An open-source altimetry data is downloaded from National Snow and Ice Data Centre (NSIDC) to evaluate the accuracy. The data includes the collection of altimetry derived from geoscience laser altimeter system (GLAS) instrument payload on National Aeronautics and Space Administration—NASA's ice, cloud, and land elevation (ICESat) satellite from 2003 to early 2010. Out of several output datasets available under this mission, global land surface altimetry data (GLAH 14) of release 34 is the latest one with respect to datum Topex/Poseidon ellipsoid EGM2008 geoid. The GLAS instrument emits laser pulses at the range of 1064 nm (infrared) and 532 nm (green). The ground laser footprint diameter is about 70 m, and the distance of each footprint in the track direction is 172 m. The vertical accuracy of GLAS elevation is < 1 m in flat regions and < 10 m in hilly terrains (https://nsidc.org). ICESat temporal repetition was 91 repeat cycles with 33 days sub-cycles.

3.2 SRTM DEM

The SRTM DEM was produced by X band and C bands interferometry during an 11-day mission from 11 to 22 February 2011 [8, 21]. The mission's geographic land coverage is between 60° N and 56° S at 90 m resolution. The elevations are referenced to a horizontal datum of WGS1984 and a vertical datum of EGM96. The absolute vertical error is calculated to be 16 m.

3.3 MERIT DEM

The MERIT DEM is derived from the existing models by removing errors that include speckle noise, stripe noise, absolute bias, and tree height bias corrections. The baseline DEMs used are SRTM DEM below 60° N, and AW3D-30 m DEM above 60° N. The data gaps present when combining the two DEMs were filled by Viewfinder Panoramas DEM (VFP–DEM), which holds data from region-specific topography maps, U.S. National Elevation data, and Canadian GeoBase DEM [21]. The measurements obtained are terrain heights as the applied error removal techniques aim to attain non-object/ground elevations.

3.4 TanDEM-X 90 DEM

The TanDEM-X mission commissioned during December 2010 and January 2015 aimed at producing 0.4 arc-second DEM of global coverage. The elevations were derived by the single-pass SAR interferometry in which a pair of SAR images that are collected from the dual satellites (TerraSAR-X and TanDEM-X) flew in a close range helix formation [22]. The advantage of SAR interferometry is that the images are unaffected by weather and day or night conditions. The overall vertical accuracy target is set to be 10 m (LE90), 2 m for flat areas with a slope less than 20%, and 4 m for slopes greater than 40% [13, 22].

3.5 CartoDEM

CartoDEM is a product of Cartosat-1 launched by the Indian Space Research Organization (ISRO) on 5th May 2005 with a mission duration of 5 years covering the Indian sub-continent. The absolute vertical accuracy was found to be 8 m. The heights are referenced to both horizontal and vertical datum WGS84 ellipsoid.

3.6 ALOS PALSAR's AW3D30

ALOS PALSAR's AW3D30 DEM is a derived DEM from SRTM and National Elevation Dataset (NED) by applying geoid corrections through radiometric terrain correction (RTC) project. The resultant DEM has a resolution of both 12.5 m and 30 m and is referenced to WGS84 ellipsoid and EGM96 geoid horizontal and vertical datums, respectively.

4 Data Processing and Descriptive Statistics

4.1 ICESat Altimetry Data Filtering

The open altimetry data is downloaded from NASA's NSIDC Website (https://search.earthdata.nasa.gov/search). The ICESat dataset is a backscatter of laser pulse recorded in the waveform. The flags in the dataset are identified and are removed with reference to the GLAH14 product data dictionary. The elevation corrections applied are saturation elevation correction and elevation bias correction. The elevation flags that are considered for data filtering are saturation correction flag, elevation use flag, saturation index, percent saturation, elevation definition flag 1, elevation cloud flag, the initial number of peaks in received echo, number of peaks found in return, and full resolution 1064 quality flag.

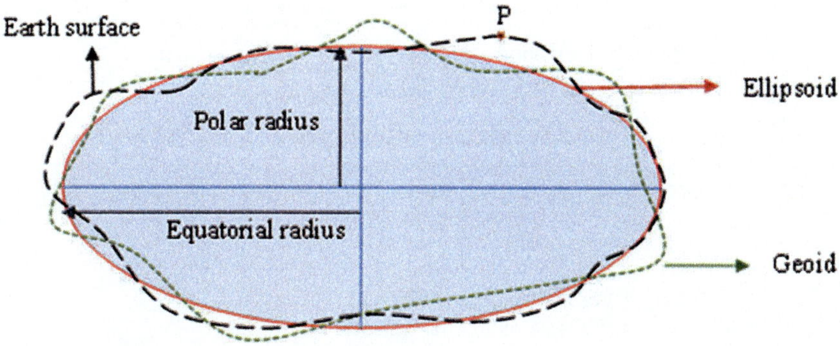

Fig. 2 Representation of ellipsoid and geoid of an Earth

4.2 Datum Transformation

The shape of the Earth is oblate spheroid/ellipsoid. The ellipsoid has a smooth surface, with the geometric parameters being the equatorial radius and polar radius. This is a place location parameter and is related to horizontal datum. The vertical datum is the one that specifies the elevation at the location. This can be obtained by developing a geoid surface of the earth. The geoid surface is defined as the shape of the ocean surface the earth would take under the influence of gravity and rotation of the earth alone (see Fig. 2). The most widely used worldwide horizontal datum is World Geodetic System (WGS), the latest model being WGS84; the vertical datum is Earth Gravitational Model (EGM) 96 and 2008.

4.3 DEM Accuracy Statistics

The descriptive statistics used to evaluate the errors in various DEMs with reference to truthful datasets are adopted from [8, 9, 13, 22] and are given in Equations from (1) to (8).

$$\text{Vertical Height Error}(\Delta h_i) = h_{i,\text{GDEM}} - h_{i,\text{ref}} \tag{1}$$

$$\text{Mean Error (ME)} = \frac{1}{n}\sum_{i=1}^{n}\left|h_{i,\text{GDEM}} - h_{i,\text{ref}}\right| \tag{2}$$

$$\text{RMSE} = \sqrt{\frac{1}{n}\sum\nolimits_{i=1}^{n}\Delta h_i^2} \tag{3}$$

$$\text{Standard Deviation} = \sqrt{\frac{1}{(n-1)}\sum_{i=1}^{n}\left(\Delta h_{i,\text{ref}} - \text{ME}\right)^2} \tag{4}$$

$$\text{LE90} \approx 1.64 \times \text{RMSE} \tag{5}$$

$$\text{LE95} \approx 1.96 \times \text{RMSE} \tag{6}$$

$$\text{Relative Vertical Accuracy (RVA)} = |\Delta_{\text{reference}} - \Delta_{\text{DEM}}| \tag{7}$$

$$\text{RMSE.RVA} = \sqrt{\frac{\sum \text{RVA}^2}{N_{\text{pairs}}}} \tag{8}$$

where

$h_{\text{i,GDEM}}$ is elevation value of a pixel "*i*" for the global digital elevation model,

$h_{\text{i,ref}}$ is elevation value of pixel "*i*" with respect to reference value/true value,

LE90 and LE95 are linear errors at 90% and 95% confidence interval,

$\Delta_{\text{reference}} = \left|\Delta_{\text{reference}}.\text{elevation}_{i+1} - \Delta_{\text{reference}}.\ \text{elevation}_{\text{i}}\right|$,

$\Delta_{\text{DEM}} = |\ \Delta_{\text{DEM}}.\ \text{elevation}_{i+1} - \Delta_{\text{DEM}}.\ \text{elevation}_{i}|$.

5 Methodology

To check the vertical accuracy of a DEM, two datasets are essential: reliable ground truth data and the DEM for which accuracy is to be assessed. As the study aims to check the efficacy of the remote sensing and GIS products, initially, remotely sensed near-ground-truth datasets are identified as ICESat altimetry. The first step in the analysis is the removal of flagged records from the altimetry. Then, the datum of the altimetry, which is Topex/Poseidon ellipsoid, is to be converted to orthometric heights referenced to WGS84 ellipsoid and EGM96 geoid. Secondly, the TanDEM-X 90 and CartoDEM representing ellipsoidal heights are to be transformed to orthometric heights matching with the ICESat altimetry data. The SRTM DEM, MERIT DEM, and ALOS PALSAR's AW3D30 DEM are default orthometric heights and require no datum transformation. The final step is to calculate descriptive statistics and critical comparison on vertical accuracies of the DEMs for flat and hilly terrain areas (LTB and UTB, respectively).

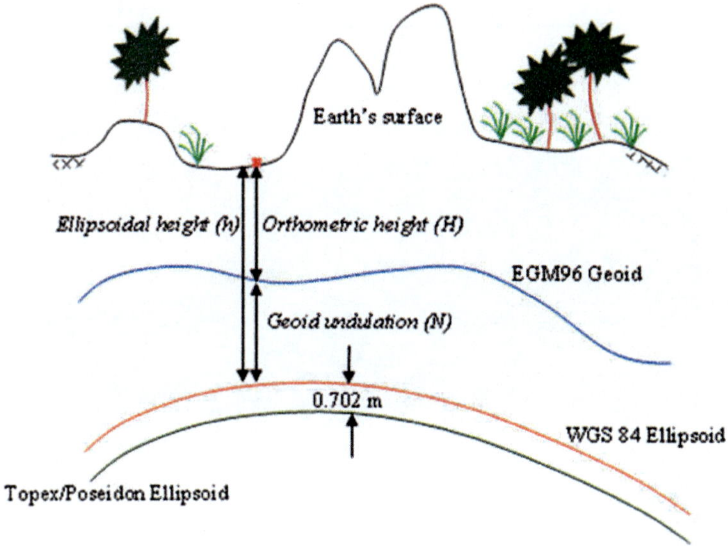

Fig. 3 Representation of orthometric heights

6 Results and Discussion

6.1 Datum Conversions

The ICESat altimetry of GLAH14 granules of release 34 is referenced to Topex/Poseidon (T/P) ellipsoid. Here, both horizontal and vertical datums are Topex/Poseidon ellipsoid. The difference between WGS84 and T/P ellipsoid is 0.702 m. Ellipsoidal heights are converted to orthometric heights using Eq. 9 (see Fig. 3).

$$H_{WGS84} = h - N - 0.702 \tag{9}$$

where H is orthometric height, h is ellipsoidal height, N is geoid height. The EGM96 geoid heights are obtained for the study area (Fig. 4 for lower Tapi basin, and Fig. 5 for upper Tapi basin), and correspondingly, the ICESat altimetry data, TanDEM-X 90 m, and CartoDEM ellipsoidal values which are originally referenced to WGS84 are transformed to orthometric heights to obtain a common datum base for comparison.

6.2 Comparison of DEMs

The DEMs are brought to common horizontal and vertical datums WGS84 and EGM96, respectively, for comparison against ICESat altimetry datasets. For each

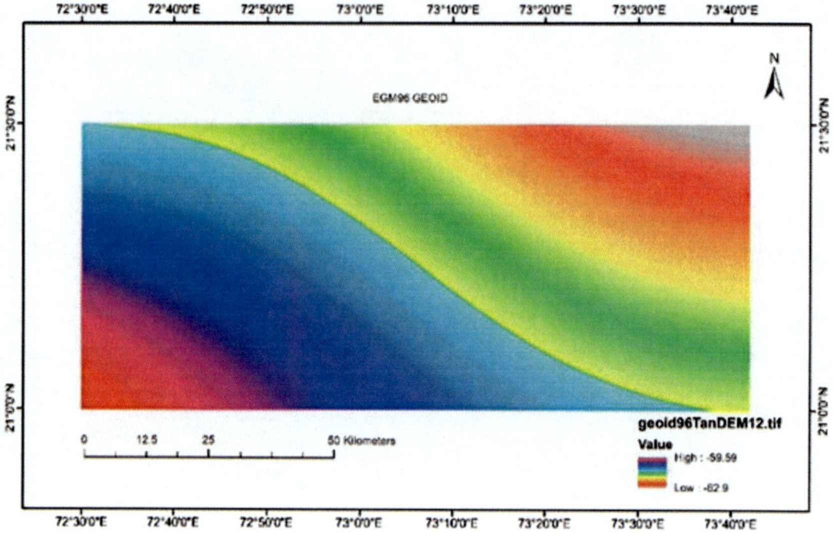

Fig. 4 Earth gravitational model 1996 for the lower Tapi basin extent

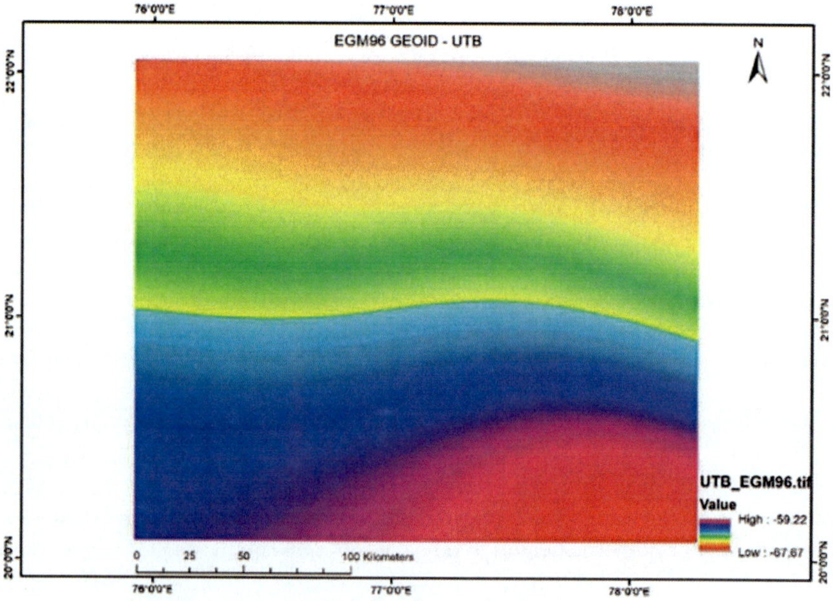

Fig. 5 Earth gravitational model 1996 for the upper Tapi basin extent

point, the difference in elevations is calculated using Eq. 9; then, the outliers are removed by applying the 3σ rule, i.e., the points beyond 3 × standard deviations are eliminated considering them as outliers [8, 14, 22]. Accuracy measures thus calculated are shown in Table 2 for LTB and Table 3 for UTB. Based on the RMSE, which is a measure of vertical accuracy of the DEM, from the Tables 2 and 3, it can be inferred that the order of DEMs ranking from superior to inferior performance is TanDEM -X 90, ALOS PALSAR's AW3D30, MERIT DEM, CartoDEM, and SRTM DEM, respectively.

From the multiplicative bias (the ratio is the sum of DEM measurements to that of the sum of observed measurements), which is a measure of model performance computed for the current dataset, the TanDEM-X 90 has perfect model performance, SRTM and AW3D30 underestimated by 0.2 times, and MERIT and CartoDEM overestimates by 0.02 and 0.01 in case of Lower Tapi basin, whereas, for upper Tapi basin, all the DEMs considered have comparable results. The fractional standard error is found to be less (0.04) in the case of TanDEM-X 90, followed by MERIT DEM and CartoDEM by a value of 0.06 for LTB, in the case of UTB, it is 0.002 for TanDEM-X

Table 2 Accuracy measures of DEMs against ICESat altimetry datasets for lower Tapi basin

Descriptive statistics	MERIT DEM	SRTM DEM	TanDEM-X 90 m	Carto DEM	ALOS AW3D30
Total points	3171	4006	3444	3240	4010
Mean error (m)	1.10	− 0.47	− 0.03	0.22	0.59
RMSE (m)	1.67	1.90	1.18	1.87	1.64
Std. deviation (m)	1.26	1.84	1.18	1.86	1.53
Skewness	− 0.87	0.01	− 0.35	0.15	0.06
ICESat maximum (m)	140.53	153.11	180.94	180.94	153.11
ICESat minimum (m)	− 1.38	− 1.38	− 1.23	− 0.18	− 1.38
Std. deviation (m)	30.92	30.35	31.16	31.40	30.27
Mean (m)	29.73	24.25	28.48	30.01	24.24
Total sum	94,284.15	97,133.5	98,074.91	97,236.7	97,206.57
DEM maximum (m)	139.97	150.00	184.48	181.06	154.00
DEM minimum (m)	− 0.01	0.00	− 7.20	− 5.13	− 2.00
Std. deviation (m)	30.96	29.72	31.34	30.76	29.76
Mean (m)	30.83	23.77	28.51	30.24	23.65
Total sum	97,770.08	95,239.0	98,161.2	97,962.6	94,838.0
Linear error					
LE 90 = 1.64*RMSE	2.74	3.12	1.94	3.07	2.69
LE 95 = 1.96*RMSE	3.27	3.73	2.31	3.67	3.21
Relative vertical accuracy (RVA) no. of pairs	3140	3975	3443	3239	3975
RVA RMSE (m)	1.204	1.856	1.280	1.957	1.290

Table 3 Accuracy measures of DEMs against ICESat altimetry datasets for upper Tapi basin

Descriptive statistics	MERIT DEM	SRTM DEM	TanDEM-X 90 m	Carto DEM	ALOS AW3D30
Total points	6358	6462	6446	6445	6428
Mean error (m)	1.19	0.90	0.03	0.69	1.30
RMSE (m)	1.71	2.12	0.89	1.97	1.48
Std. deviation (m)	1.22	1.91	0.88	1.85	1.51
Skewness	− 0.34	− 0.09	0.39	− 0.33	− 0.22
ICESat maximum (m)	801.31	801.31	801.31	801.31	801.31
ICESat minimum (m)	214.69	214.69	214.69	214.69	214.69
Std. deviation (m)	119.52	120.25	120.08	120.24	119.74
Mean (m)	362.13	363.57	363.35	364.16	362.82
Total sum	2,302,443.7	2,349,395	2,342,140.1	2,346,991	2,332,211.1
DEM maximum (m)	804.00	805	803.42	805.77	803
DEM minimum (m)	216.81	215.00	213.49	211.18	216
Std. deviation (m)	119.43	120.49	120.09	120.24	119.60
Mean (m)	363.33	364.47	363.38	364.84	364.12
Total sum	2,310,041.9	235,521	2,342,356.2	2,351,420	2,340,569
Linear error					
LE 90 = 1.64*RMSE	2.8	3.47	1.45	3.23	2.43
LE 95 = 1.96*RMSE	3.35	4.15	1.73	3.87	2.90
Relative vertical accuracy (RVA) no. of pairs	6307	6381	6363	6395	6357
RVA RMSE (m)	1.58	2.33	1.20	1.96	1.59

90, preceded by AW3D30. The relative vertical accuracy measures each observation (ICESat altimetry) point with its closest neighboring point set as a point pair (8). The RVA of TanDEM is found to be nearly consistent for both the terrains with a value of ~ 1.2 m. However, in the case of LTB, MERIT DEM showed the superior value of RVA.RMSE than TanDEM-X 90.

7 Conclusions

The present study focuses on assessing the vertical accuracy of freely available global DEMs by using ICESat altimetry. The specific findings from this current study are as follows:

- The open altimetry data of ICESat GLAH14 global land surface altimetry can be used as near-ground truth data based on literature survey.

- Most widely used global DEMs such as SRTM DEM, MERIT DEM, and AW3D30, along with CartoDEM having coverage of the entire Indian sub-continent, and the newest addition to the global DEMs the TanDEM-X 90, are considered for comparison. The RMSE is calculated (a measure of vertical accuracy) for both lower Tapi basin having a flat terrain area and upper Tapi basin having a hilly terrain. The linear error at the 90% significance level is found to be superior for the TanDEM-X 90 in both the cases with the vertical error of 1.94 m and 1.45 m for LTB and UTB, respectively. Similarly, the LE95 is found to be 2.31 m and 1.73 m.
- In both cases, i.e., flat terrain vis-à-vis hilly terrain, the performance of the DEMs with the most accuracy to the least accuracy is TanDEM-X 90, ALOS PALSAR's AW3D30, MERIT DEM, CartoDEM, and SRTM DEM. The inter-comparison of statistics for both the regions implies that the error band for SRTM DEM, MERIT DEM, and CartoDEM is more in upper Tapi basin; on the other hand, TanDEM-X 90 and CartoDEM showed more error in flat terrains.
- From the present analysis, it may be concluded that the general notion "global DEMs of hilly terrain are subjected to large errors relative to flat terrains" which did not hold true in the case of TanDEM-X 90. TanDEM-X 90 could be a potential input for various applications in the field of water resources over the other DEMs available. However, the applicability may be checked and validated through model development as the ICESat altimetry is confined to the laser tracks measurements and not spread evenly throughout the study area.

Acknowledgements The first author would like to acknowledge the Department of Science and Technology, Ministry of Science and Technology, Government of India for the financial support vide their letter no. DST/INSPIRE Fellowship/2018/[IF180589] dated July 24, 2019. The authors are grateful to the infrastructural support provided in the Centre of Excellence (CoE) on "Water Resources and Flood Management," TEQIP-II, Ministry of Human Resources Development (MHRD), Government of India, for conducting this research. The authors are thankful to various data disseminating centers for providing relevant data used in the present study.

References

1. Zhou Q (2017) Digital elevation model and digital surface model. Int Encycl Geogr People Earth Environ Technol 1–17
2. Hawker L, Bates P, Neal J, Rougier J (2018) Perspectives on digital elevation model (DEM) simulation for flood modeling in the absence of a high-accuracy open access global DEM. Front Earth Sci 6:233
3. Archer L, Neal JC, Bates PD, House JI (2018) Comparing TanDEM-X data with frequently used DEMs for flood inundation modeling. Water Resour Res 54(12):10–205
4. Ettritch G, Hardy A, Bojang L, Cross D, Bunting P, Brewer P (2018) Enhancing digital elevation models for hydraulic modelling using flood frequency detection. Remote Sens Environ 217:506–522

5. Mohanty MP, Nithya S, Nair AS, Indu J, Ghosh S, Bhatt CM, Rao GS, Karmakar S (2020) Sensitivity of various topographic data in flood management: implications on inundation mapping over large data-scarce regions. J Hydrol 590:125523
6. Liu Y, Bates PD, Neal JC, Yamazaki D (2021) Bare-Earth DEM generation in urban areas for flood inundation simulation using global digital elevation models. Water Resour Res 57(4):e2020WR028516
7. Sanders BF (2007) Evaluation of on-line DEMs for flood inundation modeling. Adv Water Resour 30(8):1831–1843
8. Du X, Guo H, Fan X, Zhu J, Yan Z, Zhan Q (2015) Vertical accuracy assessment of freely available digital elevation models over low-lying coastal plains. Int J Digital Earth 9(3):252–271
9. Jain AO, Thaker T, Chaurasia A, Patel P, Singh AK (2018) Vertical accuracy evaluation of SRTM-GL1, GDEM-V2, AW3D30 and CartoDEM-V3. 1 of 30-m resolution with dual frequency GNSS for lower Tapi Basin India. Geocarto Int 33(11):1237–1256
10. Huber M, Wessel B, Kosmann D, Felbier A, Schwieger V, Habermeyer M, Wendleder A, Roth A (2009) Ensuring globally the TanDEM-X height accuracy: Analysis of the reference data sets ICESat, SRTM and KGPS-tracks. In: 2009 IEEE international geoscience and remote sensing symposium, vol. 2, pp II–769
11. Zwally HJ, Schutz B, Abdalati W, Abshire J, Bentley C, Brenner A, Thomas R (2002) ICESat's laser measurements of polar ice, atmosphere, ocean, and land. J Geodyn 34(3–4):405–445
12. Hall AC, Schumann GJ, Bamber JL, Bates PD, Trigg MA (2012) Geodetic corrections to Amazon River water level gauges using ICESat altimetry. Water Resour Res 48(6)
13. Wessel B, Huber M, Wohlfart C, Marschalk U, Kosmann D, Roth A (2018) Accuracy assessment of the global TanDEM-X digital elevation model with GPS data. ISPRS J Photogram Remote Sens 139:171–182
14. Bhang KJ, Schwartz FW, Braun A (2007) Verification of the vertical error in C-band SRTM DEM using ICESat and Landsat-7, Otter Tail County, MN. IEEE Trans Geosci Remote Sens 45(1):36–44
15. Gruber A, Wessel B, Huber M, Roth A (2012) Operational TanDEM-X DEM calibration and first validation results. ISPRS J Photogram Remote Sens 73:39–49
16. Rizzoli P, Martone M, Gonzalez C, Wecklich C, Tridon DB, Bräutigam B, Bachmann M, Schulze D, Fritz T, Huber M, Wessel B, Krieger G, Zink M, Moreira A (2017) Generation and performance assessment of the global TanDEM-X digital elevation model. ISPRS J Photogram Remote Sens 132:119–139
17. Timbadiya PV, Patel PL, Porey PD (2015) A 1D–2D coupled hydrodynamic model for river flood prediction in a coastal urban floodplain. J Hydrol Eng 20(2):05014017
18. Vora A, Sharma PJ, Loliyana VD, Patel PL, Timbadiya PV (2018) Assessment and prioritization of flood protection levees along the lower Tapi River India. Nat Hazards Rev 19(4):05018009
19. Resmi SR, Patel PL, Timbadiya PV (2020) Impact of land use-land cover and climatic pattern on sediment yield of two contrasting sub-catchments in upper Tapi Basin, India. J Geol Soc India 96(3):253–264
20. Kale VS, Hire PS (2014) Effectiveness of monsoon floods on the Tapi River, India: role of channel geometry and hydrologic regime. Geomorphology 57(3–4):275–291
21. Yamazaki D, Ikeshima D, Tawatari R, Yamaguchi T, O'Loughlin F, Neal JC, Samspon C, Kanae S, Bates PD (2017) A high-accuracy map of global terrain elevations. Geophys Res Lett 44(11):5844–5853
22. Hawker L, Neal J, Bates P (2019) Accuracy assessment of the TanDEM-X 90 digital elevation model for selected floodplain sites. Remote Sens Environ 232:111319

GIS and RS Applications in Water Resources Management in Consumption with Crop Assessment

Suvarna Kulkarni, Sunil Gaikwad, and Makarand Kulkarni

Abstract Raigad is the district in Maharashtra located in range of Sahyadri Mountain. Most part of district contains hilly terrains. Its land covers with thick forest. Crop estimation in hilly terrain is quite difficult because of its uneven surface undulating topography. Crop acreage estimation in hilly region is challenging for remote sensing. The reflection of vegetation and crop features is seen similarly in FCC images, so it becomes hard to differentiate. Remote sensing technique is economical, faster and gives accurate result because of its higher temporal frequency and spatial resolution. Present study utilizes satellite images of Sentinel 2 satellites. The objective of this study is to identify measure, map the standing crops in Rabbi season, year 2021, in Raigad district and map taluka level area of standing crop. The objective is achieved with 67% crop accuracy, as the study area is in hilly terrain and having thick forest; hence, it is difficult to achieve reasonable accuracy. Although ground truth is difficult in hilly terrain, 150 training sets were taken during ground truth. Sentinel 2 satellite images having higher resolution 10 m are used for this study, but for achieving more accuracy in hilly terrain containing thick forest, 1–2 m higher spatial resolution images are essential with temporal resolution of 5–10 days.

Keywords Remote sensing · Erdas imagine · Satellite image · Sentinel 2 · Raigad

1 Introduction

Raigad is the district in Maharashtra, located on bank of Arabian Sea. Garden of areca nuts, cashew nuts and coconut increases its beauty. Paddy is the predominant crop in Kharip season, and production of finger millets and small millets is also taken during Kharip season in the district. Cow pea, red gram, green gram, horse gram and beans are the major crops in Rabbi season; Remote sensing technique gives higher accuracy due to repetitive coverage of area. Hence in this study, remote sensing

S. Kulkarni (✉) · S. Gaikwad · M. Kulkarni
Resources Engineering Centre, Maharashtra Engineering Research Institute, Nashik, Maharashtra 422004, India
e-mail: kulkarni.suvarna9@gmail.com

P. V. Timbadiya et al. (eds.), *Geospatial and Soft Computing Techniques*, Lecture Notes in Civil Engineering 339, https://doi.org/10.1007/978-981-99-1901-7_16

technique is used to carry out standing crop mapping in Rabbi season, year 2021, in Raigad district. For such study purpose, satellite images of Sentinel 2A and Sentinel 2B are used for analysis.

As Raigad district situated in Sahyadri mountain region, it is having hilly terrain and thick forest. Due to this, achieving reasonable accuracy for such district becomes difficult. For this study purpose, images of Sentinel 2A and Sentinel 2B satellites of European Space Agency (ESA) are utilized. This data is free downloadable and having higher resolution of 10 m. On these images, crop land and forest land reflection are seen similarly, so for this study subset of crop land is subtracted from full scene image and then classification is performed. For achieving accuracy, 150 training sets were taken during ground truth. The point, line or polygon features of every class are collected during the visit.

2 Study Area and Data Source

For standing crop assessment, Raigad district is considered. Fifteen talukas from Raigad district are covered in this study. It is located in the Konkan Region. This district comes between 18° 51′ 58″ N latitudes and 73° 18′ 22″ E longitudes. Geographical area of Raigad district is 7152 km^2. The main rivers in Raigad district are Kalu, Patalganga, Amba, Kundalika, Ghod, Ulhas, Bhogawati and Savitri. Kal and Morabe are the major dams in district while Rajanalla, Hetavane and Savatri are the medium dam (Source: Government of Maharashtra official website) (Fig. 1).

2.1 Data Used

2.2 Field Data

Crop cycle of Rabbi and summer season of Raigad district (District Agricultural Dept). List of major and medium reservoirs in the Raigad district. (Website-Mahawrd). Toposheets of Raigad district. (SOI, Survey of India official website). List of villages in Raigad district. (MRSAC, Maharashtra Remote Sensing Application Center, village maps).

2.3 Satellite Data

Sentinel satellite images are used for present study. The details of the satellite data selected for study are shown in Table 1.

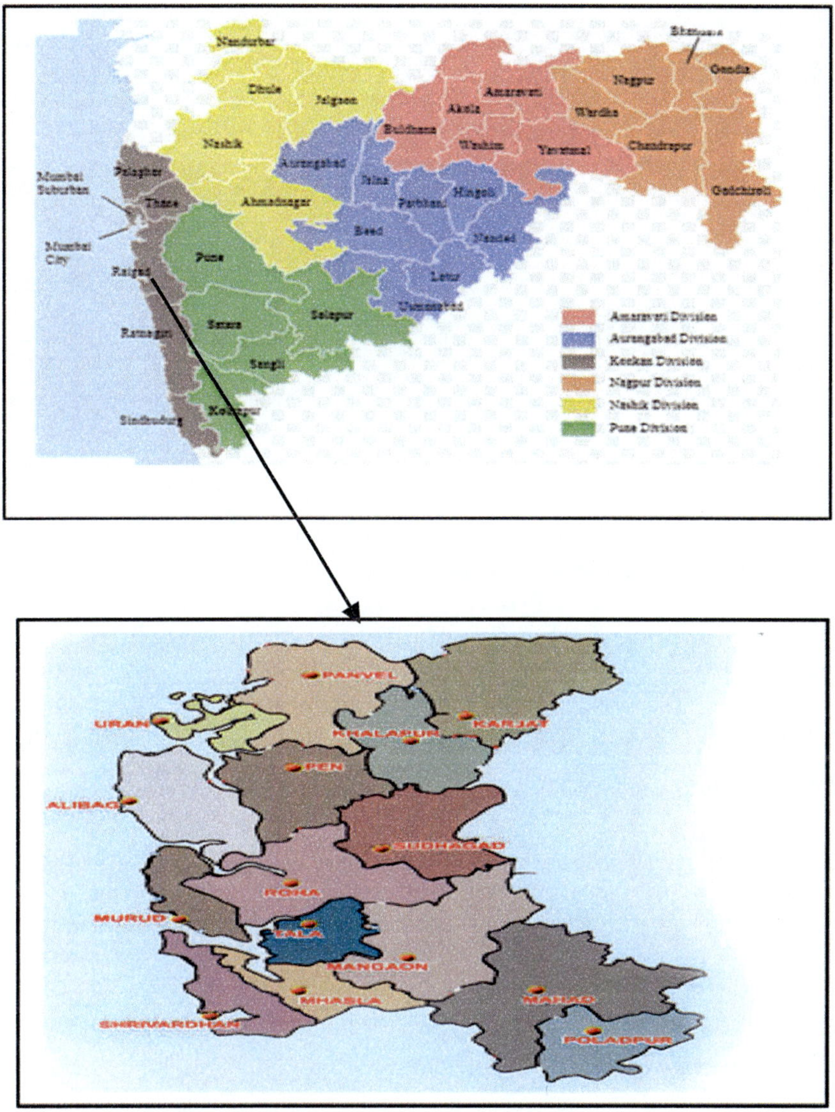

Fig. 1 Index map of Raigad district

2.4 *Mosaic of Satellite Image*

To cover all geographical area of Raigad district, six satellite images are required. Then by using above images, a mosaic image is prepared, which covers the whole 15 tahasil of Raigad district.

Table 1 Dates of pass for satellite data

Satellite	Sensor	Tile no	Date of pass for pre-monsoon	Remark
Sentinel 2	2A,2B	T 43 QBA	17/01/2021	First image
		T 43 QBA	22/01/2021	
		T 43 QCA	24/01/2021	
		T 43 QCB	16/05/2019	
		T43QCVV	24/01/2021	
		T 43 QCB	24/01/2021	Second image
		T 43 QCV	08/02/2021	
		T 43 QBA	08/02/2021	
		T 43 QBB	11/02/2021	
		T43QCBB	21/02/2021	
		T 43 QCA	21/02/2021	

2.5 *Digital Village Maps from MRSAC (Maharashtra Remote Sensing Application Centre)*

The digital village maps in vector form have been used for village-wise statistics generation.

2.6 *Methods*

The methodology adopted for analysis is described as fallows. Subset of Raigad district is taken out from the full scene of mosaic satellite image. Six satellite images are required for covering geographical area of such district. Subset is shown in Fig. 3 (Fig. 2).

2.7 *Supervised Classification*

Classification is done by using supervised classification technique in ERDAS IMAGINE software. The classes represented by pixel may be water bodies, forest, barren land, crop, vegetation urban or other land cover types.

Fig. 2 Mosaic image

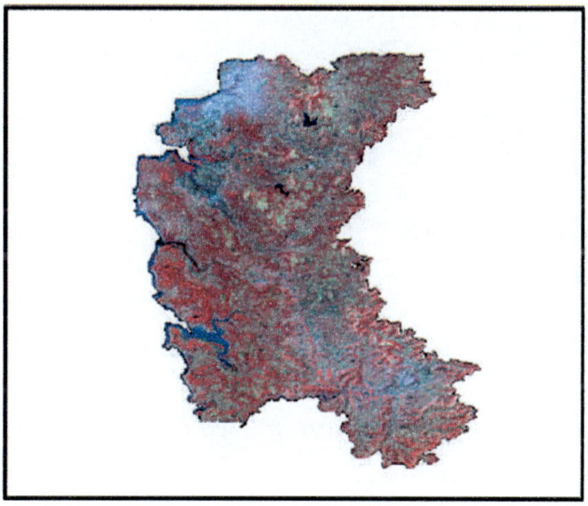

Fig. 3 Subset of Raigad district

2.8 Field Visit for Ground Truth Data Collection

For crop mapping, ground truth is very important. In ground truthing, various signature samples like, barren land, forest land, crop land, fallow land and vegetation are collected. Maximum signature samples give more accurate results. The field visit carried out from 27 to 29th January 2021 and collected signature sets are used for classification of image of Raigad district. The collected features are overlaid on the subset of satellite image as shown in Fig. 4, and supervised analysis of images is done.

Ground Truth photo 1

Ground Truth photo 2

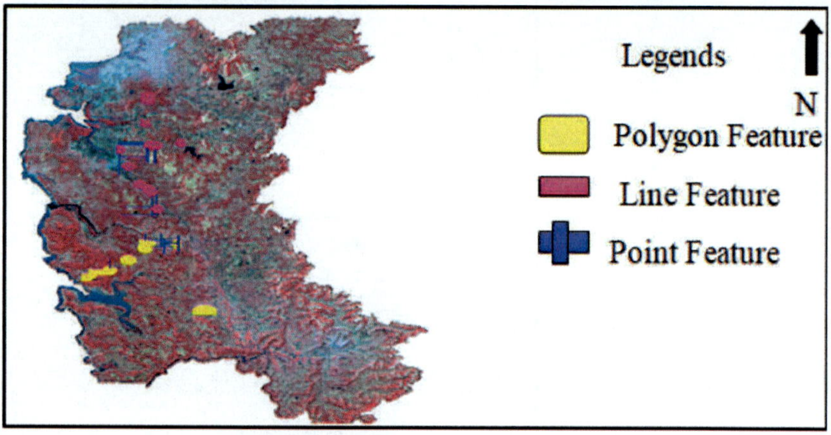

Fig. 4 Collected ground truth features

2.9 Conglomeration of Two Date Supervised Classified Images

From collection of ground truth samples, following shades are identified for the earth features in the area of interest: shades of pink-crop, cyan-barren, reddish brown-forest, gray-Fallow, mix pixel-urban, reddish pink-vegetation, blue-water in lakes and river. Then supervised classification is performed. First and second supervised classified image for standing crop is shown in Figs. 5 and 6, respectively.

MATRIX image is generated from two supervised classified images. Such image has 168 (14 × 12) probable unification of classes. Various class combinations are recoded and reduced to following seven classes. Class 1-Forest, 2-Crop, 3-Fallow, 4-Barren, 5-Water, 6-Vegetation, 7-Urban. Matrix image is shown in Fig. 7. Final recoded image of taluka-wise standing crop for Rabbi season, year 2021 is shown in Fig. 8.

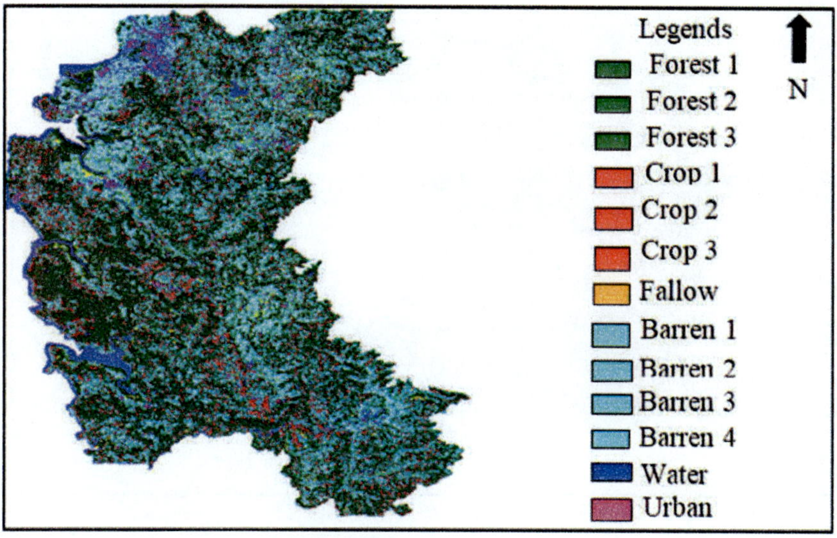

Fig. 5 First supervised classified image of Raigad district

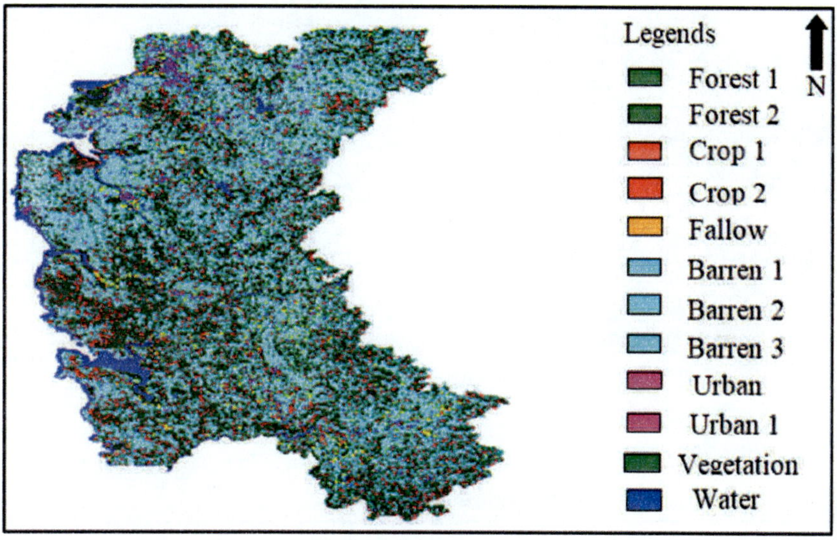

Fig. 6 Second supervised classified image of Raigad district

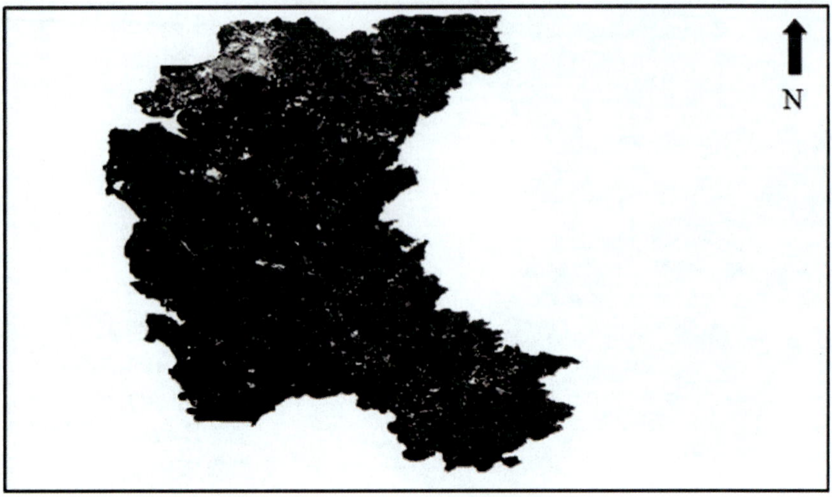

Fig. 7 Matrix image

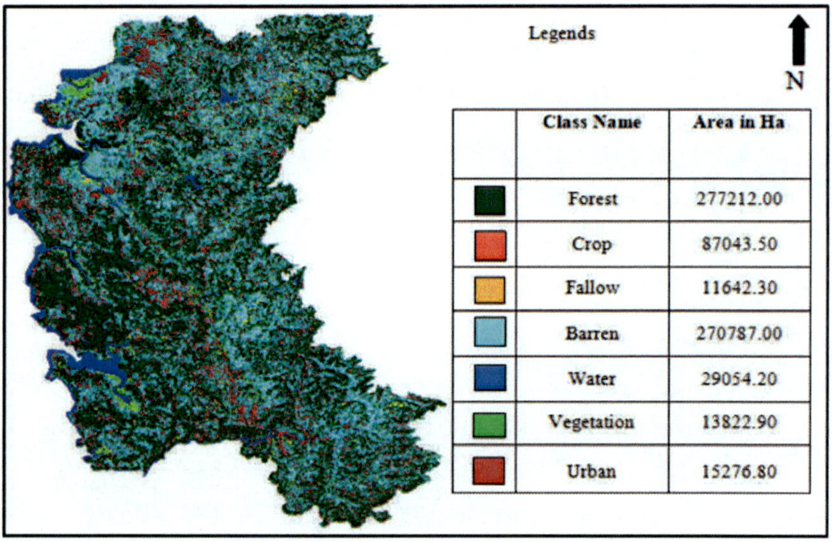

Fig. 8 Final recoded images

2.10 Creation of Area Statistics

After preparing recoded image, digital village map in vector form is superimposed on image. Taluka-wise standing crop area statistics is generated by using Summary

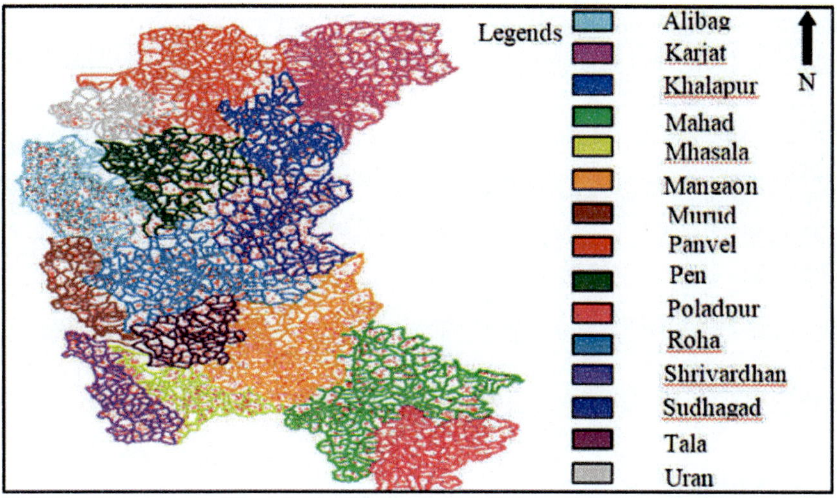

Fig. 9 Distribution of taluka-wise standing crop in Rabbi season, year 2021 of Raigad district

module in ERDAS Imagine 2010 classification software. Such distribution of Raigad district is shown in Fig. 9.

2.11 Accuracy Assessment

In order to assess the accuracy of the classification of final recode image, a following confusion matrix is generated and hence the accuracy of identification and measurement of crop in Rabbi season year 2021 for Raigad district is 67% and the overall accuracy of the supervised classification is 91.24%, which is shown in Table 2

$$\text{Total no. of samples} = 137$$

$$\text{Correct classified samples} = 125$$

$$\text{Overall Accuracy} = \frac{\text{Correct classified samples}}{\text{Total no. of samples}} = \frac{125}{137} = 91.24\%$$

Table 2 Confusion matrix ground truth reference data

Class	Crop	Fallow	Barren	Urban	Forest	Water	Vegetation	Total of row	Accuracy (%)
Crop	24	4	2	0	4	0	2	36	67
Fallow	0	13	0	0	0	0	0	13	100
Barren	0	0	31	0	0	0	0	31	100
Urban	0	0	0	10	0	0	0	10	100
Forest	0	0	0	0	10	0	0	10	100
Water	0	0	0	0	0	27	0	27	100
Vegetation	0	0	0	0	0	0	10	10	100
Total of column	24	17	33	10	14	27	12	137	91.24

3 Results and Discussion

By using remote sensing technique with sentinel satellite, two scene images, most accurate results of standing crop in Raigad district, are generated shown in tabular and graphical form in Table 3, Fig. 10, respectively.

Table 3 Abstract

S. No	Name of district	Name of taluka	No. of villages in taluka	Area of taluka in Raigad district (Ha)	Total standing crop area (Ha)	Percentage of standing crop
1		Alibaug	280	52,676.46	8416.93	15.98
2		Karjat	198	65,392.92	7937.03	12.14
3		Khalapur	146	40,826.74	4856.21	11.89
4		Mahad	187	81,665.97	9942.15	12.17
5		Mhasala	86	32,047.96	3697.4	11.54
6		Mangaon	187	68,502.23	10,130.3	14.79
7		Murud	90	26,282.17	3165.32	12.04
8		Panvel	200	60,455.38	5698.84	9.43
9		Pen	209	50,702.56	4971.51	9.81
10		Poladpur	89	36,580.76	4611.60	12.61
11		Roha	183	63,512.49	8238.07	12.97
12		Shrivardhan	79	26,092.45	3683.30	14.12
13		Sudhagad	105	46,582.32	6113.40	13.12
14		Tala	77	24,949.13	3421.08	13.71
15		Uran	73	20,794.30	1659.50	7.98
		Total	2189	697,063.84	86,542.64	12.42

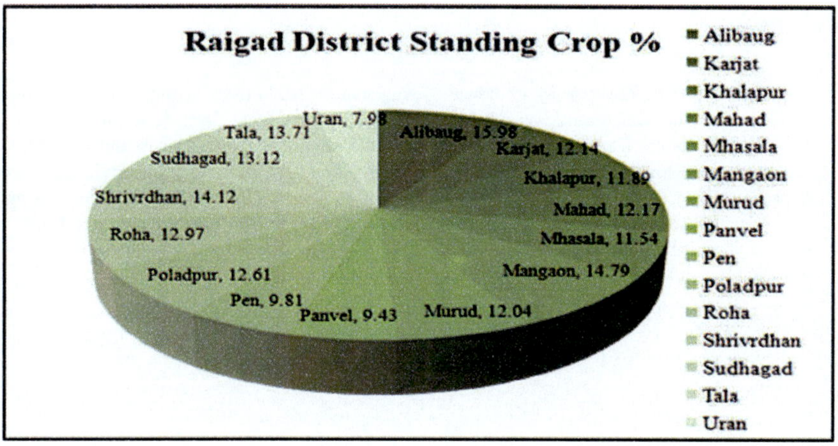

Fig. 10 Pie chart showing percentage of taluka-wise standing crop in Rabbi season, year 2021 of Raigad district

4 Conclusions

- The objective of the study is to identify measure, map the standing crops in Rabbi season, year 2021 in Raigad district and map taluka level area of standing crop. This objective has been achieved with 67% accuracy.
- The accuracy of particular study is affected due to extremely hilly area, uneven terrain and maximum forest, vegetation cover.
- It is to conclude that 12.42% of total geographical area of Raigad district is covered under standing crop for Rabbi season, year 2021.
- The creation of accurate and detailed crop maps requires high-quality ground truth and high-quality multi-temporal satellite data [1].
- The methodology adopted using the remote sensing technique with two scene images of Sentinel 2A are used to give fairly accurate results at village level [2] standing crop which has been confirmed in the field visit validation survey (Ground Truth). The village level database like area of barren land, area of fallow land, area of forest land, area of crop land and area of water can be used for periodical monitoring of land use activity.
- Remote sensing and geographic information system is the best tool for crop assessment. It is economical, cost-effective and less laborious. Remote sensing technique for assessment of standing crop in hilly area with forest cover gives reasonable accuracy; however, it is not possible to achieve accuracy around 95% due to extremely hilly terrain.

References

1. Kohirkar A, Tatu S, Kulkarni M, Gaikwad S, Kulkarni S (2021) Identification and measuring standing crops in rabbi season year 2021 of Raigad district. Using Satellite Remote Sensing Technique Resources Engineering Centre, Maharashtra Engineering Research Institutes, Nashik
2. Kohirkar A, Deshmukh S, Kulkarni M, Gaikwad S, Kuwar S (2020) Measuring village level grape crop under command area of Palked reservoir project. Using Satellite Remote Sensing Technique. Resources Engineering Centre, Maharashtra Engineering Research Institutes, Nashik
3. https://scihub.copernicus.eu
4. https://www.esa.int/Applications/Observing_the_Earth/Copernicus

A GIS-Based Multi Criteria Decision Making Technique for Groundwater Potential Zones of a Tropical River Basin, Northern Kerala, Southern India

N. P. Jesiya, M. V. Shyamkumar, and Girish Gopinath

Abstract The study encompasses the efficiency of the combination of geospatial and fuzzy-based multi-criteria decision-making (MCDM) techniques for evaluating the groundwater potential zones of Valapattanam River basin, Northern Kerala, India. The availability of groundwater (GW) mainly depends on parameters like geomorphology, geology, slope, lineament density, drainage density, soil texture, and land use land cover of the study area. First of all, the thematic data for each of these parameters was created with the help of various tools of the ArcGIS platform. As the second step, the fuzzy analytic hierarchy processes (FAHPs) were carried out by arranging the influencing parameters and their sub-criteria. Based on the assigned weights, the normalised weight of each parameter and its sub-criteria were identified. Finally, a composite groundwater potential zone (GWPZ) map was generated by weighted sum overlay analysis of thematic layers with normalised weights. According to the findings, the groundwater potential zones of the Valapattanam river basin are very good (5.3%), good (16.8%), moderate (22.5%), poor (34.7%), and very poor (20.5%). The GWPZ map was validated by using the data from India-WRIS and field data. The study also pointed out that the use of the MCDM technique with remote sensing and the GIS method has great significance in groundwater management practices.

Keywords Groundwater potential zone · Fuzzy · GIS · Valapattanam

---

N. P. Jesiya (✉) · M. V. Shyamkumar
Department of Environmental Studies, Kannur University Kannur, Kannur, Kerala, India
e-mail: jesynp@gmail.com

G. Gopinath
Department of Climate Variability and Aquatic Ecosystems, Kerala University of Fisheries and Ocean Studies (KUFOS), Kochi, Kerala, India

P. V. Timbadiya et al. (eds.), *Geospatial and Soft Computing Techniques*, Lecture Notes in Civil Engineering 339, https://doi.org/10.1007/978-981-99-1901-7_17

1 Introduction

Natural groundwater (GW) resources are the most valuable and finite source of fresh water in semi-arid and urban areas. According to the UN report, there are more than 2 billion people worldwide who depend upon GW as a major source of water for their needs like irrigation, industrial purposes, domestic usage, etc. [1]. However, increasing urbanisation, population growth, and the emergence of new water-consuming industrial sectors all place strain on GW resources. It also causes the water table to decline, springs and wells to dry up, and water chemistry to change negatively, among other effects [2–8]. Due to the lack of freshwater, approximately 0.6 million people in a semi-arid country like India were experiencing high to extremely high water stress. As a result, residents in those places must rely increasingly on groundwater resources to survive. According to a World Bank research, unless suitable efforts are made, India will become a water-stressed region by 2025 and a water-scarce region by 2050.

In view of the high-level threat to groundwater resources, valid evaluation, planning, and management have become critical and necessary phenomena. The use of remote sensing and GIS techniques allows for the rapid and cost-effective assessment of groundwater resources, which would otherwise be a very expensive, time-consuming, and laborious task [9–16]. The introduction and emergence of new technologies, and their benefits of spatial, spectral, and temporal data availability, have been found effective in gaining access to data about the factors that control the occurrence and movement of groundwater [17–19]. The application of GIS-based multi-criteria decision analysis (MCDA) is the ideal alternative for combining, analysing, and managing a wide range of geoenvironmental variables [20].

Geographical information system (GIS) techniques are particularly reliable in identifying GW prospects because they describe all important features within a spatial context and may combine data in a variety of ways [21–23]. Many hydrogeologists have turned to geospatial technology to complete their work because data is readily available and can be processed in a GIS environment to investigate various hydrogeological processes such as groundwater potential zonation, artificial recharge assessments, quality analysis, etc. with high accuracy [24–28].

The multi-criteria decision-making techniques like AHP, AHP/ANP, or fuzzy-based AHP can be used to analyse each of the geoenvironmental parameters, assessing the reliability of the result and so eliminating bias in the decision-making process. [29]. An integrated AHP-fuzzy model is more dependable than other MCDM techniques for calculating groundwater potential and vulnerability studies [8, 30–34]. It allows you to work with a large number of variables since it generates an aggregated index, handles linguistic attributes, and treats draft data without losing the vital parts of the concepts [35–38]. The integrated fuzzy AHP with GIS approach makes it feasible to evaluate the hydrological response of regions with various forms of development by including hydrogeological and anthropogenic data [8]. Furthermore, by supporting water managers in the selection of potential areas, the use of this integrated method can lower the cost of recharge operations and water management studies. The present work attempted to appraise the groundwater potential zones in the Valapattanam River basin, a tropical river basin located in northern Kerala, using the state-of-the-art GIS-based fuzzy AHP method.

2 Study Area

The Valapattanam River, which is located in a tropical monsoon climate, covers 44% of this district. The river originates from the Brahmagiri hills of Western Ghats, Kodagu, with an elevation of 1350 m. By length, Valapattanam River is the 9th longest river of Kerala, and in the case of quantum of water resources it takes 4th position. The basin located between north latitudes of 11° 49′ 30" and 12° 13′ 50″ and east longitudes of 75° 58′ 55″ and 75° 17′ 22″. The main stream has a length about 110 km and catchment area about 1900 km^2. The drainage area of the river in Kerala is 1321 km^2 (Fig. 1).

The major tributaries are Vallithodu, Aralam, Bavali, Iritty, Sreekandapuram, and Katampallipuzha. Valapattanam River is a major source of irrigation in the

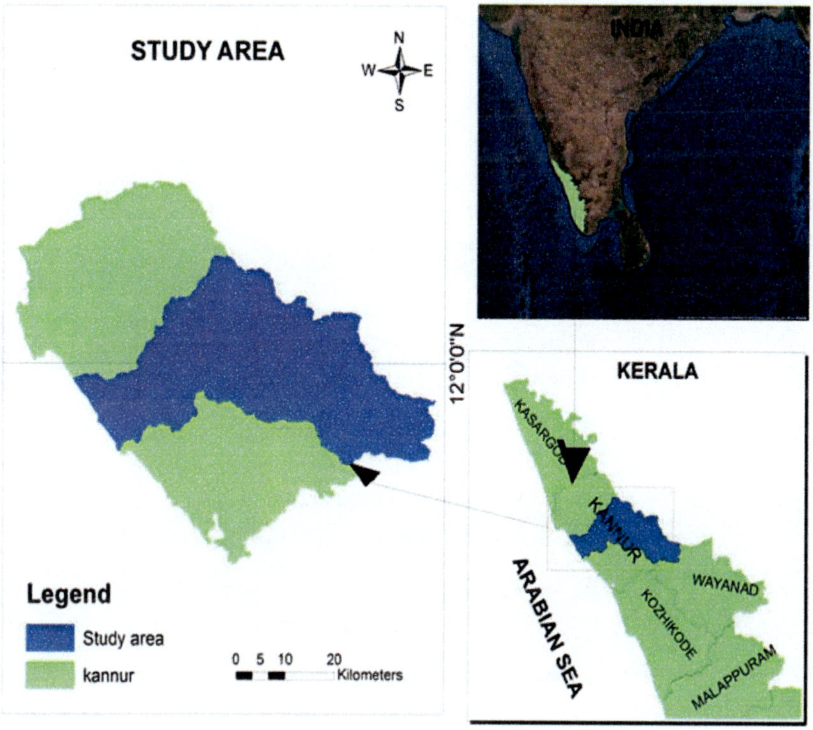

Fig. 1 Study area map of Valapattanam River basin

district (Pazhasi dam), and many wood-based industries are situated in its banks. Average annual stream flow (computed) is 4779 MCM, and average annual rainfall is 3600 mm. Water requirement for three main crops is 240 Mm^3 (ENVIS). As per the ENVIS report, water requirement for domestic use is 82$Mm^3$and that of water requirement for industrial use is 90 Mm^3.

3 Materials and Methods

The study used an integrated method using remote sensing and a GIS-based FAHPs to analyse the groundwater (GW) potential zones. The groundwater potential zonation (GWPZ) was achieved through four successive phases, including data collection, thematic layer preparation using Arc-GIS 10.8, deriving numerical index using FAHP, and the spatial and non-spatial integration of these data using ArcGIS 10.8 (Fig. 2).

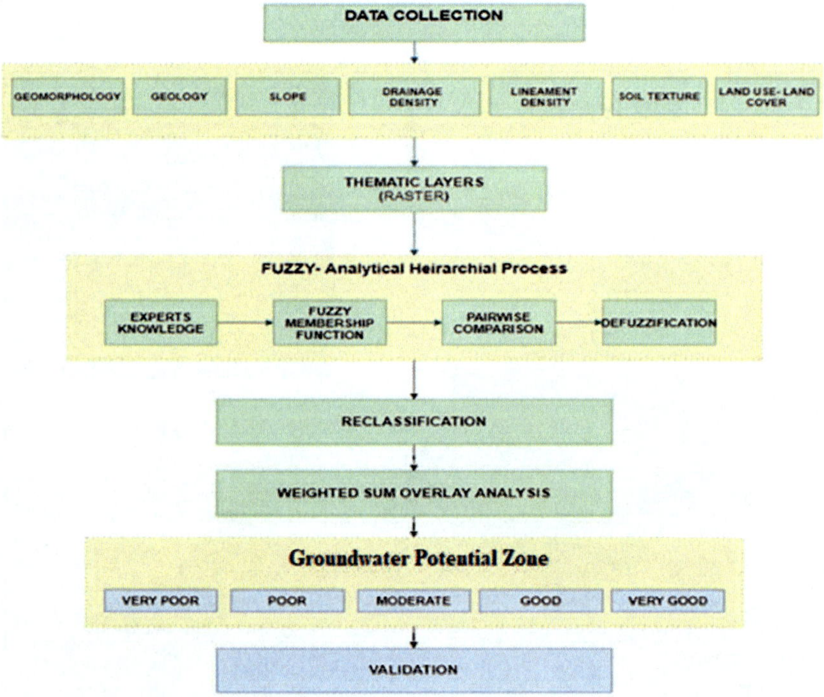

Fig. 2 Flow chart showing the methodology of groundwater potential mapping, using MCDM technique

3.1 Field Investigations and Data Collection

Valapattanam watershed boundary was delineated based on the data obtained from CWRDM-Water atlas [39]. A thorough field visit was conducted to study the environmental and geological aspects of the area. Groundwater inventory was carried out to collect hydrogeological parameters like depth to water table, total depth of well, and other primary information. Further, the collected groundwater data were used for the ground truthing of secondary water level data.

3.2 Preparation of Thematic Maps of Influencing Factors

Thematic data for geology and soil texture were developed in ArcGIS 10.8 utilising various geoprocessing features, including georeferencing, digitisation, and so on. Landuse and land cover classes of the study area were generated from IRS-P6, LISS-III data using the Image Classification tool in ArcGIS 10.8. Drainage patterns were extracted from a 1:50,000 scale survey of India's topographical maps, and lineaments were derived using the Bhuvan web portal's thematic services. In addition, point density analysis was used to generate lineament and drainage density data (Km/Km^2). Slope (in %) analysis was performed with SRTM DEM (30 m) data, which was obtained from Earth Explorer (https://earthexplorer.usgs.gov/dd) and employed surface analysis using the spatial analyst tool.

3.3 Deriving Numerical Index Using Fuzzy AHP Method

The numerical potential index is a dimensionless quantity determined by ratings and weights of influencing factors generated through the fuzzy AHP. As a first phase, each parameter was assigned a subjective score of 1–9 rely on expert opinion and scientific data. Components with the least potential are given a ranking of one, while those with the most potential are given a ranking of nine. A fuzzy AHP pairwise comparison scale is conducted using triangular fuzzy integers, and a fuzzy evaluation matrix is created for this comparison [8]. The comparison thus derived the vectors of weights for the seven major criteria and its sub-criteria. And as the output of the analysis, the normalised weightages of each parameters are displayed in Table 1.

Table 1 Rating and criteria weights evaluated after fuzzy AHP analysis

Criteria	Assigned weight	Normalised weight	Sub-criteria	Score	Normalised weight
Geomorphology	9	0.22			
			Water body	9	0.16
			Marshy	8	0.14
			Floodplain	8	0.14
			Young coastal plain	7	0.12
			Valley	7	0.12
			Pediment zone	5	0.07
			Valley fill	5	0.07
			Lower plateau (Lateritic) Dissected	3	0.05
			Residual hills	2	0.04
			Residual mount (pediment)	2	0.04
			Denudation hills	1	0.03
			Rock exposure	1	0.03
Slope	8	0.18	< 5	9	0.38
			5–10	7	0.26
			10–15	5	0.17
			15–20	3	0.10
			> 20	2	0.09
Drainage density	6	0.12	< 1	9	0.29
			1–2	8	0.24
			2–3	7	0.19
			3–4	6	0.15
			> 4	5	0.13
Lineament density	6	0.12	> 2	9	0.29
			1.5–2	8	0.24
			1–1.5	7	0.19
			0.5–1	6	0.15
			0–0.5	5	0.13
Geology	5	0.10	Fluvial costal alluvium	9	0.38
			Warkali formation	7	0.25

(continued)

Table 1 (continued)

Criteria	Assigned weight	Normalised weight	Sub-criteria	Score	Normalised weight
			Hornblende biotite gneisses	5	0.16
			Charnockite	3	0.11
			Metavolcanoes	3	0.11
Soil texture	4	0.08	Gravelly loam	7	0.49
			Gravelly clay	5	0.31
			Clay	3	0.20
Land use/land cover	3	0.06	Vegetation	9	0.39
			Water body	9	0.39
			Others	4	0.13
			Built-up area	2	0.09

3.4 Integration of Spatial and Non-spatial Data

The thematic maps were combined with non-spatial data (ratings and weights) derived from the fuzzy AHP analysis. Using the conversion tool in ArcGIS, all thematic layers were converted to raster format. Using reclassification and raster editing methods, the weightages derived from fuzzy AHP were added to the thematic layers (raster). Each of the integrated thematic layers was combined using weighted overlay analysis within the GIS platform, resulting in the spatial distribution of GW potential zones (GWPZ) for the study area.

3.5 Validation Analysis

Groundwater data were collected from CGWB for the pre- and post-monsoon seasons in the year 2019, and the groundwater fluctuation in metres was calculated. The calculated GW fluctuation data were combined with the resultant GWPZ data in order to perform validation analysis. The validation analysis was enabled by the spatial relationship between the shape file of GW fluctuation data and the GWPZ raster data.

4 Result and Discussion

The evaluation of groundwater potential in the Valapattanam River basin based on the advanced technologies point outs to the necessity to predict the potential of GW level of the area. The relative influences of geoenvironmental factors of the study area in the potentiality of GW were determined by a potential index modelling with the help of fuzzy AHP analysis. Importance of each criterion in groundwater occurrence and its analysis result was discussed below.

4.1 Groundwater Influencing Factors

4.1.1 Geology

Geology has a vital role in groundwater occurrence and distribution in any terrain. The Valapattanam River basin is underlain by charnockites, pyroxene granulites, garnetiferous gneisses, hornblende biotite gneisses, and schistose rocks, overlain by tertiaries and coastal alluvium along the coast, ranging in age from Archean to Recent [40]. Among the geological formations, the fluvial coastal sediments are assigned the relatively highest units for groundwater occurrence. Consolidated formations, viz. weathered and fractured crystallines, semi-consolidated sediments, and laterite formations and unconsolidated formations such as recent alluvium occurring along the coast are the hydrogeological units in the river basin. The weathered and fractured rocks of the consolidated formation are mostly made of charnockites, hornblende gneisses, schist, and other intrusive and constitute potential phreatic aquifers [40]. Charnockite formations are assigned a relatively low weightage because of its low degree of weathering.

The gneissic rocks are assigned moderate weightage towards groundwater potential due to highly weathered and well-jointed gneissic rocks. These formations characterised by good water potential zones with a well yield of 10–20 m^3/day [40] (Fig. 3a). The porous laterites recharge quickly, and the recharge water leaves as subsurface flow, particularly in wells on topographic highs and steep slopes. As a result, even though these formations constitute a potential aquifer in the midland regions, laterites are assigned the lowest rating. The coastal alluvium, which consists of sand, silt, and clay, has the potential to form phreatic aquifers in the area. Therefore, these geological formations are assigned the highest rating for groundwater potential.

4.1.2 Soil Texture

Soil texture influences the volume of water which can infiltrate into subsurface formations and, as a result, the rate of water infiltration to the ground. The four classes of soil

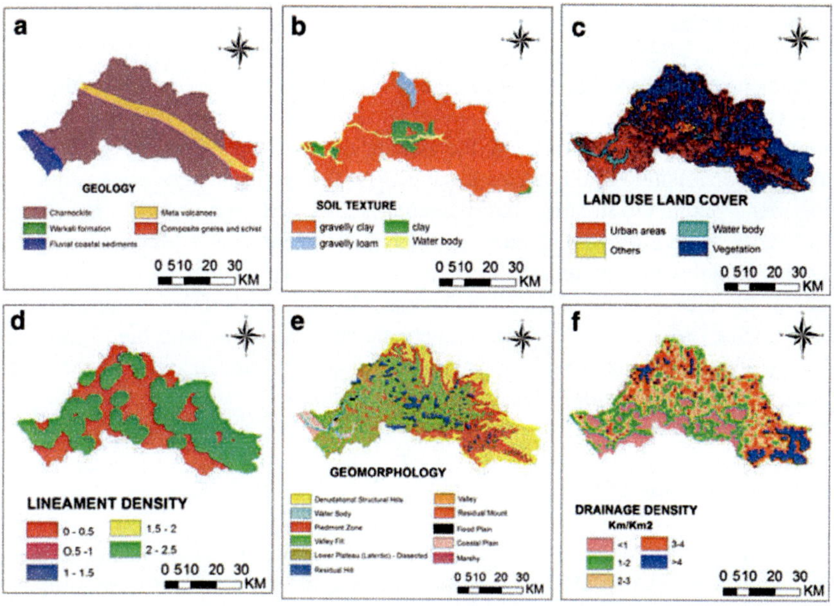

Fig. 3 Spatial distribution of **a** geology, **b** soil texture, **c** landuse/landcover, **d** lineament density, **e** geomorphology, **f** drainage density of Valapattanam River basin

occurred in the area are lateritic soil, coastal and river alluvium, brown hydromorphic soil, and forest loamy soil. The soil texture in the area is found with three kinds of soil texture: clay, gravelly clay, and gravelly loam (Fig. 3b). The predominant soil type in the Valapattanam River basin is lateritic soil, which is a weathered product derived under humid tropical conditions. Soil texture in lateritic soil ranges from sandy loam to red loam. The coastal alluvium observed in the western coastal stretch is made up of recent, mostly marine deposits. It is distinguished by being immature and having a high sand content. River alluvium with surface textures varied from sandy loam to clay is found along river valleys that cut through extensive lateritic soils. The river alluvium is characterised by being fertile, having water holding capacity.

4.1.3 Land Use/land Cover

The majority of the land in the Valapattanam River basin is used for agriculture, with urban areas ranking second [42]. The LU/LC data gives important information on moisture content of soil, infiltration, occurrence of groundwater, etc. The urban land is composed of built-up areas (commercial and residential classes) (Fig. 3c). Built-up and fallow lands inhibit groundwater recharge and are assigned the lowest rating, whereas water bodies were assigned with a highest rating towards groundwater potential.

4.1.4 Lineament Density

Lineaments denote the linear features, developed by tectonic activity. Through this major portion of water flows into the impermeable rocks. Based on the distribution in a single grid of (km/km^2), the density of lineament fractures in the study area can be classified into five classes (Fig. 3d). The intersection of lineaments and lineaments parallel to stream network regions is evidence of groundwater movement and storage. Therefore, delineation and analysis of lineaments in a hydrogeological regime provides information on the groundwater zones of that region. The presence of the intersections in the high lineament density zones favours high groundwater recharge, hence high groundwater potential. The areas with high lineament density are characterised by high groundwater potential, and 63.03% of the study area covered high lineament density zones.

4.1.5 Geomorphology

Geomorphic characteristics can be used as surface indicators to evaluate subsurface hydrologic status [19]. The three physiographic units of the study area are the coastal plains and lowlands in the west, the central undulatory midland terrain, and the high land region in the east [40]. Geomorphically, the area was classified as residual hill, lower plateau, valley fill, piedmont zone, valley, denudational structural hills, marshy areas, denudational hills, coastal plain, floodplain, and water body. The narrow coastal plains composed of alluvial deposits located parallel to the coast, and it covers an area of 2.72% of the study area. In some locations, the midland region adjacent to the east of the coastal strip constitutes a plateau land covered by a thick cover of laterite. The denudation structural hills located in the hilly tract of the eastern part makes highly rugged terrains. The residual hills and denudation hills together impart least influence on groundwater occurrence in the area. Good groundwater potential zones were recognised in valley fills that ran through the lateritic plateau. The valley fills occupied 9.70% of the area. Lateritic terrains with the thickness of ranges from 10 to 20 m were assigned with moderate potential to groundwater (Figs. 3e and 4).

Dinesh Kumar et al. [41] were studied on groundwater potential of Muvattupuzha river basin, Kerala and explained that residual hills are poor in groundwater occurrence. Hence, residual hills were assigned with least rating.

4.1.6 Slope %

Valapattanam River basin classified in to five classes based on the degree of steepness, viz. < 5%, 5–10%, 10–15%, 15–20%, and > 25% classes. Flat-to-gentle slope terrain (i.e. 0–5%) is categorised as a very good groundwater potential zone because slow surface water runoff through the terrain permits more residents time for rainwater to percolate and increases the rate of infiltration. About 11.72% of the area

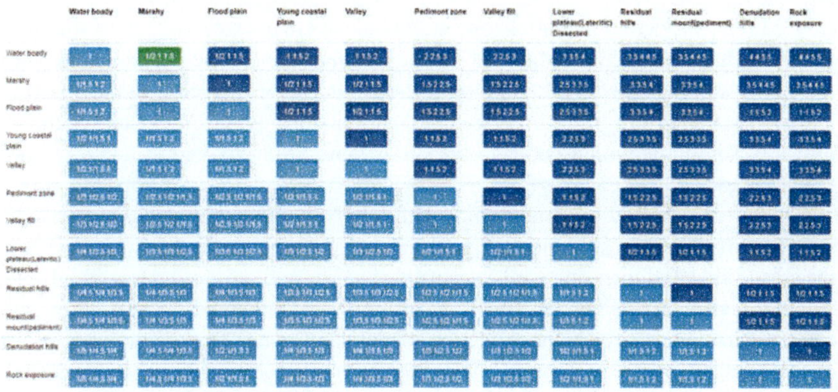

Fig. 4 Comparison matrix of geomorphology of the study area

is characterised by gentle slope terrain. Rolling lateritic hills and valley fills in the area with 5–15% and 15–25% slope terrain are characterised by a moderate infiltration rate, hence assigned a moderate rating in groundwater potential. The eastern part of the river basin consists of plateau edges and high mountain regions that have steep slopes (> 25%). This steep slope zone is characterised by poor potential for groundwater occurrence due to low infiltration rate and high surface runoff.

4.1.7 Drainage Density

The closeness of spacing of channels is expressed as drainage density, and it is expressed as length of drainage within a square grid of the area in terms of km/km^2. About five classes of drainage density zones are occurred in the Valapattanam River basin (Fig. 3f). The areas with high drainage density (> 4 km/Km^2) not suitable for groundwater potential because of the higher surface runoff. Therefore, the study area with high drainage density is assigned with least rating for groundwater potential and vice versa. In Valapattanam River basin, majority of the areas were characterised with 1–3 km/Km^2 drainage density. The less drainage density zones hinder surface runoff which consequently enhances infiltration and thereby favours groundwater recharge [23]. The intermediate drainage density classes are assigned with moderate ratings towards groundwater recharge.

4.2 Groundwater Potential Zones in Valapattanam River Basin

The GIS-based fuzzy-AHP approach delineated the Valapattanam River basin into five groundwater potential zones: very good, good, moderate, poor, and very poor

(Fig. 5). The very good potential zone mainly occurs in the western part of the study area, which covers 5.3% of the study area. The very good zone is characterised by the gently sloping coastal alluvium with low drainage density. Soil in the coastline and riverine alluvium has a high water retention capacity. Groundwater occurs in phreatic conditions with a depth to the water table of 0.5 to 5 m in the very good potential zone.

The laterite-capped midland region with relatively moderate groundwater influence factors constitutes a moderate groundwater potential zone. It covers 22.5% of the area (Table 2). Open-dug wells, according to C.G.W.B., are suitable groundwater extraction structures where the depth to groundwater is between 5 and 20 m. The eastern part comprises a poor to very poor potential zone. The area covers 55.2% of the Valapattanam River basin and consists of steep terrain, weathered rock, and high drainage density zones. Some terrains, particularly fracture planes in the east, contribute to potential groundwater zones capable of supporting bore wells [40].

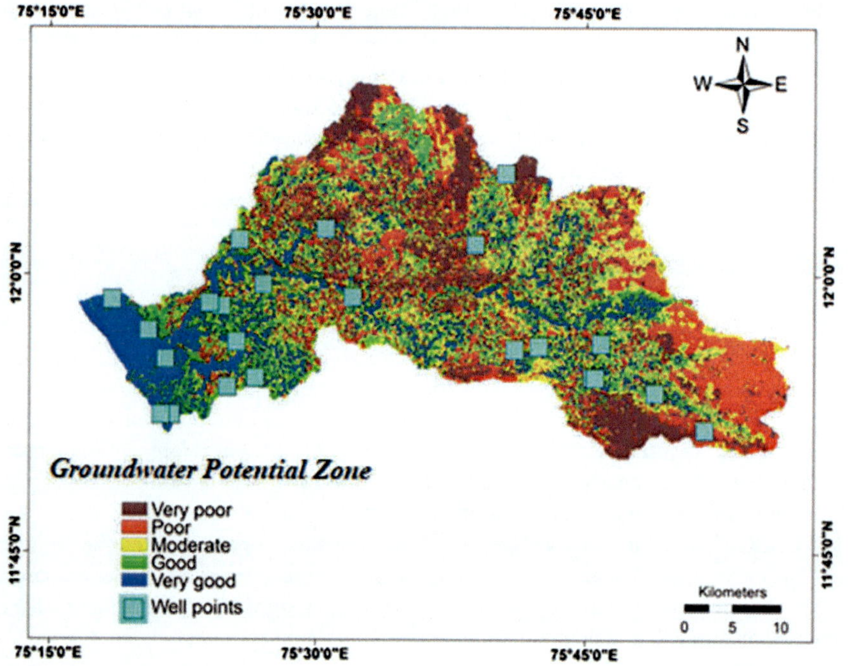

Fig. 5 Spatial distribution of groundwater potential zones in the Valapattanam River basin

Table 2 Groundwater potential zones in the Valapattanam River basin

S. No.	Ground water potential	Area in km^2	Area in %
1	Very poor	269.0	20.5
2	Poor	454.0	34.7
3	Moderate	294.8	22.5
4	Good	220.0	16.8
5	Very good	69.8	5.3

4.3 Validation of GWPZ with Available Well Data as Ground Validation

For the validation analysis, data from 23 dug wells in the Valapattanam River basin were used. The depth to water level varies from 2.6 (along the coastal area) to 22.5 m below ground level (mbgl). The validation of the outcome (GWPZs) from the study with field observation from dug wells in the Valapattanam River basin revealed that very good GWPZs are occupying the coastal area, which has a very good alluvial cover, whereas the eastern part has very little soil cover and is predominately made of hard rocks and occupies poor GWPZs. The GWPZ, derived logically from the fuzzy AHP model, has been validated with water level fluctuation data (in metres). In the validation analysis, the spatial relationship between well points and potential classes revealed that wells in very good potential zones show comparatively less fluctuation (0.7 m), indicating high groundwater yield in the area (Table 3), whereas average water level fluctuations in wells from good and moderate potential zones are 0.3 and 0.67 m, respectively. Moreover, most of the dug wells in the eastern part get dry during the summer, and these aquifers are recharged only during the monsoon. On the other hand, wells in the central and western parts of the country are perennial throughout the year. Average water level fluctuations in wells from poor and very poor potential zones are 1.6 and 1.7 m, respectively.

Table 3 Statistics of validation analysis between GWPZ classes and GW fluctuation data

Potential class	Minimum	Maximum	Average groundwater fluctuation level (m)	Std. Dev
Very good	– 1.6	2.01	0.07	0.9
Good	0.2	5.7	0.3	0.7
Moderate	0.1	1.9	0.67	0.9
Poor	0	1.1	1.6	0.5
Very poor	0	0.07	1.7	0.5

5 Conclusions

Groundwater potential analysis of the Valapattanam River basin, Kerala, with the help of remote sensing and GIS technology combined with the FAHP technique perfectly demonstrated the efficiency and reliability of the MCDM-geospatial techniques in groundwater resource management. The groundwater controlling factors and their sub-criteria were assigned a weight (on a 1–9 scale) based on their importance in hosting groundwater occurrences. The normalised weights were generated by pair-wise comparisons of features and sub features using fuzzy AHP analysis. Besides, fuzzy AHP is an excellent tool that can be used in group decision-making to eliminate ambiguity, imprecision, and uncertainty in the comparison analysis. Fuzzy weigh-tages were integrated with the particular GW controlling features, and the resulting groundwater potential zone map was generated. The Valapattanam River basin is divided into five groundwater potential zones. The very good potential zone, covering 5.3% of the area, mainly occurred in the western part of the study area. The zone is characterised by gently sloping coastal alluvium with low drainage density. The midland region of the study area constitutes a moderate groundwater potential zone, and it covers 22.5% of the area. The zone of poor to very poor potential covers 55.2% of the Valapattanam River basin and consists of steep terrain, weathered rock, and high drainage density zones. The present study proved that the MCDM methodology, geostatistical modelling, and its application in geospatial layer preparation have a great role in getting a precise and reliable picture of the current groundwater condition of the Valapattanam River basin. Furthermore, this analysis will help formulate a long-term sustainable use plan for groundwater conservation.

Acknowledgements The authors are grateful to the Department of Environmental Studies, Kannur University Campus, Mangattuparmba, Kannur, Kerala, India, for providing facilities and guidance in the preparation of the manuscript.

References

1. United Nations (2015) World population prospects. Report No. ESA/P/WP.241, United Nations, New York 59
2. Foster SSD, Morris BL, Chilton PJ (1999) Groundwater in urban development-a review Howardof linkages and concerns. In: Proceedings of impacts of urban growth on surface water and groundwater quality, IUGG 99 Symposium HS5, IAHS Publ, Birmingham vol 259, pp 3–12
3. Howard KWF, Gelo KK (2002) Intensive groundwater use in urban areas: the case of megacities. Intensive Groundwater Challenges Opportunities 484
4. Kløve B, AlaAho P, Bertrand G, Gurdak JJ, Kupfersberger H, Kværner J, Muotka T, Mykra H, Preda E, Rossi P, Uvo CB, Velasco E, Pulido- Velazquez M (2014) Climate change impacts on groundwater and dependent ecosystems. J Hydrol 518:250–266
5. Boughariou E, Allouche N, Jmal I, Mokadem N, Ayed B, Hajji S, Khanfir H, Bouri S (2018) Modeling aquifer behaviour under climate change and high consumption: case study of the Sfax region, southeast Tunisia. J Afr Earth Sci 141:118–129

6. Le Brocque AF, Kath J, Reardon-Smith K (2018) Chronic groundwater decline: a multi-decadal analysis of groundwater trends under extreme climate cycles. J Hydrol 561:976–986
7. Howard K, Gerber R (2018) Impacts of urban areas and urban growth on groundwater in the Great Lakes Basin of North America. J Great Lake Res 44(1):1–13
8. Jesiya NP, Gopinath G (2020) A fuzzy based MCDM–GIS framework to evaluate groundwater potential index for sustainable groundwater management—a case study in an urban-periurban ensemble, southern India. Groundwater Sustain Dev 11
9. Moore ID, Grayson RB, Ladson AR (1991) Digital terrain modelling: a review of hydrological, geomorphological, and biological applications. Hydrol Process 5(1):3–30
10. Krishnamurthy J, Mani A, Jayaraman V, Manivel M (2000) Groundwater resources development in hard rock terrain: an approach using remote sensing and GIS techniques. Int J Appl Earth Obs Geoinf 3(3–4):204–215
11. Jha MK, Chowdary VM, Chowdhury A (2010) Groundwater assessment in Salboni Block, West Bengal (India) using remote sensing, geographical information system and multi-criteria decision analysis techniques. Hydrogeol J 18:1713–1728
12. Arkoprovo B, Adarsa J, Prakash SS (2012) Delineation of ground- water potential zones using satellite remote sensing and geographic information system techniques: a case study from Ganjam district, Orissa India. Res J Recent Sci 1(9):59–66
13. Hammouri NA, El-Naqa A, Barakat M (2012) an integrated approach to groundwater exploration using remote sensing and geographic information system. J Water Resour Prot 4(9):717–724
14. Lee S, Kim YS, Oh HJ (2012) Application of a weights-of-evidence method and GIS to regional groundwater productivity potential mapping. J Environ Manag 96(1):91–105. https://doi.org/10.1016/j.jenvman.2011.09.016
15. Davoodi MD, Rezaei M, Pourghasemi HR, Pourtaghi ZS, Pradhan B (2015) Groundwater spring potential mapping using bivariate statistical model and GIS in the Taleghan watershed Iran. Arab J Geosci 8(2):913–929
16. Thapa R, Gupta S, Guin S, Kaur H (2017) Assessment of groundwater potential zones using multi-influencing factor (MIF) and GIS: a case study from Birbhum district West Bengal. Appl Water Sci 7(7):4117–4131
17. Bobba AG, Bukata RP, Jerome JH (1992) Digitally processed satellite data as a tool in detecting potential groundwater flow systems. J Hydrol 131(1–4):25–62
18. Meijerink AMJ (2000) Groundwater In: Schultz GA, Engman ET (eds) Remote sensing in hydrology and water management. Springer, Berlin, pp 305–325
19. Preeja KR, Joseph S, Thomas J, Vijith H (2011) Identification of groundwater potential zones of a Tropical River Basin (Kerala, India) using remote sensing and GIS techniques. J Indian Soc Remote Sens 39(1):83–94
20. Saidi S, Hosni S, Mannai H, Jelassi F, Bouri S, Anselme B (2017) GIS-based multi-criteria analysis and vulnerability method for the potential groundwater recharge delineation, case study of Manouba phreatic aquifer, NE Tunisia. Environ Earth Sci 76(15)
21. Nair HC, Padmalal D, Joseph A, Vinod PG (2017) Delineation of groundwater potential zones in river basins using geospatial tools—an example from Southern Western Ghats, Kerala, India. J Geovisualization Spat Anal 1:1–2
22. Swetha TV, Gopinath G, Thrivikramji KP, Jesiya NP (2017) Geospatial and MCDM tool mix for identification of potential groundwater prospects in a tropical river basin Kerala. Environ Earth Sci 76(12):1–17
23. Jesiya NP, Gopinath G (2019) A Customized FuzzyAHP - GIS based DRASTIC-L model for intrinsic groundwater vulnerability assessment of urban and Peri urban phreatic aquifer clusters. Groundw Sustain Dev 8:654–666
24. Gumma MK, Pavelic P (2013) Mapping of groundwater potential zones across Ghana using remote sensing, geographic information systems, and spatial modeling. Environ Monit Assess 185(4):3561–3579
25. Murthy KSR (2000) Groundwater potential in a semi-arid region of Andhra Pradesh-a geographical information system approach. Int J Remote Sens 21(9):1867–1884

26. Lone RMM, Ahmed S (2011) Integrating geospatial and ground geophysical information as guidelines for groundwater potential zones in hard rock terrains of south India. Environ Monit Assess 184:4829–4839
27. Senthil Kumar GR, Shankar K (2014) Assessment of groundwater potential zones using GIS. Front Geosci 2(1):1–10
28. Raj BS (2019) Groundwater potential zonation of Delampady Grama Panchayath, Kasaragod, Northern Kerala: a geophysical and GIS approach. Int J Res Appl Sci Eng Technol 7(5):1919–1929
29. Arulbalaji P, Padmalal D, Sreelash K (2019) GIS and AHP Techniques based delineation of groundwater potential zones: a case study from Southern Western Ghats India. Sci Rep 9(1):1–18
30. Kumar A, Krishna AP (2018) Assessment of groundwater potential zones in coal mining impacted hard-rock terrain of India by integrating geospatial and analytic hierarchy process (AHP) approach. Geocarto Int 33(2):105–129
31. Şener E, Şener Ş, Davraz A (2018) Groundwater potential mapping by combining fuzzy-analytic hierarchy process and GIS in Beyşehir Lake Basin, Turkey. Arab J Geosci 11:1–21
32. Chaudhry AK, Kumar K, Alam MA (2019) Mapping of groundwater potential zones using the fuzzy analytic hierarchy process and geospatial technique. Geocarto Int 1–22
33. Rajasekhar M, Raju GS, Sreenivasulu Y, Raju RS (2019) Delineation of groundwater potential zones in semi-arid region of Jilledubanderu river basin, Anantapur district, Andhra Pradesh, India using fuzzy logic, AHP and integrated fuzzy-AHP approaches. Hydro Res 2:97–108
34. Bhadran A, Girishbai D, Jesiya NP, Gopinath G, Krishnan RG, Vijesh, VK (2022) A GIS based fuzzy-AHP for delineating groundwater potential zones in tropical river basin, southern part of India. Geosyst Geoenvironment 100093
35. Seekao C, Pharino C (2016) Assessment of the flood vulnerability of shrimp farms using a multicriteria evaluation and GIS: a case study in the Bangpakong Sub-Basin Thailand. Environ Earth Sci 75:308
36. Rebolledo B, Gil A, Flotas X, Sanchez JA (2016) Assessment of groundwater vulnerability to nitrates from agricultural sources using a GIS-compatible logic multi-criteria model. J Environ Manage 171:70–80
37. Fernandez DS, Lutz MA (2010) Urban flood hazard zoning in Tucuman Province, Argentina, using GIS and multicriteria decision analysis. EngGeol 111:90–98
38. Aydi A (2018) Evaluation of groundwater vulnerability to pollution using a GIS-based multi-criteria decision analysis. Groundw Sustain Dev 7:204–211
39. Water Atlas of Kerala (1995) Centre for water resources development and management
40. Santhana Subramani M (2013) Ground water information booklet of Kannur district, Kerala state technical reports: series 'D'. Cent Ground Water Board, Gov India, Ministry Water Resou 1–30
41. Dinesh Kumar PK, Gopinath G, Seralathan P (2007) Application of remote sensing and GIS for the demarcation of groundwater potential zones of a river basin in Kerala, southwest coast of India. Int J Remote Sens 28(24):5583–5601
42. Jyothirmayi P, Sukumar B (2019) Role of relief and slope in agricultural land use: a case study in Valapattanam River Basin in Kannur district, Kerala using GIS and remote sensing. J Geogr Environ Earth Sci Int 21(2):1–11

Exploring Geospatial Technology in Kadiri Basin of Ananthapuramu District, A.P. for Demarcation of GWPZ and Identification of Recharge Structures

P. P. Chowdary, S. Kumar, S. Kumar, V. G. K. Villuri, and P. Srinivas

Abstract Geospatial technology (GT) has played a crucial role in identification of groundwater potential zones (GWPZ). Weighted overlay analysis (WOA) is a multicriterion study for the GWPZ under the umbrella of GT wherein investigation was carried out with multifaceted things for determining certain themes with the aid of assigning rank to the respective features class and then assign weightage to the respective parameters depending upon the weightage of the theme on the objective. For this purpose, criteria for the analysis were defined, and each parameter was assigned weightage based on its importance. In the present study, weighted overlay model in GIS environment (ArcGIS software) has been utilized to identify and demarcate the suitability for groundwater recharge zones in Kadiri basin of Ananthapuramu district, Andhra Pradesh, which was explored further for suitable recharge structures. Integration of various thematic layers was done for developing groundwater potential zones map of the study area which has four categories, i.e. poor, average, good and excellent GWPZ, respectively. Multiple thematic layers of influencing parameters were prepared and assigned features class rank as per the importance in the selection of recharge sites. Using this suitability modelling, suitable areas were identified wherein the classes with higher values indicate the most favourable zones for natural recharge in GIS platform and generated a composite map

P. P. Chowdary · S. Kumar · V. G. K. Villuri
Department of Mining Engineering, Indian Institute of Technology (ISM), Dhanbad, Dhanbad, Jharkhand 826004, India
e-mail: ppc.18DP000363@me.iitism.ac.in

S. Kumar
e-mail: satish.19pr0049@me.iitism.ac.in

V. G. K. Villuri
e-mail: vgkvilluri@iitism.ac.in

S. Kumar · P. Srinivas (✉)
Department of Civil Engineering, Indian Institute of Technology (ISM), Dhanbad, Dhanbad, Jharkhand 826004, India
e-mail: srinivas@iitism.ac.in

S. Kumar
e-mail: sachidanand.17dr000533@cve.ism.ac.in

P. V. Timbadiya et al. (eds.), *Geospatial and Soft Computing Techniques*, Lecture Notes in Civil Engineering 339, https://doi.org/10.1007/978-981-99-1901-7_18

showing proposed locations for suitable groundwater recharge structures like check dams, percolation tanks, subsurface dykes and gabion structures.

Keywords Remote sensing · GIS · Weighted overlay analysis · Kadiri basin · GWPZ map

1 Introduction

Indiscriminate exploitation of surface and subsurface water has led to severe water scarcity and environmental degradation. The spatial-temporal variation in rainfall has further aggravated the problem. To meet the challenges of the scarcity, increasing demand and depletion of groundwater levels, water resources should be developed and managed in an effective manner. Kadiri basin in Ananthapuramu region, Andhra Pradesh, India, is drought-prone because the entire Ananthapuramu district falls under rain-shadow region of A.P. The area receives scanty rainfall and has no major irrigation projects. In addition, a major part of agricultural sector in this region fully depends on groundwater irrigation. As a result of overexploitation of groundwater, water scarcity prevails in the area. Proper groundwater management in a scientific manner is very much essential for the study area. Hence, delineation of groundwater potential zones has been carried out for this area using remote sensing and GIS technologies for better and optimal utilization of this precious resource for sustainable development.

2 Materials and Methods

2.1 Study Area

Kadiri basin is in the south-eastern part of Ananthapuramu district, A.P., India is geographically located between 78° 00′–78° 20′ E longitudes and 13° 55′–14° 10′ N latitudes with a total area of 517.28 km^2 (Fig. 1). The area comprises four mandals, namely Kadiri, Gandlapenta, Nalla Cheruvu and Obuladevara Cheruvu. Agricultural land occupies the major part of basin followed by forest area.

2.2 Data Collected

Satellite imageries, toposheets, geological maps and soil maps were procured from National Remote Sensing Center, Hyderabad, Survey of India, Hyderabad, Geological Survey of India, Hyderabad and National Bureau of Soil Survey and Land

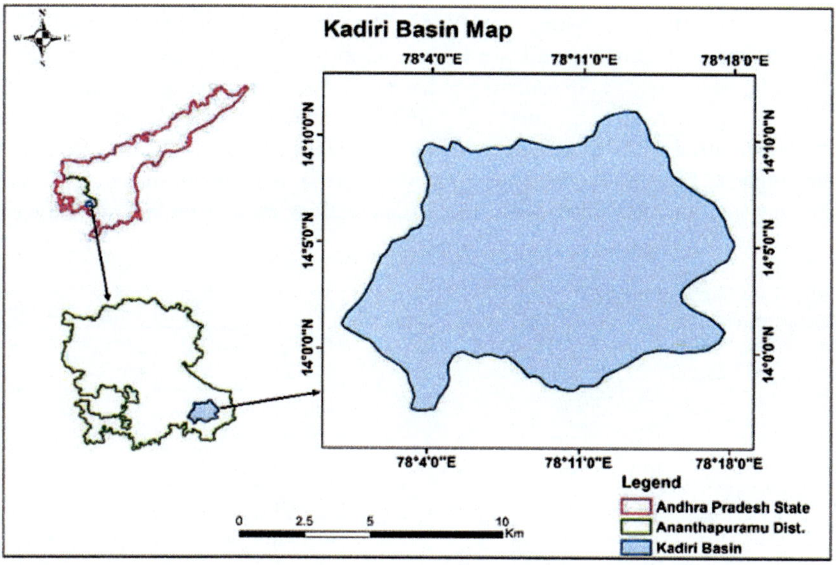

Fig. 1 Study area map

use Planning, Bangalore, respectively for the study area. Other important and required data related to study area were collected from various Andhra Pradesh State Government departments of Ananthapuramu district like Groundwater Department, District Water Management Agency, Panchayati Raj Engineering Department, Water Resources Department, Rural Water Supply and Sanitation Department, and Chief Planning Office which were utilized for various analysis in the present study.

2.3 *Methodology*

Different types of data have been used to generate the various layers of GWPZ map in Kadiri basin, Ananthapuramu district of Andhra Pradesh, India. Several steps involved to develop the six different thematic layers for GWPZ map were performed using spatial analyst tools and WOA technique in ArcGIS software. All the georeferenced maps were assigned weights. In order to identify the groundwater potential zones, several factors are needed based on their relative importance and it is achieved by rating scheme. The ratings are assigned based on the associated causative factors for prediction of ground water potential surveyed in the field and based on the knowledge by expertise on GWP causes as published in literature [9, 13]. The ranking is provided for each parameter separately for every thematic layer. In this study, ranks are given based on the rating scale of 1–20 scale and weight of each parameter assigned between 1 and 100% [7], where geology has more weight as 28 %. Higher ratings are of greater influence on GWPZ. The rating and weightage assigned for each parameter are provided in Table 1. Thus, basin-wise groundwater potential zones with four classes such as excellent, good, average and poor zones are obtained.

The weighted overlay is a method of modelling suitability considered as a simple and widely used method in different areas such as for evaluating the potential land slide area [2, 12], area for fisheries agroindustry [14], landfill site selection [3] and groundwater potential areas [8, 11, 13]. In this method, each raster layer is assigned

Table 1 Details of thematic layers created by GIS software with their subunit, weight and rating

Layer	Sub Unit	Weight	Rating
Geology	Grey granite/Pink granite	28	12
	Hornblende-biotite gneiss		28
	Meladacite		6
	Melandacite		6
	Rhyolite/Quartz		18
Soil type	Rock lands	20	6
	Fine loamy		12
	Loamy skeletal		20
Slope (°)	< 7	17	17
	7–14		14
	15–22		11
	> 22		7
Geomorphology	Denudation hills	14	8
	Granite-Gneiss		7
	Pediment-Pediplain		14
	Dyke ridge		8
Land use land cover	Agricultural land	12	12
	Barren rocky land		2
	Built up area		4
	Fallow land		5
	Forest area		8
	Scrub land		5
	Water body		10
Drainage density (Km/Km^2)	0.00–0.87	9	9
	0.88–1.74		7
	1.75–2.61		4
	2.62–3.48		2
	3.49–4.35		1

a weight and reclassified in the suitability analysis. Raster layers are overlaid, multiplying each raster cell's suitability value by its layer weight and by totaling the values to derive a suitability value. These values are written to new cells in an output layer, which are labelled in the symbology.

2.3.1 Maps Generated

Six thematic layers were generated using geospatial techniques, i.e. geology, soil type, slope, geomorphology, land use land cover and drainage density of the study area. The computed revised weights obtained for the layers were 28, 20, 17, 14, 12 and 9, respectively. A parameter assigned a higher weight value shows a major influence and similarly a lower weight value shows a minor influence on groundwater potential [1, 5, 15]. Integration of all thematic layers was done through weighted overlay technique (WOT) for developing GWPZ map of the study area using GIS software.

To ensure sustainable development of the basin, groundwater recharge structures were proposed. WOA is a multicriterion study wherein investigation was carried out with multifaceted things for determining certain themes with the aid of assigning rank to the respective features class and then assign weightage to the respective parameters depending upon the weightage of the theme on the objective. The effectiveness of this method is that the individual thematic layers and their classes are assigned weightages based on their relative contribution towards the output [4, 6]. There is no standard scale for a simple weighted overlay method. For this purpose, criteria for the analysis were defined, and each parameter was assigned weightage based on its importance [10]. Determination of weightage of each class is the most crucial in integrated analysis, as the output is largely dependent on the assignment of appropriate weightage. Consideration of relative importance leads to a better representation of the actual ground situation. In the present study, weighted overlay model in GIS environment (ArcGIS software) has been used to identify and demarcate the suitability zones for groundwater recharge which can also be utilized as sites for artificial recharge. Thus, multiple thematic layers of influencing parameters like geology, soil, slope, drainage density, lineament density and land use land cover which were prepared are assigned features class rank as per the importance in the selection of recharge sites. In this model, six parameters were converted into raster from vector base according to the weights. Each raster was assigned percentage influence based on its importance, and its features class were ranked between 1 and 6 scales (Table 2). Each input raster was weighted, and the total influence for all raster equals 100%. Moreover, individual thematic layers and their classes were assigned weightage based on their relative contribution towards the output. Using this suitability modelling, ideal areas were identified wherein the classes with higher values indicate the most favourable zones for groundwater recharge and also for location of artificial recharge structures.

Table 2 Rank and weight of different parameters for groundwater recharge zonation

Parameters	Features class	Weight	Rank
Land use land cover	Built-up area	35	1
	Barren rocky land		2
	Fallow land		3
	Forest land		4
	Scrub land		5
	Agricultural land		6
Soil	Rock lands	25	1
	Fine loamy		2
	Loamy skeletal		3
Slope	0–5%	15	5
	5–10%		4
	10–15%		3
	15–30%		2
	30–100%		1
Drainage density	Very low	15	5
	Low		4
	Medium		3
	High		2
	Very high		1
Geology	Hornblende_biotite gneiss	5	5
	Rhyolite/Quartz		4
	Grey granite/Pink granite		3
	Meladacite		2
	Melandacite		1
Lineament density	Very low	5	1
	Low		2
	Medium		3
	High		4
	Very high		5

3 Results and Discussions

3.1 Demarcation of Groundwater Potential Zones

The objective of the study was to analyse and to identify the groundwater prospect zones (GWPZ) by developing groundwater potential zone map for Kadiri basin of Ananthapuramu district in Andhra Pradesh, India. Revised weights as well as ratings to the respective subclasses (Table 1), raster layers were overlaid through WOT in GIS environment to develop GWPZ map which was further reclassified into four categories, i.e. poor, average, good and excellent groundwater potential zones, respectively. GWPZ map of the study area is shown in Fig. 2. Major portions of excellent and good groundwater potentiality occur in the eastern and central region of the study area. From GIS overlay analysis, it was inferred that groundwater potentiality of the basin is majorly good and average, except in few areas of central, eastern and south-western portion. The areal distribution of the groundwater potential zones is shown in Table 3.

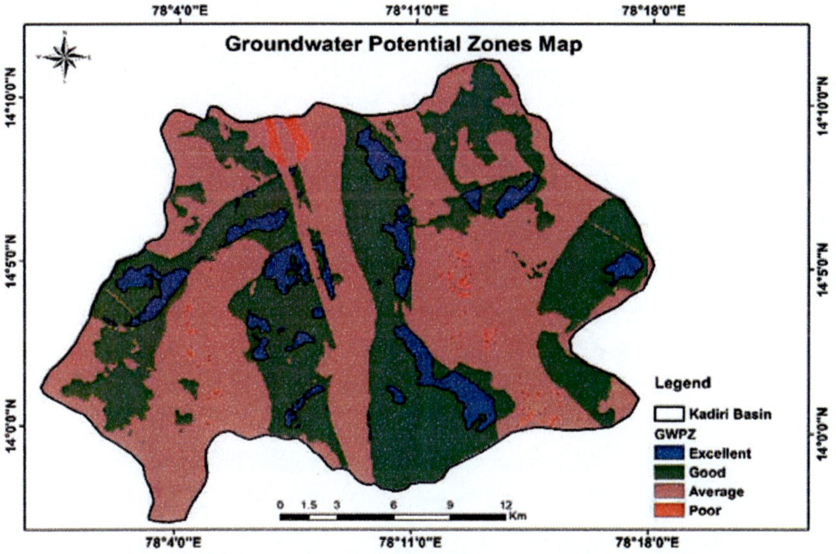

Fig. 2 GWPZ map

Table 3 Statistics of GWPZ

Groundwater potential zones	Area (Km^2)	Percentage of total area (%)
Excellent	40.29	7.79
Good	206.11	39.84
Average	266.66	51.55
Poor	4.22	0.82

3.2 Locating Suitable Sites for Groundwater Recharge Structures

The overall perspective of this study is to have a detailed study on the availability of groundwater resources in Kadiri basin and suggest suitable measures for efficient utilization of existing resources and for further improving the quantity and quality of groundwater resources in a more sustainable way. Based on the data collected and with the thematic information derived from the remote sensing data, groundwater recharge structures like check dams, percolation tanks, gabion structures and subsurface dykes were identified at various locations to improve the groundwater quantity and quality in the study area. The proposed groundwater augmenting structures in the study area are shown in Fig. 3.

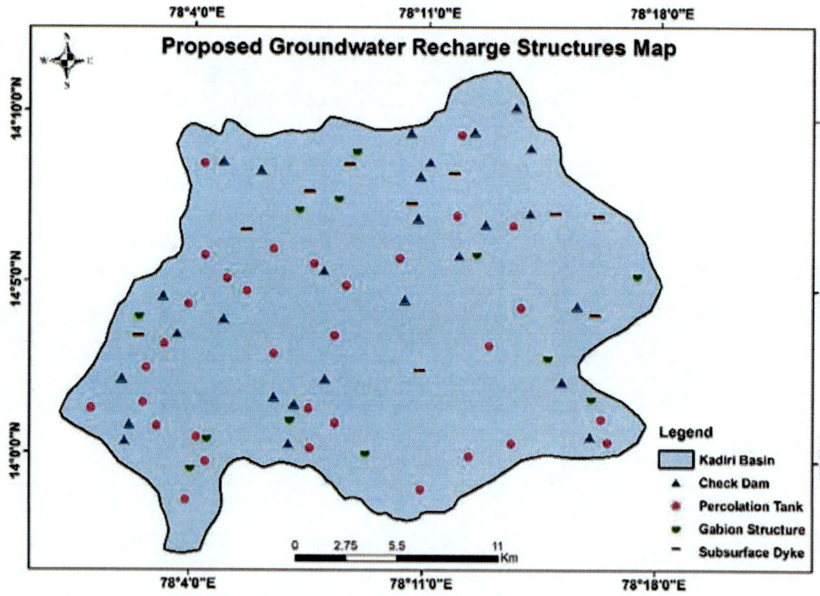

Fig. 3 Proposed groundwater recharge structures map

After assigning the weightages, each theme was overlaid by using ArcGIS, and favourable zones for artificial recharge area were delineated. Later drainage network map was superimposed over the artificial recharge zones map and considering the concern terrain conditions, groundwater augmenting structures such as percolation tanks, check dams, gabion structures and subsurface dykes were suggested accordingly. Areas suggested for the construction of 27 number of check dams were on area having 1st to 3rd order streams and for 32 numbers of percolation tanks on area having 2nd to 3rd order streams. Also, the area suggested has flat terrain for maximum storage of runoff in the proposed sites of check dam and percolation tank. Subsurface dykes (10 numbers) were proposed on the areas having shallow impervious layer with wide valley and narrow outlet. Gabion structures (12 numbers) were proposed along the small streams to conserve stream flows with practically no submergence beyond stream course.

4 Conclusions

Indiscriminate exploitation of surface and subsurface water has led to severe water scarcity and environmental degradation. With the gradual dwindling of surface sources, the role of subsurface sources is gaining momentum in the drought prone area of Kadiri basin. It is therefore, very necessary to have a quantitative and qualitative analysis of groundwater in the study area, for its planned and sustained development.

The conclusions arrived at from this research study are as follows:

- Excellent groundwater potential zones identified were mainly in pediment-pediplain with flat slope, agricultural land of loamy skeletal soil.
- Major portions of excellent and good groundwater potentiality occur in the eastern and central region of the study area.
- Areas suggested for the construction of check dams were on area having first- to third-order streams and for percolation tanks on area having second- to third-order streams. Also, the area suggested has flat terrain for maximum storage of runoff in the proposed sites of check dams and percolation tanks.
- Groundwater recharge structures of 81 no. of which 27 no. of check dams, 32 no. of percolation tanks, 10 no. of subsurface dykes and 12 no. of gabion structures were proposed at various locations for improving the quantity and quality of groundwater resources in a more sustainable way.

The following points need consideration for sustainable management of groundwater resources:

- Groundwater basins used to provide drinking water supplies should be protected from depletion by other uses as well as from contamination.
- In drought-prone areas like Kadiri basin, certain groundwater basins should be earmarked for drinking purposes only.

- Implement the proposed groundwater recharge structures like check dams, percolation tanks, gabion structures, subsurface dykes to improve the groundwater resources and retain soil fertility for sustainable development of the study area.
- Severe unemployment problem in the study area and migration of farmers for works due to scarcity of water and prevailing drought conditions can be averted by improving the groundwater levels by implementing proposed groundwater recharge structures which increases crop yield there by living standards of people will be raised.
- The groundwater potential zones map generated could be useful for optimal utilization of the groundwater resources and for identification of suitable locations for extraction of water, preparation of better management plans that will help in improving the socio-economic conditions and for sustainable development in the Kadiri basin.

Acknowledgements Authors acknowledge the grant received from SERB, DST, vide research project no: (SR/FTP/ETA-486/2012 dated 16.12.2016) for carrying out the research work. The authors are also thankful to organizations like NRSC, NBSSLUP, GSI, SOI, APSGWB, PRED, CPO, RWSSP, DWMA, Ananthapuramu, for providing the necessary data utilized in the present work.

References

1. Agarwal R, Garg PK (2016) Remote sensing and GIS based groundwater potential & recharge zones mapping using multi-criteria decision making technique. Water Resour Manage 30(1):243–260
2. Ahmed MF, Rogers JD, Ismail EH (2014) A regional level preliminary landslide susceptibility study of the upper Indus river basin. Eur J Remote Sens 47(1):343–373
3. Al-Anbari MA, Thameer MY, Al-Ansari N (2018) Landfill site selection by weighted overlay technique: case study of Al-Kufa Iraq. Sustainability 10(4):999
4. Chowdhury A, Jha MK, Chowdary VM (2010) Delineation of groundwater recharge zones and identification of artificial recharge sites in West Medinipur district, West Bengal, using RS, GIS and MCDM techniques. Environ Earth Sci 59:1209–1222
5. Jhariya DC, Kumar T, Gobinath M, Diwan P, Kishore N (2016) Assessment of groundwater potential zone using remote sensing, GIS and multi criteria decision analysis techniques. J Geol Soc India 88(4):481–492
6. Karthick S (2017) Semi supervised hierarchy forest clustering and KNN based metric learning technique for machine learning system. J Adv Res Dyn Control Syst 9(1):2679–2690
7. Nagarajan M, Singh S (2009) Assessment of groundwater potential zones using GIS technique. J Indian Soc Remote Sens 37(1):69–77
8. Pani S, Chakrabarty A, Bhadur S (2016) Groundwater potential zone Identification by analytical hierarchy process (AHP) weighted overlay in GIS environment—a case study of Jhargram Block, Paschim Medinipur. Int J Remote Sens Geosci 5(3):1–10
9. Pasupuleti S, Sandilya DK, Singha S, Singha SS, Saha S (2019) Delineation of groundwater potential zones utilising geospatial techniques in Kadiri watershed of Anantapur district, Andhra Pradesh India. J Environ Biol 40(1):61–68
10. Saraf AK, Choudhury PR (1998) Integrated remote sensing and GIS for groundwater exploration and identification of artificial recharge sites. Int J Remote Sens 19(10):1825–1841

11. Saranya T, Saravanan S (2020) Groundwater potential zone mapping using analytical hierarchy process (AHP) and GIS for Kancheepuram District, Tamilnadu, India. Model Earth Syst Environ 1–18
12. Shit PK, Bhunia GS, Maiti R (2016) Potential landslide susceptibility mapping using weighted overlay model (WOM). Model Earth Syst Environ 2(1):21
13. Singha SS, Pasupuleti S, Singha S, Singh R, Venkatesh AS (2021) Analytic network process based approach for delineation of groundwater potential zones in Korba district, Central India using remote sensing and GIS. Geocarto Int 36(13):1489–1511
14. Teniwut WA, Hamid SK, Makailipessy MM (2019) Using spatial analysis with weighted overlay on selecting area for fisheries agroindustry in Southeast Maluku, Indonesia. In: Journal of physics: conference series, vol 1424, no 1. IOP Publishing, p 012016
15. Thapa R, Gupta S, Guin S, Kaur H (2017) Assessment of groundwater potential zones using multi-influencing factor (MIF) and GIS: a case study from Birbhum district West Bengal. Appl Water Sci 7(7):4117–4131

Comparison of Spatial Interpolation Methods for Mapping Seasonal Groundwater Levels

Akash Singh Raghuvanshi and H. L. Tiwari

Abstract Proper knowledge of spatio-temporal variation of groundwater level are very important for efficient planning and management of groundwater level. In sparsely monitored basins, levels of groundwater are generally monitored at random points which may be far away from each other. Interpolation of groundwater level over a region in an accurate manner is a pre-requisite for modeling as well as management of water resources that can be achieved by adopting accurate and reliable techniques of interpolation of scattered data over a region. In the present study, two deterministic interpolation methods, i.e., inverse distance weighted (IDW), radial basis functions (RBFs), and three geostatistical interpolation methods, i.e., ordinary kriging (OK), simple kriging (SK), and ordinary cokriging (OCK) are evaluated to predict the spatial and temporal variation of groundwater levels. Cross-validation procedure is adopted to evaluate the prediction performance of adopted interpolation methods. Groundwater level data for the year 2019 from 31 sparsely located monitored observation wells over Sagar district of Madhya Pradesh, India are used to evaluate the performance of interpolation methods with respect to different statistical indicators of cross-validation. Interpolation maps of estimate groundwater level are produced using all five spatial interpolation techniques. Results of the analysis are presented and discussed.

Disclaimer: The presentation of material and details in maps used in this chapter does not imply the expression of any opinion whatsoever on the part of the Publisher or Author concerning the legal status of any country, area or territory or of its authorities, or concerning the delimitation of its borders. The depiction and use of boundaries, geographic names and related data shown on maps and included in lists, tables, documents, and databases in this chapter are not warranted to be error free nor do they necessarily imply official endorsement or acceptance by the Publisher or Author.

A. S. Raghuvanshi (✉) · H. L. Tiwari
Department of Hydrology, Indian Institute of Technology Roorkee, Roorkee 247667, Uttarakhand, India
e-mail: akash_sr@hy.iitr.ac.in

A. S. Raghuvanshi
Department of Civil Engineering, Maulana Azad National Institute of Technology, Bhopal 462003,, Madhya Pradesh, India

P. V. Timbadiya et al. (eds.), *Geospatial and Soft Computing Techniques*, Lecture Notes in Civil Engineering 339, https://doi.org/10.1007/978-981-99-1901-7_19

Keywords Groundwater level · Spatial interpolation · Geostatistics · Cross-validation · ArcGIS 10.8

1 Introduction

Groundwater is one of the most important water sources. Groundwater management is important for their sustainable development. Therefore, we need appropriate information about the spatio-temporal behavior of the water table in a region. However, monitoring groundwater levels is inherently costly and time consuming, especially during installation stages that require the drilling of wells and piezometers. As a result, the number of surveillance sites available in a particular region is relatively small and often does not reflect the actual range of variation that may exist. Therefore, accurate groundwater level spatial interpolation at unsampled sites is required for better groundwater management.

Among various interpolation methods, there is no clear optimal method, thus results need to be compared depending on specific situation [10]. This study presents the use of GIS tools to generate the groundwater level surface for a sparsely monitored region from groundwater levels monitored at random locations. Geostatistical analyst tool in ArcGIS 10.8 is used to explore the spatial variability in groundwater levels for the Sagar district region located in Madhya Pradesh, India.

2 Study Area and Data Source

2.1 Sagar District

The Sagar district is located in the north central part of Madhya Pradesh, India, and occupies an area of 10,252 km^2. The district extends between the latitude of 23° 10′ and 24° 27′N, longitude of 78° 04′ and 79° 21′E. Figure 1 shows the index map for the Sagar district.

2.2 Data Used

The pre (April–June) and post-monsoon (Oct–Dec) seasonal groundwater level data of 31 central ground water board (CGWB) monitoring wells for 2019 are used. The location and groundwater level of observation wells are collected directly from the India Water Resources Information System (WRIS). Traditionally, geostatistical studies are performed on at least 100 samples [11]. However, the sample size is small in this study.

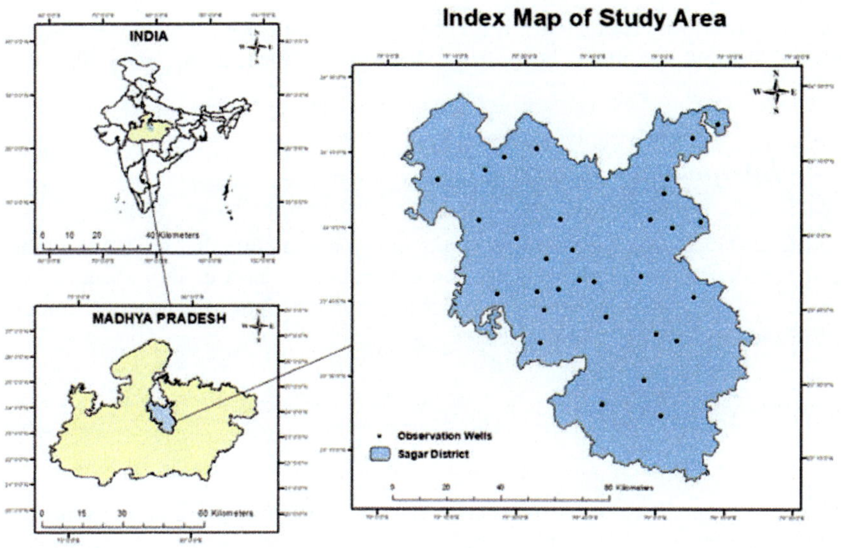

Fig. 1 Index map of study area

3 Materials and Method

3.1 Exploratory Analysis

The datasets are initially visualized in order to identify incorrect coordinate information and illogical data points. The screened datasets are then subjected to exploratory information evaluation to identify the outliers which can be unfavorable to spatial prediction. The variogram, in particular, may be very touchy to outliers due to the fact it's far primarily based totally at the squared differences among information [8]. Description of the information values is done through fundamental precis of statistics, inclusive of means, medians, variances, and skewness.

The geostatistical method mainly kriging is taken into consideration the high-quality unbiased linear prediction (BLUP) if the information meets the situations of normality, variance uniformity, and stationarity [7]. However, spatial information, specifically weather information, violates those situations. High asymmetries and outliers have undesired consequences on variogram shape and kriging estimates [6]. For spatio-temporal data that follow a Gaussian distribution, the effects of extremes are reduced, and more stable variograms are obtained, making it easier to model spatial variability [5]. Data transformations is required prior to kriging to standardize data distribution, eliminate outliers, and improvise data stationarity [4]. In this study, the normality of groundwater level data is visualized using tools such as histograms and boxplots, and the mean and median, symmetry (skewness) and flattening (kurtosis) coefficients. It is checked numerically by comparing it with the normal distribution and also according to the formal statistical Shapiro–Wilk test.

3.2 Interpolation Methods

Inverse distance weighting directly works on the assumption that the point closest to the predicted position is weighted more heavily, and the weight is reduced as a function of distance, hence the name inverse distance weighting. The exact formula for this interpolator is

$$z(s_0) = \frac{\sum_{i=1}^{n} \frac{s_i}{d_i^p}}{\sum_{i=1}^{n} \frac{1}{d_i^p}} \tag{1}$$

where $z(s_0)$ is predicted value of unsampled point, n is total no. of sampling points, d_i is separation distance between unsampled point and ith sampled point, and p denotes the weighing power.

Radial basis functions (RBFs) are an exact set of interpolation methods. That is, the surface must pass through each measurement sample value. RBF is used to generate smooth surfaces from a large number of data points. There are five different functions: thin-plate spline (TPS), spline with tension (SPT) [2], fully regularized spline (CRS) [2], multiquadric function (MQ), and inverse multiquadric function (IMQ) [12]. Each basis function has a different form and has a different interpolation plane.

Kriging belongs to a family of generalized least squares regression methods in geostatistics which uses observations sampled in a particular search environment as a linear combination to estimate values in unsampled locations [5, 7].

$$\widehat{Z}(s_o) = \sum_{\alpha} \omega_\alpha Z(s_\alpha) \tag{2}$$

$\widehat{Z}(s_o)$ is the estimated variable of interest (groundwater level) at the unsampled location $s_{o,}$ and $Z(s_\alpha)$ is the observed values at the sampled locations in the vicinity of s_o.

OK is most commonly used form of kriging, in which mean is considered unknown and fluctuates locally, which makes it possible to maintain stationarity in the local neighborhood.

Simple kriging is mathematically the simplest but the least common. It is assumed that the expected value (mean) of the random field is known and depends on the covariance function. SK assumes second-order stationary that is constant mean, variance, and covariance across the domain or the region of interest [11].

The OCK method is a modification of the OK method. The main advantage of OCK is that it uses multiple variables in the estimation process. The OCK method is used to improve the predictability of the primary variable by using the auxiliary variable, assuming that both primary and auxiliary variables are in good correlation [7]. This method is especially suitable when the main attributes of interest are sparse, but the relevant secondary information is abundant.

3.3 Variograms (Semivariogram)

The existence of spatial structures (spatial autocorrelation) in which nearby observations are more similar than distant observations is a pre-requisite for the application of geostatistics [5]. Experimental variograms measure the average dissimilarity between unsampled values and nearby data values [4] and can therefore represent autocorrelation at various distances. The kriging method requires a model of functions that characterize spatial variability, variograms, and key characteristic parameters such as nugget effects, thresholds, and ranges [5]. The experimental variogram is calculated based on the following formula:

$$\hat{\gamma}(d) = \frac{1}{2N(d)} \sum_{\alpha=1}^{N(d)} (Z(s_\alpha + d) - Z(s_\alpha))^2 \tag{3}$$

with $Z(s_\alpha)$ and $Z(s_\alpha + d)$ being the values observed at the locations u_α and u_α + d separated by the distance d and N(d) being the number of such pairs. If the value at $Z(s_\alpha)$ and $Z(s_\alpha + d)$ is autocorrelated, the result of Eq. (3) will be small, relative to an uncorrelated pair of points. From an experimental variogram analysis, the appropriate model (Gaussian, spherical, etc.) is usually fitted by the weighted least squares method, and the parameters (range, threshold, and nugget) are used in kriging. Exponential, Gaussian, and spherical are the most commonly used (theoretical) variogram models in hydrological kriging applications [1] and are also used to model experimental variograms.

3.4 Cross-Validation

The performance of the various interpolation methods (IDW, RBF, SK, OK, and OCK) is evaluated through a cross-validation process. Cross-validation is a validation technique that removes observations one by one from the dataset and re-estimates from the remaining sampled data using the selected model. If the sample size of the data is very small, such as when there are only 31 observations, the method comparison is done by cross-validation [7]. This is a common way to verify the accuracy of the interpolation method [1].

The overall performance of the interpolation methods for groundwater level estimation is conducted using correlation and error-based measures. The correlation includes the coefficient of determination (R^2), Nash–Sutcliffe efficiency coefficient (E), and Willmott agreement index (d), whereas the error measures include the mean relative error (MRE), the root mean square error (RMSE), and the mean error (ME).

4 Results and Discussion

4.1 *Exploratory Data Analysis*

Seasonal groundwater data from 31 observation wells are examined to understand the pattern in the data. Histogram (Fig. 2) and standard statistics (Table 1) are used to describe the data.

The histogram shows symmetry indicating normality of distribution. In fact, seasonal groundwater elevation data have mean and median close enough and skewness value close to zero, thus suggesting this distribution as symmetrical distribution. Finally, the Shapiro–Wilk test confirms the normality of the original data ($p = 0.83 > 0.05$, $p = 0.71 > 0.05$). Since original data follow a normal distribution hence, it is decided to work on original data without transformation.

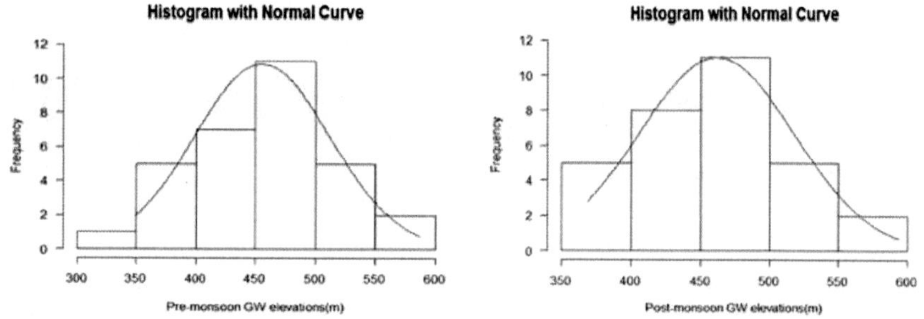

Fig. 2 Histogram of seasonal groundwater levels (amsl). Curve represents the fitting of a normal distribution

Table 1 Descriptive statistics for seasonal groundwater elevation level (m) data

Season	Min	Max	Mean	SE	Median	Q1	Q3	SD	CV	Skew	Kurt	SW
Pre-monsoon	348.9	586.1	455.3	10.2	460.1	416.1	490.9	57.1	12.53	0.10	– 0.64	0.83
Post-monsoon	368.4	592.9	461.7	10.1	464.0	420.1	496.6	56.2	12.17	0.13	– 0.61	0.71

Min. minimum, Max. maximum, SE. standard error, Q1 1st quartile, Q3 3rd quartile, SD standard deviation, Skew. skewness, Kurt. kurtosis, CV coefficient of variation, SW probability corresponding to the Shapiro–Wilk test of normality

4.2 Spatial Analysis of Groundwater Level Data

The spatial variation of seasonal groundwater elevation is considered to be isotropic ignoring the separation direction because the size of the sample (31) is limited and would not possibly detect anisotropy [7].

Table 3 and Figs. 3 and 4 show the variogram model and its parameters tuned for seasonal groundwater levels. The nugget effect to nugget to threshold ratio is used to classify the spatial dependence of the variable [3]. Variables have strong spatial dependence when the nugget-threshold ratio is less than 0.25, and moderate spatial dependence when the ratio is 0.25 to 0.75 [8]. Otherwise, the variables are less spatially dependent. Therefore, in our case, the groundwater level was strongly spatially correlated (Table 2).

Elevation, as auxiliary information, has decreased semi-variances. It may be visible that the sill is higher for OK and SK (4507.4 m^2 and 4,295.7 m^2) than for OCK (3713.8 m^2 and 3577 m^2). This is predicted due to the fact the covariate, elevation which become taken into consideration for OCK, however, now no longer for OK and SK variogram, in part explains the variability of the groundwater level data.

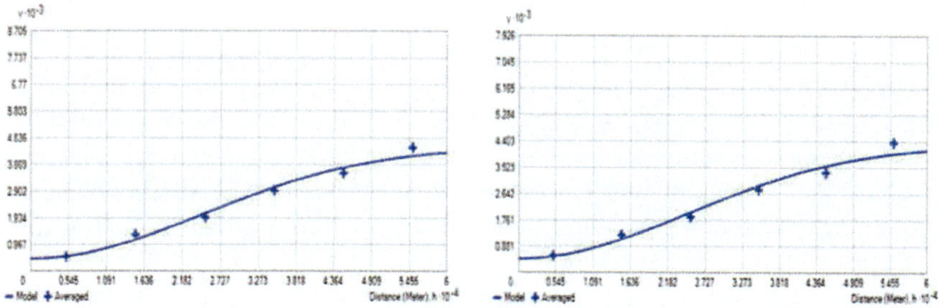

Fig. 3 Experimental (points) and fitted theoretical (curve) variograms of pre-monsoon (left) and post-monsoon (right) groundwater levels for OK and SK

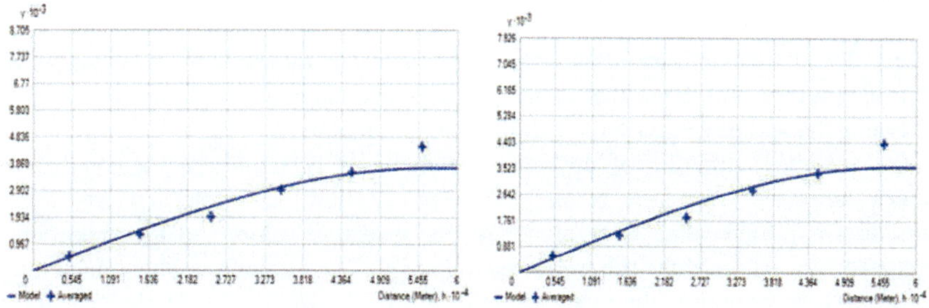

Fig. 4 Experimental (points) and fitted theoretical (curve) variograms of pre-monsoon (left) and post-monsoon (right) groundwater levels for OCK

Table 2 Summary of semivariogram parameters of best-fitted theoretical model to predict seasonal groundwater level (amsl)

Interpolation	Season	Best-fit model	Nugget (m)	Sill (m)	Range (km)	Nugget/Sill
OK & SK	Pre-monsoon	Gaussian	449.5	4507.4	60	0.1
	Post-monsoon	Gaussian	465.0	4,295.7	60	0.11
OCK	Pre-monsoon	Spherical	0	3713.8	56	0
	Post-monsoon	Spherical	0	3577	56	0

Table 3 Summary of descriptive statistics for observed and predicted pre-monsoon groundwater level

Pre-monsoon	Mean	Max	Min	Median	C5	C25	C75	C95	SD	CV
Observed	455.31	586.18	348.93	460.15	369.9	416.15	490.9	539.44	57.09	12.53
IDW	459.20	524.98	389.37	454.86	408.47	426.34	493.52	505.80	36.82	8.02
RBF	455.96	531.87	355.74	463.84	383.73	419.81	491.70	524.13	46.29	10.15
OK	456.16	525.41	367.55	452.66	383.88	424.14	493.96	520.94	43.97	9.64
SK	456.34	526.9	373.19	452.65	384.13	426.06	494.13	521.84	43.90	9.62
OCK	456.10	573.65	369.1	464.45	382.38	414.46	491.36	528.43	51.58	11.31

For IDW method, an optimal power value *(p)* as well as the number of the closest neighbors to include are determined, whereas for RBF method, the choice of radial basis function, their kernel parameter, and number of the closest neighbors to include are determined by minimizing the root mean square error (RMSE) statistics obtained from a cross-validation procedure. In this study, for IDW method, the optimizing parameter of the weight function *(p)* is taken as 2.0, whereas for RBF method, multiquadric radial basis function is found to be optimal among all the functions.

4.3 Groundwater Mapping

Tables 3 and 4 show various descriptive statistical parameters for the measured seasons (before and after the monsoon) and the parameters predicted by two deterministic and three geostatistical interpolation methods.

One of the hallmarks of geostatistical methods is smoothing, as predicted values are less variable than measured values. In other words, the minimum expected value is greater than the measured value, and the maximum expected value is less than the measured value [9]. This smoothing phenomenon is least for OCK while it is the most accentuated for SK which has 373.19, 376.34 m as minimal and 526.9, 533.14 m as maximal values compared to 348.93, 368.43 m as minimal and 586.18, 592.98 m as maximal for the measured groundwater values during pre- and post-monsoon periods, respectively.

Table 4 Summary of descriptive statistic for observed and predicted post-monsoon groundwater level

Post-monsoon	Mean	Max	Min	Median	C5	C25	C75	C95	SD	CV
Observed	461.68	592.98	368.43	464.05	373.57	420.1	496.65	547.28	56.18	12.17
IDW	465.25	531.62	390.41	460.35	417.72	431.36	500.74	513.31	37.29	8.01
RBF	461.87	538.51	358.2	468.28	394.93	426.49	497.84	529.61	45.84	9.92
OK	462.21	525.41	367.55	454.70	395.24	429.62	499.88	527.09	43.78	9.47
SK	462.51	533.14	376.34	454.70	395.56	431.98	500.88	528.03	43.60	9.42
OCK	462.14	580.27	377.8	469.88	384.66	420.20	497.92	535.21	51.61	11.16

Min. Minimum, Max. maximum, C5 5th centile, C25 25th centile or 1st quartile, C75 75th centile or 3rd quartile, C95 95th centile, SD standard deviation, CV coefficient of variation, OK ordinary kriging, SK simple kriging, OCK ordinary cokriging

This phenomenon is confirmed by the standard deviation, especially the decrease in the estimated variance for the measured data variance of 43, 39.7% of SK; 40.6, 39.2% for OK; 18.4, 15.6% for OCK during pre- and post-monsoon periods, respectively, and also by the coefficient of variation which is minimal for SK (9.62, 9.42%) followed by OK (9.64,9.47) and OCK (11.31,11.16%) compared to measured values (12.53, 12.17%) for pre- and post-monsoon periods.

Smoothing phenomenon among deterministic methods is least for RBF and most accentuated for IDW which has 389.37, 390.41 m as minimal and 524.98, 531.62 m as maximal values compared to 348.93, 368.43 m as minimal and 586.18, 592.98 m as maximal for the measured groundwater values during pre- and post-monsoon periods, respectively.

In geostatistical methods, estimates of SK are higher than those of OCK and OK with SK, OK, and OCK mean values of 456.34, 462.51 m; 456.16, 462.21 m; 456.10, 462.14 m for pre- and post-monsoon periods, respectively. Similar results can be seen in minimal values and for different percentiles. Similar analysis is carried out for deterministic interpolation methods, and overall estimates of IDW are found to be higher among all five interpolation methods used in this study.

Figures 5 and 6 represent the groundwater maps obtained from five methods of spatial interpolation (IDW, RBF, SK OK, and OCK) for pre- and post-monsoon periods for the year 2019.

4.4 Performance Evaluation Study

To deepen the comparative study of spatial interpolation, the performance indicators for cross-validation are shown in Figs. 7 and 5. The boxplot of groundwater level prediction error (Fig. 7) shows that interpolation generally corrects the predicted groundwater level of 31 observation wells. The perfect match between the predicted and measured values is represented by 0 in the Fig. 7. Comparing the five methods, the residual (error) between the measured and predicted groundwater levels of OCK

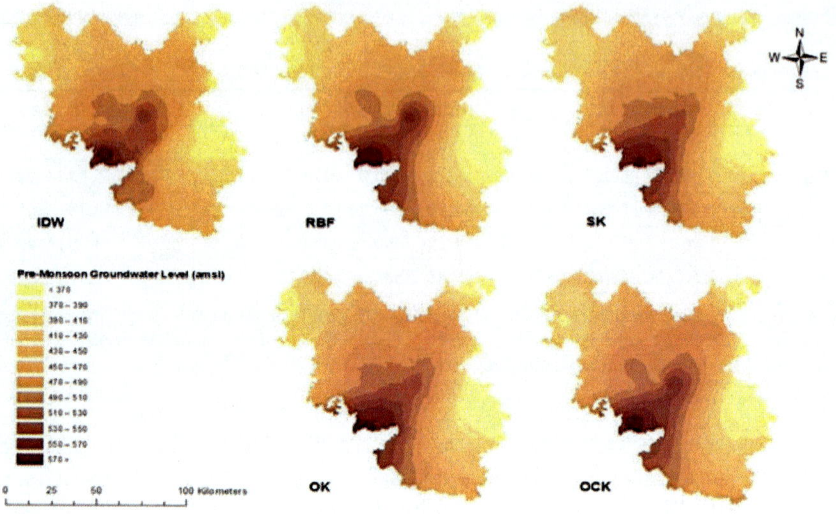

Fig. 5 Maps of pre-monsoon groundwater level (m) amsl estimated by IDW, RBF, SK, OK, and OCK

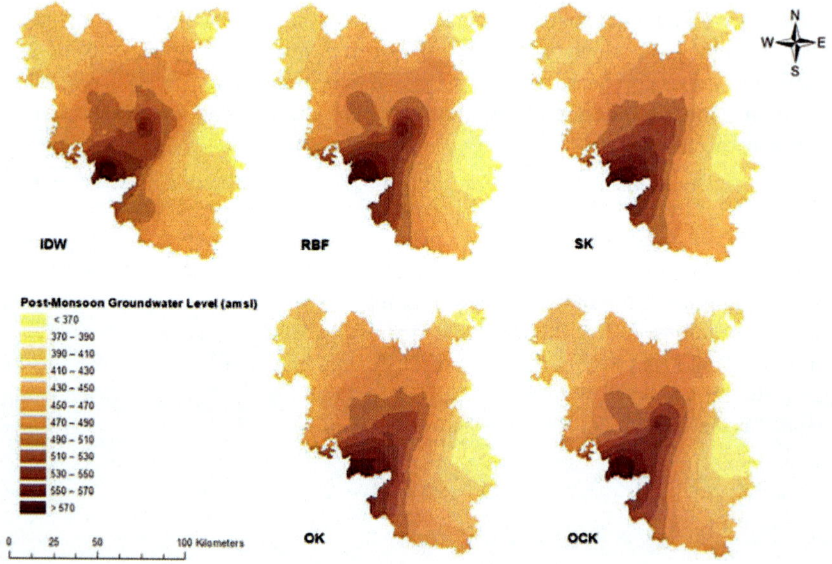

Fig. 6 Maps of post-monsoon groundwater level (m) amsl estimated by IDW, RBF, SK, OK, and OCK

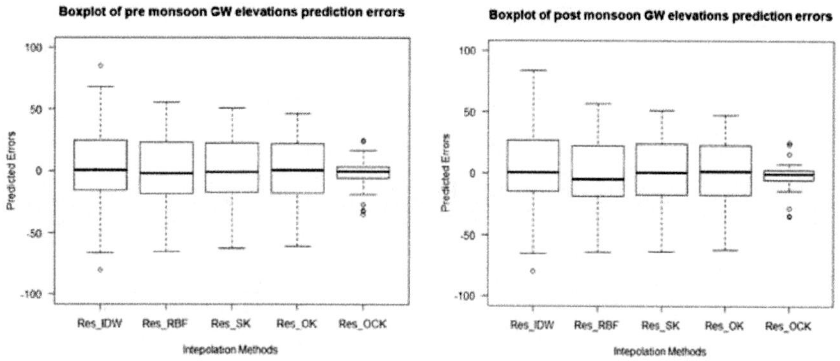

Fig. 7 Boxplots of seasonal groundwater level prediction errors using IDW, RBF, SK, OK, OCK

is significantly reduced. This shows that OCK is the best interpolator, though there exists small underestimation and overestimation.

Because aim of this study was to compare various methods, first the impact of various interpolation method on accuracy was considered. The 3 kriging methods performed better than the deterministic approach to estimate groundwater levels for each season. Performance measures of interpolation methods are summarized in Table 5. High values of coefficients of determination, Nash–Sutcliffe efficiency, and Willmott agreement index suggested an amazing fit among measured and predicted water level depth. Of the five interpolation methods, OCK had the best overall performance, with OK, SK, and RBF significantly superior to IDW. Not only the performance indicators of the model, but also the errors confirmed the above facts. The low RMSE and ME of all interpolation methods showed applicability to the prediction of groundwater level, and the superiority of OCK over all other methods was fully demonstrated by its minimum error value.

The second approach to assess the accuracy of a method is done by adopting regression coefficients (intercept and slope). The best model performance is represented by small intercept and large gradient. Among the five interpolation methods, the OCK, method which considered elevation as an auxiliary variable for predicting groundwater levels, showed best results.

5 Conclusions

The following conclusions are derived from the foregoing study:

- Introduction of elevation information improved the performance of covariate kriging method, OCK in particular, in sparsely monitored region.

Table 5 Performance evaluation of interpolation methods to predict groundwater levels

Interpolation	Season	Efficiency			Error			Intercept	Slope
		R^2	E	d	RMSE	ME	MRE		
IDW	Pre-monsoon	0.62	0.60	0.83	35.51	3.88	0.14	226.73	0.51
	Post-monsoon	0.63	0.61	0.84	34.49	3.56	0.15	221.66	0.52
RBF	Pre-monsoon	0.72	0.72	0.90	29.44	0.65	0.12	141.14	0.69
	Post-monsoon	0.72	0.72	0.90	29.20	0.18	0.13	141.82	0.69
SK	Pre-monsoon	0.75	0.74	0.91	28.58	1.02	0.12	152.95	0.66
	Post-monsoon	0.75	0.74	0.91	28.01	0.83	0.12	151.89	0.67
OK	Pre-monsoon	0.77	0.76	0.92	27.56	0.85	0.11	148.23	0.67
	Post-monsoon	0.76	0.76	0.92	27.17	0.52	0.12	146.98	0.68
OCK	Pre-monsoon	0.98	0.98	0.99	5.82	0.25	0.02	49.587	0.89
	Post-monsoon	0.99	0.99	0.99	4.79	− 0.01	0.02	42.536	0.90

- Smoothing phenomenon in geostatistical method is least for OCK while it is the most accentuated for SK whereas smoothing phenomenon among deterministic methods is least for RBF and most accentuated for IDW.
- In geostatistical methods, estimates of SK were higher than OCK and OK, whereas overall estimates of IDW are found to be higher among all five interpolation methods.
- Geostatistical methods performed better than the deterministic methods for predicting groundwater levels.

References

1. Adhikary PP, Dash CJ (2017) Comparison of deterministic and stochastic methods to predict spatial variation of groundwater depth. Appl Water Sci 7:339–348
2. Arslan H (2014) Estimation of spatial distrubition of groundwater level and risky areas of seawater intrusion on the coastal region in Carsamba Plain, Turkey, using different interpolation methods. Environ Monit Assess 186:5123–5134
3. Cambardella CA, Moorman TB, Novak JM, Parkin TB, Karlen DL, Turco RF, Konopka AE (1994) Field-scale variability of soil properties in Central Iowa Soils. Soil Sci Soc Am J 58:1501–1511
4. Deutsch CV, Journel AG (1998) GSLIB: Geostatistical Software Library and user's guide, 2nd edn. Oxford University Press, New York
5. Goovaerts P (1997) Geostatistics for natural resources evaluation. Oxford University Press, New York
6. Gringarten E, Deutsch CV (2001) Teacher's aide: variogram interpretation and modeling. Math Geol 33:507–534
7. Isaaks EH, Srivastava RM (1989) An introduction to applied geostatistics. Oxford University, New York

8. Liu D, Wang Z, Zhang B, Song K, Li X, Li J, Li F, Duan H (2006) Spatial distribution of soil organic carbon and analysis of related factors in croplands of the black soil region, Northeast China. Agric, Ecosyst Environ 113(1–4):73–81
9. Rata M, Douaoui A, Larid M (2020) Comparison of geostatistical interpolation methods to map annual rainfall in the Chéliff watershed, Algeria. Theoret Appl Climatol 141:1009–1024
10. Sun Y, Kang S, Li F, Zhang L (2009) Comparision of interpolation methods for depth to groundwater and its temporal and spatial variation in the Minqin oasis of Northwest China. Environ Model Softw 24:1163–1170
11. Webster R, Oliver MA (2007) Geostatistics for environmental scientists, 2nd edn. Wiley, Chichester
12. Xie Y, Chen T, Lei M, Yang J (2011) Spatial distribution of soil heavy metal pollution estimated by different interpolation methods: accuracy and uncertainty analysis. Chemosphere 82:468–476

Waterlogging Risk Assessment of Patiala City, Punjab Using Analytical Hierarchy Process and GIS Analysis: A Case Study

S. Gorai, A. Dhir, and D. Ratha

Abstract Waterlogging is one of the most critical environmental problems of an urban area and requires proper management. Urbanization without proper planning causes waterlogging, resulting in significant built-up damages, economic losses, and negative impacts on health. To manage this issue properly, waterlogging mapping is the preliminary task to identify the probable waterlogging areas. In this study, we carried out a waterlogging risk analysis and prepared a waterlogging risk zone for the Patiala city of Punjab. Analytical hierarchy process (AHP) incorporated with GIS techniques has been used to prepare waterlogging risk mapping. Four criteria are selected for AHP based on the best data availability. These criteria are elevation, slope, land use land cover, and hydrologic soil group. These criteria are further classified into different importance levels relating to flood risk. The pair-wise comparison matrix of AHP is prepared using the importance level suggested by experts, and the weights of four different criteria are calculated. The waterlogging map is produced using the weights and representative factors of the respective criteria. The results of the waterlogging risk map are split up into five risk zones: severe, high, moderate, low, and very low. These findings of the water logging map will demonstrate the areas under flood risk for which decision-makers may take necessary action to prevent such environmental problems.

Keywords Waterlogging · Analytical hierarchy process · GIS analysis · Patiala city

---

S. Gorai · A. Dhir
School of Energy and Environment, Thapar Institute of Engineering and Technology, Patiala 147004, India
e-mail: amit.dhir@thapar.edu

D. Ratha (✉)
Department of Civil Engineering, Thapar Institute of Engineering and Technology, Patiala 147004, India
e-mail: dnath.ratha@thapar.edu

P. V. Timbadiya et al. (eds.), *Geospatial and Soft Computing Techniques*, Lecture Notes in Civil Engineering 339, https://doi.org/10.1007/978-981-99-1901-7_20

1 Introduction

Waterlogging, the accumulation of rainwater in an urban area, became a serious issue to handle. Rapid urbanization is the primary cause behind waterlogging in a metropolitan area. Urbanization replaces natural surfaces with built-up areas such as roads and buildings, preventing rainwater from entering the soil. Therefore, the rainwater accumulated on the surface produces runoff immediately. But other factors such as elevation, slope, and soil group play a crucial role in transporting the water to rivers or water bodies. The disturbances on these factors will directly impact stormwater's transportation, which will be stopped, and the accumulation will start. The proliferation of rainwater will lead to waterlogging in an urban area. For mitigation of waterlogging problem, a waterlogging risk map should be executed.

The analysis of areas prone to waterlogging in an urban area is conducted worldwide, and numerous studies are available on the execution of waterlogging risk map [1–4]. Execution of waterlogging risk map is the integration of thematic layers statistical modeling in GIS. Statistical modeling, analytical hierarchy process (AHP), frequency ration, fuzzy logic, machine learning techniques, etc., are used [5]. AHP is one the most popular methods to assess waterlogging maps and used by several authors [1–4].

AHP is a statistical model based on multi-criteria decision-making developed by Saaty in 1977 [6]. The criteria of AHP are not limited to any numbers but should be sufficient to execute the risk map. The importance level of selected criteria for waterlogging of an area is given from surveys or works of literature. The importance level will form a pair-wise comparison matrix, and the weight of each criterion will be analyzed by normalizing the pair-wise comparison matrix. The weight of each criterion will be integrated with that criterion's class to evaluate the risk map.

Patiala city is the fourth largest and populated city of Punjab state and is located in the Ghaggar river basin of northern India. Frequent flood events are occurred in the basin, mainly due to extreme rainfall. In the last 22 years (1988–2010), five significant floods occurred in the basin [7]. The city is located at a distance of 50 km from the foothill of the Himalayas, and therefore, a higher probability of flash flood exists. In context to frequent floods in the Ghaggar river basin and nearest Himalaya, it is essential to determine the waterlogging risk zone over Patiala city. Therefore, the study's objective is to prepare the waterlogging risk map of Patiala city of Punjab using AHP analysis and GIS analysis. The study's findings will help to take the initiative for waterlogging preventive measures for future flood events.

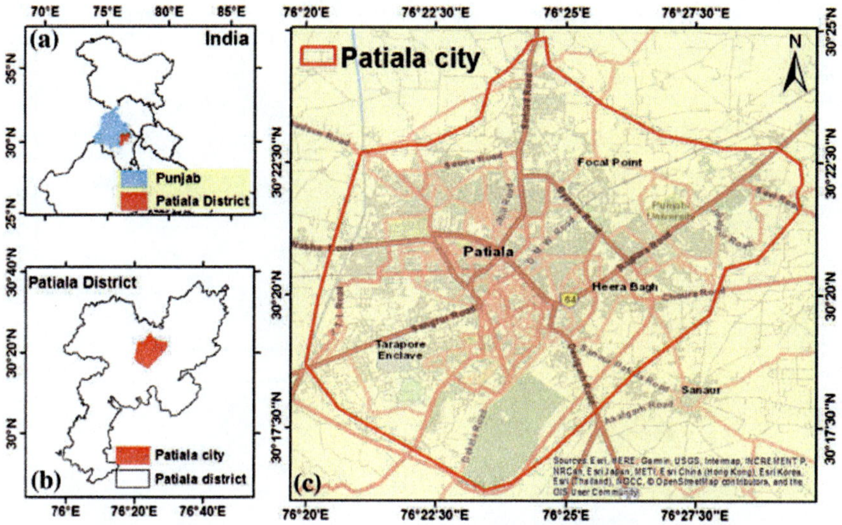

Fig. 1 Study area of Patiala city (**a** The northern part of India shows the state of Punjab, and Patiala district is showing inside Punjab state; **b** Patiala district is showing the Patiala city; **c** The study area of Patiala city)

2 Study Area and Data Source

2.1 Patiala City

Patiala is the fourth largest city of Punjab state in India and is located in the latitude of 30° 20′ 24″ N and longitude of 76° 22′ 47.98″ E. This city lies at an elevation of 255 m above the mean sea level. The annual average temperature of Patiala is 24.5 °C, where the highest temperature is observed in June, and the lowest temperature is observed in January. The Patiala city is located under the semi-arid region's climatic condition, and southwest monsoon rainfall is the primary source of rainwater. The average annual rainfall of Patiala is 698 mm, where the average southwest monsoon rainfall is 575 mm [8]. The city's total area is 160 km^2, and the city's total population is 763280 (projected to 2020). Therefore, the density of Patiala city is 4800/km^2. Figure 1 shows the study area in the Patiala district of Punjab state of India.

2.2 Criteria Used for AHP

The criteria used for AHP analysis are elevation, slope, soil group, and land use land cover (LULC) based on their data availability for Patiala city (Fig. 2).

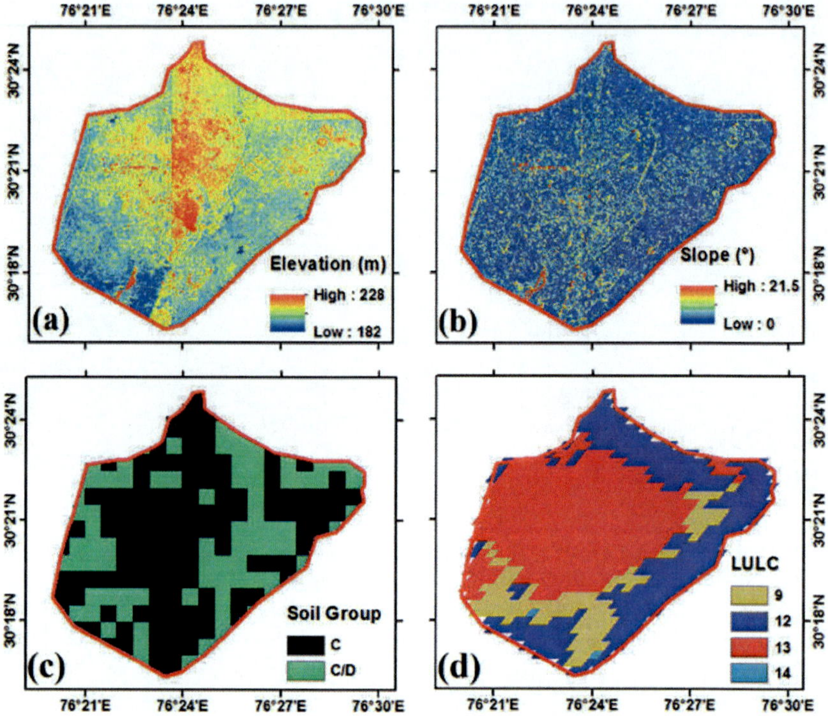

Fig. 2 Thematic layers of **a** elevation, **b** slope, **c** hydrologic soil group, and **d** land use land cover (LULC) [9: savannas; 12: cropland; 13: urban builtup; 14: cropland/natural vegetation]

Elevation: The 30 m resolution elevation data was collected from Shuttle Radar Topography Mission (SRTM) dated 23.09.2014. Figure 3a shows the elevation profile of Patiala city. The elevation of Patiala city ranges between 182–228 m. The average elevation of Patiala city is 206 m.

Slope: The slope data is calculated from the elevation data. The conversion factor for slope calculation in ArcGIS is 9.26×10^{-06}. Figure 3b shows the slop profile of Patiala city. The slope of Patiala city lies in the range of 0–21.5°. The average slope of Patiala city is 1.89°.

Soil group: The soil group data is collected from Global Hydrologic Soil Groups (HYSOGs250m) for curve number-based runoff modeling dated 22.04.2020. Figure 3c shows the soil group of Patiala city. The hydrologic soil group of Patiala city is covered by C (62.13%) and C/D group (37.87%).

Land use land cover: The 500 m resolution land use land cover is collected from MODIS land cover V6, 2019. Figure 3f shows the land use land cover pattern of Patiala city. The land use land cover of Patiala city is covered by savannas (15.12), cropland (37.64%), urban builtup (46.92%), and cropland/natural vegetation (0.32%).

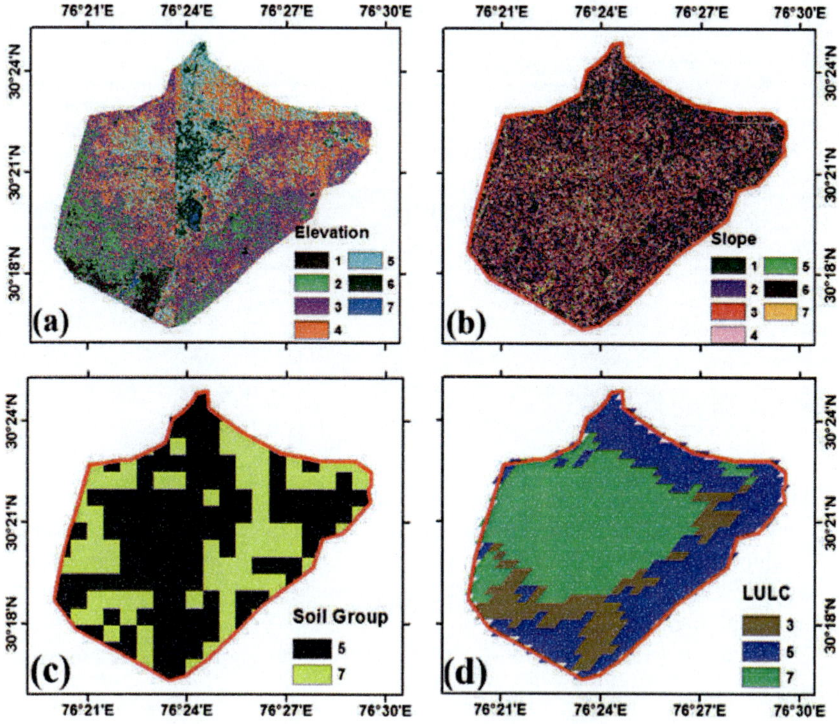

Fig. 3 Classification of thematic layers of **a** elevation, **b** slope, **c** soil group, and **d** LULC

2.3 Materials and Method

The waterlogging risk map is analyzed using the AHP method, and the waterlogging risk can be stated as follows [9].

$$\text{Watelogging Risk} = \sum_{i=1}^{n} W_i \times R_i \tag{1}$$

where W_i is the weightage of each criterion and R_i is the classified values of each criterion. The weightage of each criterion is calculated using AHP.

The AHP method is selected for waterlogging risk mapping using different criteria. A pair-wise comparison matrix is constructed based on the relative importance (Eq. 2). The importance values are shown in Table 1. The values are given based on the importance of each row of the matrix compared with the matrix column. In the pair-wise comparison matrix, the rows follow the inverse value of each criterion and its significance with others. The diagonal value of the matrix will be equally important.

Table 1 Level of importance for pair-wise comparison matrix [9]

Level of importance	*Definition*	*Level of importance*	*Definition*
1	Equally important	*7*	Very strong important
3	Moderate important	*9*	Extreme important
5	Strong important	*2,4,6,8*	Intermediate values

$$\text{Pair}-\text{wise comparison matrix} = \begin{bmatrix} P_{11} & P_{12} & . & P_{1n} \\ P_{21} & P_{22} & . & P_{2n} \\ . & . & . & . \\ P_{n1} & P_{n1} & . & P_{nn} \end{bmatrix}, \; P_{11} = P_{22} = P_{nn} = 1 \quad (2)$$

After constructing the pair-wise comparison matrix, the matrix is transformed to a normalized matrix calculating the weightage in each row. After that, the mean of each column is calculated from the normalize matrix, and weightage fraction (W_i) of each factor will be derived. This weightage factor can be transformed to a percentage form multiplying by 100.

The consistency ratio (*CR*) of the pair-wise matrix is analyzed using the following equation

$$CR = \frac{CI}{RI} \quad (3)$$

where *RI* is the random index given in Table 2, and *CI* can be calculated as

$$CI = \frac{\lambda_{\max} - n}{n - 1} \quad (4)$$

where $\lambda_{\max}$ is the principal eigenvalue of the comparison matrix, and *n* is the total number of criteria.

Table 2 Random index (RI) value [10]

n	*3*	*4*	*5*	*6*	*7*
RI	0.58	0.90	1.12	1.24	1.32

3 Results and Discussion

3.1 AHP Model Formulation

Four criteria are selected for waterlogging risk mapping which are elevation, slope, soil group, and land use land cover. A pair-wise comparison matrix is constructed upon experts' decisions, as shown in Table 3 for waterlogging risk in the city. The pair-wise comparison matrix is transformed to a normalized matrix calculating weightage by taking each relation value upon its column summation (Table 4). Now, calculating the average of each row value will give the weightage of each criterion (W_i), and later, this weightage is converted in percentage multiplying by 100. The weightage of elevation is 5.04%; the slope is 44.50%; land use land cover is 40.75%, and the soil group is 9.71%. Therefore, Eq. 1 can be revised as

$$\text{Waterlogging Risk} = (5.04 \times \text{Elevation}) + (44.50 \times \text{Slope}) + (40.75 \times \text{Landuseland cover}) + (9.71 \times \text{Soil group}) \quad (5)$$

The maximum eigenvalue (λ_{max}) is found 4.121, and the total number of criteria is 4. Therefore, CI is calculated using Eq. 4 and found 0.040. The RI value is taken from Table 3 and found 0.90. Now, the consistency ratio value is found 0.0448 using Eq. 3. The normalized matrix is found consistent as its value is less than 0.10.

Table 3 Pair-wise comparison matrix for AHP flood risk mapping

	Elevation	Slope	LULC	Soil group
Elevation	1	0.14	0.14	0.33
Slope	7	1	1	7
LULC	7	1	1	5
Soil group	3	0.14	0.2	1

Table 4 Normalized matrix of pair-wise comparison matrix and weightage of criteria for flood risk mapping

	Elevation	Slope	LULC	Soil group	W_i	W_i(%)
Elevation	0.0556	0.0614	0.0598	0.0248	0.0504	5.04
Slope	0.3889	0.4386	0.4274	0.5251	0.4450	44.50
LULC	0.3889	0.4386	0.4274	0.3751	0.4075	40.75
Soil group	0.1667	0.0614	0.0855	0.0750	0.0971	9.71

Table 5 Classification table of different criteria for waterlogging risk mapping

Parameter	Classification	Importance level	Weightage
Elevation	< 200 m	1	0.038
	200–203	2	0.146
	203.1–206	3	0.338
	206.1–208	4	0.235
	208.1–211	5	0.183
	211.1–215	6	0.052
	> 215	7	0.008
Slope	< 1.067	1	0.329
	1.067–2.052	2	0.330
	2.053–3.109	3	0.204
	3.110–4.432	4	0.092
	4.433–6.527	5	0.034
	6.528–10.556	6	0.009
	> 10.556	7	0.002
LULC	Savannas	3	0.151
	cropland	5	0.380
	Cropland/natural vegetation		
	Urban builtup	7	0.469
Soil	C	5	0.621
	C/D	7	0.379

3.2 Criteria Factor Selection

Four criteria decided for waterlogging map are classified into different categories based on its importance level for waterlogging in the city. Table 5 shows the classification based on its importance level along with the respective weighted area. Figure 3 shows the spatial classification of four criteria's thematic layers of the study area. The elevation and slope of the city are classified into 1–7 using Jenks natural break classification in the ArcGIS module. Land use land cover (LULC) is classified into three categories based on its pattern. The soil group is classified based on infiltration capacity. Soil group C and C/D have an ability of low and very low infiltration rates, respectively, and therefore, the classification is based on this.

3.3 Waterlogging Risk Analysis

The waterlogging analysis is analyzed using Eq. (5) and processed in the ArcGIS 10.5 raster calculator module. The relative weight of the factor is used in Table 5.

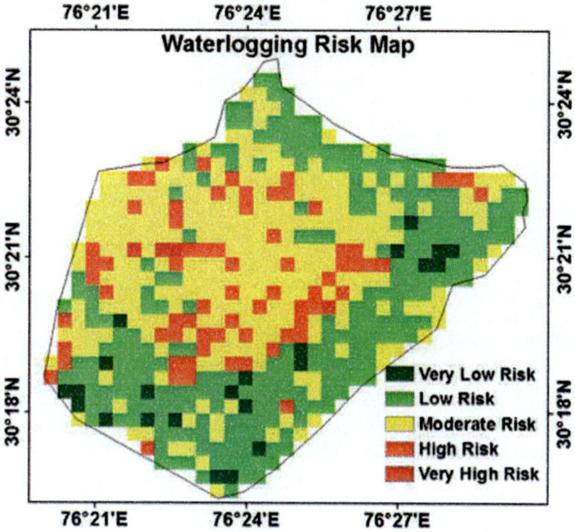

Fig. 4 Waterlogging risk map of Patiala city

Figure 4 shows the waterlogging risk map of Patiala city. The highest importance level is used for the analysis which is 1 to 7; therefore, the risk analysis value will be lie between 100 and 700. Based upon the range, the five types of risk zone are classified which are very low (100–250), low (251–400), moderate (401–500), high (501–600), and very high (601–700). It is found from risk analysis that a significant portion of the city is under moderate risk to waterlogging, i.e., 42.37% and low risk, i.e., 41.02%. The very low, high, and very high area are 3.90%, 12.03%, and 0.68%.

4 Conclusions

This study analyzes the waterlogging risk map of Patiala city of Punjab district. Analytical hierarchy process integrating with ArcGIS is used for the analysis. Four thematic layers or criteria, such as elevation, slope, land use land cover, and soil group, are used for the analysis. The city's elevation lies in the range of 182–228, and the slope lies in the range of 0–21.5°. The LULC pattern of the city is savannas (15.12), cropland (37.64%), urban builtup (46.92%), and cropland/natural vegetation (0.32%). The soil cover of the city is covered by C (62.13%) and C/D (37.87%). The pair-wise comparison matrix is prepared based on expert advice, and the weighted of each criterion is analyzed by normalizing the pair-wise comparison matrix. The weightage of elevation is 5.04%; the slope is 44.50%; land use land cover is 40.75%, and the soil group is 9.71%. These criteria are classified for the relative weight of its category. The waterlogging risk map is analyzed and found that the area under very low, low, moderate, high, and very high risk is 3.90%, 41.02%, 42.37%, 12.03%, and 0.68%, respectively. The major portion of the city is under low to moderate risk, but a part of the city is under high risk. Some preventive measures such as climate-resilient infrastructure, construction and management of sustainable drainage systems, rainwater harvesting, conservation of water bodies, and early warning system might be adopted to reduce waterlogging risk and impacts social, economic, and environmental.

References

1. Quan RS, Liu M, Lu M, Zhang LJ, Wang JJ, Xu SY (2010) Waterlogging risk assessment based on land use/cover change: A case study in Pudong New Area Shanghai. Environ Earth Sci 61(6):1113–1121
2. Sar N, Chatterjee S, Das Adhikari M (2015) Integrated remote sensing and GIS based spatial modelling through analytical hierarchy process (AHP) for water logging hazard, vulnerability and risk assessment in Keleghai river basin, India. Model Earth Syst Environ 1(4):1–21
3. Zeng J, Huang G (2018) Set pair analysis for karst waterlogging risk assessment based on AHP and entropy weight. Hydrol Res 49(4):1143–1155
4. Roy S, Bose A, Singha N, Basak D, Chowdhury IR (2021) Urban waterlogging risk as an undervalued environmental challenge: An Integrated MCDA-GIS based modeling approach. Environ Challenges 4:100194
5. Souissi D, Zouhri L, Hammami S, Msaddek MH, Zghibi A, Dlala M (2020) GIS-based MCDM—AHP modeling for flood susceptibility mapping of arid areas, southeastern Tunisia. Geocarto Int 35(9):991–1017
6. Saaty TL (1977) A scaling method for priorities in hierarchical structures. J Math Psychol 15(3):234–281
7. Bhattacharya S, Kumar V (2010) Unprecedented floods in Ghaggar basin. Dams Rivers People (July):1–9. Available from: https://sandrp.files.wordpress.com/2018/03/an_analysis_of_the_flood_disaster_in_ghaggar_basin_in_july_2010.pdf
8. Gorai S, Ratha D, Dhir A (2021) Adapting rainfall variability to flood risk: a case study of the Ghaggar River Basin. J Geoll Soc India 97(11):1347–1354
9. Dahri N, Abida H (2017) Monte Carlo simulation-aided analytical hierarchy process (AHP) for flood susceptibility mapping in Gabes Basin (southeastern Tunisia). Environ Earth Sci 76(7):302
10. Luu C, Von Meding J, Kanjanabootra S (2018) Assessing flood hazard using flood marks and analytic hierarchy process approach: a case study for the 2013 flood event in Quang Nam Vietnam. Nat Hazards 90(3):1031–1050

Analysis of Hazipur Village Water Distribution Network by Using EPANET

Akhilesh Sonker, Tuhin Mukherjee, and Ganesh D. Kale

Abstract Water is our fundamental need. The provision of safe and sufficient amount of water for both rural and urban areas is one of the crucial tasks. The importance of a water supply network is to provide safe drinking water to its community with sufficient quantity and quality and also with satisfactory pressure head. EPANET software is public domain software, which can be competently used for designing any type of water distribution network (WDN). None of the reviewed studies have analyzed WDN of the Hazipur Village, Mau District, Uttar Pradesh, India. Thus, in the present study, WDN of the Hazipur Village is analyzed for the year 2051 by using EPANET software. From the analysis, it is observed that head is varying between 150 m and 143.53 m, demand is varying between 0.41 lps and 0.01 lps, and pressure is varying from 49.2 m to 43.05 m. Also, it is observed that flow is varying from 9.91 lps to 0.01 lps, velocity is varying from 0.82 m/s to 0.01 m/s, and head loss is varying from 13.86 m to 0.01 m. Each node is found to be receiving water with adequate pressure, and demand at each node is also found to be fulfilled.

Keywords Hazipur village · Mau district · Uttar Pradesh · Water distribution network analysis · EPANET

1 Introduction

Water is our fundamental need. The provision of safe and sufficient amount of water for both urban and rural areas is one among the important tasks [8]. The importance of a water supply network is to provide safe drinking water to its community with adequate quantity and bare minimal quality and also with adequate pressure head with achieving financial constraint [15]. Therefore, analysis and design of pipe networks are vital as water availability is a vital economic development parameter [11].

A. Sonker · T. Mukherjee · G. D. Kale (✉)
Department of Civil Engineering, Sardar Vallabhbhai National Institute of Technology, Surat, 395007, India
e-mail: gdk@ced.svnit.ac.in

P. V. Timbadiya et al. (eds.), *Geospatial and Soft Computing Techniques*, Lecture Notes in Civil Engineering 339, https://doi.org/10.1007/978-981-99-1901-7_21

Software used for data management of hydraulic properties of network and design of water distribution networks (WDNs) consists of public domain software such as EPANET, branch and loop software, and commercial software such as Aquis, WaterGEMS, and WaterCAD. EPANET software is a public domain software, which can be efficiently used for designing any kind of WDN. It provides many advantages such as analysis of water quality, extended period simulation, and calculations of residual chlorine for disinfection. It can also be used to modernize or reinstate the existing water supply systems [14]. Thus, in the present study, EPANET software is used for the analysis of WDN.

Analysis/design of WDN is an important topic. Therefore, literature review is carried out on the topic of design/analysis of WDN. Various studies were carried out on analysis/design of WDNs in India [2, 3, 8, 9, 12, 16, etc]. Also various studies were carried out on analysis/design of WDNs outside of India [1, 5, 7, 10, 13, etc]. From the reviewed literature, it is inferred that none of the reviewed studies have analyzed WDN of the Hazipur Village, Mau District, Uttar Pradesh, India. Thus, in the present study, WDN of the Hazipur Village is analyzed for the year 2051 by using EPANET software.

2 Study Area and Data Source

2.1 Hajipur Gram Panchayat

Hazipur is a Village in Ghosi Block and Ghosi Tehsil of the Mau District, Uttar Pradesh, India. The population of the Hazipur Village was 2605 No's (according to census 2011). After that population was forecasted for crucial year (2051) as 5785 No's. Tube well is adopted as the source of water supply. The expected maximum yield is about 800 to 1200 L per minutes. The capacity of overhead tank is 150 KL. Hence, for the aforesaid population, by contemplating the per capita demand as 55 L per capita per day (lpcd) to which additional 15% extra quantity of water is included toward unaccounted for water to get total rate of water supply as 63.25 lpcd [17]. Index plan of Hazipur Gram Panchayat water supply scheme (WSS) is shown in Fig. 1.

2.2 Data Used

To analyze WDN of the Hazipur Village, subsequent data is collected from Uttar Pradesh (UP) Jal Nigam, Mau District [18]. Collected data consists of node-to-node connectivity of pipe, pipe length, and roughness coefficient of pipe in terms of Hazen William's C, elevations of storage tank or service reservoir, and layout map of proposed pipe distribution network (PDN) of WSS that is prepared in AutoCAD

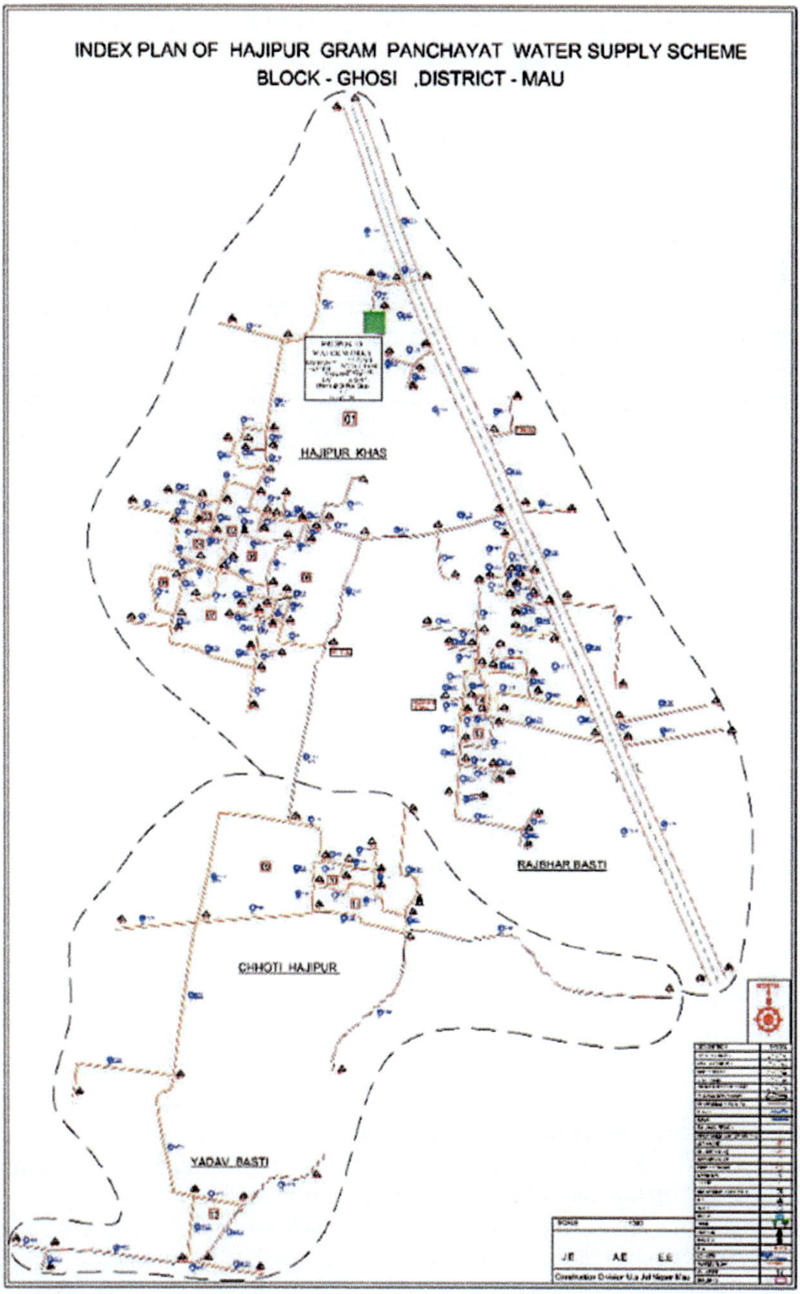

Fig. 1 Index plan of the Hajipur Gram Panchayat WSS. *Source* Uttar Pradesh Jal Nigam, Mau District) [18]

by conducting level surveying. For Hazipur Village WSS, material of pipe used is DI.

2.3 *Methodology*

EPANET is public domain software used for WDS modeling, and it is developed by Water Supply and Water Resource Division of the United States Environmental Protection Agency (EPA). This software provides a hydraulic analysis, which can handle systems of any size. EPANET monitors water flow in every pipe, water level in every tank and also calculates pressure at each node all over the network. EPANET is a WDN modeling software that executes extended period simulation of hydraulic as well as water quality behavior [3]. Thus, EPANET software is used in the present study for the analysis of WDN.

Todini's global gradient algorithm (GGA) is employed in EPANET to solve system of equations. The GGA utilizes a linearization of the conservation equations in an iterative Newton–Raphson method that consequences in a two-step solution procedure at every iteration [19].

Gradient method uses a linearized equation [4] given below.

$$
\begin{aligned}
&{}_{t+1}H_i - {}_{t+1}H_j - ({}_nR_{ox}[{}_tQ_{ox}]^{n-1})_{t+1}Q_x = (1-n)R_{ox}{}_tQ^n_{ox} \\
&\text{where,}\quad x = 1, 2 \ldots X
\end{aligned}
$$

where

R_{ox}	Known resistance constant of pipe x
${}_tH_{oi}$ and ${}_tH_{oj}$	Known or assumed nodal heads for the tth iteration
${}_t\Delta H_i$, ${}_t\Delta H_j$ and ${}_t\Delta Q_x$	The unknown corrections for the tth iteration

3 Results and Discussion

In the present study, WDN of the Hazipur Village is analyzed for the year 2051 by using EPANET software. After simulation of WDN, the following results are obtained, which includes estimates of demand, head, and pressure available at each node as shown in the Table 1. Also results consist of estimates of flow, velocity, unit head loss, and friction factor for each pipe as shown in the Table 2.

From junction report (Table 1), it is observed that head is varying between 150 m and 143.53 m, demand is varying between 0.41 lps and 0.01 lps, pressure is varying from 49.2 m to 43.05 m. From pipe report (Table 2), it is observed that flow is varying from 9.91 lps to 0.01 lps, velocity is varying from 0.82 m/s to 0.01 m/s, and head loss is varying from 13.86 m to 0.01 m. As per Central Public Health and Environmental Engineering Organisation [6], minimum residual pressure needed for the two-story

Table 1 Global climate models used in the present analysis

Node ID	Demand (LPS)	Head (m)	Pressure (m)	Node ID	Demand (LPS)	Head (m)	Pressure (m)
Resvr J/1	– 12.73	150	0	Junc J/17	0.22	148.73	48.44
Junc J/2	0.13	149.61	49.2	Junc J/18	0.05	148.73	46.99
Junc J/3	0.22	149.38	47.54	Junc J/19	0.03	148.29	46.97
Junc J/4	0.2	149.05	48.03	Junc J/20	0.02	148.09	46.31
Junc J/5	0.2	148.28	46.61	Junc J/21	0.04	147.89	46.24
Junc J/6	0.15	147.86	47.7	Junc J/22	0.01	147.3	45.51
Junc J/7	0.17	147.45	46.51	Junc J/23	0.02	147.25	45.99
Junc J/8	0.1	147.26	46.32	Junc J/24	0.02	147.21	46.92
Junc J/9	0.03	147.21	46.33	Junc J/25	0.05	147.26	46.62
Junc J/10	0.04	147.2	46.4	Junc J/26	0.04	147.86	46.95
Junc J/11	0.05	147.25	46.45	Junc J/27	0.05	149.05	48.93
Junc J/12	0.05	147.3	45.52	Junc J/28	0.02	149.05	49.13
Junc J/13	0.08	147.41	45.54	Junc J/29	0.02	149.05	48.25
Junc J/14	0.08	147.89	46.69	Junc J/30	0.14	149.35	47.73
Junc J/15	0.04	148.09	46.81	Junc J/31	0.27	149.09	47.77
Junc J/16	0.1	148.29	47.03	Junc J/32	0.14	149.06	47.39
Junc J/33	0.23	147.01	46.14	Junc J/71	0.02	146.65	45.66
Junc J/34	0.04	147.01	46.65	Junc J/72	0.02	146.65	45.69
Junc J/35	0.17	146.33	45.53	Junc J/73	0.31	145.78	45.4
Junc J/36	0.03	146.33	45.87	Junc J/74	0.08	145.19	44.76
Junc J/37	0.27	145.7	44.4	Junc J/75	0.09	145.1	44.48
Junc J/38	0.1	145.7	44.74	Junc J/76	0.18	145.08	45.11
Junc J/39	0.18	145.33	44.78	Junc J/77	0.41	145.11	45.14
Junc J/40	0.07	145.32	44.88	Junc J/78	0.06	145.11	45.03
Junc J/41	0.25	145.3	44.87	Junc J/79	0.05	145.15	44.31
Junc J/42	0.08	145.29	45.96	Junc J/80	0.06	145.09	44.58
Junc J/43	0.15	145.26	43.91	Junc J/81	0.1	145.08	44.26
Junc J/44	0.12	146.98	46.98	Junc J/82	0.02	145.09	44.95
Junc J/45	0.1	146.9	46.2	Junc J/83	0.01	145.15	44.22
Junc J/46	0.07	146.93	46.23	Junc J/84	0.14	144.98	45.84
Junc J/47	0.08	147	45.87	Junc J/85	0.41	144.97	44.68
Junc J/48	0.04	146.93	45.61	Junc J/86	0.13	144.95	46.21
Junc J/49	0.05	146.96	45.64	Junc J/87	0.21	144.89	44.04
Junc J/50	0.02	146.96	45.22	Junc J/88	0.34	144.33	45.11
Junc J/51	0.1	146.84	45.71	Junc J/89	0.1	144.32	45.3

(continued)

Table 1 (continued)

Node ID	Demand (LPS)	Head (m)	Pressure (m)	Node ID	Demand (LPS)	Head (m)	Pressure (m)
Junc J/52	0.07	146.83	45.55	Junc J/90	0.2	143.62	44.34
Junc J/53	0.1	146.9	45.62	Junc J/91	0.17	143.58	44.28
Junc J/54	0.04	146.9	45.91	Junc J/92	0.11	143.57	43.64
Junc J/55	0.1	146.81	46.6	Junc J/93	0.16	143.57	43.64
Junc J/56	0.24	146.51	46.1	Junc J/94	0.15	143.53	44.73
Junc J/57	0.05	146.61	46.2	Junc J/95	0.02	143.53	44.43
Junc J/58	0.04	146.65	46.01	Junc J/96	0.03	143.53	44.25
Junc J/59	0.05	146.75	46.2	Junc J/97	0.05	143.57	45.44
Junc J/60	0.02	146.75	46.21	Junc J/98	0.08	143.57	42.99
Junc J/61	0.03	146.6	45.66	Junc J/99	0.08	148.14	46.47
Junc J/62	0.05	146.65	46.12	Junc J/100	0.06	148.14	46.7
Junc J/63	0.05	146.65	45.69	Junc J/101	0.01	148.14	48.08
Junc J/64	0.12	146.65	45.72	Junc J/102	0.02	148.14	47.78
Junc J/65	0.11	146.73	45.86	Junc J/103	0.06	147.93	46.13
Junc J/66	0.11	146.74	44.98	Junc J/104	0.02	147.93	46.84
Junc J/67	0.11	146.78	44.89	Junc J/105	0.01	147.93	46.82
Junc J/68	0.03	146.84	45.11	Junc J/106	0.01	147.93	46.26
Junc J/69	0.03	146.73	45.99	Junc J/107	0.07	147.29	45.62
Junc J/70	0.03	146.65	45.82	Junc J/108	0.03	147.29	45.54
Junc J/109	0.12	146.6	45.12	Junc J/125	0.02	143.92	43.32
Junc J/110	0.04	146.6	44.96	Junc J/126	0.08	143.92	43.72
Junc J/111	0.13	145.04	44.08	Junc J/127	0.04	143.92	43.65
Junc J/112	0.12	145	44.68	Junc J/128	0.03	143.92	43.92
Junc J/113	0.06	145	44.05	Junc J/129	0.01	143.93	44.16
Junc J/114	0.09	144.99	44.88	Junc J/130	0.07	143.91	43.51
Junc J/115	0.03	144.99	44.44	Junc J/131	0.04	143.91	43.45
Junc J/116	0.04	144.99	44.22	Junc J/132	0.01	143.91	43.15
Junc J/117	0.3	144.63	43.8	Junc J/133	0.01	143.91	43.21
Junc J/118	0.13	143.96	43.39	Junc J/134	0.1	143.9	44.51
Junc J/119	0.08	143.93	43.16	Junc J/135	0.02	143.9	44.05
Junc J/120	0.05	143.93	43.16	Junc J/136	0.09	143.89	44.39
Junc J/121	0.06	143.95	43.9	Junc J/137	0.01	143.89	44.52
Junc J/122	0.05	143.93	43.05	Junc J/138	0.02	143.89	44.01
Junc J/123	0.08	143.93	42.94	Junc J/139	0.19	144.56	43.26
Junc J/124	0.05	143.92	43.73	Junc J/140	0.09	144.97	44.43

Table 2 Simulated flows, velocities, unit head losses, and friction factors

Link ID	Flow (LPS)	Velocity (m/s)	Unit head loss m/km	Friction factor	Link ID	Flow (LPS)	Velocity (m/s)	Unit head loss m/km	Friction factor
P-1	12.73	1.02	7.89	0.019	P-20	0.02	0.01	0	0.048
P-2	9.9	0.79	4.96	0.019	P-21	0.04	0.02	0.01	0.039
P-3	7.55	0.61	3	0.02	P-22	0.01	0.004	0	0.045
P-4	7.26	0.58	2.79	0.02	P-23	0.02	0.01	0	0.046
P-5	4.5	0.59	3.73	0.021	P-24	0.02	0.01	0	0.046
P-6	4.32	0.56	3.45	0.021	P-25	0.05	0.02	0.02	0.038
P-7	0.91	0.36	2.85	0.025	P-26	0.04	0.02	0.01	0.04
P-8	0.76	0.3	2.02	0.026	P-27	0.1	0.04	0.04	0.035
P-9	0.71	0.28	1.78	0.026	P-28	0.02	0.01	0	0.043
P-10	0.62	0.24	1.39	0.026	P-29	0.02	0.01	0	0.045
P-11	0.69	0.27	1.68	0.026	P-30	0.14	0.05	0.09	0.033
P-12	0.75	0.29	1.99	0.026	P-31	1.99	0.78	12.14	0.022
P-13	2.1	0.82	13.41	0.022	P-32	0.14	0.06	0.09	0.033
P-14	2.22	0.63	6.77	0.022	P-33	1.58	0.62	7.87	0.023
P-15	2.29	0.65	7.12	0.022	P-34	0.04	0.02	0.01	0.039
P-16	2.42	0.47	3.14	0.023	P-35	1.3	0.51	5.52	0.024
P-17	2.69	0.52	3.82	0.022	P-36	0.03	0.01	0.01	0.041
P-18	0.05	0.02	0.01	0.039	P-37	1.1	0.43	4.03	0.024
P-19	0.03	0.01	0	0.042	P-38	0.1	0.04	0.05	0.035
P-39	0.73	0.29	1.9	0.026	P-75	0.03	0.01	0.01	0.04
P-40	0.07	0.03	0.03	0.036	P-76	0.03	0.01	0.01	0.041

(continued)

Table 2 (continued)

Link ID	Flow (LPS)	Velocity (m/s)	Unit head loss m/km	Friction factor	Link ID	Flow (LPS)	Velocity (m/s)	Unit head loss m/km	Friction factor
P-41	0.48	0.19	0.87	0.027	P-77	0.03	0.01	0.01	0.04
P-42	0.08	0.03	0.03	0.036	P-78	0.02	0.01	0	0.044
P -43	0.15	0.06	0.11	0.033	P-79	0.02	0.01	0	0.044
P-44	1.29	0.5	5.41	0.024	P-80	3.76	0.49	2.67	0.022
P-45	0.42	0.17	0.69	0.028	P-81	1.37	0.54	6.03	0.024
P-46	0.42	0.17	0.69	0.028	P-82	0.7	0.28	1.76	0.026
P-47	0.61	0.24	1.35	0.027	P-83	0.38	0.15	0.57	0.028
P-48	1.27	0.5	5.27	0.024	P-84	0.22	0.09	0.2	0.031
P-49	0.12	0.05	0.07	0.034	P-85	2.09	0.4	2.38	0.023
P-50	0.5	0.2	0.94	0.027	P-86	0.06	0.02	0.02	0.037
P-51	0.58	0.23	1.23	0.027	P-87	0.58	0.23	1.24	0.027
P-52	0.02	0.01	0	0.044	P-88	0.52	0.2	1	0.027
P-53	0.75	0.29	1.98	0.026	P-89	0.43	0.17	0.72	0.028
P-54	0.35	0.14	0.48	0.029	P-90	0.23	0.09	0.23	0.031
P-55	0.44	0.17	0.73	0.028	P-91	0.02	0.01	0	0.042
P-56	0.58	0.23	1.22	0.027	P-92	0.01	0.004	0	0.062
P-57	0.04	0.02	0.01	0.039	P-93	0.57	0.22	1.19	0.027
P-58	0.74	0.29	1.95	0.026	P-94	0.33	0.13	0.43	0.029
P-59	0.27	0.1	0.29	0.03	P-95	0.42	0.17	0.68	0.028
P-60	3.24	0.63	5.38	0.022	P-96	0.13	0.05	0.08	0.033
P-61	0.76	0.3	2.05	0.026	P-97	0.21	0.08	0.19	0.031

(continued)

Table 2 (continued)

Link ID	Flow (LPS)	Velocity (m/s)	Unit head loss m/km	Friction factor	Link ID	Flow (LPS)	Velocity (m/s)	Unit head loss m/km	Friction factor
P-62	0.85	0.33	2.48	0.025	P-98	1.4	0.4	2.86	0.024
P-63	0.84	0.33	2.47	0.025	P-99	0.1	0.04	0.05	0.035
P-64	0.91	0.36	2.84	0.025	P-100	0.96	0.37	3.11	0.025
P-65	0.02	0.01	0	0.044	P-101	0.38	0.15	0.57	0.028
P-66	0.03	0.01	0	0.042	P-102	0.14	0.05	0.08	0.033
P-67	0.04	0.02	0.01	0.04	P-103	0.02	0.01	0	0.042
P-68	0.12	0.05	0.06	0.034	P-104	0.37	0.15	0.55	0.029
P-69	0.18	0.07	0.15	0.032	P-105	0.19	0.08	0.16	0.031
P-70	0.34	0.13	0.45	0.029	P-106	0.02	0.01	0	0.047
P-71	0.49	0.19	0.89	0.027	P-107	0.03	0.01	0	0.042
P-72	0.36	0.14	0.5	0.029	P-108	0.05	0.02	0.01	0.038
P-73	0.71	0.28	1.8	0.026	P-109	0.08	0.03	0.03	0.036
P-74	0.24	0.09	0.24	0.03	P-110	2.55	0.5	3.47	0.022
P-110	2.55	0.5	3.47	0.022	P-132	0.32	0.12	0.41	0.029
P-111	0.09	0.03	0.04	0.035	P-133	0.61	0.24	1.37	0.027
P-112	0.01	0.002	0	0.097	P-134	0.13	0.05	0.08	0.033
P-113	0.02	0.01	0	0.045	P-135	0.07	0.03	0.03	0.036
P-114	2.38	0.46	3.05	0.023	P-136	0.23	0.09	0.23	0.031
P-115	0.04	0.02	0.01	0.038	P-137	0.22	0.09	0.21	0.031
P-116	0.01	0.004	0	0.056	P-138	0.02	0.01	0	0.045
P-117	0.01	0.002	0	0.111	P-139	0.15	0.06	0.11	0.033

(continued)

Table 2 (continued)

Link ID	Flow (LPS)	Velocity (m/s)	Unit head loss m/km	Friction factor	Link ID	Flow (LPS)	Velocity (m/s)	Unit head loss m/km	Friction factor
P-118	0.28	0.9	15.62	0.022	P-140	0.04	0.02	0.01	0.04
P-119	0.03	0.01	0	0.041	P-141	0.03	0.01	0.01	0.041
P-120	2.19	0.86	14.4	0.022	P-142	0.01	0.004	0	0.056
P-121	0.04	0.02	0.01	0.039	P-143	0.37	0.14	0.53	0.029
P-122	2.03	0.79	12.53	0.022	P-144	0.06	0.02	0.02	0.037
P-123	0.34	0.13	0.46	0.029	P-145	0.01	0.002	0	0.097
P-124	0.06	0.02	0.02	0.037	P-146	0.01	0.004	0	0.059
P-125	0.16	0.06	0.11	0.032	P-147	0.24	0.09	0.24	0.03
P-126	0.03	0.01	0	0.041	P-148	0.02	0.01	0	0.044
P-127	0.04	0.02	0.01	0.039	P-149	0.12	0.05	0.07	0.034
P-128	1.55	0.61	7.66	0.023	P-150	0.01	0.004	0	0.043
P-129	1.06	0.42	3.77	0.024	P-151	0.02	0.01	0	0.044
P-130	0.32	0.12	0.4	0.029	P-152	0.19	0.08	0.16	0.031
P-131	0.13	0.05	0.08	0.033	P-153	0.09	0.04	0.04	0.035

building is 12 m, and pressure head available at all nodes is more than minimum residual pressure requirement of 12 m.

4 Conclusions

- In the current study, WDN of the Hazipur Village is analyzed for the year 2051 by using EPANET software.
- As per Central Public Health and Environmental Engineering Organisation [6], minimum residual pressure needed for the two-story building is 12 m, and pressure head available at all nodes is more than minimum residual pressure requirement of 12 m.
- From the study, it is concluded that each node is having adequate pressure, and there is no shortage in supply of water needed for fulfilling the demand.

Acknowledgements Authors are thankful to Uttar Pradesh (UP) Jal Nigam, Mau District for providing necessary data required for this study.

References

1. Alkali AN, Yadima SG, Usman B, Ibrahim UA, Lawan AG (2017) Design of a water supply distribution network using EPANET 2.0: a case study of Maiduguri Zone 3, Nigeria. Arid Zone J Eng Technol Environ 13(3):347–355
2. Anisha G, Kumar A, Kumar JA, Raju PS (2016) Analysis and design of water distribution network using EPANET for Chirala Municipality in Prakasam District of Andhra Pradesh. Int J Eng Appl Sci 3(4)
3. Athulya T, Ullas AK (2020) Design of water distribution network using EPANET software. Int Res J Eng Technol (IRJET) 7(3):1774–1778
4. Bhave PR, GuptaR (2009) Analysis of water distribution networks. Narosa Publishing House Pvt Ltd New Delhi
5. Cai L, Wang R, Ping J, Jing Y, Sun J (2015) Water supply network monitoring based on demand reverse deduction (DRD) technology. Procedia Engineering 119:19–27
6. Central Public Health and Environmental Engineering Organization (1999) Manual on water supply and treatment. Ministry of Urban Development New Delhi India
7. Kara S, Karadirek IE, Muhammetoglu A, Muhammetoglu H (2016) Hydraulic modeling of awater distribution network in a tourism area with highly varying characteristics. Procedia Eng 162:521–529
8. Koradiya D, Khokhani VH (2018) Water distribution network analysis and design using LOOPand EPANET software. Int J Sci Res Develop 6(2):2275–2279
9. Kumar A, KumarK BB, Matial N, Dey E, Singh M, Malhotra N (2015) Design of water distribution system using EPANET. Int J Adv Res 3(9):789–812
10. Martin-Candilejo A, Santillán D, Iglesias A, Garrote L (2020) Optimization of the design of water distribution systems for variable pumping flow rates. Water 12(2):359
11. Mehta VN, Joshi GS (2017) Design of rural water supply system using Branch 3.0-a case study for Nava-Shihora region, India. Int J Civ Eng Technol 8(2):618–630

12. Mehta VN, Joshi GS (2019) Design and analysis of rural water supply system using loop 4.0 and water gems V8i for Nava Shihora Zone I. Int J Eng Adv Technol 9:2258–2266
13. Saminu A, Abubakar N, Sagir L (2013) Design of NDA water distribution network using EPANET. Int J Emerg Sci Eng (IJESE) 1(9):5–9
14. Sonaje NP, Joshi MG (2015) A review of modeling and application of water distribution networks (WDN) softwares. Int J Tech Res Appl 3(5):174–178
15. Srivastava H, Singhal A (2018) Design and analysis of optimized water distribution network at BITS Pilani, Pilani Campus. In: Proceeding of 3rd world congress at civil structural and environmental engineering (CSEE'2018) Budapest Hungary
16. Sumithra RP, NethajiMVE AJ (2013) Feasibility analysis and design of water distribution system for Tirunelveli Corporation using LOOP and WaterGEMS software. Int J Appl Bioeng 7(1):61–70
17. Technical Statement Hazipur Gram Panchayat W/S Scheme, Dist. Mau, 2019–20
18. Uttar Pradesh Jal Nigam, Mau District
19. https://epanet22.readthedocs.io/en/latest/12_analysis_algorithms.html

Comparison of Heuristic and Metaheuristic Evolutionary Algorithms on Optimal Design of Water Distribution Networks

Prerna Pandey, Devang Singh, Shilpa Dongre, and Rajesh Gupta

Abstract Complexity in the design of water distribution networks (WDNs) is primarily due to the discrete nature of available pipe sizes and the nonlinear relationship between pipe discharge and head loss through it. Several evolutionary algorithms have been suggested by researchers for the optimal design of WDNs in the last two-decades considering different complexities. These have been tested and evaluated on many benchmark networks. Most of these algorithms are inspired by natural phenomenon of selection and called *metaheuristic* evolutionary algorithms. Metaheuristic algorithms are parameter-based, and thus fixing their values for a particular problem is always a challenge. To overcome this issue, *heuristic* evolutionary algorithms like Jaya, Rao-I, and Rao-II have been suggested by Rao (2016, 2019) which are parameter-free algorithms. How best these three algorithms are applicable to design WDNs of various complexities which have been explored through their applications on various bench mark problems. Further, the efficiency and efficacy of these algorithms are compared with a metaheuristic algorithm 'particle swarm optimization' (PSO) which is based on the phenomenon of bird flocking and require few parameters to be defined. PSO is successfully used by many researchers and found to be better in comparison with other metaheuristic algorithms. In this study, Rao-I algorithm is found to be faster in providing optimal solution as compared to Jaya and Rao-II. In comparison with PSO, Rao-I is observed to provide similar solution in more or less same number of evolutions. Even though these heuristic evolutionary algorithms do not mimic any natural phenomenon, they are observed to provide better optimal solutions for WDNs.

Keywords Water distribution networks (WDNs) · Optimization · Heuristic · Metaheuristic · Parameter-free algorithm

P. Pandey (✉) · D. Singh · S. Dongre · R. Gupta
Department of Civil Engineering, Visvesvaraya National Institute of Technology (VNIT), Nagpur, Maharashtra 440010, India
e-mail: prerna.pandey1203@gmail.com

P. V. Timbadiya et al. (eds.), *Geospatial and Soft Computing Techniques*, Lecture Notes in Civil Engineering 339, https://doi.org/10.1007/978-981-99-1901-7_22

1 Introduction

The water distribution networks (WDNs) are crucial and costly infrastructure and hence constitute a major financial drain in all urban societies. Because urban areas continue to grow as their networks age, new networks, and network improvements or upgrades remain critical issues. Of all the money spent on water supply system facilities, distribution network expenses can be somewhat substantial, sometimes even exceeding 70%. Therefore, the network should be optimized. Minimizing the cost of the network, whilst fulfilling demand and pressure restrictions at all nodes, is an optimal design of the WDN.

The optimal WDN design for a known layout and demand pattern comprises of minimizing the network cost by selecting pipe sizes for various links. As a result, pipe sizes are regarded as design variables. The optimal design of a network can be accomplished using different algorithms, including deterministic, stochastic or heuristic, and metaheuristic algorithms. Despite their simplicity, deterministic algorithms such as linear programming gradient (LPG) [1] and nonlinear programming (NLP) [2] have more chances of being trapped at local optimal solution. As a result, the necessity for improved evolutionary algorithms (EAs) becomes apparent.

The genetic algorithm (GA) [3], the simulated annealing (SA) method [4], the scrambled frog leap algorithm (SFLA) [5], and the differential evolution (DE) [6] algorithm are some of the metaheuristic techniques. Particle swarm optimization (PSO) [7] is inspired by the social behaviour of birds flocking, and ant colony optimization (ACO) [8] is inspired by the social behaviour of ants which are some nature inspired algorithms. El-Ghandour and Elbeltagi [9] compared five different optimization models for the design and rehabilitation of WDNs, namely the GA, PSO, ACO, memetic algorithm (MA), and modified SFLA. They found that, of these five models, the PSO model provided the best solution for design or rehabilitation, using speedy convergence and the efficiency rate metric as performance criteria for the selected networks. Hybrid algorithms were proposed to improve the efficiency of metaheuristic algorithms. Kadu et al. [10] used critical path method to select candidate pipe sizes for each link, thereby reducing search space drastically in GA-based search method. Zheng et al. [11] introduced DE-NLP, where NLP approximates optimal solutions and DE polishes them. Sedki and Ouazar [12] created PSO-DE, combining DE's exploration power with PSO's local search power. Moreover, [13] proposed CSHS, a hybrid cuckoo search (CS), and harmony search (HS) algorithm. The CSHS created to counter CS's poor convergence rate and lack of information flow between HS people.

Metaheuristic EAs, as discussed above, require fine-tuning of around 4 to 7 parameters as depicted in Table 1 [14]. Further, these algorithms include algorithm-specific constants. These algorithm-specific parameters need to be adjusted for each optimization problem through a trial-and-error procedure, making their application difficult. On the other hand, the heuristic EAs, Jaya, Rao-I, and Rao-II method, does not require any algorithm-specific parameter values to achieve its goal and mostly require

Table 1 Comparison of different algorithms based on its parameter requirement

Algorithm	Parameters requirement
GA	Crossover probability, mutation probability
PSO	Cognition weight, social weight, inertia weight
ACO	Pheromone evaporation parameter, pheromone weighting parameter
SA	Annealing rate, initial temperature
HS	HM size, pitch adjustment rate, HM considering rate
CSHS	Scaling factor, discovering probability of alien eggs
DE	Mutation probability, crossover probability
CSA	Step size control parameter 'α' and discovering probability parameter 'Pa'
Jaya, Rao-I, Rao-II	N.A

common parameters such as population size, iteration count, and, in some situations, a penalty parameter to covert the constrained problem to an unconstrained problem.

This paper focuses on demonstrating less complicated optimization algorithm applications on WDNs design, with no or minimal interaction of algorithm-specific parameters. The parameter-free algorithms are also compared with the PSO, which is considered to be the best amongst all metaheuristic EAs.

2 Methodology

2.1 Mathematical Model for Optimal Design of WDN

A WDN optimization model [15] can be formulated as a cost minimization problem subjected to satisfying the constraints of node-flow continuity and loop head loss and assuring availability of pressure more than the desired one at each node.

Objective Function

$$\text{Min } f(D_1 \dots D_i) = \sum C(D_i) * L_i \tag{1}$$

Constraints

$$\sum_{i \in j} Q_i + q_j = 0; \quad j = 1 \dots n_d \tag{2}$$

$$\left(\sum_{i=1}^{y} HL_i + \sum_{p=1}^{yp} E_p \right)_l = 0; \quad l = 1, 2, \dots nl \tag{3}$$

$$H_j \geq H_j^{\min}; \quad j = 1 \ldots n_d \tag{4}$$

$$D_i \in \{D_{\min} \ldots D_{\max}\}; \quad i = 1, \ldots I \tag{5}$$

$$HL_i = \frac{\alpha * L_i * Q_i^{1.852}}{C_{HW}^{1.852} * D_i^{4.87}} \tag{6}$$

where $i = 1 \ldots I$ (number of links in the pipe network); $C(D_i)$ = unit cost of pipe having diameter D (commercial), q = nodal demand, n_d = number of demand nodes, L = length of the pipe, E_p = energy added to water by pump, Q = flow in any pipe, HL = head loss in pipe, C_{HW} = Hazen-Williams coefficient, α = numerical constant, H_j = actual head available at node j, $H_j^{\min}$ = minimum head required at node j, $D_{\min}$ and D_{max} = minimum and maximum diameter of commercially available pipes, respectively.

Modified Pipe Cost Problem According to Optimization Method

The objective function as in the Eq. (1) is modified using the penalty function. The penalty-based model of [10] is used. Using that Eq. (1) can be modified as

$$\text{Min} f(D_1 \ldots D_i) = \sum_{i=1}^{I} C(D_i) * L_i + \sum_{j=1}^{n_d} p * q_j * \left\{\max\left(H_j^{\min} - H_j, 0\right)\right\} \tag{7}$$

where p = penalty function.

2.2 Computer Codes

The optimization model discussed in above section can be solved using various metaheuristic and heuristic EAs. In this work, codes in MATLAB are developed using EPANET programmer's toolkit DLL files [16] for Jaya algorithm, Rao-I, Rao-II, and PSO. Hydraulic network solver EPANET solves Eqs. (2) and (3) for any member of population selected randomly based on Eq. (5). EPANET provides the nodal heads as output and used to check availability of pressure as per Eq. (4). Penalty is considered if constraints are not satisfied at one or more number of nodes.

2.3 *Solution Techniques*

Jaya Algorithm

Jaya is a simple yet powerful optimization algorithm proposed by [17]. This algorithm is based on the concept that the solution obtained for a given problem should move towards the best solution and avoid the worst solution.

Let the function to be optimized be *f(x)*, 'm' be the number of design variables (that is, $j = 1, 2, \ldots, m$), '*n*' be the number of candidate solutions (i.e. *k*, population size = 1,2,...,*n*) and *i* indicates the iterations. The procedure is as follows:

- At the start, population size, number of design variables, and the criteria for termination are required to be initialized. From the given set of population, the best ($f(x)_{\text{best}}$) and worst ($f(x)_{\text{worst}}$) solutions are obtained by solving the objective function for all candidate solutions available.
- Let $X_{j,k,I}$ be representing the value of the *j*th variable for the *k*th population in the *i*th iteration. Then, $X_{j,best,i}$ will indicate the best value, and $X_{j,worst,i}$ exhibits the worst value of the *j*th variable with respect to the best and worst solutions of the function, respectively, during the *i*th iteration.
- The value for the decision variable $X_{j,k,i}$ is then modified for the next iteration using Eq. (8), which assures that the solution always moves towards the best one.

$$X'_{j,k,i} = X_{j,k,i} + r_{1,j,i}\left(X_{j,\text{best},i} - \left|X_{j,k,i}\right|\right) + r_{2,j,i}\left(X_{j,\text{worst},i} - \left|X_{j,k,i}\right|\right) \quad (8)$$

- Where $X'_{j,k,i}$ is the modified value of the variable, $r_{1,j,i}$ and $r_{2,j,i}$ are two randomly generated numbers between 0 and 1 for the *j*th variable in the *i*th iteration.
- The modified variables are then estimated, if it is found better than the previous solution, then the modified solution is accepted, and the previous value is discarded. Else, the previous solution is kept as the current solution. If termination criteria (e.g. number of iterations) are reached, the solution is reported, otherwise the procedure is repeated till the best solution is observed.

Rao-I and Rao-II

The Rao-I and Rao-II [8] algorithms are based on the similar concept as used in Jaya algorithm, except the difference in the equations required to update the solutions as given in Eqs. (9) and (10).

$$X'_{j,k,i} = X_{j,k,i} + r_{1,j,i}\left(X_{j,\text{best},i} - X_{j,\text{worst},i}\right) \quad (9)$$

$$X'_{j,k,i} = X_{j,k,i} + r_{1,j,i}\left(X_{j,\text{best},i} - X_{j,\text{worst},i}\right) + r_{2,j,i}\left(\left|X_{j,k,i} \text{ or } X_{j,l,i}\right| - \left|X_{j,l,i} \text{ or } X_{j,k,i}\right|\right) \quad (10)$$

Particle Swarm Optimization (PSO)

[7] developed PSO algorithm which replicate natural swarming behaviour. A particle represents a swarming bird, bee, fish, or other natural agent. PSO provides a population-based search strategy. The process flow of the algorithm is as follows:

- Set of population and design variables at the starting are required to be initialized. Each particle is tested for pressure constraints at each node using EPANET. If any node fails to satisfy the minimum pressure head constraints, the fitness index for the particular particle is calculated as in Eq. (11):

$$\text{Fitness index} = \text{Cost} + (W * \text{penalty}) \tag{11}$$

- Fitness index is the cost of the network, when pressure at all the nodes is satisfied, the 'W' becomes zero. The lower fitness index resembles the best or feasible solution. Later, the local and global best for each particle is identified.
- The initial velocity for each particle is set using Eq. (12);

$$V_{m,n} = x_{\min,n} + r_1 * \left[x_{\max,n} - x_{\min,n}\right] \tag{12}$$

where $x_{\min,n}$ is the minimum value of individual in population, r_1 is random number (0,1), and $x_{\max,n}$ is the maximum value of individual.

New position of the particle is find using Eq. (13);

$$x_{m,n}(t+1) = x_{m,n}(t) + V_{m,n}(t+1) \tag{13}$$

where $x_{m,n}$ is position matrix, whilst $v_{m,n}$ is the velocity matrix.

- The fitness index of every particle is identified using new position, followed by updating local and global best.
- Using Eq. (14), the velocities are updated, and further, the new positions of particle updated again using Eq. (12).

$$V_{m,n}(t+1) = V_{m,n}(t) * w + C_1 * r_1 * \left(p_{m,n} - x_{m,n}\right) + C_2 * r_2 * \left(g_n - x_{m,n}\right) \tag{14}$$

where C_1 and C_2 are positive constants defined [0,2], w is inertia weight (0.5 + (r_1/2)), r_1 and r_2 are the random numbers ranges between [0,1], $p_{m,n}$ is local best, and g_n is global best.

- The fitness index, position of the particle, local, and global best are updated till the termination criteria met.

As can be observed, the PSO approach works similarly to the Jaya, Rao-I, and Rao-II algorithms, substituting the current solution for the prior one if the present solution is infeasible or vice versa. Additionally, the parameters C_1 and C_2 must be

Table 2 Description of all the five WDN used for the application of proposed algorithms

Case study	Name	Description	Reference
3.1	Two-loop network	The network consisting of a single source node, 6 demand nodes, and 8 pipes as shown in Fig. 1a	[1]
3.2	Hanoi network	The network consists of a single source, three loops, 31 demand nodes, and 34 pipes primary pipes as shown in Fig. 1b	[2]
3.3	Kadu network	The network consists of two-reservoir, 34 pipes, and 26 nodes arranged in nine loops which is shown in Fig. 1c	[10]
3.4	New York tunnel	The network consists of one source node and 19 demand nodes and 21 pipes as shown in Fig. 1d	[18]
3.5	Bengali camp (Chandrapur) zone network	The network is single sourced gravity feed and consists of 34 nodes, 38 pipes as shown in Fig. 1e	[19]

predefined for PSO, and their values have to be between [0,2], whilst the random numbers r_1 and r_2 have to be between [0,1], and the inertia must be proportional to the random number. As a result, the PSO only requires a few parameters, whose range has little effect on the solution. Thus, the application of both algorithms on WDNs is compared.

3 Case Studies

Five case studies as described in Table 2 are considered to compare different EAs.

4 Applications and Result

All the four algorithms are used to solve all five WDN problems. The performance evaluation of proposed algorithms was done by running the algorithms for ten times with varied population sizes to see how well they performed, and how well they converged. In order to minimize the number of function evaluations required, the best population size for each WDN is established. Determinants of the number of choice factors and the average number of function evaluations necessary to reach the optimal solution were also discovered. Thus, as the network expands in size, so does the number of function evaluations required to achieve optimality. When

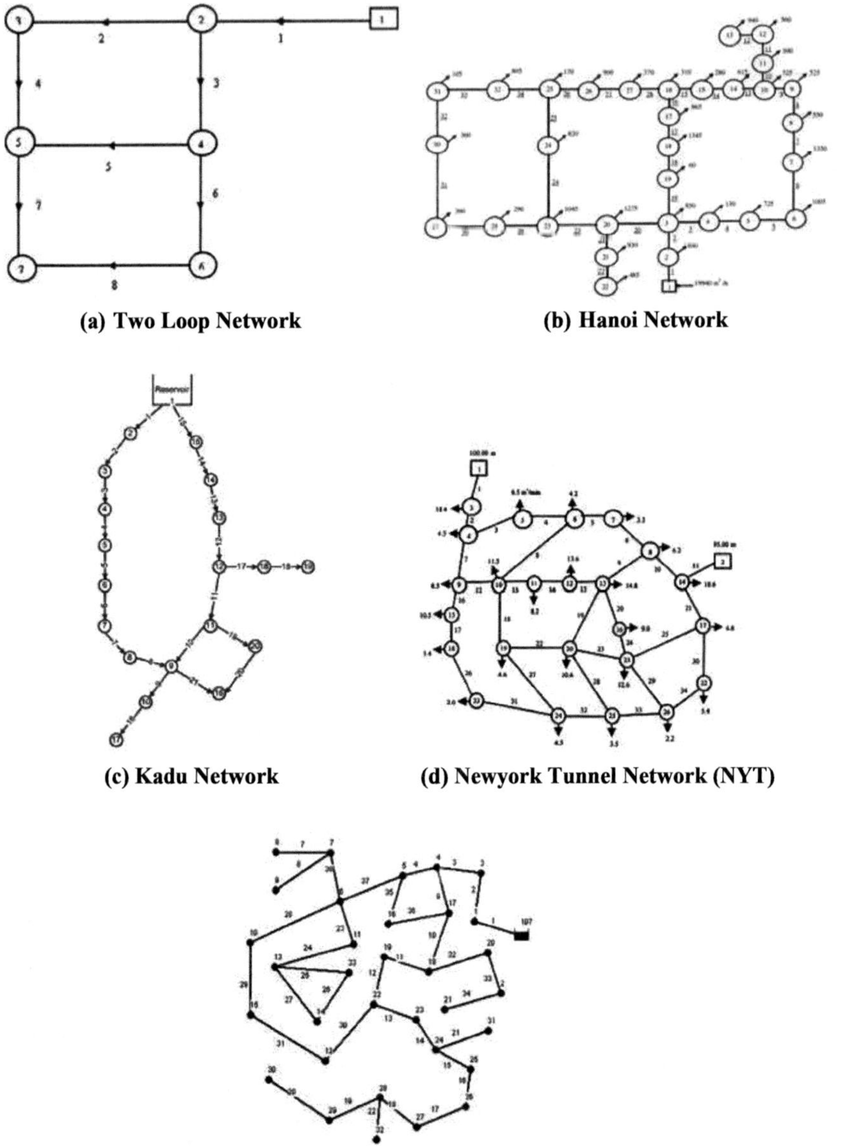

(a) Two Loop Network

(b) Hanoi Network

(c) Kadu Network

(d) Newyork Tunnel Network (NYT)

(e) Bengali Camp Chandrapur Network

Fig. 1 Layout of benchmark networks

solving WDN problems using various methods, it is typical to observe that the ideal population size is almost ten times the number of decision variables.

4.1 Two-Loop WDN

The optimal design of a two-loop WDN is performed using the diameter set same as [1]. All the four algorithms provided the same diameters and final cost of $419,000. The number of evaluations required by these and few other algorithms are compared in Table 3. Rao-I and PSO algorithms outperformed other algorithms in terms of number of evaluations. Rao-I and PSO outperformed Jaya significantly, whereas Jaya outperformed Rao-II by an enormous margin. A variation of cost with number of generations is shown in Fig. 2. Rao-II that required 10 times more evaluations, as a result, Rao-II is omitted from the subsequent benchmark network design.

Table 3 Comparison of heuristic EAs with metaheuristic EAs

Algorithm	Evaluation	Algorithm	Evaluation
Rao-I	1880	SA [4]	25,000
Rao-II	36,960	SFLA [5]	11,323
Jaya	3120	GA [20]	2200
PSO	1875		

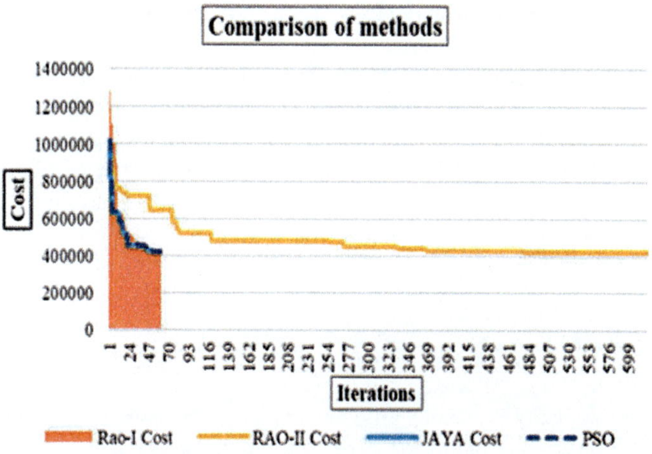

Fig. 2 Comparison of proposed algorithms on the basis of cost and iterations

4.2 Hanoi Network

The Hanoi WDN uses 6 commercial pipe sizes, having a 2.865×10^{26} search space. The results of the proposed approach are compared to earlier solutions. It is observed that PSO and Rao-I have optimal costs of $\$6.0811 \times 10^6$ and $\$6.067 \times 10^6$, respectively, which is less costly than GA [10] and SA [4]. Compared to other EA, the computational convergence is relatively minimal at 42,480 and 32,651 Kadu et al. [10] got early convergence by limiting the candidate diameters for each link using the critical path method. However, time required by the PSO and Rao-I is almost equivalent to GA. As shown in Fig. 3a, the Rao-I and PSO provide better optimal solutions with lesser time consumption, function evaluation, and cost than Jaya.

4.3 Kadu Two-Source Network

The Kadu network, which was initially designed using GA, achieved an optimal cost of Rs.123,268,864 using 14 commercially available pipe sizes. Kadu et al. [10] used a hybrid critical path—GA algorithm to find the best solution in the shortest time. However, EPANET analysis indicates deficiencies at nodes 6 and 26. Haghighi et al. [21] used hybrid GA–ILP and provided best solution costing Rs.131,312,815 in 4440. The heuristic technique of [22] provided the best solution having cost of Rs. 140,100,000. Siew et al. [23] used penalty free NSGA-II algorithm. Whilst considering full search space, they provided the best solution of Rs.125,460,980 in 436,000 evaluations, and with reduced search space, the best solution has a cost of 125,826,425. Sayyed et al. [24] used GA with reduced solution space and combined head and flow deficit-based penalty to provide best solution of 125,754,310 in 7600 evaluations. In this study, PSO, Rao-I, and Jaya algorithms provided optimal cost of Rs.125,457,320, Rs. 126,162,570, and Rs. 127,207,360, respectively, in 176,000, 180,000, and 185,000 function evaluations as shown in Table 4. The best solution obtained using PSO is better than that obtained by both Rao-I and Jaya algorithms. Further, it is better than any other solution reported in the literature.

4.4 New York Tunnel

This WDN is a capacity expansion problem. Only the diameters of the new pipes are used as decision variables in this WDN design, and the rest of the pipe diameters are kept constant. Each additional pipe can be one of 15 potential sizes, or ‘do nothing’, giving a search space of 1.93×10^{25}. The optimal solution obtained using Rao-I, PSO, and Jaya is $\$38.76 \times 10^6$, which is similar to GA, ACO, and SFLA [9]. However, computational convergence required to obtain the optimal cost by Rao-I is

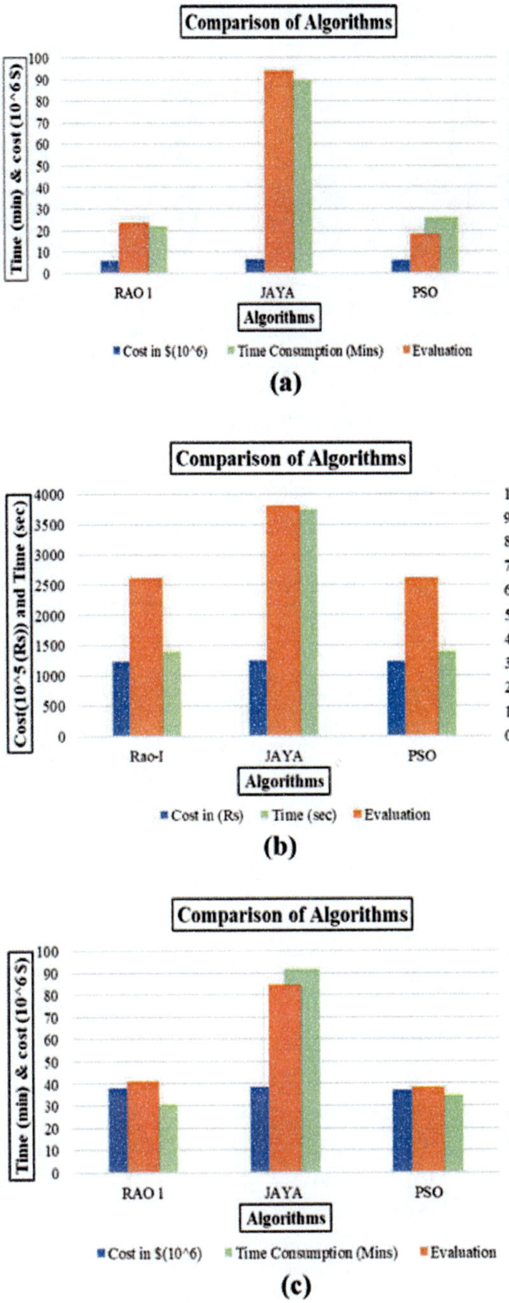

Fig. 3 Comparison of proposed algorithms on the bases of time, cost, and number of evaluations for **a** Hanoi, **b** Kadu, **c** New York tunnel, **d** Bengali camp (Chandrapur) zone WDN

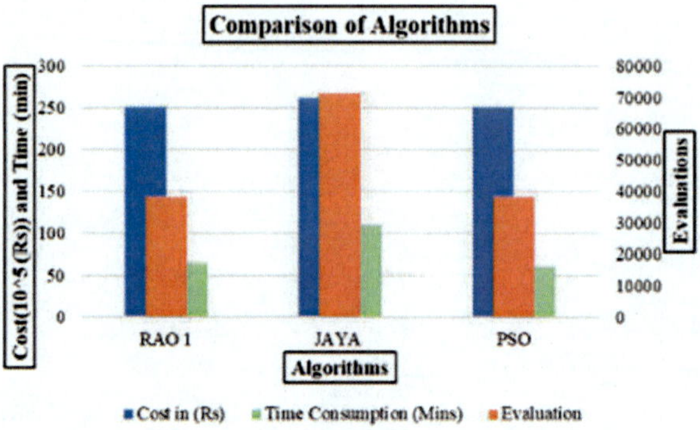

(d)

Fig. 3 (continued)

Table 4 Optimal solutions (diameter and available head) of Kadu network

Pipe No.	Length (m)	Optimal diameter (mm)			Pipe No.	Length (m)	Optimal diameter (mm)		
		PSO	Rao-I	Jaya			PSO	Rao-I	Jaya
1	300	900	1000	900	18	650	400	350	450
2	820	900	800	1000	19	760	150	150	150
3	940	350	400	350	20	1100	150	150	150
4	730	300	300	300	21	660	750	700	700
5	1620	150	150	150	22	1170	150	150	150
6	600	250	200	250	23	980	500	450	450
7	800	800	800	800	24	670	400	350	350
8	1400	150	150	150	25	1080	600	700	700
9	1175	450	450	450	26	750	250	250	200
10	750	500	500	500	27	900	250	300	300
11	210	900	750	800	28	650	300	300	300
12	700	700	700	700	29	1540	200	200	200
13	310	500	700	500	30	730	300	300	250
14	500	500	500	450	31	1170	150	150	150
15	1960	150	150	150	32	1650	150	150	150
16	900	500	500	500	33	1320	150	150	150
17	850	350	400	300	34	3250	150	150	150
Cost (Rs.) (10^6)		125.45	126.16	127.20					
Evaluation		176,000	180,000	185,000					

Table 5 Optimal solutions (Available head) of Kadu network

Node No.	Required head (m)	Available head (m)		
		PSO	Rao-I	Jaya
1	85	98.29	98.97	98.26
2	85	95.08	93.21	96.30
3	85	87.43	88.19	88.32
4	85	85.21	85.19	85.83
5	82	85.71	82.05	85.76
6	82	88.78	87.92	88.41
7	85	91.20	89.43	92.35
8	85	88.42	86.75	89.38
9	85	86.56	86.39	87.52
10	85	85.28	85.10	85.38
11	82	83.15	82.39	83.06
12	82	94.15	92.95	93.52
13	85	88.09	86.33	89.55
14	82	82.88	82.13	82.83
15	82	91.74	89.53	90.19
16	85	85.54	85.01	85.55
17	82	86.24	82.80	87.74
18	82	82.67	82.15	83.10
19	82	85.04	86.29	87.00
20	80	87.14	85.06	80.44
21	82	83.11	82.65	82.29
22	80	80.33	80.23	83.31
23	80	80.77	80.39	81.55
24	80	80.11	80.11	80.05

6500, and with PSO, it is 6100. The similar in case of GA is 23070, ACO is 55950, and SFLA is 7963. Although, when PSO, Rao-I, and Jaya are compared with each other as shown in Fig. 3c, the PSO and Rao-I have shown remarkable performance in comparison with Jaya.

4.5 Bengali Camp (Chandrapur) Zone WDN

[19] designed this WDN using cross-entropy (CE). The network uses 12 commercial pipe diameters, with a search space of 1.02×10^{41}. In 38,400 function assessments, the CE gave an ideal cost of Rs. 25,235,630. PSO and Rao-I are able to obtain better

optimal solutions with cost of Rs. 25,114,762 and Rs. 25,121,644, which is better than CE. Also, PSO requires 37,500 function evaluations, whilst Rao-I requires 38,000. As can be seen in Fig. 3d, Rao-I and PSO outperform Jaya.

5 Summary and Conclusions

A comparison amongst parameter-free heuristic EAs, Jaya, Rao-I, and Rao-II for minimum cost design of WDNs is carried out. Also, these heuristic EAs are compared with metaheuristic PSO which is observed to require comparatively less parameters as compared to other metaheuristic algorithms. Also, PSO does not require any such parameters to be defined, whose values are based on trial and error.

The performances of different algorithms are examined by considering four benchmark networks and one real-life network in a single objective optimization scenario. The performance is also compared with the recent state-of-the-art algorithms such as GA, ACO, SFLA, and SA. From the results of example networks, it is clear that parameter-free heuristic EAs converge much faster to optimal solution than other algorithms. For all the example networks, Rao-I and PSO showed best convergence with better optimal solutions, whereas the performance of Jaya is also comparatively good. However, Rao-II has shown very slow convergence than these three algorithms and also could not converge to known optimal solution. Therefore, it is concluded that both heuristic and metaheuristic EAs, i.e. Rao-I and PSO have shown almost similar and remarkable performance in optimal design of WDNs.

References

1. Alperovits E, Shamir U (1977) Design of optimal water distribution systems. Water Resour Res 13(6):885–900
2. Fujiwara O, Khang DB (1990) A two-phase decomposition method for optimal design of looped water distribution networks. Water Resour Res 26(4):539–549
3. Savic DA, Walters GA (1997) Genetic algorithms for least-cost design of water distribution networks. J Water Resour Plan Manag 123(2):67–77
4. Cunha MDC, Sousa J (1999) Water distribution network design optimization: simulated annealing approach. J Water Resour Plan Manag 125(4):215–221
5. Eusuff M, Lansey K, Pasha F (2006) Shuffled frog-leaping algorithm: a memetic meta-heuristic for discrete optimization. Eng Optim 38(2):129–154
6. Vasan A, Simonovic SP (2010) Optimization of water distribution network design using differential evolution. J Water Resour Plan Manag 136(2):279–287
7. Eberhart R, Kennedy J (1995) Particle swarm optimization. In: Proceedings of the IEEE international conference on neural networks, vol 4, 1942–1948.
8. Rao R (2020) Rao algorithms: Three metaphor-less simple algorithms for solving optimization problems. Int J Ind Eng Comp 11(1):107–130
9. El-Ghandour HA, Elbeltagi E (2018) Comparison of five evolutionary algorithms for optimization of water distribution networks. J Comput Civil Eng 32(1): 04017066–1–10
10. Kadu MS, Gupta R, Bhave PR (2008) Optimal design of water networks using a modified genetic algorithm with reduction in search space. J Water Resour Plan Manag 134(2):147–160

11. Zheng F, Simpson AR, Zecchin AC (2014) Coupled binary linear programming–differential evolution algorithm approach for water distribution system optimization. J Water Resour Plan Manag 140(5):585–597
12. Sedki A, Ouazar D (2012) Hybrid particle swarm optimization and differential evolution for optimal design of water distribution systems. Adv Eng Inf 26(3):582–591
13. Sheikholeslami R, Zecchin AC, Zheng F, Talatahari S (2016) A hybrid cuckoo–harmony search algorithm for optimal design of water distribution systems. J Hydrol 18(3):544–563
14. Tolson BA, Asadzadeh M, Maier HR, Zecchin A (2009) Hybrid discrete dynamically dimensioned search (HD-DDS) algorithm for water distribution system design optimization. Water Resour Res 45(12)
15. Bhave P R (2003) Optimal design of water distribution networks. Alpha Science Int'l Ltd.
16. Rossman LA (2000) Epanet 2 users manual, us environmental protection agency
17. Rao R (2016) Jaya: A simple and new optimization algorithm for solving constrained and unconstrained optimization problems. Int J Ind Eng Comp 7(1):19–34
18. Schaake JC, Lai FH (1969) Linear programming and dynamic programming application to water distribution network design. MIT Hydrodynamics Laboratory
19. Shibu A, Reddy MJ (2011) Least cost design of water distribution network by cross entropy optimization. In: 2011 world congress on information and communication technologies. IEEE, pp 302–306
20. Siew C, Tanyimboh TT (2012) Penalty-free feasibility boundary convergent multi-objective evolutionary algorithm for the optimization of WDS. Water Resorces Manage 26(15):4485–4507
21. Haghighi A, Samani HM, Samani ZM (2011) GA-ILP method for optimization of water distribution networks. Water Resour Manage 25(7):1791–1808
22. Suribabu CR (2012) Heuristic-based pipe dimensioning model for water distribution networks. J Pipeline Syst Eng Pract 3(4):115–124
23. Siew C, Tanyimboh TT, Seyoum AG (2014) Assessment of penalty-free multi-objective evolutionary optimization approach for the design and rehabilitation of water distribution systems. Water Resour Manag 28(2):373–389
24. Sayyed MAHA, Gupta R, Tanyimboh TT (2015) Noniterative application of EPANET for pressure dependent modelling of water distribution systems. Water Resour Manag 29(9):3227–3242

Analysis of Sarangpur City's Zone-3 Water Distribution Network by Using LOOP 4.0 and EPANET Software

Vinod Kumar Malviya and Ganesh D. Kale

Abstract The water distribution system (WDS) is designed for supplying sufficient water for domestic, industrial, commercial, and firefighting processes. It has become difficult to get a constant fresh water supply because of rapid growth of population. Based on demand quantities and required pressures, the water distribution network (WDN) is utilized to supply water to customers. Recently, several computer programmes like EPANET, WADISO, UNWB-LOOP, KYPIPE, and WATER have been developed and made available for analysis of WDNs. In pressurized pipe networks, EPANET performs extended time simulation for water flow and quality parameters. LOOP can be utilized to design and simulate new, partial, or fully existing gravity and pumped WDSs. None of the reviewed studies have examined existing WDN of the Zone-3 in Sarangpur City, Rajgarh District, Madhya Pradesh State, India. The WDN of the Sarangpur City is designed for 30 years (i.e. from 2012 to 2042). Thus, analysis of aforesaid WDN is performed in the present work corresponding to the year 2042 by using LOOP 4.0 and EPANET software. The aforesaid WDN assessed in this study is found to be capable of satisfying the demand of the year 2042.

Keywords Zone-3 · Sarangpur City · Water distribution network · EPANET · LOOP 4.0

1 Introduction

The water distribution system (WDS) is designed for supplying adequate water for domestic, industrial, commercial, and firefighting processes. It has become hard to get a constant supply of fresh water due to rapid growth of population. Around 2.5 billion people on the earth do not have access to safe drinking water [1]. Depending on demand quantities and desired pressures, the water distribution network (WDN)

V. K. Malviya · G. D. Kale (✉)
Department of Civil Engineering, Sardar Vallabhbhai National Institute of Technology, Surat, Gujarat 395007, India
e-mail: gdk@ced.svnit.ac.in

P. V. Timbadiya et al. (eds.), *Geospatial and Soft Computing Techniques*, Lecture Notes in Civil Engineering 339, https://doi.org/10.1007/978-981-99-1901-7_23

is used to supply water to customers [2]. In general, WDSs consist of a variety of outlets, storage tanks to meet customer node demands, and interconnected pipeline networks [3].

For commercial and educational use, many modelling programmes are now available for analysis of WDNs. Recently, many computer programmes such as EPANET, WADISO, UNWB-LOOP, WATER, and KYPIPE have been developed and made available for analysis of WDNs [4]. Inside pressurized pipe networks, EPANET conducts extended time simulation of the flow of water and the quality parameters. EPANET monitors flow of water in every pipe/link, pressure at every node, height of water in every tank, and form of chemical concentration across the network throughout the simulation time period, source, and water age [1]. LOOP can be used to design and simulate new partial or fully prevailing gravity as well as pumped WDSs [5]. Thus, in the present study, EPANET and LOOP software are used for the analysis of WDN.

To meet the continuously increasing population's water demand, it is necessary to provide sufficient and uniform amount of water by means of designed pipe network [6]. This can be assured through analysis/design of WDN. Thus, analysis/design of the WDN is necessary.

Many studies have been carried out on analysis/design of WDNs in India [1, 2, 5–12], etc. Apart from aforesaid studies, many studies have been carried out on analysis/design of WDNs outside of the India also [13–15], etc. None of the reviewed studies carried out in India have analyzed existing WDN of the Zone-3 in Sarangpur City, Rajgarh District, Madhya Pradesh State, India. Thus, in the present study, aforesaid WDN is evaluated by using EPANET and LOOP software for the end year of its life cycle, i.e. 2042.

2 Study Area and Data Source

2.1 Study Area

Sarangpur is the city and tehsil in the Rajgarh District, Madhya Pradesh, India. Sarangpur is positioned at latitude 23.57° N and longitude 76.47° E, on the bank of the River Kali Sindh [16]. It is governed by the Bhopal Division. It is situated 65 km south of the Rajgarh district headquarters. It is the headquarter of the tehsil. The Agra Mumbai Highway NH-3 serves Sarangpur. It is situated 126 km away from the Indore, which is Madhya Pradesh's business capital. It is 160 km away from the state capital, i.e. Bhopal [16]. Zone-3 area of the Sarangpur City is selected as study area in the present study. Sarangpur has four zones. By contemplating population in 2011 (as per census 2011) as base, the forecasted population of the study area in 2012 was 9463 [17]. Layout of existing WDN of the study area is shown in Fig. 1.

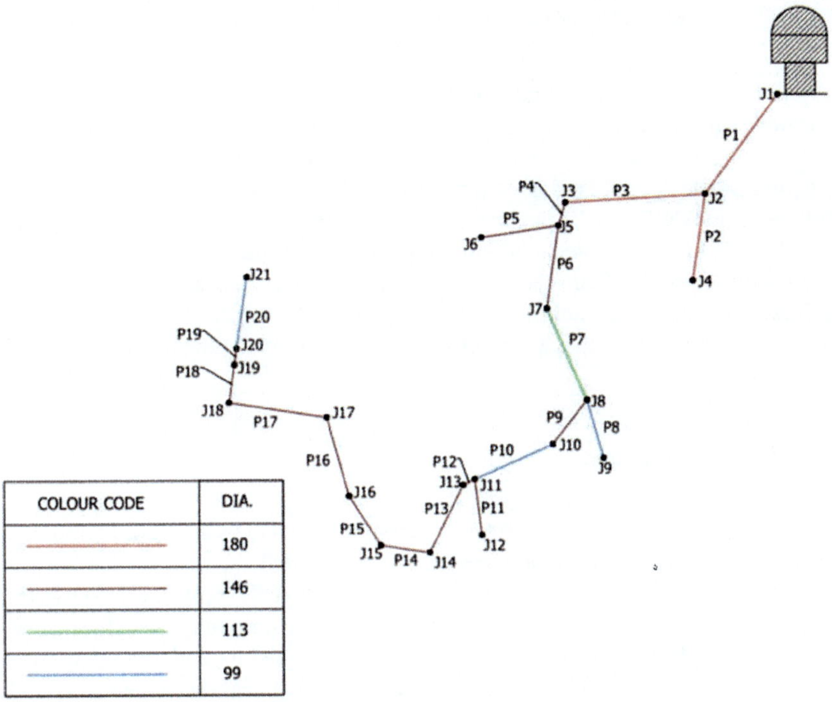

Fig.1 Layout of Zone-3 WDN in the Sarangpur City

2.2 *Data Collection*

To analyze the WDN of the Zone-3 in Sarangpur City, required data is collected from the Jal Praday Vibhag, Nagar Palika, Sarangpur [17]. The collected data consisted of pipe details (diameter, length, C value, material), junction details (node elevation, mean demand at node), network design in AutoCAD file, pump details (discharge, head, pumping hours, power, no. of pump), elevated storage reservoir details (capacity, tank-top level, bottom level, plinth level, water demand, existing and forecasted population, diameter, supply timings), daily and hourly demand, total area, and design period.

3 Methodology

In the present study, EAPNET and LOOP 4.0 software are used for the analysis of existing WDN of the study area.

3.1 EPANET Software

EPANET is a computer programme that simulates water quality and hydraulic behaviour within the pressurized pipe network system over an extended time period. In the WDN, nodes also known as pipe junctions, pipes, valves, pumps, and tanks/ reservoirs are consisted. During the simulation period, which comprises of multiple time steps, EPANET tracks the water flow in every pipe, pressure at every node, water level in every tank, and concentration of chemical species across the network. As an addition to chemical species, source tracing as well as water age can also be simulated. EPANET can help in determining alternative management methods to enhance water quality across a system [7].

3.2 LOOP Software

In 1990, Dr. Prasad Modak and Juzer have developed LOOP software with the assistance of World Bank. LOOP software is computer-aided software, which is used in developing countries to plan and design low-cost water supply and waste-water disposal systems. This programme uses Newton–Raphson technique and flow equations of Hazen-Williams or Darcy-Weisbach for looped distribution network design. LOOP version 4.0 manages up to 1000 pipes and 750 nodes. It can handle several sources with fixed or variable heads, fixed or unknown flows, check valves, booster pump, and pressure control valves. This software system displays hydraulic grade line along the selected section that estimates head losses, operating status of the valve, velocities, cost, and pumping heads. Programme is structured for easy data entry, editing, and updating. It is provided in a quick BASIC form that complies with programme execution with speed [18].

4 Results and Discussion

In the present study, WDN of the Zone-3 in Sarangpur City is evaluated by using EPANET and LOOP 4.0 software for the end year of its life cycle, i.e. 2042. The results of the analysis are given in following paragraphs. Table 1 given below is showing simulated results for WDN of the Zone-3 in Sarangpur City corresponding to the demand of the year 2042. The results in Table 1 are obtained from the EPANET software, which include velocity (m/s), flow (LPS), unit head loss (m/km), and friction factor values for different links of the Zone-3 WDN.

From the above Table 1, it is observed that values of velocities in links are ranging from 0.06 to 2.53 m/s, whilst values of flow are varying from 1.06 L/s (LPS) to 30 LPS for the given WDN. The negative sign of the flow indicates flow in the direction opposite to the direction considered in the method of analysis. Values of unit head

Table 1 Simulated values of velocity, flow, unit head loss, and friction factor obtained from the EPANET software for WDN of the Zone-3 in Sarangpur City corresponding to the year 2042

Link No.	Velocity (m/s)	Flow (LPS)	Unit head loss (m/km)	Friction factor
P/1	1.31	33.32	8.81	0.018
P/2	0.06	1.46	0.03	0.029
P/3	1.18	30.00	7.26	0.018
P/4	1.65	27.66	17.31	0.018
P/5	0.11	1.92	0.12	0.027
P/6	1.43	24.00	13.31	0.019
P/7	2.31	23.13	43.30	0.018
P/8	0.28	2.16	1.02	0.025
P/9	1.21	20.31	9.77	0.019
P/10	2.53	19.44	59.77	0.018
P/11	0.10	1.69	0.10	0.027
P/12	0.94	15.74	6.09	0.020
P/13	0.76	12.77	4.14	0.020
P/14	0.58	11.32	3.31	0.021
P/15	0.57	9.56	2.42	0.021
P/16	0.43	7.18	1.42	0.022
P/17	0.35	5.84	0.97	0.023
P/18	0.21	3.60	0.40	0.025
P/19	0.12	1.99	0.13	0.027
P/20	0.14	1.06	0.27	0.028

loss are varying from 0.03 to 59.77 m/km. Also, the values of friction factor are varying from 0.018 to 0.029. Pressure head values (in metre) at various nodes of the Zone-3 WDN in Sarangpur City corresponding to the year 2042 are shown in Table 2, and these are obtained from the EPANET software.

Table 2 shows that values of pressure head at various nodes are varying from 15.52 and 27.08 m. As per Central Public Health and Environmental Engineering Organization [19], the minimum residual pressure head needed for the two-storey building is 12 m. So, head available at all nodes is more than minimum residual pressure requirement of 12 m. Simulated results obtained from the LOOP 4.0 software for the Zone-3 WDN in Sarangpur City corresponding to the demand of the year 2042 are shown in Table 3. The results in Table 3 include velocity (m/s), flow (LPS), and unit head loss (m/km) values corresponding to links of the given WDN.

From Table 3, it is observed that values of velocities in links are varying from 0.06 m/s to 2.53 m/s, whilst values of flow are varying from 1.06 LPS to 33.32 LPS. Also, values of unit head loss are varying from 0.031 m/km to 59.79 m/km. Pressure head values (in metre) at various nodes of WDN of the Zone-3 in Sarangpur City

Table 2 Simulated values of pressure head obtained from the EPANET software for WDN of the Zone-3 in Sarangpur City corresponding to the year 2042

Node ID	Pressure head (m)
J/2	26.08
J/3	25.22
J/4	27.08
J/5	25.87
J/6	25.87
J/7	24.94
J/8	21.35
J/9	20.30
J/10	19.89
J/11	15.53
J/12	15.52
J/13	15.46
J/14	15.20
J/15	16.06
J/16	16.94
J/17	16.84
J/18	16.76
J/19	16.75
J/20	15.75
J/21	15.73
Tank J/1	22.50

corresponding to the year 2042 are shown in Table 4, and these are obtained from the LOOP 4.0 software.

Table 4 shows that values of pressure heads are varying from 15.18 m to 26.07 m. As per Central Public Health and Environmental Engineering Organization [19],

Table 3 Simulated values of velocity, flow, and unit head loss obtained from the LOOP 4.0 software for WDN of the Zone-3 in Sarangpur City corresponding to the year 2042

Link No.	Velocity (m/s)	Flow (LPS)	Unit head loss (m/km)
P/1	1.31	33.32	8.81
P/2	0.06	1.46	0.031
P/3	1.18	30.00	7.26
P/4	1.63	27.52	17.26
P/5	0.12	1.97	0.18
P/6	1.43	24.00	13.32
P/7	2.31	23.21	43.32
P/8	0.31	2.23	1.59

(continued)

Table 3 (continued)

Link No.	Velocity (m/s)	Flow (LPS)	Unit head loss (m/km)
P/9	1.21	20.31	9.79
P/10	2.53	19.47	59.79
P/11	0.08	1.52	0.095
P/12	0.94	15.74	6.12
P/13	0.71	12.69	3.98
P/14	0.51	11.31	3.21
P/15	0.57	11.02	1.39
P/16	0.51	7.26	1.59
P/17	0.41	5.91	1.02
P/18	0.21	3.60	0.39
P/19	0.11	1.99	0.15
P/20	0.13	1.06	0.27

Table 4 Simulated values of pressure head obtained from the LOOP 4.0 software for WDN of the Zone-3 in Sarangpur City corresponding to the year 2042

Node ID	Pressure head (m)
J/2	26.07
J/3	25.20
J/4	27.01
J/5	25.89
J/6	25.89
J/7	24.96
J/8	21.41
J/9	20.35
J/10	19.94
J/11	15.63
J/12	15.49
J/13	15.41
J/14	15.18
J/15	16.11
J/16	16.98
J/17	16.88
J/18	16.76
J/19	16.79
J/20	15.89
J/21	15.69

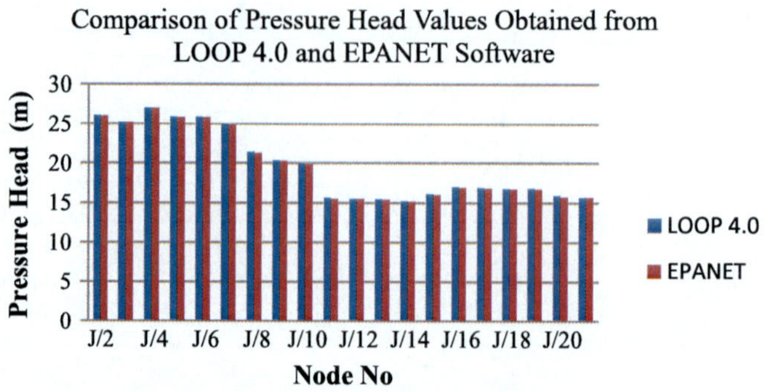

Fig. 2 Variation of pressure head values at various nodes derived from LOOP 4.0 and EPANET software

minimum residual pressure head needed for the two-storey building is 12 m. So, head available at all nodes is more than minimum residual pressure requirement of 12 m.

The comparison of pressure head values at various nodes obtained from LOOP 4.0 and EPANET software is shown in Fig. 2. From Fig. 2, it is observed that pressure head values obtained from the EPANET and LOOP 4.0 software are found to be close to each other for the given WDN.

5 Conclusions

- The existing WDN assessed in this study is found to be capable of satisfying the demand of the year 2042.
- As per Central Public Health and Environmental Engineering Organization [19], minimum residual pressure head needed for the two-storey building is 12 m, and pressure head available at all nodes is more than minimum residual pressure requirement of 12 m.
- J-21 is the farthest point in the network; even there also the pressure head is sufficient.
- Velocity, flow, unit head loss, and pressure head values obtained from the EPANET and LOOP 4.0 software are found to be close to each other for the given WDN.

Acknowledgements The authors are thankful to Jal Praday Vibhag, Nagar Palika, Sarangpur, Rajgarh District, Madhya Pradesh, India for providing necessary data required for this study.

References

1. Bhoyar RD, Mane MS (2017) Modelling and optimization of water distribution system site: Nagpur. Int J Adv Eng Res Dev 4(7):2348–4470
2. Kathrotiya C, Malek AM (2018) Design and cost estimation of water distribution network using EPANET software for village Dudhala-Gujarat. Int J Sci Res Dev (IJSRD) 6(2), 2321-0613
3. Sadafule VV, Hiremath RB, Tuljapure SB (2013) Design and development of optimal loop water distribution system. Int J Appl Innov Eng Manag, (IJAIEM) 2(11):374–378
4. Adeniran AE, Oyelowo MA (2013) An EPANET analysis of water distribution network of the University of Lagos, Nigeria. J Eng Res 18(2):69–83
5. Sumithra RP, Nethaji MVE, Amaranath J (2013) Feasibility analysis and design of water distribution system for Tirunelveli Corporation using Loop and Water GEMS software. Int J Appl Bioeng 7(1):61–70
6. Kumar A, Kumar K, Bharanidharan B, Matial N, Dey E, Singh M, Thakur V, Sharma S, Malhotra N (2015) Design of water distribution system using EPANET. Int J Adv Res 3(9):789–812
7. Arunkumar M, Mariappan VN (2011) Water demand analysis of municipal water supply using EPANET software. Int J Appl Bioeng 5(1):9–19
8. Athulya T, Ullas AK (2020) Design of water distribution network using EPANET Software. Int Res J Eng Technol (IRJET) 7(3), 2395-0056
9. Joshi M, Chavda M, Rajyaguru D, Sarvaiya S (2014) Design of water supply distribution network for Kuchhadi Village. Indian J Res 3(2), 2250-1991
10. Lungariya P, Katharotiya N, Mehta D, Waikhom S (2016) Analysis of continuous water distribution in Surat city using EPANET: a case study. Glob Res Dev J Eng 1(7):2455–5703
11. Mehta VN, Joshi GS (2019) Design and analysis of rural water supply system using Loop 4.0 and WaterGEMSV8i for Nava Shihora Zone I. Int J Eng Adv Technol 9(1), 2249-8958
12. Yengale MM, Wadhai PJ, Khode BV (2012) Analysis of water distribution network for Karanja village—a case study. Int J Eng 2(3):2352–2355
13. Motevalizadeh M, Irandoust M (2016) Hydraulic analysis of water supply networks and controlling the leak using WaterGEMS model. Int J Adv Biotechnol Res (IJBR) 7(2), 0976-2612
14. Sarbu I (2009) Design of optimal water distribution systems. Int J Energy 3(4):59–67
15. Terlumun UJ, Ekwule OR (2019) Evaluation of a municipal water distribution networks using Watercard and Watergems. J Eng Sci 5(2):147–156
16. https://en.wikipedia.org/wiki/Sarangpur,_Madhya_Pradesh. Accessed 06 Dec 2020
17. Jal Praday Vibhag, Nagar Palika, Sarangpur, Rajgarh district, Madhya Pradesh, India
18. Ajudiya B, Bhagde S, Hathaliya H (2017) Planning and designing of rural water supply system. Int J Tech Innov Mod Eng Sci (IJTIMES) 3(5):2455–2584
19. Central Public Health and Environmental Engineering Organization (1999) Manual on water supply and treatment. Ministry of Urban Development, New Delhi, India

Improved Design Solutions for Benchmark Networks Using Genetic Algorithm Involving Penalty Based on Combined Flow and Pressure Deficit

Laxmi Gangwani, Shilpa Dongre, Rajesh Gupta, and Mohd Abbas H. Abdy Sayyed

Abstract Genetic algorithms (GAs) have been used extensively for optimal design of water distribution networks (WDNs) by many researchers in the past. The simple GA is modified from time to time through various measures, e.g. type of coding, values of GA parameters, search space reduction, penalty methods, type of analysis, etc. to improve the efficiency of GA in identifying minimum cost solution with minimum computational efforts. All such improvements in GA are tested on some benchmark problems of different sizes and varying complexities to show the advantages of proposed improvement. Most of the improved GA methodologies worked well to identify global optimal solutions for small networks and sometimes suggested some improved solutions for moderate to large WDNs. Most recent development in GA is the use of combined flow and pressure deficit-based penalty approach with reduced search space. In this study, a computer code based on the above penalty approach is developed by coupling GA algorithm with EPANET 2.0 to design various benchmark networks. However, global search is carried out instead of search in reduced space to explore the possibility of better solutions. Out of the several networks considered, the optimal solutions for two networks are presented herein which are found cheaper than the solutions obtained by earlier studies. Improved solutions for these benchmark networks are presented in this paper.

L. Gangwani (✉)
Department of Civil Engineering, Shri Ramdeobaba College of Engineering and Management, Nagpur 440013, India
e-mail: gangwanilr@rknec.edu

S. Dongre · R. Gupta
Department of Civil Engineering, Visvesvaraya National Institute of Technology (VNIT), Nagpur 440010, India
e-mail: shilpadongre@civ.vnit.ac.in

R. Gupta
e-mail: rajeshgupta@civ.vnit.ac.in

M. A. H. Abdy Sayyed
Centre for Urban Science and Engineering, Indian Institute of Technology Bombay (IITB), Powai, Mumbai 400076, India
e-mail: abbas_cuse@iitb.ac.in

P. V. Timbadiya et al. (eds.), *Geospatial and Soft Computing Techniques*, Lecture Notes in Civil Engineering 339, https://doi.org/10.1007/978-981-99-1901-7_24

Keywords Flow and pressure deficit-based penalty · Genetic algorithm · Optimization · Pressure-dependent analysis · Water distribution network

1 Introduction

A water distribution network (WDN) consists of various components such as pumps, links, hydrants, reservoirs, tanks, valves and other appurtenances. The design of WDN should be such that the desired quantity of water at sufficient pressure is supplied to the consumers [4]. Since last five decades and more, researchers have developed various optimization methods to obtain minimum cost design of WDNs based on diverse mathematical programming techniques. These techniques search for the optimal solution by starting search from a point (a feasible solution) in search space and move towards better solution using some deterministic rules. Often such search terminates at local optimal solution. In the last two to three decades, several evolutionary algorithms have been developed that optimizes an objective function by exploiting the information by starting from several initial randomly selected solutions. These methods are inspired by the biological or other natural phenomenon and use computational methods based on them for searching the entire search space and terminate usually when sufficient search is made. It is observed that evolutionary techniques have better capacity to reach global optimal solution and provide several near optimal solutions.

Genetic algorithms (GAs) which derive its base from Darwin's philosophy of survival of the fittest have been used for design of WDNs by several researchers [11, 15, 19]. Since its first application on WDN design, there has been a lot of developments in GA by various researchers to improve the efficiency and effectiveness of GA by suggesting improvements in basic GA parameters, coding system and string representation schemes, fitness functions, penalty approach and search space reduction. Besides others, the two important parameters are: (1) the search space reduction that allows searches in limited area by restricting the candidate pipe sizes for each link in the network; and (2) a self-adaptive penalty that helps in improving the infeasible solutions on the boundary of the feasible-infeasible front.

The most recent improvement in GA was suggested by Abdy Sayyed et al. [2]. The penalty for not meeting the constraints was applied based on the deficiency in meeting demands as well as pressure instead of the usual approach that is based on deficiency in pressure only. Efficiency of the methodology was tested with several networks including a two-source network from Kadu et al. [11] used by many researchers. Even with several modifications in GA-based methodology for WDN design, their application to large network problem is limited and research for faster convergence is continued.

Kadu et al. [11] provided two different solutions for the two-source network with full and reduced search space, respectively, to show the impact of reduction in search space on the cost of optimal solution and required number of evaluations. Haghighi et al. [10] used GA combined with ILP (GA-ILP) to provide the optimal solution for the same network considering the full search space and provided a better solution. Siew et al. [17] used penalty-free GA and obtained two different solutions, one considering full search space and the other considering the reduced search space for the Kadu's network. Several better solutions compared to the best solutions of Kadu et al. [11] were reported by Siew et al. [17]. Abdy Sayyed et al. [2] modified the penalty and obtained a better solution for the reduced search space, without exploring the full search space. Thus, in each of the subsequent research, one or several better solutions than the previous best were reported. The main objective of this paper is to show the application of methodology of Abdy Sayyed et al. [2] on the two-source network of Kadu et al. [11] and other benchmark networks by considering full search space as solutions in the full search space were not explored.

2 Optimization Problem and Methodology

2.1 Water Distribution Network Design Optimization

A general optimization problem formulation for a WDN with number of demand nodes (ND), number of pipes (X) and number of loops (Y) can be described as follows:

$$\text{Minimize } f(D_1, \ldots, D_X) = \sum_{x=1}^{X} c(D_x) \times L_x \tag{1}$$

Subject to constraints:

$$\sum_{x \in j} Q_x + q_j = 0; \; j = 1, \ldots, (\text{ND}) \tag{2}$$

$$\sum_{x \in y} h_x + \sum E_p = 0; \quad y = 1, \ldots, Y \tag{3}$$

$$H_j^{\text{avl}} \geq H_j^{\text{des}}; \; j = 1, \ldots, (\text{ND}) \tag{4}$$

$$D_x \in \{D_{\min}, \ldots, D_{\max}\}; \quad x = 1, \ldots, X \tag{5}$$

in which, $c(D)$ = per unit cost of pipe having diameter 'D'; Q = pipe discharge; L = length of pipe; q = demand at nodes; h = head loss in pipe; Ep = energy added by a pump to water; Hj^{avl} and Hj^{des} = available head and desirable head at node j, respectively; and $D_{\min}$ and $D_{\max}$ = minimum and maximum diameter of available pipes, respectively. Equations (2) and (3) are general flow continuity and loop head loss equations, respectively. Equation (4) assures available head more than the desirable head at each demand node, and Eq. (5) allows selection of commercial pipe sizes only.

In GA, constrained optimization problem is converted to an unconstrained one by using a penalty approach. Equations (2) and (3) are handled using hydraulic solver like EPANET. Abdy Sayyed et al. [2] suggested using pressure-dependent analysis (PDA) to obtain deficiency in not meeting the demands and pressures. In PDA, nodal outflows are dependent on available pressure and represented through node-head-flow relationship. Several node-head-flow relationships are available, and any one can be used [5]. The most commonly used relationship is as suggested by Wagner et al. [18] and Chandapillai [6] and can be written as

$$q_j^{\text{avl}} = q_j^{\text{req}}, \text{ if } H_j^{\text{avl}} \geq H_j^{\text{des}} \tag{6}$$

$$q_j^{\text{avl}} = q_j^{\text{req}} \left(\frac{H_j^{\text{avl}} - H_j^{\text{min}}}{H_j^{\text{des}} - H_j^{\text{min}}} \right)^{\frac{1}{n_j}}, \text{ if } H_j^{\text{min}} < H_j^{\text{avl}} < H_j^{\text{des}} \tag{7}$$

$$q_j^{avl} = 0, \; if\, H_j^{\text{avl}} \leq H_j^{\text{min}} \tag{8}$$

where q^{avl} = available flow; q^{req} = required flow; H^{avl} = available head; H^{min} = minimum required head; and H^{des} = desirable head at a node. The exponent n in Eq. 7 is usually taken as 2; however, it varies with pressure head requirements at demand nodes [9]. The subscript j refers to the node number.

The objective function in Eq. (1) can be modified to include penalty cost as

$$\text{Minimize } f(D_1, \ldots, D_x) = \sum_{x=1}^{X} c(D_x) L_x + \sum_{j=1}^{M-S} p \times (q_j^{\text{req}} - q_j^{\text{avl}}) \times \{\max(H_j^{\text{des}} - H_j^{\text{avl}}, 0)\} \tag{9}$$

The penalty cost is an equivalent energy cost to lift the deficit water by the deficit head and p is a penalty multiplier [2]

2.2 GA Methodology

GA-based design methodology for WDN consists of the following steps:

1. Generation of initial population of random solution of suitable size considering available pipe diameters;
2. Computation of system cost for each solution in the population;
3. Hydraulic analysis of each solution using EPANET 2.0 or any other solver;

4. Computation of penalty cost for non-feasible solutions;
5. Computation of objective function value;
6. Computation of fitness;
7. Production of new population for subsequent generations using reproduction, crossover, mutation and elitism operator; and
8. Repeating the process till the terminating criteria is met.

2.3 Computer Software Development

GA code available on Kanpur GA Lab, IIT Kanpur (https://www.egr.msu.edu/~kdeb/codes.shtml), is suitably modified and clubbed with hydraulic simulator EPANET 2.0 [14] for optimal design of WDNs on C platform. It does not have PDA facility and therefore required a series of additional fictitious components as suggested by Abdy Sayyed et al. [1] in the network at each node to get a solution in single run.

3 Application of Methodology on Benchmark Networks

Several networks of varying sizes and complexities like two-loop network of Alperovits and Shamir [3], three-loop network of Fujiwara and Khang [7], New York Tunnel expansion problem of Schaake and Lai [16], Bakryun network of Lee and Lee [13], two-source network problem of Kadu et al. [11] and Pumped Source GoYang Network of Kim et al. [12] were considered. The GA methodology of Abdy Sayyed et al. [2] could identify optimal solution in all the cases. Results of only two networks are discussed herein as better than previously reported solutions are obtained for them.

3.1 Two-Source Network of Kadu *et al.* (2008)

A two-reservoir network from Kadu et al. [11] with 34 links, 26 nodes and 9 loops is shown in Fig. 1. Nodes 1 and 2 are the source node and have reservoir water levels as 100.00 and 95.00 m, respectively. A set of 14 commercial pipe diameters is used in this design optimization problem. Link data, node data, available pipe sizes and their cost can be referred from Kadu et al. [11].

The GA results for the network were obtained considering the following GA parameters: population size of 320, crossover probability of 0.72 and mutation probability of 0.003. The obtained diameters of different links are shown in Col. (3) and column (4) of Table 1. The network cost is obtained as Rs. 125,209,860 in 157,760 evaluations. This solution is 0.2% cheaper than the previous best solution of 125,460,980 obtained by Siew et al. [17] in 436,000 evaluations. Thus, the combined

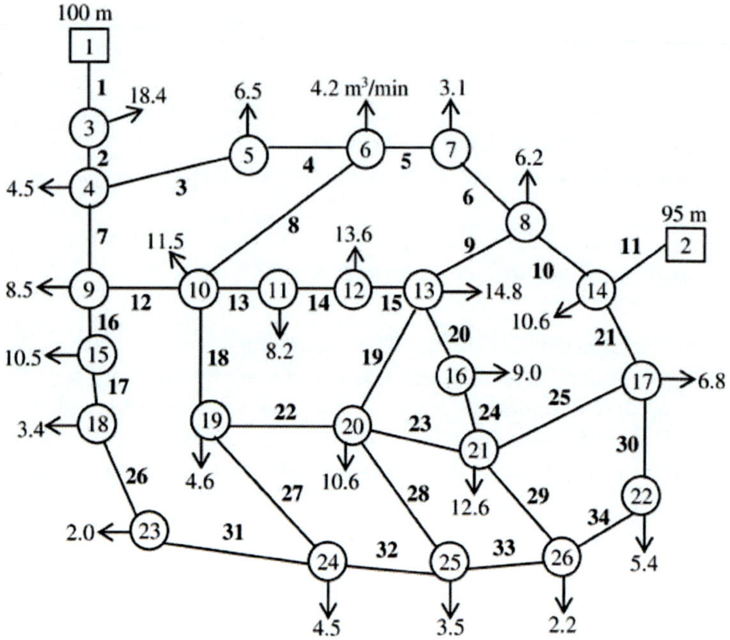

Fig. 1 Layout of two-source network (Kadu et al. [11])

flow and deficit-based penalty approach is found better than penalty-free approach in terms of the cost of final solution and the required number of evaluations. The results reported by other researchers are also provided in Table 1 for comparison. The obtained pressure heads with the final solution are given in Table 2. It can be observed that available heads at all the nodes are satisfied at all the demand nodes.

3.2 GoYang Pumped Source Network

GoYang network was first presented by Kim et al. [12]. It includes thirty pipes, twenty-two demand nodes and one constant pump of 4.52 kW linking to one reservoir with a constant head of 71 m as shown in Fig. 2. The Hazen–Williams roughness coefficient for each new pipe is 100. The minimum required pressure head above the ground elevation at each node is 15 m. Link data, node data, available pipe sizes and their cost can be referred from Geem [8].

The design solution for the network was obtained by considering the GA parameter as: population size of 400, crossover probability of 0.84 and mutation probability of 0.0025. The results obtained are shown in Tables 3 and 4. The best solution as obtained in present study has a cost of 177,010,355 Won obtained in 11,200 evaluations. The

Table 1 Diameters of alternative solution for two-source network by Kadu et al. [11]

Pipe	Length (m)	Diameter in mm							
		Full solution space				Reduced solution space			
		Present work	Kadu et al. [11]	Haghighi et al. [10]	Siew et al. [17]	Kadu et al. [11]	Siew et al. [17]	Abdy Sayyed et al. [2]	Abdy Sayyed et al. Altered reduce space [2]
(1)	(2)	(3)	(4)	(5)	(6)	(7)	(8)	(9)	(10)
1	300	900	1000	1000	900	1000	900	900	900
2	820	900	900	900	900	900	900	900	900
3	940	350	400	400	350	350	400	350	350
4	730	300	350	350	300	250	250	300	300
5	1620	150	150	150	150	150	150	150	150
6	600	250	250	250	250	250	200	200	250
7	800	800	800	800	800	800	800	800	800
8	1400	150	150	150	150	150	150	150	150
9	1175	450	400	400	450	600	600	600	450
10	750	500	500	500	500	700	600	700	500
11	210	750	1000	1000	900	900	900	800	800
12	700	700	700	700	700	700	700	700	750
13	310	600	800	800	500	500	500	500	500
14	500	450	400	400	500	450	500	450	450
15	1960	150	150	150	150	150	150	150	150
16	900	500	500	500	500	450	500	500	500

(continued)

Table 1 (continued)

		Diameter in mm							
Pipe	Length (m)	Full solution space				Reduced solution space			
		Present work	Kadu et al. [11]	Haghighi et al. [10]	Siew et al. [17]	Kadu et al. [11]	Siew et al. [17]	Abdy Sayyed et al. [2]	Abdy Sayyed et al. Altered reduce space [2]
(1)	(2)	(3)	(4)	(5)	(6)	(7)	(8)	(9)	(10)
17	850	350	350	350	350	350	350	350	350
18	650	400	350	350	400	400	350	300	350
19	760	150	150	150	150	450	450	450	150
20	1100	150	200	150	150	150	150	200	150
21	660	700	700	700	700	600	600	700	700
22	1170	150	150	150	150	150	150	150	150
23	980	450	400	450	450	150	150	150	450
24	670	350	400	400	350	400	350	350	350
25	1080	700	700	700	700	500	600	500	700
26	750	250	250	250	250	200	250	250	250
27	900	250	250	250	250	350	300	300	250
28	650	300	200	200	300	250	300	300	300
29	1540	200	300	300	200	150	200	200	200

(continued)

Table 1 (continued)

		Diameter in mm							
Pipe	Length (m)	Full solution space				Reduced solution space			
		Present work	Kadu et al. [11]	Haghighi et al. [10]	Siew et al. [17]	Kadu et al. [11]	Siew et al. [17]	Abdy Sayyed et al. [2]	Abdy Sayyed et al. Altered reduce space [2]
(1)	(2)	(3)	(4)	(5)	(6)	(7)	(8)	(9)	(10)
30	730	300	300	300	250	300	300	300	300
31	1170	150	200	200	150	150	150	200	200
32	1650	150	150	150	150	150	150	150	150
33	1320	150	250	200	150	150	150	150	150
34	3250	150	150	150	150	200	150	150	150
Cost (rupees)		125,209,860	131,678,935	131,312,815	125,460,980	126,368,865	125,826,425	126,365,955	125,754,310
Evaluations		157,760	120,000[a]	4440[a]	436,000	25,200[a]	82,400	7000	7600

[a]Infeasible solutions (based on EPANET 2.0 with $\alpha = 1.85$; $\beta = 4.87$; $\omega = 10.68$)

Table 2 Available heads at nodes for two-source network by Kadu et al. [11]

Node	Min HGL Reqd. (m)	Available head (m) (based on EPANET 2 with $\alpha = 1.85$; $\beta = 4.87$; $\omega = 10.68$)							
		Full solution space				Reduced solution space			
		Present work	Kadu et al. [11][a]	Haghighi et al. [10][a]	Siew et al. [17]	Kadu et al. [11][a]	Siew et al. [17]	Abdy Sayyed et al. [2]	Abdy Sayyed et al. altered reduce space [2]
1 (Res.)	100.00	100.00	100.00	100.00	100.00	100.00	100.00	100.00	100.00
2 (Res.)	95.00	95.00	95.00	95.00	95.00	95.00	95.00	95.00	95.00
3	85.00	98.28	98.95	98.96	98.28	98.98	98.26	98.32	98.29
4	85.00	95.04	95.65	95.66	95.04	95.76	94.98	95.17	95.06
5	85.00	87.33	90.85	90.85	87.47	88.79	90.68	87.28	87.51
6	85.00	85.35	89.4	89.41	85.63	85.28	85.07	85.18	85.64
7	82.00	85.21	87.75	87.73	85.83	87.99	82.95	84.08	85.62
8	82.00	87.82	89.99	89.96	88.82	91.62	89.35	90.91	88.31
9	85.00	91.12	91.77	91.79	91.12	91.83	91.05	91.41	91.17
10	85.00	88.30	89.05	89.08	88.30	88.88	88.22	88.81	89.19
11	85.00	87.53	88.85	88.88	86.42	87.10	86.38	87.03	87.33
12	85.00	85.38	84.98*	85.01	85.13	85.13	85.12	85.05	85.19
13	82.00	82.39	82.02	81.88*	83.25	86.73	84.85	85.68	82.85
14	82.00	92.95	94.49	94.49	94.14	94.13	94.15	93.46	93.49
15	85.00	88.01	88.44	88.46	87.97	87.11	87.92	88.15	87.93
16	82.00	82.18	84.53	84.81	83.11	82.05	83.04	82.04	82.67
17	82.00	89.56	90.88	90.88	90.69	90.29	90.04	91.67	90.09
18	85.00	85.45	85.46	85.47	85.39	85.24	85.39	85.25	85.11

(continued)

Table 2 (continued)

Node	Min HGL Reqd. (m)	Available head (m) (based on EPANET 2 with $\alpha = 1.85$; $\beta = 4.87$; $\omega = 10.68$)							
		Full solution space				Reduced solution space			
		Present work	Kadu et al. [11][a]	Haghighi et al. [10][a]	Siew et al. [17]	Kadu et al. [11][a]	Siew et al. [17]	Abdy Sayyed et al. [2]	Abdy Sayyed et al. altered reduce space [2]
19	82.00	86.11	85.11	85.24	86.14	85.93	83.82	82.07	85.39
20	82.00	82.34	82.10	83.78	83.15	83.72	82.07	82.40	82.77
21	82.00	86.35	87.39	87.38	87.37	83.98	87.00	85.15	86.86
22	80.00	85.09	86.45	86.55	80.69	84.80	85.50	87.07	85.62
23	82.00	83.01	82.09	82.07	82.96	82.17	83.05	82.03	82.06
24	80.00	**80.16**	79.94*	79.89*	80.28	83.63	80.86	80.12	**80.44**
25	80.00	80.53	79.96*	79.77*	81.10	80.15	80.54	80.54	80.95
26	80.00	80.22	82.87	84.04	**80.04**	78.38*	**80.39**	**80.03**	80.67

[a] Infeasible solutions; * indicates a shortfall in head; Bold face denotes the nodes with critical HGL

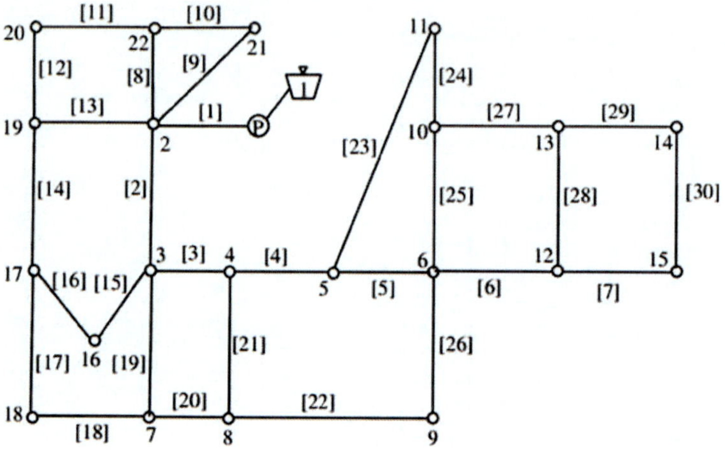

Fig. 2 Layout of GoYang network

previous best solution reported in the literature by Geem et al. [8] was obtained using harmony search technique as shown in Table 3.

4 Discussion and Conclusions

The solutions for various benchmark networks considering full search space were not explored by Abdy Sayyed et al. [2]. They obtained optimal solutions with reduced search space consisting of five candidate pipe sizes for each link in Kadu's network, which were decided using a heuristic method and the search space remained constant from iteration to iteration. The design solution for Kadu's network as initially obtained by Abdy Sayyed et al. [2] (Table 1, Column 9) was inferior to the solutions reported by Siew et al. [17] (Table 1, Column 8), wherein dynamic search space was considered. Upon analysing the problem, it was found that due to reduction in search space, candidates pipe sizes for some of the links were over-restricted. Abdy Sayyed et al. [2] further altered the reduced search space and got better solution (Table 1, Column 10) in lesser evaluations as compared to the solution provided by Siew et al. [17].

In the present study, global search resulted in either same or better solutions for all the benchmark networks. The least-cost solution for the two-source network of Kadu et al. [11] is obtained herein in 157,760 evaluations only. which is 0.2% cheaper than the solution obtained earlier by Siew et al. [17] in 436,000 evaluations.

It can be concluded that the performance of any network depends greatly on the GA parameters and specifically on the penalty approach. Herein, the self-organizing penalty as suggested by Abdy Sayyed et al. [2] based on both, the deficiency in meeting the demand and pressure using PDA, is used by considering full search

Table 3 Diameters of alternative solution for the GoYang network

Pipe number	Pipe length (m)	Diameter (mm)			
		Present work	Kim et al. [12] (original design)	Kim et al. [12] (NLP)	Geem [8] (HS)
1	165	200	200	200	150
2	124	125	200	200	150
3	118	125	150	125	125
4	81	100	150	125	150
5	134	80	150	100	100
6	135	80	100	100	100
7	202	80	80	80	80
8	135	80	100	80	100
9	170	80	80	80	80
10	113	80	80	80	80
11	335	80	80	80	80
12	115	80	80	80	80
13	345	80	80	80	80
14	114	80	80	80	80
15	103	80	100	80	80
16	261	80	80	80	80
17	72	80	80	80	80
18	373	80	80	100	80
19	98	80	80	125	80
20	110	80	80	80	80
21	98	80	80	80	80
22	246	80	80	80	80
23	174	80	80	80	80
24	102	80	80	80	80
25	92	80	80	80	80
26	100	80	80	80	80
27	130	80	80	80	80
28	90	80	80	80	80
29	185	80	80	100	80
30	90	80	80	80	80
Cost (Won)	–	177,010,355	179,428,600	179,142,700	177,135,800
Evaluations		11,200	–	–	10,000

(1000 Won» 1 US Dollar)

Table 4 Node data and computational results for GoYang network

Node number	Water demand (cmd)	Ground level (m)	Pressure head in m			
			Present work	Kim et al. [12] Original	Kim et al. [12] (NLP)	Geem [8] (HS)
1	– 2550.00	71.00	15.62	15.61	15.61	15.61
2	153.00	56.40	28.93	28.91	28.91	24.91
3	70.50	53.80	28.73	31.18	31.15	26.32
4	58.50	54.90	26.58	29.53	29.10	24.11
5	75.00	56.00	24.2	28.16	27.47	22.78
6	67.50	57.00	21.51	26.91	25.44	20.67
7	63.00	53.90	27.72	30.46	30.75	25.34
8	48.00	54.50	26.70	29.80	29.48	24.41
9	42.00	57.90	21.21	26.05	24.84	20.01
10	30.00	62.10	16.17	21.50	20.17	15.43
11	42.00	62.80	16.03	20.92	19.79	15.06
12	37.50	58.60	18.16	24.34	22.95	18.16
13	37.50	59.30	17.46	23.54	22.07	17.38
14	63.00	59.80	15.33	21.43	20.84	15.27
15	445.50	59.20	15.48	21.59	20.78	15.42
16	108.00	53.60	28.31	31.06	30.65	25.88
17	79.50	54.80	26.75	29.05	28.97	24.29
18	55.50	55.10	26.44	28.76	28.87	23.99
19	118.50	54.20	27.36	29.49	29.14	24.89
20	124.50	54.50	26.68	28.80	27.96	24.43
21	31.50	62.90	19.74	21.06	20.18	16.89
22	799.50	61.80	19.36	21.47	20.07	17.21

space and by developing a separate design code. The optimal solutions obtained for Kadu's network and GoYang network are found to be cheaper and obtained in lesser number of evaluations than the best solutions reported in the literature. The proposed work thus establishes the superiority of combined flow and head deficit-based penalty approach over other penalty approaches used in GA.

GA-based search in full space is possible only for the moderate size network. Therefore, reduction in search space is desirable for large networks. However, for not missing the chance of obtaining the global optimum solution, dynamic search space reduction or repetition of run with new reduced search space is recommended, in case pipe size of any of the links in final solution is observed on extreme of the selected candidate pipe sizes for that link.

References

1. Abdy Sayyed MAH, Gupta R, Tanyimboh TT (2015) Noniterative application of EPANET for pressure dependent modelling of water distribution systems. Water Resour Manage 29(9):3227–3242
2. Abdy Sayyed MAH, Gupta R, Tanyimboh TT (2019) Combined flow and pressure deficit-based penalty in GA for optimal design of water distribution network. ISH J Hydraul Eng. https://doi.org/10.1080/09715010.2019.1604180
3. Alperovits E, Shamir U (1977) Design of optimal water distribution systems. Water Resour Res 13(6):885–900
4. Bhave PR (2003) Optimal design of water distribution networks, Narosa Publishing House Pvt. Ltd. New Delhi, India; and Alpha Science International Ltd., Pangbourne, UK
5. Bhave PR, Gupta R (2006) Analysis of water distribution networks. Narosa Publishing House Pvt. Ltd. New Delhi, India; and Alpha Science International Ltd., Pangbourne UK
6. Chandapillai J (1991) Realistic simulation of water distribution system. J Transp Eng 117(2):258–263
7. Fujiwara O, Khang DB (1990) A two-phase decomposition method for optimal design of looped water distribution networks. Water Resour Res 26(4):539–549
8. Geem ZW (2006) Optimal cost design of water distribution networks using harmony search. Eng Optimizat 38(03):259–277. https://doi.org/10.1080/03052150500467430
9. Gupta R, Bhave PR (1996) Comparison of methods for predicting deficient network performance. J Water Resour Plann Manage 122(3):214–217
10. Haghighi A, Samani HM, Samani ZM (2011) GA-ILP method for optimization of water distribution networks. Water Resour Manage 25(7):1791–1808
11. Kadu MS, Gupta R, Bhave PR (2008) Optimal design of water networks using a modified genetic algorithm with reduction in search space. J Water Resour Plann Manage 134(2):147–160
12. Kim JH, Kim TG, Kim JH, Yoon YN (1994) A study on the pipe network system design using non-linear programming. J Korean Water Resour Ass 27(4):59–67
13. Lee S-C, Lee S-I (2001) (2001) Genetic algorithms for optimal augmentation of water distribution networks. J Korean Water Resour Ass 34(5):567–575
14. Rossman LA (2000) EPANET 2: User's Manual
15. Savic D, Walters G (1997) Genetic algorithms for least-cost design of water distribution networks. J Water Resour Plann Manage 123(2):67–77
16. Schaake J, Lai D (1969) Linear programming and dynamic programming applications to water distribution network design. Rep. No. 116, Dept. of Civil Engineering, Massachusetts Institute of Technology, Cambridge, MA
17. Siew C, Tanyimboh TT, Seyoum AG (2014) Assessment of penalty-free multi-objective evolutionary optimization approach for the design and rehabilitation of water distribution systems. Water Resour Manage 28(2):373–389. https://doi.org/10.1007/s11269-013-04888
18. Wagner JM, Shamir U, Marks DH (1988) Water distribution reliability: simulation method. J Water Resour Plann Manage 114(3):276–294
19. Wu ZY, Simpson AR (2001) Competent genetic-evolutionary optimization of water distribution systems. J Comput Civ Eng 15(2):89–101

Optimum Design of Rural Water Supply System Using JalTantra and Evolutionary Algorithms

Vidhi N. Mehta and H. M. Patel

Abstract Water is a precious resource, and distribution of safe and sufficient drinking water to each consumer is the prime concern of the authority. The design of water distribution system includes planning of topology, selection of pipe material and sizing of pipes to fulfill the demands and to maintain other hydraulic properties of the system. For feeding rural pockets, branched systems are most suitable expanding in a large area by following road network. Many softwares are available for optimum design to achieve the least cost of pipes. Evolutionary algorithms (EA) such as genetic algorithm (GA) or PSO are also useful for such nonlinear problems. In this paper, the optimum design is carried out using JalTantra and GA. The main objective of the study is to design and analyze the water distribution system (WDS) economically and to provide water supply of desired quantity and pressure which may provide water to the community in accordance with their requirements. JalTantra tool has been used to develop the topology of pipeline system and for designing optimum size of pipes. The evolutionary algorithms are also used to design pipeline. The fitness function and constraints are developed, and least cost solution is obtained. The results obtained EA are compared with that from JalTantra tool. EA is found to have more flexibility to test the different operating conditions.

Keywords Water distribution network · JalTantra · Optimization · Evolutionary algorithm

V. N. Mehta (✉) · H. M. Patel
Civil Engineering Department, Faculty of Technology and Engineering, The Maharaja Sayajirao University of Baroda, Vadodara, Gujarat 390001, India
e-mail: mehta.vidhi-ced@msubaroda.ac.in

H. M. Patel
e-mail: hmpatel-ced@msubaroda.ac.in

P. V. Timbadiya et al. (eds.), *Geospatial and Soft Computing Techniques*, Lecture Notes in Civil Engineering 339, https://doi.org/10.1007/978-981-99-1901-7_25

1 Introduction

Water distribution networks are among the most important issues facing society; without water, humans cannot survive. Therefore, it is of great interest to have water distribution networks that satisfy the needs of users and do not cause economic loss. Water distribution networks play an important role in improving the standard of living in a community [2]. Here, this paper is focused on optimum design of drinking water supply network to the rural area. Water supply network consists of a primary network and secondary network. Source to Water treatment Plant (WTP) and WTP to sump and then sump to elevated service reservoir (ESR) are the primary network of the water supply system which is known as gravity main, and further ESR to the household is the secondary network of water supply system. In regional rural water supply scheme, the primary network consists of long pipeline with branches to feed the sumps. The optimum design of such network considerably reduce the cost of the project.

There are many softwares like WaterGEMS, EPANET, BRANCH, etc. which are used to design water supply network. For this study, JalTantra and genetic algorithm (GA)-based program have been used for the design of water supply system. JalTantra system has been developed at IIT Bombay. JalTantra is an open-source freeware software for optimization of water supply network. In JalTantra, the problem of integer linear programing optimization problem is solved by using the linear solver library CBC [7] and its open-source Java interface by Google.

For evolutionary algorithm, genetic algorithm (GA) has been used in this study. Genetic algorithms are search and optimization algorithms based on the principles of natural evolution, which were first introduced by John Holland in 1970 [3, 5]. Genetic algorithms also implement the optimization strategies by simulating evolution of species through natural selections. GAs have been extensively applied to solve the problem of designing the optimal water distribution network [6, 8]. Multi-objective heuristic approaches have also been formulated not only to minimize the network cost but to take into consideration other conflicting objectives as well, such as the reliability of the system [4]. A simple model of the evolutionary process will contain the following features:

(i) A population, the individuals within which die off and are replaced by offspring.
(ii) A breeding process, involving the formation of offspring, by combining the genetic information (genes) of parents selected from the population.
(iii) A selection process by which fitter individuals in a population are more likely to breed (and successfully rear their young) than less fit individuals. Fit individuals are those well suited to their environment [10].

To design gravity-based water supply networks with minimum cost, the constraints involve nonlinear friction loss terms. In the case of pumped water supply, the cost function also becomes nonlinear due to pump and energy costs. The decision variables of pipe size are discrete in nature. In such a problem, GA can be applied to obtain optimum solution. In GA, each solution is evaluated using an objective function called a fitness function, and this process is repeated until some form of convergence in fitness is achieved [9].

In the present study, one gravity network and two pumped networks of rural water supply scheme are selected. The objective of the study is to apply the JalTantra and GA-based computer program for the field problem to obtain optimum solutions and to evaluate the results obtained by both the methods.

2 Study Area and Data Source

Study area for the research is Tilakwada, Garudeshwar Taluka, which is located in Narmada District of Gujarat State. Narmada District is situated in Western India and located at Latitude-21.8 and Longitude-73.5. The ground elevation in the district ranges from 45 to 241 m above m.s.l. The Gujarat Water Supply and Sewage Board (GWSSB) is developing rural water supply schemes in the Garudeshwar Taluka of the Narmada District.

Figure 1 shows flow diagram of headwork (HW) to HW of Tilakwada Taluka. There are seven numbers of existing ESR to supply water to the community of that area. Each ESRs gets supply from the sump via pump, and each sump gets supply of water from main headwork (MHW). In this project, there is a requirement of augmentation of water supply as usage of water increases with increased amenities and increase in the population of the city. The main source of the supply is Narmada main canal as shown in figure.

Now, to get the optimum size of pipes in JalTantra, the network has been divided into three parts as shown in Fig. 2. One is from WTP to Devaliya chokadi sump which is named as gravity network as shown in Fig. 2. Second from Devaliya chokadi sump to Namalpur Secondary Headwork (SHW) which is named as pumped network-1 as pump is required to supply water at far end as shown in Fig. 3. Third from Devaliya chikadi sump to Naliya SHW which is named as pumped network-2 as pump is required to supply water at far end as shown in Fig. 4. End node of the network is sump of each ESRs to design network.

Population data and available commercial diameter of pipe with cost data for the research has been collected from GWSSB regional office Rajpipala. Population data collected from census 2011 and population estimated for the 30 years. The system data of all the three networks are shown in result tables (Tables 1, 2, 3, 4, 5, 6 and 7).

3 Methodology

Design procedure of water supply system in JalTantra is shown in Fig. 5.

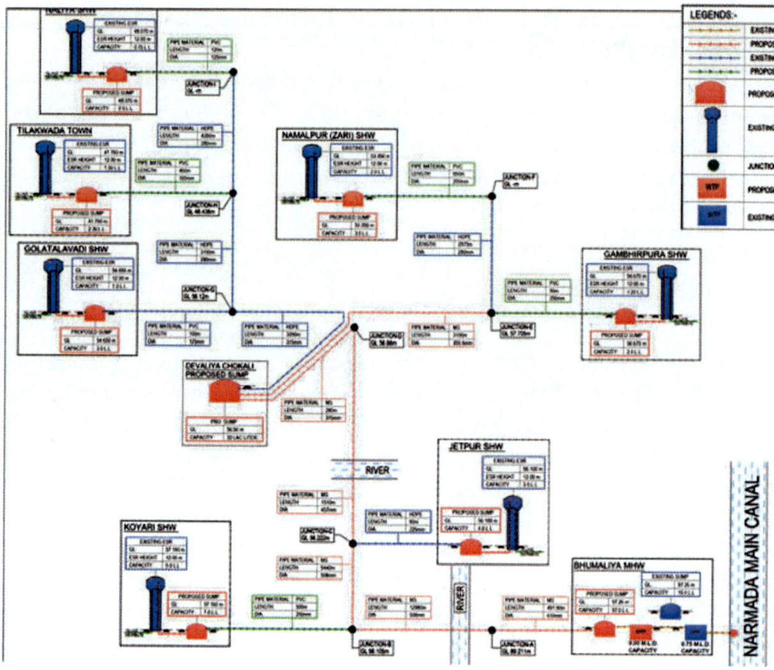

Fig. 1 Flow diagram of HW to HW of Tilakwada Taluka. *Source* GWSSB

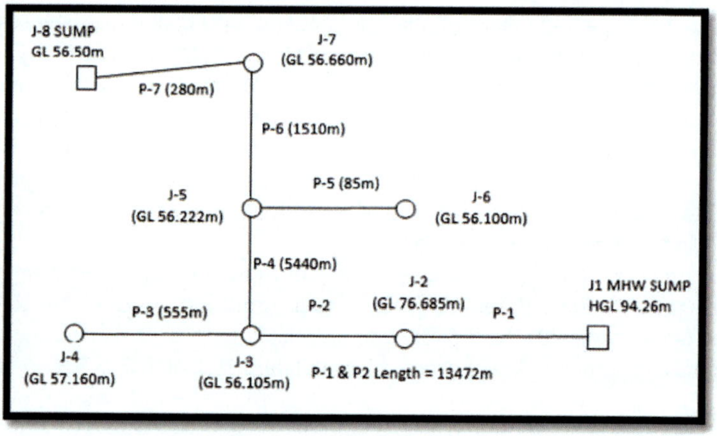

Fig. 2 Gravity network

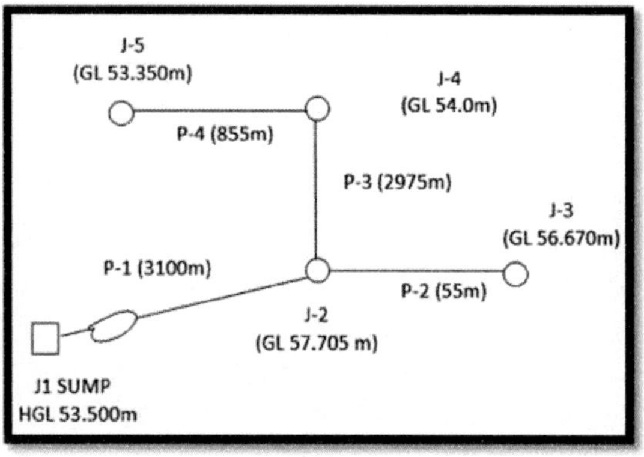

Fig. 3 Pumped network-1

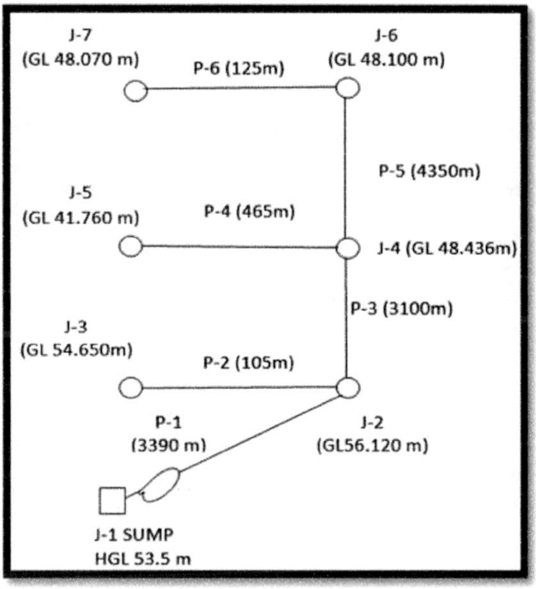

Fig. 4 Pumped network-2

3.1 JalTantra and GA Input Parameters

The following input parameters are required for JalTantra tool:

1. General data: Minimum residual pressure, roughness coefficient, minimum head loss/km, maximum head loss/km, number of supply hours, source head and source elevation.
2. Pipe data: Node to node connectivity, length of pipe.
3. Node data: Elevation, demand of node, minimum pressure required at node.

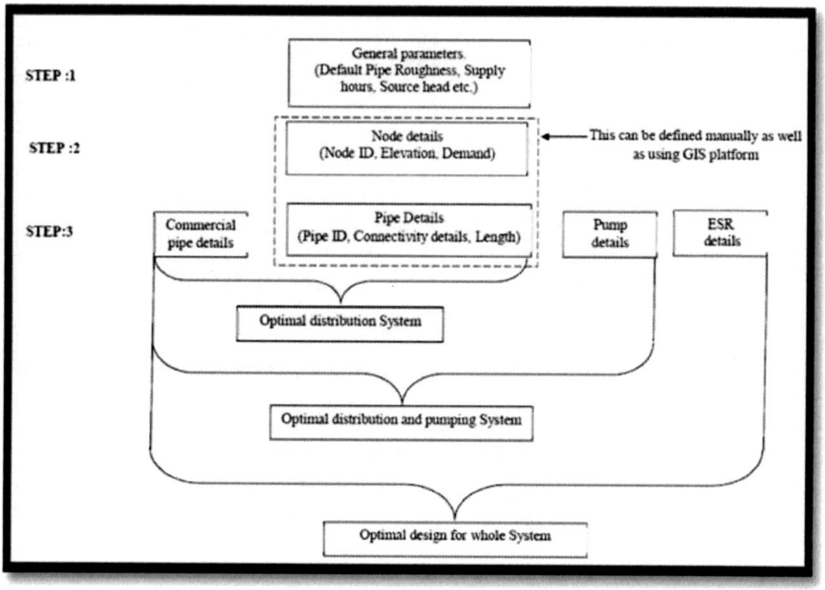

Fig. 5 Design procedure of water supply system in JalTantra [1]

4. Commercial pipe size data: Cost of pipe, roughness of pipe material, available commercial size of pipe.
5. Pump data: Efficiency of pump, minimum pump size, capital cost per KW, energy cost/ Kwh, design life time, discount rate, inflation rate, pipes without pump.

The objective function is JalTantra is to minimize total cost which includes pipe cost, pump cost and energy cost. The constraints are the residual pressures at the demand nodes [7]. GA program is developed by authors in MATLAB by interfacing its optimization toolbox. The population size is set as 40. The objective function is to minimize total cost which includes pipe cost, pump cost and energy cost (Eq. 1). The constraints are the pressure values at the demand nodes which should be greater than the required residual head (Eq. 2). A total of 21 commercial pipe data are used. The decision variables are index number (Ii) for each pipe which are integer values, fraction of length (x) for pipe with multiple diameter and head generated by the pump. The last two variables are continuous. The choice of index number of commercial pipe array enables the inclusion of two or more pipes of same internal diameter but different material or roughness and different cost per unit length. The index number variable also defines pipe parameters such as cost per unit length and friction coefficient. As the problem is of mixed integer programming type, the integer variables are specified in GA function argument. In MATLAB optimization toolbox, integer GA algorithm uses the real coded array for members of population and minimize the penalty function. The penalty function value of a member of a population is fitness function or objective function if the member is feasible. If the member is infeasible, the penalty function is the maximum fitness function among feasible members of the population, plus a sum of the constraint violations of the (infeasible) point. Thus, the penalty function value for optimum solution is the objective function value, and it indicates total cost for the problems presented here.

3.2 *Objective Function*

$$\text{Minimize } Z = \sum_{i=1}^{NP} C_i(I_i)L_i + \mathrm{CP} \cdot \mathrm{P_p}\left(\mathrm{h_p}\right) + \mathrm{DF} \cdot \mathrm{PH} \cdot \mathrm{EP} \cdot \mathrm{P_p}\left(\mathrm{h_p}\right) \quad (1)$$

If first pipe length L0 has multiple diameter length L1 and L2, then L1 = xL0 and L2 = (1 − x)L0.

3.3 *Constraints*

For each end demand node,

$$\left(P_n - (H_o - E_n) - h_p + \sum_{j=1}^{NL} HF_j\right) w \le 0 \quad (2)$$

where: NP = Total pipes in network; Ci = Cost of pipe per unit length for pipe i; Ii = Index number of pipe i in the array of commercial diameters; Li = Length of pipe i; CP = Cost of pump per kW; Pp = Pump capacity in kW; hp = Pump head in meter, DF = Depreciation factor for energy cost (14.040115 for the problem presented); PH = Total yearly pumping hours (5840 for the problems presented); EP = Energy charges per unit in Rs (Rs 5 for the problem presented); x = Fraction of length in multiple diameter pipe with length L0; Pn = Minimum residual head required at the end demand node; H0 = Elevation of HGL at source, En = Ground elevation at end demand node; HFj = Head loss in pipe j which is a part of NL pipes in series from source to end demand nodes; w is the weight applied to constraints.

4 Results

4.1 Rainfall Characteristics Results of Gravity Network with JalTantra and GA

Table 1 shows the results of gravity network with JalTantra (Case-1). In JalTantra input, the total length of pipe-1 and pipe-2 is given as combined length. But the solution of JalTantra splits the combined length into two pipes. Table 2 shows the results of gravity network with GA (Cases 2, 3 and 4). In Case-2, the length of first pipe is adopted same as that obtained from JalTantra result (Case-1). In Cases 3 and 4, the length of pipe-1 is optimized along with diameters of all the pipe. Length of pipe-2 is adjusted to achieve total length of pipe-1 and pipe-2. Figure 6 shows the penalty values with generation for Case-4. The best penalty value in the figure represents the value of the objective function which is the total cost of the project.

4.2 Results of Pumped Network-1 with JalTantra and GA

Table 3 shows the results of pumped network-1 with JalTantra (Case-1). Two results are obtained using GA for pumped network-1 which are shown in Tables 4 and 5 (Case-2 and Case-3). Figure 7 shows the penalty values with generations for Case-3.

4.3 Results of Pumped Network-2 with JalTantra and GA

Table 6 shows the results of pumped network-2 with JalTantra (Case-1). One result is obtained using GA for pumped network-1 is shown in Table 7 (Case-2). Figure 8 shows the penalty values with generations for Case-2.

Table 1 Results of gravity network with JalTantra (Case-1) (P1 = 5390 m, P2 = 8082 m)

Pipe ID	Start node	End Node	Flow (lps)	Diameter (m)	Roughness 'C'	Head loss (m)	HGL end (m)	Pressure (m)	Pipe cost (Rs./m)	Cost (Rs.)
1	1	2	137.726	406.4	140	12.455	81.805	5.120	2680	14,443,909
2	2	3	137.726	457	140	10.547	71.258	15.153	3017	24,384,546
3	3	4	37.928	198.3	150	3.412	67.846	**10.686**	1256	697,080
4	3	5	99.798	406.4	140	6.923	64.334	8.112	2680	14,579,200
5	5	6	20.754	158.7	150	0.506	63.828	**7.728**	802	68,170
6	5	7	79.044	355.6	140	2.391	61.943	5.283	2341	3,534,910
7	7	8	79.044	355.6	140	0.443	61.500	**5.000**	2341	655,480
								Cost		58,363,295

Table 2 Results of gravity network with GA (Case-2, Case-3 and Case-4)

Pipe ID	Case-2			Case-3			Case-4		
	Length (m)	Dia. (m)	Pressure (m)	Length (m)	Dia. (m)	Pressure (m)	Length (m)	Dia. (m)	Pressure (m)
1	5390	406.4	5.120	8081	457	7.029	4232	406.4	7.796
2	8082	457	15.153	5391	406.4	15.152	9240	457	16.317
3	555	176.3	**8048**	555	158.7	**4001**	555	158.7	**5.166**
4	5440	406.4	8.112	5440	406.4	8.111	5440	406.4	9.277
5	85	110.1	**5.231**	85	110.1	**5.230**	85	110.1	**6.396**
6	1510	355.6	5.283	1510	355.6	5.282	1510	323.9	5.071
7	280	355.6	**5.000**	280	355.6	**4.999**	280	406.4	**5.000**
Cost (Rs)		58,181,200			58,075,970			58,244,288	

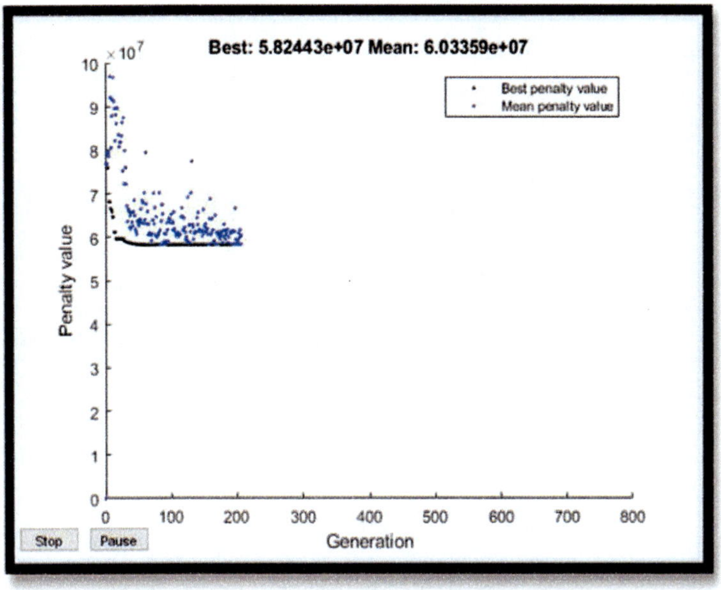

Fig. 6 Fitness values with generations for Case-4 of gravity network

5 Discussion

The rural water supply scheme for Tilakwada Taluka of Narmada District consists of one gravity network and two pumped networks. JalTantra software is used to obtain economical diameters of pipes. GA is also used to obtain the optimum solution. The performance of the algorithms is evaluated based on the total cost and residual pressures at demand nodes. Both the techniques use different approaches. JalTantra uses LP whereas GA is evolutionary algorithm based on random search. The problem formulation is also different in terms of decision variables. JalTantra considers lengths for each commercial diameter for each link. For example, for gravity network of 7 pipes and 21 commercial diameters, total constraints are $7 \times 21 = 147$. For a similar problem in GA presented here, the unknowns are 7 integer values. As the feasible region for this integer variable is limited by available commercial diameter values, the GA-based techniques are also found efficient in obtaining solution. Figures 6, 7 and 8 indicate that the solution process converges to generate optimum solution.

5.1 Gravity Network

For gravity network, it is observed that the first pipe length supplied as input is divided into two segments (5390 m and 8082 m) with different diameters. JalTantra

Table 3 Results of pumped network with JalTantra (Case-1)

Pipe ID	Start node	End node	Flow (lps)	Diameter (m)	Roughness 'C'	Head loss (m)	Pump head (m)	HGL end (m)	Pressure (m)	Pipe cost (Rs./m)	Pipe cost (Rs.)	Pump capacity (KW)	Pump cost (Rs.)	Energy cost (Rs.)	Total cost (Rs.)
1	1	2	37.5465	273.0	140	4.480	18.587	67.607	9.902	1792	6E+06	9.777	29,331	4,008,253	
2	2	3	20.7300	198.3	150	0.110		67.496	**10.826**	1256	69,080				
3	2	4	16.8165	176.3	150	7.190		60.416	6.416	991	3E+06				
4	4	5	16.8165	176.3	150	2.066		58.350	**5.000**	991	847,305				
											9E+06		29,331	4,008,253	13,457,394

Table 4 Results of pumped network-1 with GA (Case-2)

Pipe ID	Start node	End node	Flow (lps)	Diameter (m)	Roughness 'C'	Head loss (m)	Pump head (m)	HGL end (m)	Pressure (m)	Pipe cost (Rs./m)	Pipe cost (Rs.)	Pump capacity (KW)	Pump cost (Rs.)	Energy cost (Rs.)	Total cost (Rs.)
1	1	2	37.5465	273	140	4.480	14.551	63.571	5.866	1792	6E+06	7.654	22,962	3,137,895	
2	2	3	20.7300	123.4	150	1.113		62.458	**5.788**	486	26,730				
3	2	4	16.8165	198.3	150	4.055		59.515	5.515	1256	4E+06				
4	4	5	16.8165	198.3	150	1.166		58.350	**5.000**	1256	1E+06				
											1E+07		22,962	3,137,895	13,553,267

Table 5 Results of pumped network-1 with GA (Case-3)

Pipe ID	Start node	End node	Flow (lps)	Diameter (m)	Roughness 'C'	Head loss (m)	Pump head (m)	HGL end (m)	Pressure (m)	Pipe cost (Rs./m)	Pipe cost (Rs.)	Pump capacity (KW)	Pump cost (Rs.)	Energy cost (Rs.)	Total cost (Rs.)
1	1	2	37.5465	273	140	4.480	18.587	67.607	9.902	1792	6E+06	9.777	29,331	4,008,251	
2	2	3	20.7300	110.1	150	1.939		65.668	**8.998**	390	21,450				
3	2	4	16.8165	176.3	150	7.190		60.416	6.416	991	3E+06				
4	4	5	16.8165	176.3	150	2.066		58.350	**5.000**	991	847,305				
											9E+06		29,331	4,008,251	13,409,762

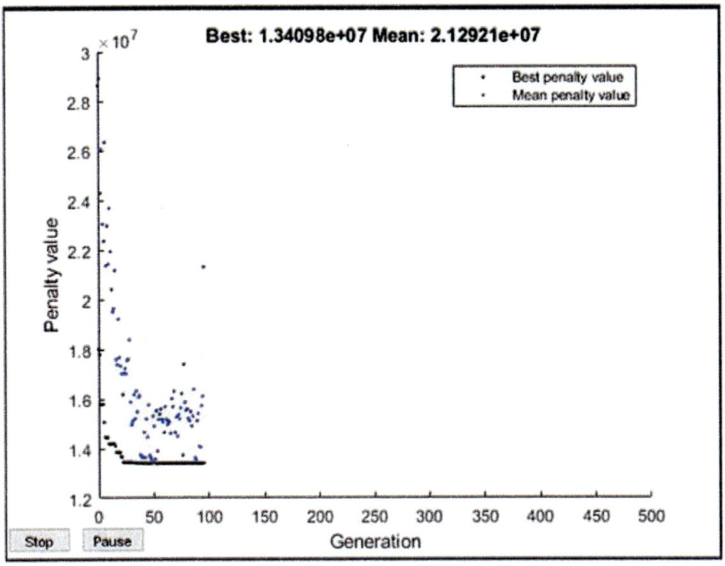

Fig. 7 Fitness values with generation for Case-3 of pumped Network-1

uses integer linear techniques to obtain a solution. JalTantra solution provides total cost of pipeline as 583.6 lakh (Case-1). GA-based program is used for same lengths of pipe in Case-2. In this case, GA-based solution has resulted total cost of 581.8 lakh which is marginally less than that obtained in Case-1. The GA-based solution is improved by introducing first two pipe lengths as variables in Case-3 and Case-4. It is observed that in Case-3, the total cost is further reduced but the residual pressure at the end of pipe-3 falls below the specified value of 5 m. To improve the performance of GA, the weights are introduced to the value of constraint terms to make them more sensitive. The result of improved GA is obtained with cost of Rs 582.4 lakh and residual pressure within limit in Case-4. The cost is marginally higher than that obtained by JalTantra (Case-1). However, the GA provides residual pressure near to the expected values.

5.2 *Pumped Network-1*

JalTantra solution provides total cost of pipeline as 134.6 lakh (Case-1). GA-based solution as per Case-2 has resulted total cost of 135.5 lakh which is marginally more than that obtained in Case-1 but provides better control of residual pressures. Trials were made to check the possibility to reduce the cost. The solution of Case-3 is obtained with lesser cost of Rs 134.1 lakh. The performance of GA-based algorithm is found satisfactory.

Table 6 Results of pumped network-2 with JalTantra (Case-1)

Pipe ID	Start node	End node	Flow (lps)	Diameter (m)	Roughness 'C'	Head loss (m)	Pump head (m)	HGL end (m)	Pressure (m)	Pipe cost (Rs./m)	Pipe cost (Rs)	Pump capacity (KW)	Pump cost (Rs.)	Energy cost (Rs.)	Total cost (Rs.)
1	1	2	41.4945	273.0	140	5.896	20.025	67.629	11.509	1792	6,074,880	11.641	34,923	4,772,502	
2	2	3	17.5515	176.3	150	0.275		67.355	**12.705**	991	104,055				
3	2	4	23.9430	198.3	150	8.130		59.499	11.063	1256	3,893,600				
4	4	5	11.2440	158.7	150	0.890		58.609	**16.849**	802	372,930				
5	4	6	12.699	176.3	150	6.250		53.250	5.150	991	4,310,850				
6	6	7	12.699	176.3	150	0.180		53.070	**5.000**	991	123,875				
											14,880,190		34,923	4,772,502	19,687,616

Table 7 Results of pumped network-2 with GA (Case-2)

Pipe ID	Start node	End node	Flow (lps)	Diameter (m)	Roughness 'C'	Head loss (m)	Pump head (m)	HGL end (m)	Pressure (m)	Pipe cost (Rs./m)	Pipe cost (Rs.)	Pump capacity (KW)	Pump cost (Rs.)	Energy cost (Rs.)	Total cost (Rs.)
1	1	2	41.4945	273.0	140	5.896	20.025	67.629	11.509	1792	6,074,880	11.641	34,923	4,772,511	
2	2	3	17.5515	110.1	150	2.720		64.909	**10.259**	390	40,950				
3	2	4	23.9430	198.3	150	8.130		59.500	11.064	1256	3,893,600				
4	4	5	11.2440	110.1	150	5.281		54.219	**12.459**	390	181,350				
5	4	6	12.699	176.3	150	6.250		53.250	5.150	991	4,310,850				
6	6	7	12.699	176.3	150	0.180		53.070	**5.000**	991	123,875				
											14,625,505		34,923	4,772,511	19,432,940

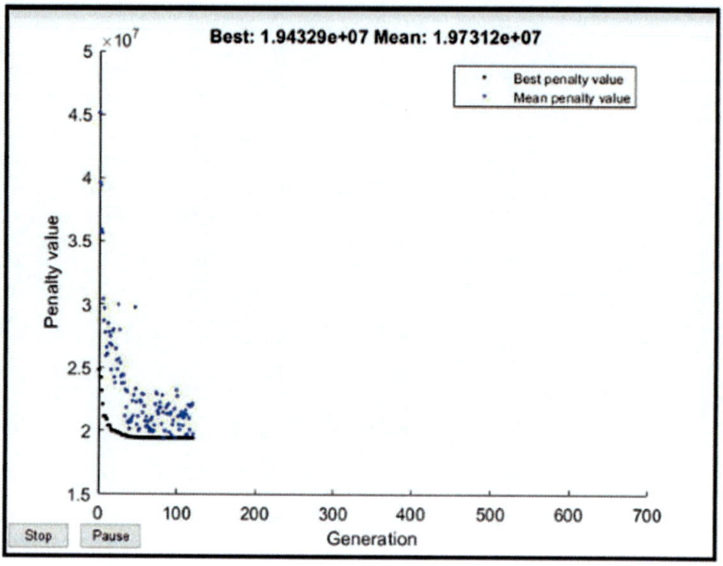

Fig. 8 Fitness values with generation for Case-2 of pumped network-2

5.3 *Pumped Network-2*

Testing the performance of GA-based program with different networks with varied configurations provides the knowledge of its applicability and robustness in comparison with JalTantra. For pumped network-2, JalTantra solution provides total cost of system as 196.9 lakh (Case-1). GA-based solution as per Case-2 has resulted total cost of 194.3 lakh which is marginally less than that obtained in Case-1. In this case also, the performance of GA-based algorithm is found satisfactory.

So, the best results obtained by GA for the total cost are compared to that obtained from JalTantra tool for all the three networks which are shown in Fig. 9. GA-based algorithm has shown equal performance with JalTantra. GA provides flexibility to achieve better control of residual pressures for pressure-driven demands in a network. In GA-based program, head loss can be found using different formula such as Darcy–Weisbach equation. JalTantra does not provide such flexibility as it is not an open-source platform.

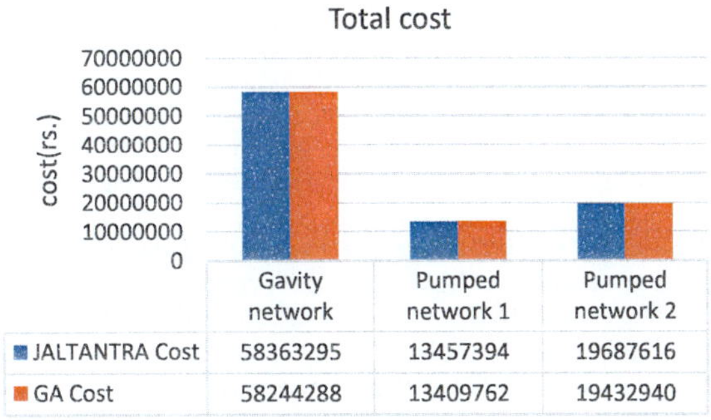

Fig. 9 Comparison of JalTantra and GA total cost

6 Conclusion

The following conclusions are derived based on the study presented in this paper.

1. JalTantra software is found very versatile to obtain economical solution for both the gravity and pumping network. It also considers to divide the long pipeline into multiple segments of different diameters to generate least cost solution. Its free availability, user interface and solution approach show the potential of its use for real life design problems.
2. GA has a capability to generate the optimum solution for the design of pipe system with the proposed strategy of formulating objective function and constraints. The results obtained by GA for all the three networks show good agreement with results obtained by using JalTantra. Thus, GA-based program is found efficient to obtain optimum solution and to provide flexibility in comparison with software like JalTantra.
3. The proposed GA problem formulation has index number of commercial pipe array as a decision variable. This enables the inclusion of two or more pipes of same internal diameter but different material or roughness and different cost per unit length.
4. GA program results are improved by introducing weight to constraints. In certain simulations, GA-based algorithms produced better result than that obtained by JalTantra. But, in general, both the JalTantra and GA-based model generate solution with minimum cost with marginal difference.
5. The JalTantra and GA both can handle constrained network to maintain minimum residual pressure at demand nodes.

Acknowledgements The authors are grateful to GWSSB regional office, Rajpipala, to provide data for the study.

References

1. Sinha A, Ghorpade A, Hooda N, Damani O, Kalbar P (2019) JalTantra: a web-based free-for-all platform for water network optimal design. Indian Water Works Assoc, 1–8
2. Beatriz M, Marco Antonio C et al (2018) Using a genetic algorithm with a mathematical programming solver to optimize a real water distribution system. Water J 10:1318
3. Holland J (1975) Adaptation in natural and artificial systems. University of Michigan Press
4. Reca J, Martínez J, Rafael L (2017) A hybrid water distribution networks design optimization method based on a search space reduction approach and a genetic algorithm. Water J 9(110):845
5. Haldurai L, Madhubala T, Rajalakshmi R (2016) A study on genetic algorithm and its applications. Int J Comput Sci Eng 4(10):139–143
6. Montesinos P, Garcia-Guzman A, Ayuso JL (1999) Water distribution network optimisation using modified genetic algorithm. Water Resour Res 35:3467–3473
7. Hooda N, Damani O (2019) JalTantra: a system for the design and optimization of rural piped water networks. Informs J Appl Anal 1–12
8. Reca J, Martínez J (2006) Genetic algorithms for the design of looped irrigation water distribution networks. Water Resour Res 42:W05416
9. Savic AD, Walters GA, Atkinson RM, Smith MR (1999) Genetic algorithm optimization of large water distribution system expansion. Meas Control 32:104–109
10. Walters GA, Savic DA (1996) Recent applications of genetic algorithms to water system design. Trans Ecol Environ 12:143–152

Comparison of a Long Short-Term Memory Model with Statistical-Based Water Demand Prediction Models on a Case Study of Spain

Prityush K. Sahu, Prerna Pandey, Shilpa Dongre, and Rajesh Gupta

Abstract Accurate prediction of future water demand is desirable in both design and operation of water distribution networks (WDNs). While the long-term forecasting is helpful in planning and designing the system, a short- to medium-term forecasting contributes in better operation, maintenance practices and calibration of the system. The automations in WDNs have emphasised on more accurate prediction of daily and weekly demands for online scheduling of pump operations and valve adjustments. In this paper, a deep learning algorithm called recurrent neural network (RNN) has been used to train the model and forecast hourly water demand, using long short-term memory (LSTM) layer for a city of Spain. The performance of LSTM model is compared with some statistical hybrid models such as ensemble empirical mode decomposition (EEMD), difference pattern sequence forecasting (DPSF) (EEMD-DPSF), and the EEMD with DPSF and autoregressive integrated moving average (ARIMA) (EEMD-DPSF-ARIMA), etc. by means of root mean squared error (RMSE) and mean absolute error (MAE) and mean percentage absolute error (MAPE). Results of this study show that the LSTM-based model can make predictions with improved accuracy than the other models that are being compared when dealing with data with higher time resolutions, data points with abrupt changes, and data with a relatively high uncertainty level. It is also observed that with respect to EEMD-DPSF, LSTM-based models provide better performance in predicting multiple successive water demands.

Keywords Water demand prediction · LSTM · Soft computing approach · Errors

P. K. Sahu (✉) · P. Pandey · S. Dongre · R. Gupta
Department of Civil Engineering, Visvesvaraya National Institute of Technology (VNIT), Nagpur, Maharashtra 440010, India
e-mail: prityushsahu96@gmail.com

P. Pandey
e-mail: prerna.pandey1203@gmail.com

S. Dongre
e-mail: shilpa_dg@rediffmail.com

R. Gupta
e-mail: drrajeshgupta123@hotmail.com

P. V. Timbadiya et al. (eds.), *Geospatial and Soft Computing Techniques*, Lecture Notes in Civil Engineering 339, https://doi.org/10.1007/978-981-99-1901-7_26

1 Introduction

Water has always been an important component to support life on earth. With the advancement in civilisation, water and sanitation system has become an integral part of urban infrastructure. Due to exponential growth in the world population, the water demand exerted by the system has been increased drastically. Only 3% of the water existing on earth is fresh water, out of which two-thirds is captured in frozen glaciers or otherwise unavailable for our use. As a result, around 1.1 billion people on earth have lack of access to potable water, and a total of 2.7 billion face scarcity for at least for a month in a year. Already, there are many regions that have been identified in the world where water shortage has become a severe condition. South Africa, Brazil, England, Middle East Countries, China, and India are the major ones. In India, cities like Delhi, Chennai, Bangalore, Hyderabad, Nashik, and many other are under extreme water crisis as per the recent studies [4]. Hence, it is really important to plan and manage such a scarce resource to keep the integrity of this ecosystem.

The prime goal of any water utility is to provide safe and reliable potable water to its consumers all the time. The water distribution system is generally being designed to support the future 30 years population. This long-term demand forecasting is done considering the population growth of an area, which is usually obtained through various methods considering the past growth pattern of population. However, migration of people from rural to urban area, climate change that introduces a lot of uncertainties, and other social factors may limit the accuracy of these conventional methods, as historical trends will no longer be reliable for predicting future water demand. Factors like weather conditions, seasonal changes, festive, or emergency events do contribute to the short-term variation in demand. The demand pattern gets changed so frequently that, without short-term demand forecasting, it becomes really difficult to desired water supplies. The necessity of automation in operation of WDS also emphasise on estimation of water demand accurately. Several soft computing and advanced statistical methods are being developed for both long-term and short-term prediction.

Artificial neural networks (ANNs), recurrent neural networks (RNNs), support vector machine (SVM), adaptive neuro-fuzzy interface system (ANFIS) models are the examples of various soft computing techniques. Apart from the conventional methods and time series analysis, various advanced statistical approaches have also been made to forecast water demand. The time series models are useful for short- and medium-term forecast, wherein fluctuation in demand is observed with time. It usually requires understanding the past trend, calculating some statistical parameters and projection based on calculated parameters.

The autoregressive integrated moving average (ARIMA) is one of the widely used linear model in time series forecasting. Arandia et al. [2] presented a seasonal autoregressive integrated moving average (SARIMA) model with data assimilation. The tailoring process was adopted for identification, estimation, and validation of the models and exploring how the length of demand history can be used for improving the forecast performance.

The hybrid models consist of combination of two or more forecasting models, which sometimes helps to overcome the drawbacks of the original techniques. The authors combine the model effectively and calibrate various parameters in it to get the improved results. Wavelet–bootstrap–artificial neural network (WBANN) modelling approach was proposed by Tiwari and Adamowski [9] to forecast medium-term urban water demand with limited data. The bootstrap technique is a data-driven simulation methodology that uses intensive resampling with replacement which reduces the uncertainties.

Empirical mode decomposition (EMD) which was introduced by Norden et al. [7] is found to be best suited for both nonlinear and non-stationary time series analysis. In this process, the whole time series is decomposed to a number of intrinsic mode function (IMF) and one residual. Wu and Huang [10] added finite white noise to the original data, to overcome the problem of mode mixing that is inherent in EMD and suggested ensemble empirical mode decomposition (EMDD). Wu suggested random white noise having zero mean and specified standard deviation [8], to be summed to available time series to decompose the time series by EEMD. Two hybrid models EEMD-DPSF and EEMD-DPSF-ARIMA were proposed by Pandey et al. [8] to improvise the prediction results as compared to the stand-alone methodologies. In EEMD-DPSF model, the time series was decomposed into finite numbers of IMFs and a residual using EEMD. Then with the help of differenced pattern sequence-based forecasting (DPSF) method, the future values of all IMFs and residuals were predicted, and with the help of these values, final prediction outcome was obtained. The hybrid EEMD-DPSF-ARMIA model is a modification of the EEMD-DPSF model by incorporation of ARIMA into it. It was difficult to achieve accurate predictions for all types of IMFs using the DPSF method as DPSF provides superior performance for the stationary, seasonal, and cyclic time series. Hence to take the advantage of auto-regression-based methods in hybrid EEMD-DPSF, the non-stationary time series is processed using ARIMA method.

Most of the approaches available in the literature are used on different data bases, and therefore, easy comparison of different approaches is not possible. Herein, the main objective of the paper is to compare the performance of a RNN model using long short-term memory (LSTM) layer with some of the hybrid statistical models mentioned earlier.

The model is briefly explained, and its application is shown on prediction of water demand using dataset of Spain. This dataset is used earlier in some of the recent studies.

2 Theory

2.1 Recurrent Neural Network (RNN)

A recurrent neural network (RNN) is a classification of artificial neural networks. In deep learning, RNN models are a special kind of neural networks that are designed to process sequential data. RNNs can use their internal state (memory) to process sequences of inputs unlike feed-forward neural networks. By doing so, it not only receives new chunk of information at each time step, but also it adds this new information as a weighted version of the previous output. The recurrent networks allow to operate over sequence of vectors in the input, output, or generally in both cases (Fig. 1).

Loss Function: In recurrent neural network, the loss function L for all time steps is defined on the basis of the loss occurring at every time step as shown below:

$$L(\hat{y}, y) = \sum_{t=1}^{T_y} L(\hat{y}^{\langle t \rangle}, y^{\langle t \rangle})$$

where $\hat{y}$ = Estimated output; y = Target output; t = Time steps

Activation Function: These are critical part of any neural network. The choice of activation function in hidden layer controls how well the model learns about the training dataset. While the choice of activation function in output layer defines the type of prediction, the model can make. Activation functions in neural network define how the weighted sum of input is transformed into an output from a node or nodes in a layer of the network. Some commonly used activation functions in RNN are (Fig. 2):

- Sigmoid
- Tanh
- RELU.

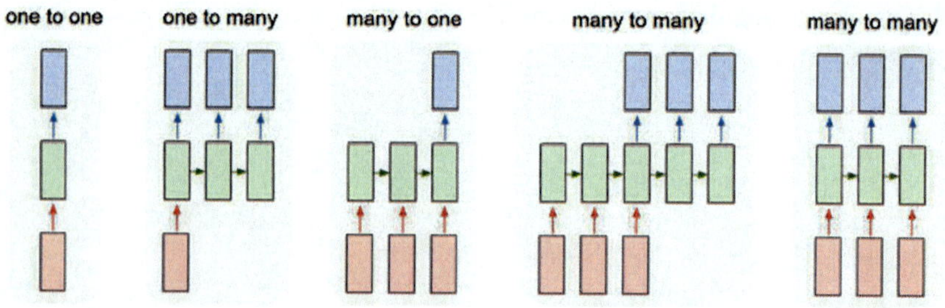

Fig. 1 Types of RNN models

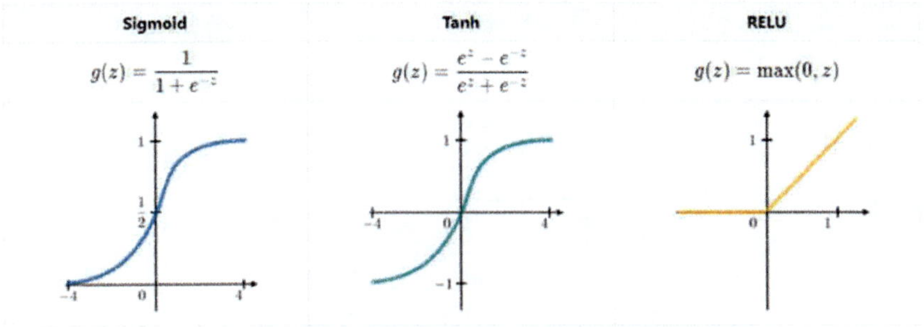

Fig. 2 Activation functions used in neural networks

The issue with RNN models is that, with time as the model gets fed more new data, it tends to "forget" about the previous data it had seen, as it gets dissolved between the new data, the transformation from activation function, and the weight multiplication. It is, called vanishing gradient problem. To solve this problem LSTM is being introduced to enhance memory of conventional RNN.

2.2 *Long Short-Term Memory (LSTM)*

It is an artificial RNN architecture used in the field of deep learning. Unlike feed-forward neural networks, LSTM has got feedback connection. It has a complex cell structure than normal recurrent neuron which allows it to regulate how to learn and unlearn effectively with great efficiency from the different types of input sources. A common LSTM unit consists of a cell involving an input gate, an output gate, and a forget gate. The major component of LSTM is the cell state (cell memory). The cell state encodes the information that has been observed up to each step. Because of an explicit gating mechanism, the cell can figure out what should be done with the state vector: whether to read from it, write to it, or discard it. The LSTM cell is a specially designed unit of logic that helps to reduce the vanishing gradient problem sufficiently to make recurrent neural networks more useful for long-term memory tasks, i.e. sequence predictions (Fig. 3).

- **Input Gate**: It decides the information that should enter the cell state.
- **Forget Gate**: Decides whether cells can erase its memory to make space for new information to be updated.
- **Output Gate**: Decides whether to make output information available or not.

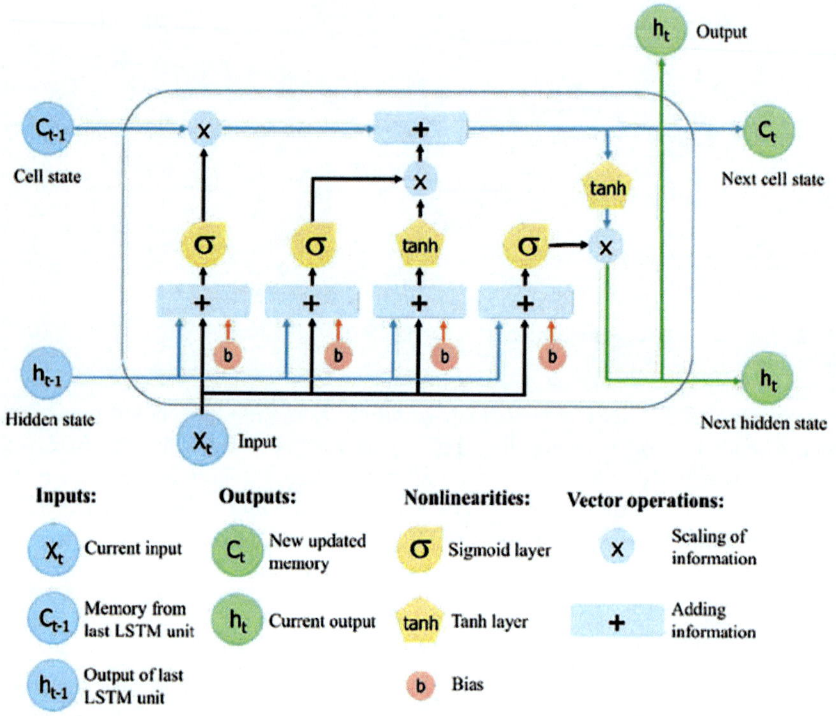

Fig. 3 Basic LSTM structure

2.3 *Core LSTM Equations*

The equations used inside LSTM network right from input to output are shown as follows [6]:

$f_t = \sigma_g\left(W_f \times x_t + U_f \times h_{t-1} + b_f\right) f_t$ *is the forget gate*

$i_t = \sigma_g(W_i \times x_t + U_i \times h_{t-1} + b_i)$ i_t *is the input gate*

$o_t = \sigma_g(W_o \times x_t + U_o \times h_{t-1} + b_o)$ o_t *is the output gate*

$c_t' = \sigma_c(W_c \times x_t + U_c \times h_{t-1} + b_c)$

$c_t = f_t * c_{t-1} + i_t * c_t'$ c_t *is the cell state*

$h_t = o_t * \sigma_c(c_t)$ h_t *is the hidden state*

σ_g and σ_c are sigmoid and tanh functions.

* represents element wise multiplication. where x_t = Input vector; h_{t-1} = Previous cell output; c_{t-1} = Previous cell memory; W and U = Weight vectors for all the gates, b = Bias at each gate.

2.4 Model Performance

The performance of the model has been determined using the commonly used indices: root mean square error (RMSE), mean absolute error (MAE), and mean absolute percentage error (MAPE). These parameters are mathematically expressed as:

$$\text{RMSE} = \sqrt{\frac{1}{N}\sum_{i=1}^{N}|O_i - P_i|^2}$$

$$\text{MAE} = \frac{1}{N}\sum_{i=1}^{N}|O_i - P_i|$$

$$\text{MAPE} = \frac{1}{N}\sum_{i=1}^{N}\frac{|O_i - P_i|}{O_i} \times 100\%$$

where O_i = Observed/Actual demand at ith hour; P_i = Predicted demand at ith hour; n = Number of total demand observation points.

3 Case Study

3.1 Description of the Case Study

This study has been performed on a water demand dataset of a city in south-eastern Spain. The district metered area (DMA) under consideration has approximately 5000 consumers distributed over an area of approximately 8 km^2. The dataset consists of hourly demand data from 1 Jan to 8 April of the year 2005 and has been chosen for water demand forecasting due to its appropriate seasonal variations. It has been used in many renowned similar case studies such as Herrera et al. [3]; Anele et al. [1] and Pandey et al. [8]. The dataset has got 2352 number of data points. In the study, 1882 data points (around 80% of whole) have been used for training the forecasting model. The forecasting has been done, and the model performance is validated using rest of the data. The motive of this study was to check the performance of LSTM over advanced statistical demand prediction methods like EEMD-DPSF, EEMD-DPSF-ARIMA, EEMD-ARIMA, EEMD-PSF, EEMD-PSF-ARIMA, ARIMA, etc. (Table 1 and Fig. 4).

Table 1 Summary of the summary of the dataset (m^3/h)

Min.	Median	Mean	Max.	Std. Dev.
0.22	17.25	18.95	55.39	8.154

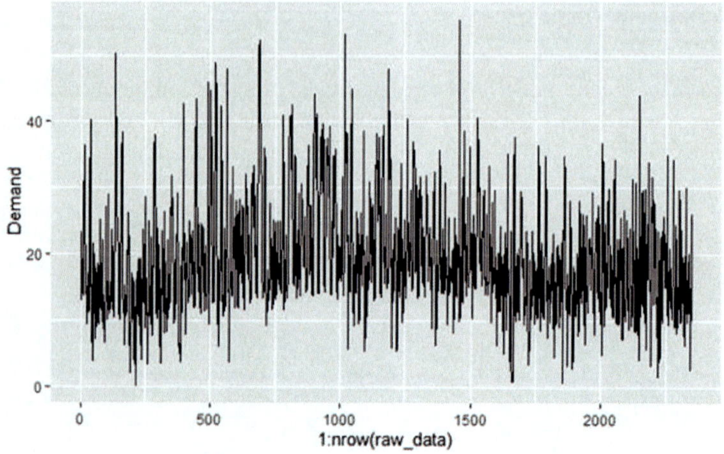

Fig. 4 Water demand dataset plot (2352 h)

3.2 *Computational Methodology*

Various R and Python packages are commonly used to develop prediction models. The RNN-based LSTM model was basically developed in python environment with the help of the functions defined in Keras Library [5]. Keras is an application programming interface (API). It is an open-source software library which provides python interface for working in deep learning algorithms like neural networks. It contains many neural network building blocks such as activation functions, layers, optimisers, and many more things which makes it easy to simplify the code which are required to run deep neural networks and recurrent neural networks. Here, the reference has been taken from a source code that was used to predict temperature by an R programmer BrandonYEO. The dataset also included various other parameters such as pressure, vapour pressure, wind velocity, wind temperature, wind flow directions to constitute a multivariate prediction model.

In this study, the real demand dataset is first normalised to help the model train data in an efficient way. Choosing the appropriate methodology is an essential step while working with forecasting or prediction models. Hence, a part of the dataset (in general 70–80% data of whole data) is used to train the model on its pre-defined training block. While training the dataset, the MSE value should reduce with increasing number of epochs, but over doing it leads to an over-trained model resulting in poor solutions. In this study, the epochs was set to 10 which gave the best result. The MSE value on 10th epoch was found to be 0.008 on normalised dataset, which can be seen in the *R* plot (Fig. 5) indicating a well-trained model.

Once the model is trained, forecasting is done and validated with rest of the dataset. The actual and predicted normalised demands are plotted against timeline, and the variations are checked. If the plot indicates less variations in actual and

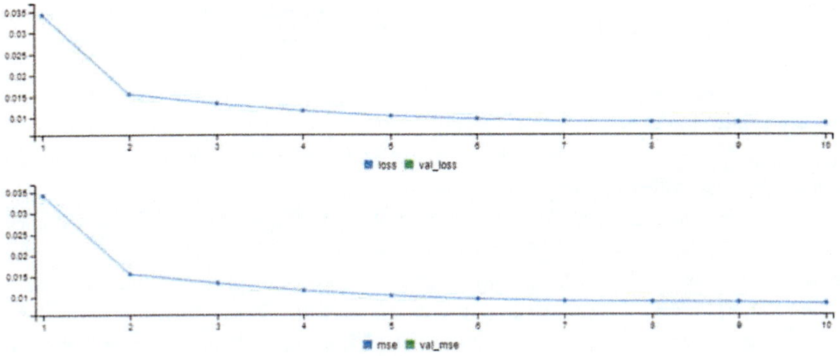

Fig. 5 Model training error validation plot

predicted values, the forecasting results can be considered as acceptable results. The normalised demands are de-normalised and plotted against timeline.

3.3 *Results Obtained by LSTM*

It can be observed from the prediction plot (Fig. 6) that the predicted demand is very much close to the actual demand. The demands are plotted for seven days (i.e. 168 h) ahead. The MSE, RMSE, and MAE have been calculated using de-normalised actual and predicted demands for a time horizon of next 168 h (7 days) from 1882nd data point onwards. The RMSE, MAE, and MAPE value for the same have been found to be 2.1616 m^3/h, 1.8208 m^3/h, and 17.4142%, respectively.

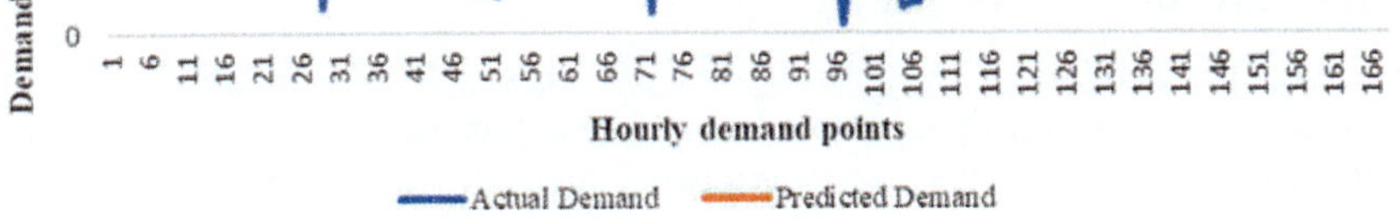

Fig. 6 Actual and predicted demand plot (168 h)

3.4 Comparison of Results with Other Statistical Models

All of the models were able to predict the overall trend at every time step, with errors mostly producing at the extreme values. A detailed comparative analysis has been done to judge the performance of various models taken into consideration, which can be seen in Tables 2, 3 and 4.

It can be observed that for all the prediction horizon (6 h, 12 h, and 24 h), the value of RMSE, MAE, and MAPE is less for LSTM model as compared the statistical models. The MAPE value for 12 h and 24 h prediction horizon was found to be close for LSTM, EEMD-DPSF, and EEMD-DPSF-ARIMA, but LSTM performed remarkably well in the 6-h prediction horizon. It is important to compare various models' performance at different time resolutions to facilitate better informed decision making since different models perform their best in different scenarios.

Datasets with abrupt changes are generally very difficult to predict. From Fig. 6, it can be seen that the LSTM model was able to train the Spain dataset with frequent abrupt changes and was able to forecast short-term demand pretty accurately. For making a prediction, LSTM looks back at a series of past data, and after prediction, it takes up the newly predicted data as an input for forecasting demand at next time step. This helps LSTM model to perform with improved accuracy.

Since water demand is affected by multiple factors like seasonality, temperature, rainfall, humidity, and many more, the LSTM model can incorporate all these additional factors to carryout multivariate analysis to improve the prediction results. In this study, additional factors like temperature, day of the week, pressure, wind speed, and rainfall were considered to make the short-term demand prediction which showed significantly better performance as compared to other statistical models.

4 Conclusion

From the above study, it could be seen that short-term water demand prediction by LSTM model was found to be significantly better, which was inspired by many LSTM-based models that were used to forecast various other parameters in other research domain. LSTM models do require a dataset with a huge number of data points to train it properly. The following conclusions can be drawn considering all the outcomes of the current study,

- LSTM performs best with predicting data with very high time resolution with the help of a dataset having large number of data points. Hence, it may be used to predict water demand at urban areas where larger side data is available thanks to the advanced operation and maintenance systems.
- LSTM has got the potential to do multivariate analysis, which helps in improving the performance of the prediction model as compared to univariate ones.

Table 2 Comparison of performance of proposed method using RMSE (m3/hr)

Prediction horizon (h)	LSTM	Statistical approaches					
		EEMD-DPSF	EEMD-DPSF-ARIMA	EEMD-ARIMA	EEMD-PSF	EEMD-PSF-ARIMA	ARIMA
6	1.245	4.472	4.958	8.724	7.426	8.187	6.81
12	1.875	5.357	6.687	9.598	8.018	8.858	7.997
24	2.152	7.269	7.79	9.127	8.834	8.883	8.547

Table 3 Comparison of performance of proposed method using MAE (m^3/h)

Prediction horizon (h)	LSTM	Statistical approaches					
		EEMD-DPSF	EEMD-DPSF-ARIMA	EEMD-ARIMA	EEMD-PSF	EEMD-PSF-ARIMA	ARIMA
6	0.945	3.088	3.291	6.654	6.108	7.401	5.321
12	1.614	3.685	4.186	7.607	6.681	7.277	6.093
24	1.936	4.352	5.661	8.191	6.89	7.528	6.941

Table 4 Comparison of performance of proposed method using MAPE (%)

Prediction horizon (h)	LSTM	Statistical approaches					
		EEMD-DPSF	EEMD-DPSF-ARIMA	EEMD-ARIMA	EEMD-PSF	EEMD-PSF-ARIMA	ARIMA
6	5.211	13.604	13.723	18.637	18.112	31.168	17.223
12	13.005	13.23	14.493	20.819	19.021	30.653	18.724
24	15.754	16.155	17.293	22.735	24.032	33.909	27.099

- For longer prediction horizons (12 h, 24 h and more), LSTM performs more or less similar as EEMD-DPSF and EEMD-DPSF-ARIMA, since they have got nearly equal MAPE values.
- LSTM is a RNN-based model, which comes under deep learning and soft computing category. Such kind of models operates through multiple levels of hidden layers making us unable to understand the exact processes of the algorithm.

References

1. Anele AO, Hamam Y, Abu-Mahfouz AM, Todini E (2017) Overview, comparative assessment and recommendations of forecasting models for short-term water demand prediction. Water 9(11):887
2. Arandia E, Ba A, Eck B, McKenna S (2016) Tailoring seasonal time series models to forecast short-term water demand. J Water Res Plann Manag 142(3). https://doi.org/10.1061/(ASCE)WR.1943-5452.0000591
3. Herrera M, Torgo L, Izquierdo J, Pérez-García R (2010) Predictive models for forecasting hourly urban water demand. J Hydrol 387(1–2):141–150
4. https://www.thehindu.com/news/cities/mumbai/indias-large-cities-staring-at-water-crisis/article28764312.ece
5. Ketkar N (2017) Introduction to keras. Deep learning with Python. Apress, Berkeley, CA, pp 97–111
6. Mu L, Zheng F, Tao R, Zhang Q, Kapelan Z (2020) Hourly and daily urban water demand predictions using a long short-term memory based model. J Water Res Plann Manag 146(9):05020017
7. Norden EH, Zheng S, Steven RL, Manli CW, Hsing HS, Quanan Z, Nai-Chyuan Y, Chi CT, Henry HL (1998) The empirical mode decomposition and the Hilbert spectrum for nonlinear and non-stationary time series analysis. In: Proceedings of the royal society of London. series A: Math phys eng sci 454:903–995. https://doi.org/10.1098/rspa.1998.0193
8. Pandey P, Bokde ND, Dongre S, Gupta R (2021) Hybrid models for water demand forecasting. J Water Res Plann Manag 147(2):04020106
9. Tiwari MK, Adamowski JF (2015) Medium-term urban water demand forecasting with limited data using an ensemble wavelet–bootstrap machine-learning approach. J Water Res Plann Manag 141(2). https://doi.org/10.1061/(ASCE)WR.1943-5452.0000454
10. Wu Z, Huang NE (2009) Ensemble empirical mode decomposition: a noise-assisted date analysis method. Adv Adapt Data Anal 01(01):1–41. https://doi.org/10.1142/S1793536909000047

Developing Leak Detection Strategies in Water Distribution Networks Using Machine Learning Techniques

Kushang V. Shah and H. M. Patel

Abstract Data availability and monitoring facilities enable the application of machine learning (ML) and artificial intelligence (AI) in the field of environmental hydraulics. Determination of anomalies is the complex imposition in the hydraulic modeling of water distribution systems. Leakage detection in water distribution networks (WDNs) becomes a challenging problem to be solved in the direction of meeting time and efficiency constraints. In this study, hydraulic model was developed with known demands and fictitious leakages introduced to treat the model for probable leakage. Conditional probabilities were generated for all nodes using simulation results obtained by EPANET. Prediction models are developed by applying support vector machine (SVM) and artificial neural network (ANN). Both models are compared by keeping the same input parameters of classifying models for training–testing datasets and aiming with conditional probabilities. Results are quite encouraging to resolve complexities in the leak detection for water distribution systems.

Keywords Artificial neural network · EPANET · Leak detection · Support vector machine · Water distribution system

1 Introduction

The loss of water in water distribution systems due to leakage is a major concern for city authorities, and it has posed a challenge for the research community for quick and precise detection of leakage spots with limited available hydraulic data. Traditional philosophies are intended for fractions of major losses for consideration of leakages in water distribution systems. Traditional works of excavating buried

K. V. Shah (✉) · H. M. Patel
Department of Civil Engineering, Faculty of Technology and Engineering, The Maharaja Sayajirao University of Baroda, Vadodara 390001, India
e-mail: kushang.shah-ced@msubaroda.ac.in

H. M. Patel
e-mail: hmpatel-ced@msubaroda.ac.in

P. V. Timbadiya et al. (eds.), *Geospatial and Soft Computing Techniques*, Lecture Notes in Civil Engineering 339, https://doi.org/10.1007/978-981-99-1901-7_27

pipes at trial locations are associated with many problems such as cost, delay in restoration of supply and traffic obstructions. Existing practice needs to be improved to determine the quantities and location of leakages in WDNs with adequate precision for its repair. Adequate flow and pressure monitoring within WDNs can provide important inputs for leak detection. Use of advanced technologies such as machine learning (ML) techniques can enhance the prediction capabilities in WDNs. These advanced techniques can help to secure the zero leakage of water and to prevent contamination in WDNs by providing predicting capabilities into the system. ML techniques provide appropriate classification by minimizing leakage in WDNs with better accuracy rather than onsite trials.

Roots of the leak detection studies originated in the early 90s when most of the hydraulic models have been worked on the basis of mathematical framework. Mukherjee and Narasimhan [10] investigated multiple leakages and calculated efficiency for a large hypothetical model of WDNs using the generalized likelihood method which gives the nonlinear detection strategy for WDNs. Poulakis et al. [11] developed genetic algorithm and inverse transient method to detect leaks and friction factor of water distribution network. Mashford et al. [6] developed the SVM model. Model is trained and tested with respect to results obtained from the EPANET simulation and pressure sensors. Performance assessment of the SVM showed the leak size and location. Outcomes are predicted with a reasonable degree of accuracy. Yang et al. [15] selected a neural network approach as a classifier, which identifies self-similarity degrees as the network inputs for leakages. System achieves the correct detection range but requires validation accuracy. Romano et al. [13] used several AI techniques including wavelets for de-noising of the recorded pressure and flow signals, ANN selected for the short-term forecasting of future pressure and flow signal values, statistical process control analysis implemented for the analysis of discrepancies between the predicted (i.e., expected) and the actually observed signal values. These approaches gave better results for real-time studies but defended SCADA units for online distribution systems. Mounce et al. [8] used support vector regression (SVR) for identifying sampling interval for event detection in WDNs. Improvements required for pressure gain instead of averaging pressure data for models. Mohammad et al. [9] conducted measurement error sensitivity analysis for detecting and locating leaks in pipelines using ANN and SVM. Sensor's error is quite deviating for such analysis because observation relies on sensors. Gustavo et al. [2] created an ANN model to estimate pressure on all nodes of a network. The calibration of pipe roughness followed particle swarm optimization (PSO) to minimize the objective function represented by the difference between simulated and forecasted pressure values. Przystałka [12] created a neural network autoregressive model with an exogenous input to develop performance optimization of leak detection in WDNs. Methodology requires maximum observations from sensors. The applicability of the technique in practice depends on the ability of pressure sensors. Mohammadreza et al. [7] represent graph theory and ANN on the basis of pressure logger data, and they developed the scenarios to identify leakages. Ivana et al. [4] used a random forest classifier to detect leak locations on two different sized WDNs with sparse sensor placement. A great number of leak scenarios were simulated with Monte Carlo method, and leak parameters (leak location and emitter coefficient) have been determined. Model accuracy is required to be checked for multiple leakage spots and optimized numbers of features.

In Indian scenarios, multiple online sensors are not encouraged due to economic and feasibility aspects. Examination of leakages requires observation in terms of pressures and flows within the network. Physical measurement of pressure and flow at multiple spots could be a monotonous task for municipal authorities. Hydraulic models should be treated with prediction capabilities through minimum numbers of observations. Application of ML techniques can improve such capabilities to model but creates some challenges such as adequate validation, confusion in decision making, inconsistent sample inputs and real-time computations. There is limited research of ML applications in leak detection in terms of minimum observations and its combination with statistical approaches. The objective of the present study is to develop strategies for investigation of the probable locations of leakage in WDNs. Two ML techniques are used for the studies; those are SVM and ANN. Conditional probabilities are applied to each node within the network to create the biases. Model is grouped for input attributes of simulation results and tested with biases. Results of both the techniques are compared for its validation and its capabilities to locate the leakages in WDNs.

2 Methods for Hydraulic Modeling, Leakage Classification and Prediction Model

2.1 Pressure Driven Analysis (PDA)

PDA method allows the actual demand delivered at a node corresponding to available pressure. PDA is the one way to avoid having designed demands at nodes with inadequate pressure. In PDA approach, the governing equation for leakages is as follows [14]:

$$d = kD\left(\frac{P}{P_{\text{ref}}}\right)^{P_{\text{exp}}} \tag{1}$$

where d is the leakage demand, D is the reference leakage demand, k is the coefficients, p worked as a pressure within stretch, and P_{ref} is the referenced pressure along which leakage is supposed to be measured. Pexp could refer to the material.

2.2 Conditional Probability Approach for Classification Model

Datasets need to be segregated to set up input parameters as features and probabilities as labels for classification models. Conditional probability approaches can be applied to obtain a framework of probable biases and non-biases based on the observations or hypothesis [3].

$$P(A) = \frac{P(A \text{ and } B)}{P(A)} \tag{2}$$

where $P(B|A)$ = The probability of B given A, $P(A$ and $B)$ = Probability of occurrence A and B, $P(A)$ = Probability of occurrence.

2.3 Support Vector Machine as Prediction and Leak Detection Model

Datasets need SVM to work well for higher dimensional features and find the supporting vectors to divide the data. SVM defines the hyperplane using "kernel trick." Kernel needs to be defined as linear for the classifying model [1]. Linear relationship can be presented as follows:

$$w.x + b = 0 \tag{3}$$

where w is related to hypothesis made for the programming and x refers to 2D array. Gamma function is assigned for scaling the data frame into bounding limits.

2.4 Artificial Neural Network (ANN) as Prediction and Leak Detection Model

ANN requires less prescribed statistical training and detects all possible interactions between predictor variables [5]. ANN provides sequential analysis of multiple inputs and classifies the problem toward decision making output. ANN provides significance for validations of leakage models. In the present study, rectified linear unit (ReLU) is used as an activation function. However, "Sigmoid" activation function is also used for probability-based outputs when outcomes are classified in binaries [1]. Optimizer is introduced into the neural network to reduce the losses which can change the attributes for optimization. Stochastic gradient descent (SGD) algorithm is selected as an optimizer for adjustment of weights in the prediction model.

3 Hydraulic Modeling and Creation of Datasets for Classification–Prediction Model

Hydraulic model developed in EPANET for baseline scenario. Model comprises WDN of 151 junctions and 172 distribution pipes which resembles the part of Vesu area water supply network in Surat city. Network takes water supply from the elevated service reservoir (ESR 1. in Fig. 1). ESR 1 is considered as a tank with a type of mixing model in EPANET. Internal diameter of pipes ranges from 100 to 800 mm. Hazen—William's roughness coefficient for DI pipe assigned as 130. Demands are estimated from the population served by the demand node. Nodal demand schedule is considered uniform for supply hours. PDA method is selected for model analysis.

Two emitters are introduced for leakage at non demand nodes (J-2180 and J-2171. in Fig. 1) with discharge coefficient of 0.2. Tanks are attached to collect leakages and to compute cumulative volume of leakage. Pressure exponent considered as 0.5 [14] for hydraulic-leakage modeling.

Datasets have been prepared for classification and prediction models using simulation results of hydraulic model. Conditional probability approach is applied on each junction to obtain a framework of probable biases and non-biases based on hypothesis. In the present study, a conditional probabilistic model is formulated with respect to the following conditionality.

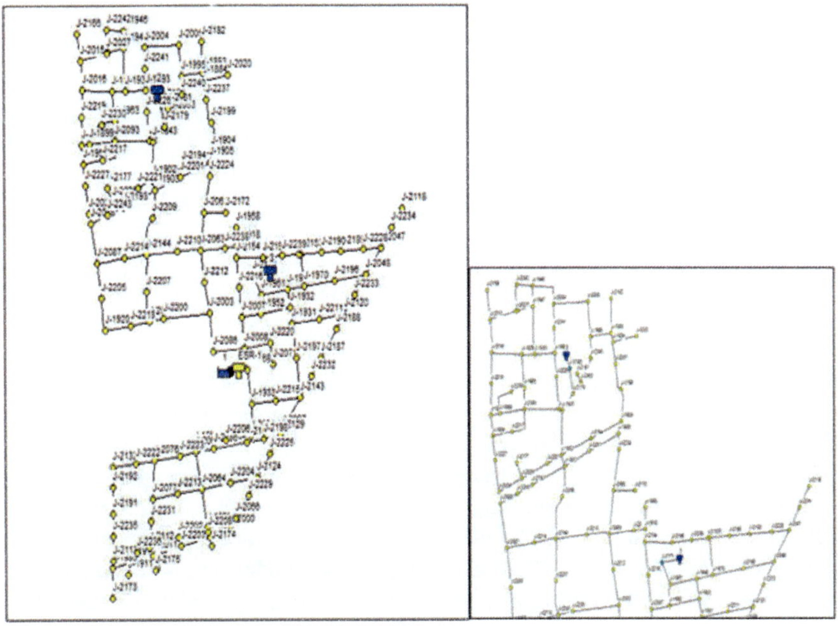

Fig. 1 Hydraulic model of WDN in EPANET and introducing leakage pressure points

Table 1 Partial datasets for classification and prediction model

Pressure (m)	Base demand (LPS)	Actual demand (LPS)	Head (m)	Probability 1	Probability 2	Probability 3
14.60	1.582	1.540	24.57	1	0	1
14.51	1.140	1.100	24.57	1	0	1
35.77	0.150	0.150	46.20	0	0	0
16.91	0.795	0.800	27.16	0	0	0
17.24	0.832	0.830	27.43	0	0	0
35.40	1.640	1.640	46.01	0	0	0
35.04	0.455	0.460	46.01	0	0	0
14.73	0.270	0.270	25.10	1	0	1
14.79	0.816	0.810	24.92	1	0	1
16.59	1.444	1.440	26.91	0	0	0
16.66	1.345	1.350	27.16	0	0	0

$$L(P)_{\text{Pressure}} = \frac{p_{\text{observed}}}{p_{\text{preferenced}}}, \quad L(P)_{\text{Demand}} = \frac{D_{\text{base}}}{D_{\text{actual}}}, \quad L(P)_{\text{Head}} = \frac{hf_{\text{atobs}}}{hf_{\text{atref}}} \tag{4}$$

Subject to:

$$1 \leq L(P)_{\text{Pressure}} \geq 0.70, \quad L(P)_{\text{Demand}} \leq 0.95, \quad L(P)_{\text{Head}} \geq 0.50 \tag{5}$$

Otherwise, 0

Table 1 shows sample datasets of 11 nodes out of 151 which are generated using hydraulic simulation and combined with conditional probabilities.

4 Classification and Prediction Model

Model is classified into training and testing data from the datasets. Among the total sets of (151, 7) (rows, columns), sets of (120, 7) considered as training datasets and sets of (31, 7) considered as testing datasets. Input parameters are assigned in terms of array (*X*, *y*). Here, *X* includes attributes as 2D array and *y* includes variables as 1D array. The first six columns are treated as attributes through which parameters can be analyzed and the last column is treated as variable, i.e., (120, 6) as attributes and (120, 1) as expectations for decisions on leakages. The model procedure is coded using Python programming with Python libraries by introducing SVM and ANN to work as leak detection models for water distribution systems. Models are compared for prediction capabilities and validation to identify leakage in selected network.

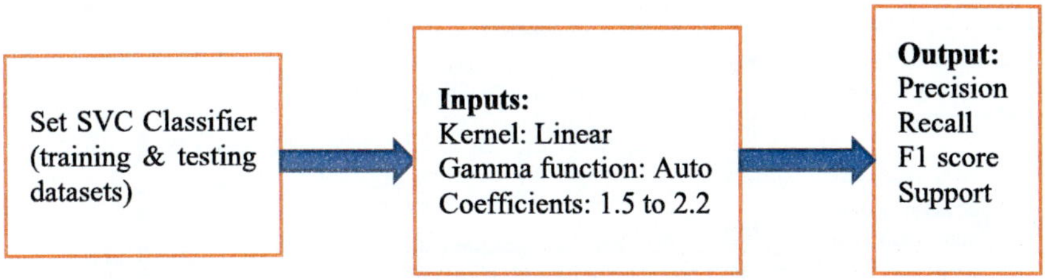

Fig. 2 SVM model in programming tool for leak detection

4.1 Support Vector Machine (SVM)

It is always been a question for the advance and trend algorithms to justify complexities with better model outputs. SVM needs to be checked for solving the complexities of leakage prediction in selected network. Classifier is applied to set leakage predictions by defining training features and model fitness with respect to testing labels. Support vector classifier (SVC) with linear kernel activation is selected as classifier for training and testing datasets used for this study and gamma function is applied as "auto." Coefficients are applied to SVC as penalties for wrongly identified points.

4.2 Artificial Neural Network (ANN)

ANN is increasingly used for detection of complex relationships (nonlinear and linear) between dependent and independent variables. In the present ANN model, there are six different features which can be treated as input, and one could be the decision making for leakage. Dense layer is added with 16 neurons, and ReLU is selected as the activation function for the model. Model is treated toward one output as this is a binary classification problem. Further, a dropout layer is added by keeping the fraction rate of inputs between 0 and 1. Dense layer is repeated by reducing neurons by eight and the same activation function "ReLU." Again, a dropout layer is applied to the model, and training is updated. Finally, dense layer of one neuron is applied as a single final output and its activation function changed to "Sigmoid" which is used to predict probability-based outputs. Model is configured for training and prediction by setting up "optimizer" with SGD algorithm.

5 Results and Discussion

Results obtained from SVM and ANN models are quite encouraging to set up predictive capabilities of leak detection. Figure 4 indicates classification of datasets with respect to conditional probabilities using SVM.

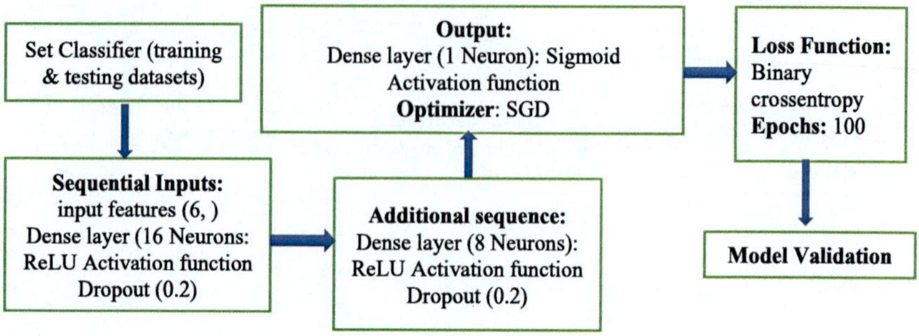

Fig. 3 ANN model in programming tool for leak detection

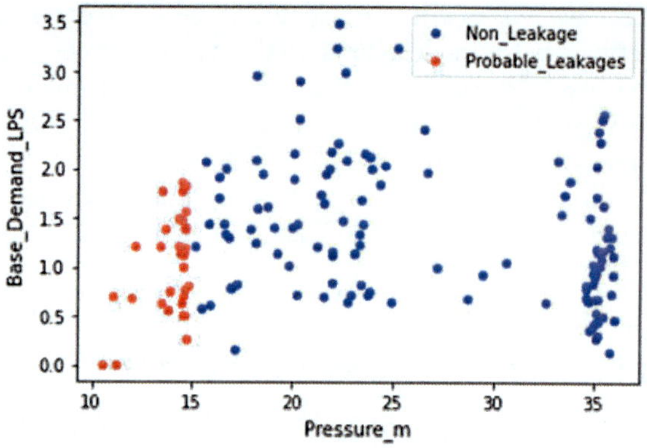

Fig. 4 Classification of datasets with respect to conditional probabilities using SVM

SVM model shows 95% precision for non-leakage spots and 100% for leakage spots. This value indicates correctly predicted results out of predicted possible values. Recall shows that the model's maximum survival in extreme conditions as non-leakages zones is 100%. $F1$ Score of SVM model is 97% for non-leakage spots and 96% for leakage spots. "Support" shows that out of 31 testing samples, twelve could be the spot of leakages and 19 samples are classified as non-leakage zones. SVM shows 97% of model accuracy for this problem.

Table 2 SVM model output

	Precision	Recall	$F1$-score	Support
Class 0	0.95	1	0.97	19
Class 1	1	0.92	0.96	12
Accuracy	–	–	0.97	31
Macro avg	0.97	0.96	0.97	31
Weighted avg	0.97	0.97	0.97	31

Table 3 ANN results (last four epoch)

Epoch	Loss	Accuracy	Validation loss	Validation accuracy
97	0.2714	0.8956	0.3247	0.8710
98	0.4201	0.8087	0.2919	0.9032
99	0.3762	0.8242	0.2921	0.9032
100	0.3949	0.8504	0.3805	0.8065

The ANN model validates the system with the validation accuracy of 80.65% at 100th training "epoch" value. Model accuracy of results is 85.04% with loss values of 0.3949, and validation loss value is 0.3805.

Complexity is resolved related to judgment of leakage spots as features (120, 6) are conditions by the biases (120, 1) with the help of conditional probability and binary classification which segregates leakage spots (1) from non-leakage spots (0). Predictions are presented with 97% accuracy in SVM and confirmed by the ANN for the leakages with 85.04% accuracy. Prediction models show losses as they have made decisions from such varied possibilities of leakages in selected hydraulic model.

6 Conclusions

Motive of the study was finding out useful strategies for leak detection within WDNs. Inputs are generated using EPANET simulation results and conditional probabilities for the prediction models. SVM and ANN techniques of ML are applied as prediction models for the leak detection of selected network. ML is the only possible solution which is applied to resolve complexity for prediction of leakages. Model treatment requires biases for leakages. Biases cannot be applied directly to the model due to independence of attributes. However, conditional probabilities prove to be an effective solution to classify probable leakages to individual nodes. Statistical approach is combined with ML by binary classification of the decision variables. Model performance of ANN confirms the leakage event and appeared less challenging than the SVM. SVM model presents fair results and assembles the group of the leakage event in a better form of presentation. Moreover, SVM model is applicable for prediction of leakages even in the intricate environments. SVM is good for solving confusion metrics which generates better predictions. Both models are presented as best practices for this study that can be utilized to explore the potential applications with varied complexities in system improvement of WDNs.

References

1. Géron A (2019) Hands-on machine learning with Scikit-Learn, Keras, and TensorFlow, concepts, tools, and techniques to build intelligent systems, 2nd edn. O'Reilly Media, Inc. https://www.oreilly.com/catalog/errata.csp?isbn=9781492032649
2. Gustavo M, Daniel M, Bruno B, Thaisa G, Luvizotto E (2017) Calibration model for water distribution network using pressures estimated by artificial neural networks. Water Resour Manage 31(4):4339–4351
3. Guttag JV (2016) Introduction to computation and programming using python with application to understanding data, 2nd edn. The MIT Press Cambridge, Massachusetts, London, England. https://lccn.loc.gov/2016019367
4. Ivana L, Bože L, Zoran C, Sikirica A (2021) Data-driven leak localization in urban water distribution networks using big data for random forest classifier. Mathematics, MDPI, 9(6)672:1–14
5. Jack VT (1996) Advantages and disadvantages of using artificial neural networks versus logistic regression for predicting medical outcomes. J Clin Epidemiol 49(11):1225–1231
6. Mashford J, De Silva D, Donavan M, Stewart B (2009) An approach to leak detection in pipe networks using analysis of monitored pressure values by support vector machine. Third Int Conf Netw Syst Secur 38(1):534–539
7. Mohammadreza S, Mohammadreza JG, Jafar Y (2020) A methodology for leak detection in water distribution networks using graph theory and artificial neural network. Urban Water J 17(6):1–9
8. Mounce SR, Mounce RB, Boxall JB (2012) Identifying sampling interval for event detection in water distribution networks. J Water Resour Plann Manage ASCE 138(2):187–191
9. Mohammad TN, Mysorewala M, Lahouari C, Siddiqui B, Sabih M (2014) Measurement error sensitivity analysis for detecting and locating leak in pipeline using ANN and SVM. IEEE 7(14):1–4
10. Mukherjee J, Narasimhan S (1996) Leak detection in networks of pipelines by the generalized likelihood ratio method. Am Chem Soc 35(6):1886–1893
11. Poulakis Z, Valougeorgis D, Papadimitriou C (2003) Leakage detection in water pipe networks using a Bayesian probabilistic framework. Probab Eng Mech 18(4):315–327
12. Przystałka P (2018) Performance optimization of a leak detection scheme for water distribution networks. IFAC Papers on Line 51(24):914–921
13. Romano M, Kapelan Z, Savic DA (2010) Real-time leak detection in water distribution systems. water distribution system analysis 2010—WDSA2010, Tucson, AZ, USA, 97(425):1074–1082
14. Rossman LA (2000) EPANET 2: User's manual: EPA/600/R-00/057. United States Environmental Protection Agency, Cincinneri, Ohio
15. Yang J, Wen Y, Li P (2010) Approximate entropy-based leak detection using artificial neural network in water distribution pipelines. In: 11th International Conference Control, Automation, Robotics and Vision, pp 1029–1034

Leakage Management in WDN System Using Optimization Technique

Ashwini Singh and A. B. Mirajkar

Abstract The elimination of leaks in the water distribution network (WDN) system is one of the major issues facing the water industry. Pressure reduction valves (PRVs) installation in the water network is a successful method for reducing leaks since leaks rely on pressure. The quantity of leakage depends on the water distribution network's (WDN) operating pressure. As a method of avoiding water system leaks, pressure control is progressively gaining favors. In this study, pressure reduction valves are used to demonstrate how successful they are in reducing leakage (PRVs). A variable speed pump works to reduce pressure shortages during periods of high demand while reducing excess pressure and limiting leakage during periods of low demand. Three PRVs were employed to further reduce leakage. This paper proposes a modified reference pressure approach for locating valves in the WDN. A multiobjective genetic algorithm is used to determine the appropriate control value of the pressure reduction valve in terms of variations in demand patterns and a reduced rate of leakage in the WDN.

Keywords WDS/WDN · PRVs · Genetic algorithm · Leakage reduction · Modified reference pressure

1 Introduction

Water management is a prior priority for this running world as it is a major source of requirement. In this day and age of widespread, long-term droughts, impacts from climate change, and rapidly increasing rates of urbanization as cities around the world explode in size and number, water loss is a critical issue. To fulfill the increased water demand, WDNs require innovative technical solutions. The age of the infrastructure

A. Singh (✉) · A. B. Mirajkar
Department of Civil Engineering, Visvesvaraya National Institute of Technology Nagpur, Maharashtra 440010, India
e-mail: singhthesingh61@gmail.com

A. B. Mirajkar
e-mail: abmirajkar@civ.vnit.ac.in

P. V. Timbadiya et al. (eds.), *Geospatial and Soft Computing Techniques*, Lecture Notes in Civil Engineering 339, https://doi.org/10.1007/978-981-99-1901-7_28

exacerbates one of the most serious issues with water network systems. One of the major concerns is the leakage reduction as the actual demand which is required to be delivered tampers. The leakages found in water network is due to pressure variation as demand increases according to that pressure is set and if demand suddenly gets less but still the pressure is same which in return will cause breaks in the pipeline, in short, causing leakage hence, we can say that leakage is directly proportional to pressure.

Water leakages are a type of physical water loss that is not remunerated. A multitude of causes contribute to water losses in water distribution systems (WDS). Faulty pipeline connections, old pipes that break when working under high pressure, and other factors can all lead to leaks. The rate of pipe replacement has been and continues to be, insufficient. Twenty percent or more of the pipelines in one-third of the world's water utilities are nearing the end of their usable lives. This indicates that there is a high need for pipe replacement. This also implies that many pipes will reach the end of their useful life before being replaced, putting them in danger of fracture and fatigue. These problems get enlarged when found in the bigger system like a water distribution network for a city as their demand will be altered due to the defect being caused by faulty measures. WDN leakage is known to reduce system hydraulic capacity and increase the rate of pipe breaks, putting consumers at risk of inadequate water supply, service outages, and damage from failure occurrences.

Water balance measurements can occasionally be used to confirm the presence of a leak. This method could be used to identify a source such as water mains, for example. As a result, it is a useful tool for keeping track of water supply and use. The strategy, however, will not work if the source only accounts for a small portion of the water balance or if the data is highly questionable. In this trending world, pressure managements have many techniques such as pressure and flow control valves used in pipeline network, tank water storage level optimization, usage of pipe schedules, etc. The pressure management techniques are less efficient but still, they manage to achieve some reduction in the leakage rate.

A genetic algorithm is an adaptable heuristic search algorithm based on natural selection and genetics as evolutionary notions. It is based on Darwin's 'Survival of the Fittest' theory of evolution. GA was first studied in the mid-1970s and has since evolved into a potent optimization technique. Genetic algorithms are heuristic strategies for exploring through a problem's solution space in order to identify the best answer or combination of solutions. Natural genetics is made up of three main elements: reproduction, crossover, and mutation.

2 Study Area and Data Source

In this section, a benchmark problem for leakage control has been discussed utilizing appropriate optimization techniques. The network analysis was broken down into three sections: discharge recalculation, leakage calculations, and PRV localization. The losses due to leakage might be greatly minimized with this proposed strategy.

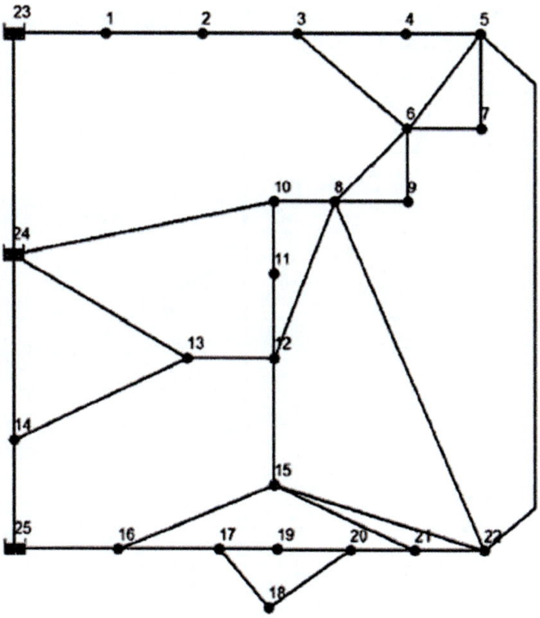

Fig. 1 Reference network: showing the node numbering

Rather of opting for a wider infrastructure, choose one benchmark problem to calculate all of the approach with suitable algorithm implementation. The intricacy of mistake is readily mastered and handled this way. There are 22 nodes and 37 links in the network (pipelines) (Figs. 1 and 2 and Table 1).

2.1 Data Used

See Table 2.

2.2 Mathematical Formulation

2.2.1 Law of Conservation

First the simulation was performed in the EPANET where the input parameters were assigned and was run further to which optimization model will be coupled. The initiation is done by applying the law of conservation represented by usage of sets of equation.

The mathematical equation of conservation of mass, continuity at node is given by

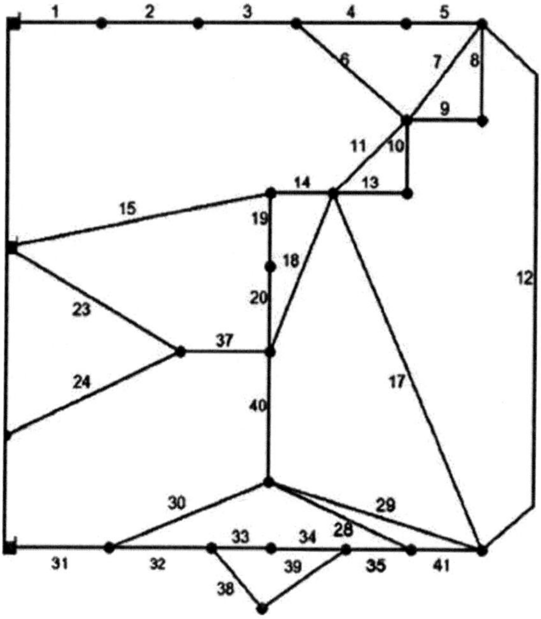

Fig. 2 Reference network: showing the pipe numbering

$$\sum_j Q_{ij,k} - K^* Q_{\text{req},i} - L_{i,k} = 0 \tag{1}$$

where $Q_{ij,k}$ is given the load condition K, the flow across the pipeline between nodes i and j.

$Q_{\text{req},i}$ is the required flow at node i (l/s).

$L_{i,k}$ is for load condition K, the leakage correlated with node i.

2.3 *Pressure Driven Analysis (PDA)*

In this part determination of the optimal flow, demand and losses of water in WDS are performed by pressure driven analysis so the better flow through the network is used for the analysis. The model is solved by as follows.

$$Q_{\text{req},i} = \left\{ Q_{i,\text{des}} * \begin{cases} Q_{i,\text{des}} \\ \left(\frac{P_{i,k}-P_{m,i}}{P_{\text{ser}}-P_{m,i}}\right)^{0.5} \\ 0 \end{cases} \right. \quad \begin{cases} \text{for} P_{i,k} > P_{\text{ser}} \\ \text{for} P_{m,i} \le P_{i,k} \le P_{\text{ser}} \\ \text{for} P_{i,k} \le P_{m,i} \end{cases} \tag{2}$$

where $Q_{i,\text{des}}$ is the nodal (i) desired demand.

$P_{i,k}$ is during the load condition K, the pressure at node i.

Table 1 Details of nodes and their elevation

Node ID	Elevation (m)	Minimum total head (m)	Demand (l/s)
N-01	18	48	5
N-02	18	48	10
N-03	14	44	0
N-04	12	42	5
N-05	14	44	30
N-06	15	45	10
N-07	14.5	44.5	0
N-08	14	44	20
N-09	14	44	0
N-10	15	45	5
N-11	12	42	10
N-12	15	45	0
N-13	23	53	0
N-14	20	50	5
N-15	8	38	20
N-16	10	40	0
N-17	7	37	0
N-18	8	38	5
N-19	10	40	5
N-20	7	37	0
N-21	10	40	5
N-22	15	45	20

P_{ser} is the minimal pressure necessary to meet the desired demand.
$Q_{req,i}$ is the required flow at node i (l/s).
$P_{m,i}$ is the pressure below which there is no water supply (For PDA analysis). The value $P_{m,i} = 0$ m is used for all the nodes [17].

After the calculation of the PDA, the demands are recalculated for every node, and hydraulic simulations are re-performed in the EPANET.

2.4 *Leakage Calculation*

The leakage rate is normalized using the observed pressure for all nodes. The pressure-based leakage formula is used to predict the water network breakdown rate. Assume that leakage is dispersed across the pipeline network. Equation 3 may be used to calibrate under the load condition K, the leakage corresponding with node i ($L_{i,k}$).

Table 2 Pipeline dimension of benchmark problem

Link ID	Start node	End node	Length (m)	Diameter (m)
P-01	23	1	606.0	0.457
P-02	23	24	454.0	0.457
P-03	24	14	2782.0	0.229
P-04	25	14	304.0	0.381
P-05	10	24	3383.0	0.305
P-06	13	24	1767.0	0.475
P-07	14	13	1014.0	0.381
P-08	16	25	1097.0	0.381
P-09	2	1	1930.0	0.457
P-10	3	2	5150.0	0.305
P-11	12	13	762.0	0.457
P-12	15	16	914.0	0.229
P-13	17	16	822.0	0.305
P-14	18	17	411.0	0.152
P-15	20	18	701.0	0.229
P-16	19	17	1072.0	0.229
P-17	20	19	864.0	0.152
P-18	21	20	711.0	0.152
P-19	21	15	832.0	0.152
P-20	22	15	2334.0	0.152
P-21	12	15	1996.0	0.229
P-22	11	12	777.0	0.229
P-23	10	11	542.0	0.229
P-24	8	12	1600.0	0.457
P-25	8	10	249.0	0.305
P-26	9	8	443.0	0.229
P-27	6	8	743.0	0.381
P-28	22	8	931.0	0.229
P-29	22	21	2689.0	0.152
P-30	4	3	326.0	0.152
P-31	5	4	844.0	0.229
P-32	6	3	1274.0	0.152
P-33	5	6	1115.0	0.299
P-34	7	6	615.0	0.381
P-35	5	22	1408.0	0.152
P-36	5	7	500.0	0.381
P-37	6	9	300.0	0.229

$$L_{i,k} = C_L * L_I * P_{i,k}^{\gamma} \tag{3}$$

where C_L is the leakage coefficient per unit length of the service pressure connection.

L_i is the total pipeline length (in meters) linked with node *i*.

$P_{i,k}$ is the node *i* pressure (in m) for demand *K* and is the leakage exponential used to create relationships between flow from orifice and head difference. For fractures in pipe or joints caused by a differential in pressure between the internal and exterior pipe, a leakage exponential value of 1.18 [13] is found. As a result, for this investigation, the authors used the same number, 1.18, as the leakage exponential.

Equation 4 [19] may be used to compute the entire length of pipeline associated with node *i* which is represented as

$$L_i = 0.5 * \sum_j L_{ij} \tag{4}$$

The entire length of pipeline linked to node *i* is $L_{i,j}$.

2.5 Localization of PRVs

In the water pipeline network, the updated reference pressure algorithm enhances PRV localization and removes the drawback. Given that '*G*' indicates the range of pipelines and G_v ($G_v \in G$) is a subset of it, the pipeline linking nodes *i* and *j* will be a PRV candidate site if:

$$N_j > P_{\text{ref}} \text{ and } N_i < P_{\text{ref}} \quad \text{(Rule 1)} \tag{5}$$

$$N_j - N_i > 0.1 \times P_{\text{ref}} \quad \text{(Rule 2)} \tag{6}$$

The pipeline pressures at nodes *j* and *i* are N_j and N_i, respectively, while the reference pressure is P_{ref}. During valve localization (Eq. (3.9 and 3.10)), P_{ref} is chosen. P_{ref} is varied throughout a range to find different values of $E_{v,n}$ ($E_{v,n}$ yields the total number of potential valve sites for a current value of P_{ref}) [15]. The P_{ref} is the pressure value that corresponds to $E_{v,n}$, minimal value. The process of localization chooses average load conditions.

The updating part is the objective function with the cos function which is given below as this is explained more precisely in the result and discussion part, this is generated in the MATLAB.

```
function value = Objective Function(x)
k1 = 5;
k2 = 7;
% Objective Function:
for i = 1:3
```

```
% value = Q−K−L
value = (((1/k1) * cos ((k1 * x (1)) ^2)) / cos (x (1))) + (((1/k2) * cos ((k2*x (1))
^2)) / cos (x (1)));
end
```

where k1 and k2 are the random iteration value taken.

X when one particular node is giving the water to other branches.

3 Results and Discussions

In order to eliminate excessive pressure in a WDS, our optimization aims to identify the ideal PRV pressure settings. The ideal PRV pressure settings in a water distribution system are first determined using our PRV model, as illustrated in Fig. 4. With 37 links, 22 nodes, and 3 reservoirs, this WDS served as a case study for PRV localizations in Liberatore and Sechi [15], Araujo et al. [3], and Gupta et al. [11]. They gave the information for the demand pattern, reservoir heads, discharge coefficient C_L, and leakage exponent parameter (2006). Lower limits of 30 apply to the node pressure (m). The ideal pressure regulation issue is defined for 24 h. The test may generally be applied with any factor values. The check valve mode (the longer working mode in our system) will, however, be more relevant if we take into account lower demands. Additionally, by accounting for these changing demand components, we are able to demonstrate how the enlarged PRV model works better than the original one in handling pressure regulation difficulties.

In EPANET, the simulation of the WDN system is done, and the corresponding hydraulic parameters (pressure, head and so on) are saved. Before reaching the actual network, the processes started simulating the small networks like with six nodes and six pipes. After that, the process took two three runs for the benchmark problem as it requires the feeding of data done properly so as to get the flow direction which will justify the continuity equation (Fig. 3).

Work of this study starts with considering the two scenarios: one in which all nodes are given maximum water to achieve maximum leakage, and another in which all nodes are given zero water to achieve zero leakage, i.e., now an upper and lower bound has been established, and only the results between these are to be obtained. The cumulative distribution function is used to demonstrate how water is distributed in each node. The coding begins by guiding the input data, and additional commands are allocated for generating nodes, reservoirs, and other details. As we have 50 iterations, we need to offer one random input data, which may be anything, before starting the first iteration; therefore, the values chosen were 5 and 7 (K_1 and K_2). The second iteration will now receive a response from the first and so on. The distinction between the rest of the researchers was that the objective function was the same, but the cos function was added. In trigonometry, the cos function is one in which if input is maintained at zero and the output is maximized. That is why, maximum and zero leakage are used to stop leaking value between these two points. For a dynamic system, GA is used. In MATLAB GA, an object is needed to assign the values;

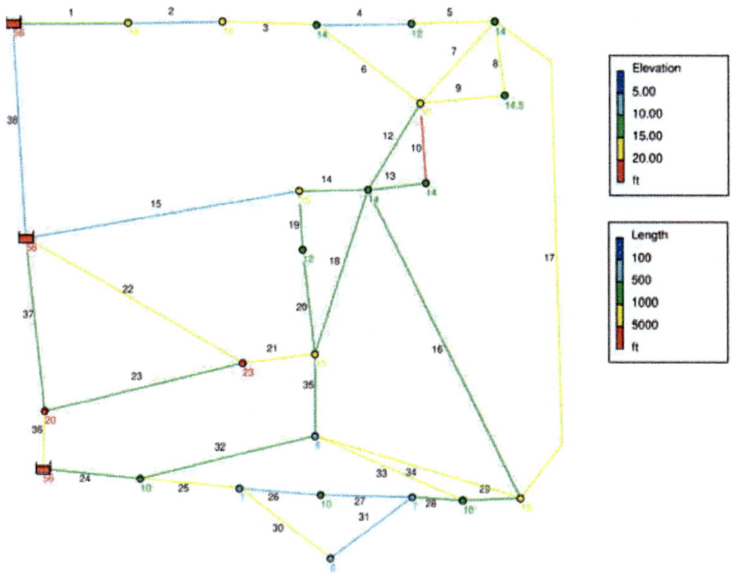

Fig. 3 Spatial map of hydraulic parameters

therefore, the code has one object named s, which is multivariable and contains all of the methodologies used. For individual evolution, the method employs a population of 50 for 200 generations. For this optimization procedure, the crossover and mutation probability were 0.9 and 0.1, respectively.

MATLAB was used to calculate the PRVs optimal operating value. The hydraulic parameters like head, elevation, and other characteristics at each node which were analyzed in EPANET Software, were used in MATLAB as input (using 5–6). The input parameter was assigned in the MATLAB to get the graphical view of the data which is clearly mentioned in the figures.

After completing the adjusted reference pressure together with the requirements of rules 1 and 2, the selection was completed to not only reduce leakage but also to consider the PRVs' economic factors. Initial concurrence can be achieved as the number of PRVs increases; however in a functional WDN system, an increase in PRVs causes the maintenance cause to increase. The valve located for the study is at 37, 40, 41 links, as the specified reference nodes, and the minimum allowable pressure requirements are stated as 30 m above ground level (Fig. 5).

As a consequence, the GA was run three times, and the optimum solution was found by averaging the results. After pressure control, the average leakage rate was reduced to 337.32 l/s from 417.34 l/s. The leakage reduction rate was higher at each phase than in studies with the three number of PRVs. The leakage rate was lowered by 19.2%, a significant improvement above the other studies' findings. This demonstrates the system's long-term viability. It takes 3–5 s to calibrate the PRV's ideal operational setting for the present load state. As a result, the proposed approach

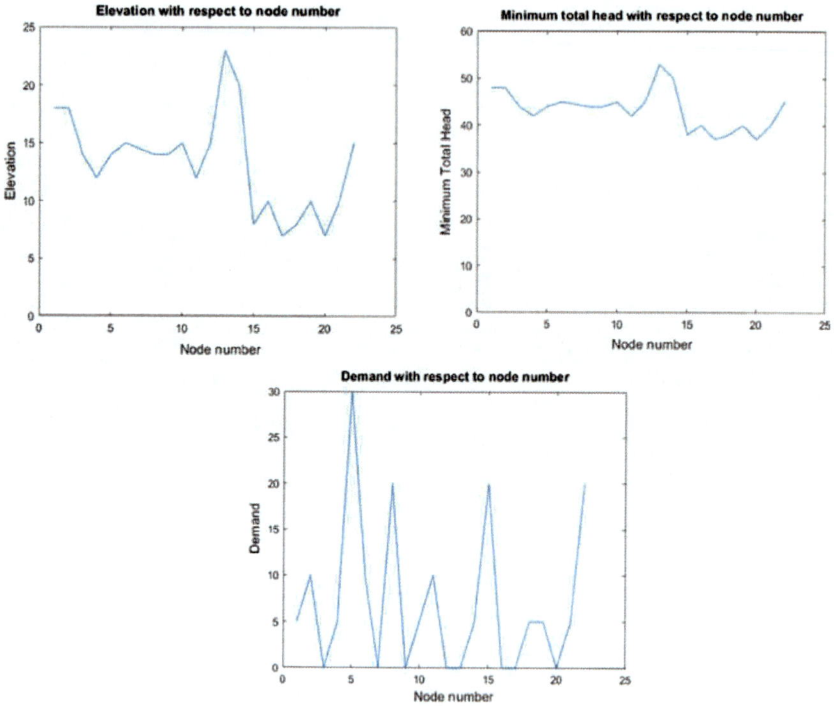

Fig. 4 Representation of the elevation data, minimum total head and demand

Fig. 5 Final network output

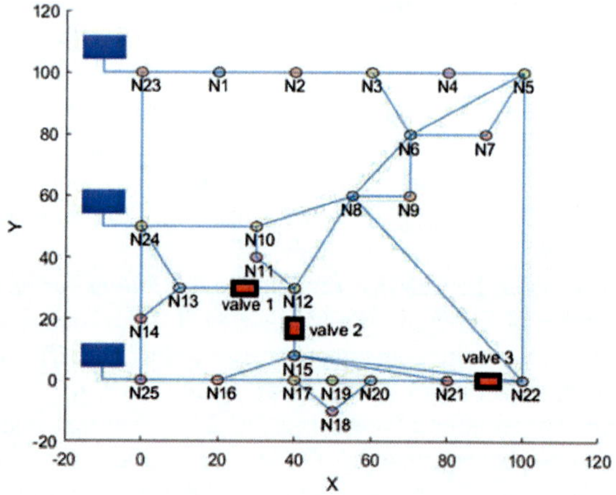

might potentially be used in real-time. According to prior studies, PRV has never been discovered at this site. This is because the investigation's new localization rule (Rule 2, 6) resulted in a lower leakage rate than previous studies.

4 Conclusions

Water distribution companies are responsible for reducing leakage and controlling it, and this can only be done with an integrated leakage control system that comprises leak characterization, detection, location, and repair, as well as the development of a continuous monitoring system. For a benchmark situation, this research provides a pressure-based leakage reduction strategy. Despite the fact that this has increased the rate of leakage in WDS, a variable speed pump and flow-controlled pressure lowering values, or PRVs, have been added for even further pressure reduction to lessen pressure shortages during times of increasing water demand. The access pressure decreases as the pump speed is slowed down during low load situations. As a result, the rate of leakage has also decreased.

The following conclusions are derived from the foregoing study:

- Finding valve settings that cause the least amount of pressure exceedance can be done using formal optimization approaches.
- Compared to earlier researchers' methods, the new optimization process which employs a cumulative approach is more reliable and effective.
- The capacity to accommodate slight pressure variations around the target, which are managed by a distinct goal function with less pressure restrictions, is a crucial aspect of the suggested technique.

References

1. Abdel Meguid H, Ulanicki B (2010) Pressure and leakage management in water distribution systems via flow modulation PRVs. In: Water distribution systems analysis 2010, pp 1124–1139
2. Adedeji KB, Hamam Y, Abe BT, Abu-Mahfouz AM (2017) Leakage detection and estimation algorithm for loss reduction in water piping networks. Water 9(10):773
3. Araujo LS, Ramos H, Coelho ST (2006) Pressure control for leakage minimisation in water distribution systems management. Water Resour Manage 20(1):133–149
4. Ávila CAM, Sánchez-Romero FJ, López-Jiménez PA, Pérez-Sánchez M (2021) Leakage management and pipe system efficiency. Its influence in the improvement of the efficiency indexes. Water 13(14):1909
5. Bello AD, Alayande WA, Otun JA, Ismail A, Lawan UF (2015) Optimization of the designed water distribution system using MATLAB. Int J Hydra Eng 4(2):37–44
6. Dai PD, Li P (2014) Optimal localization of pressure reducing valves in water distribution systems by a reformulation approach. Water Resour Manage 28(10):3057–3074
7. Eliades DG, Kyriakou M, Vrachimis S, Polycarpou MM (2016) EPANET-MATLAB toolkit: an open-source software for interfacing EPANET with MATLAB. In: Proceedings—14th international conference on computing and control for the water industry (CCWI), vol 8

8. Farley M (2001) Leakage management and control: a best practice training manual (No. WHO/SDE/WSH/01.1). World Health Organization
9. Gupta AD, Kulat K (2018) Leakage reduction in water distribution system using efficient pressure management techniques. Case study: Nagpur, India. Water Supply 18(6):2015–2027
10. Gupta A, Bokde N, Kulat K, Yaseen ZM (2020) Nodal matrix analysis for optimal pressure-reducing valve localization in a water distribution system. Energies 13(8):1878
11. Gupta A, Bokde N, Marathe D, Kulat K (2017) Leakage reduction in water distribution systems with efficient placement and control of pressure reducing valves using soft computing techniques. Eng Technol Appl Sci Res 7(2):1528–1534
12. Houcque D (2005) Introduction to Matlab for engineering students. Northwestern University 1
13. Jowitt PW, Xu C (1990) Optimal valve control in water-distribution networks. J Water Resour Plan Manag 116(4):455–472
14. Lambert AO, Fantozzi M (2010) Recent developments in pressure management. In: IWA conference water loss 2010
15. Liberatore S, Sechi GM (2009) Location and calibration of valves in water distribution networks using a scatter-search meta-heuristic approach. Water Resour Manage 23(8):1479–1495
16. Maskit M, Ostfeld A (2021) Multi-objective operation-leakage optimization and calibration of water distribution systems. Water 13(11):1606
17. Meirelles G, Manzi D, Brentan B, Goulart T, Luvizotto E (2017) Calibration model for water distribution network using pressures estimated by artificial neural networks. Water Resour Manage 31:4339–4351
18. Mosetlhe TC, Hamam Y, Du S, Monacelli E (2020) A survey of pressure control approaches in water supply systems. Water 12(6):1732
19. Nicolini M, Zovatto L (2009) Optimal location and control of pressure reducing valves in water networks. J Water Resour Plan Manag 135(3):178–187
20. Paluszczyszyn D (2015) Advanced modelling and simulation of water distribution systems with discontinuous control elements
21. Purohit GN, Sherry AM, Saraswat M (2013) Optimization of function by using a new MATLAB based genetic algorithm procedure. Int J Comput Appl 61(15)
22. Puust R, Kapelan Z, Savic DA, Koppel T (2010) A review of methods for leakage management in pipe networks. Urban Water J 7(1):25–45
23. Roshani E, Filion Y (2014) WDS leakage management through pressure control and pipes rehabilitation using an optimization approach. Proc Eng 89:21–28
24. Rossman LA (2000) EPANET 2: user's manual
25. Sahu RC, Gupta A (2020) Genetic algorithm based pressure management technique for leakage reduction in the water distribution system. In: 2020 3rd international conference on intelligent sustainable systems (ICISS). IEEE, pp 1464–1470
26. Van Zyl JE, Savic DA, Walters GA (2004) Operational optimization of water distribution systems using a hybrid genetic algorithm. J Water Resour Plan Manag 130(2):160–170
27. Zarei N, Azari A, Heidari MM (2022) Improvement of the performance of NSGA-II and MOPSO algorithms in multi-objective optimization of urban water distribution networks based on modification of decision space. Appl Water Sci 12(6):1–12

Optimum Placement of Pressure and Acoustics Sensors for Leak Detection in Ramnagar GSR Water Distribution Network of Nagpur City

N. Poojitha and Rajesh Gupta

Abstract Leakages in water distribution networks (WDNs) not only result in loss of water but also in revenue loss. Further, it may lead to contamination which results in an increase of risk to human health. Leaks should be controlled before they occur, and also, emergency measures are to be taken to control the losses through leaks as soon as they are noticed. Detecting leaks from pipes laid underground is a difficult task. Pressure and acoustic sensors are in use to help locating leaks at the earliest. While pressure sensor notices any abnormal decrease in pressure caused by a leak, acoustic sensor notices change in noise pattern caused by a leak to identify it. As deployment and maintenance of these sensors are costly, it is desirable to keep their numbers to a minimum. This paper focuses on the application of methodology for obtaining locations of both types of sensors individually for the Ramnagar GSR network of Nagpur city. A multi-criteria decision-making method named fuzzy Decision-Making Trial and Evaluation Laboratory (DEMATEL () approach is used to determine the optimum locations of pressure sensors, and an entropy-based approach is used to obtain optimal locations of the acoustic sensors.

Keywords Water distribution networks · Leak detection · DEMATEL · Sensor placement optimization · Entropy

1 Introduction

Water distribution networks (WDNs) in most of the cities are laid several years ago. With the deterioration of the pipes and other appurtenances, it becomes more prone to leakages and bursts. Rapid temperature changes, abrupt pressures change in the pipe, incorrect use of materials and poor installation practice are some of the other reasons for causing leaks. Advancement in science and technology has led us to exposure

---

N. Poojitha (✉) · R. Gupta
Department of Civil Engineering, Visvesvaraya National Institute of Technology, Nagpur 440010, India
e-mail: npoojitha35@gmail.com

P. V. Timbadiya et al. (eds.), *Geospatial and Soft Computing Techniques*, Lecture Notes in Civil Engineering 339, https://doi.org/10.1007/978-981-99-1901-7_29

of many creative and innovative techniques for efficient management of leaks. For instance, sensor-based WDN is one of the smart technologies that has been evolved in the field of water leakage management [1–8, 10, 12, 14].

Fuzzy Decision-Making Trial and Evaluation Laboratory (DEMATEL) technique is an emerging multi-criteria decision-making approach used to analyse a design model that includes complex factors and derive a cause-and-effect relationships among the various factors involved in a system. It deals with many real-world models that include lot of uncertainties and imprecise factors and helps in developing the interdependencies among the various components involved in a system. However, still the research is going on to deal with a real-world WDN problem [9, 11].

Entropy-based approach is used to address the problem of optimal placement of acoustic sensor in WDNs. In this approach, entropy is assigned to every pipe segment, and eventually, sensor location is derived from maximizing the total entropy in the pipe network [5, 13, 15]. This method uses sensing range of the sensor and mainly focuses on the area that a sensor could cover, so that it helps in reducing the overlap of coverage areas of two sensors and subsequently helps in redundancy of number of sensors used. Present study is focused on application of fuzzy DEMATEL approach for placing pressure sensors and application of entropy-based approach for placing acoustic sensors in Ramnagar GSR WDN of Nagpur city.

2 Materials and Methods

2.1 Fuzzy DEMATEL Approach

As mentioned earlier, fuzzy DEMATEL approach is used for optimal placement of pressure sensors in a network. It uses a fuzzy logic and helps in building matrices for comparison. The main motive of this study is considering and evaluating the relationships among the pressure sensors placed in the network, in order to get the complete knowledge of the mutual influence of one sensor over other. Further, it helps in identifying the nodes for hosting a sensor that can have a bigger impact on a network. A step-by-step method for optimum placing of pressure sensors is followed.

Sensitivity matrix for leak detection. To install a sensor in the network, there is need to find the peculiar points in the network. These points should be more sensitive to the anomalies occurring in the network. Initially, consider the normal operation and find the pressure at a particular node j at a time step t (P_J^N). Now, simulate the leak at node i by raising the demand with certain amount (q_i) at node i, and find the pressure at node j at the same time step t (P_j^i). Increasing the demand at node i represents that, there is anomaly included at node i. The primary concern of this is to know how much sensitive is node j for an anomaly occurring at node i. Sensitivity of a particular node ($s_{i,j}$) is found using the formula given in Eq. 1.

$$s_{i,j} = \frac{P_J^N - P_j^i}{q_i} \tag{1}$$

Now, sensitivity matrix is calculated by simulating the leak at every node individually and finding the sensitivities of all the other nodes towards the simulated leak. Row i of the matrix signifies the sensitivity of node j (column nodes) towards the leak simulated at node i. Generally, the amount of leak simulated is modelled through an emitter using Eq. 2.

$$q_i = \beta P_i^{\alpha} \tag{2}$$

where α is for emitter exponent and β is for emitter coefficient. α and β values vary from 0 to 1 based on the pipe environmental condition, geometry of leakage and few other parameters. Mostly, using these values results in a minute decimal variation between the pressure at normal condition and leak condition. So, it would be difficult to observe the sensitivity. In order to avoid this problem, here a constant leak of 10 LPS was introduced in the present study.

Fuzzy DEMATEL method. In the classical DEMATEL approach, crisp values are used to evaluate the mutual influence of one factor over other. However, in modern world applications, human knowledge is often unclear in given judgements because of lot of uncertainties involved in a model. So exact crisp values cannot be used to assess a model while finding the mutual dependencies of the complex factors. In order to solve this problem, triangular fuzzy sets are introduced along with DEMATEL. This fuzzy logic helps in managing the effect on input assessments for a DEMATEL because of impreciseness involved in the problem. The main uncertainty considered in WDN model is leakages. Steps included in fuzzy DEMATEL approach are as follows.

Step 1—Fuzzy linguistic scale: A fuzzy linguistic scale was created to in order to evaluate the interdependencies among the different factors in the system. In the present study, sensitivity matrix obtained is used as an input for fuzzy DEMATEL approach to collect the judgement. Pair-wise comparison of all the elements of the sensitivity matrix is done to state the mutual influence of one node i over the node j and vice-versa. Fuzzy linguistic terms and its corresponding triangular fuzzy number used in the present study are shown in Table 1. The limits of a triangular fuzzy number in fuzzy linguistic scale can be changed according to the user's convenience. Now, each element of a normalized sensitivity matrix is assigned with the corresponding code based on the value of the element resulting in a matrix with each element having a triangular fuzzy set.

Step 2—Defuzzification to obtain a direct relation matrix: To convert this fuzzy matrix into a crisp matrix, defuzzification is applied. In DEMATEL approach, Converting Fuzzy to Crisp Scores (CFCS) defuzzification method is used. Let us consider a triangular fuzzy set $f_{ij} = \left(l_{ij}, m_{ij}, r_{ij}\right)$ where i is a criterion having j alternatives.

1. Normalization: $r_i^{\max} = \max\left(r_{ij}\right)$ and $l_i^{\min} = \min\left(l_{ij}\right)$

Table 1 Fuzzy linguistic scale

Linguistic evaluation	Triangular fuzzy number	Representing code used
No influence (NI)	0, 0, 0.25	1
Low influence (LI)	0, 0.25, 0.5	2
Medium influence (MI)	0.25, 0.5, 0.75	3
High influence (HI)	0.5, 0.75, 1	4
Extreme influence (EI)	0.75, 1, 1	5

$$\Delta_{\min}^{\max} = r_i^{\max} - l_i^{\min} \tag{3}$$

Now, calculate for every alternative,

$$x_{lj} = \frac{l_{ij} - l_i^{\min}}{\Delta_{\min}^{\max}} \tag{4}$$

$$x_{mj} = \frac{m_{ij} - l_i^{\min}}{\Delta_{\min}^{\max}} \tag{5}$$

$$x_{rj} = \frac{r_{ij} - l_i^{\min}}{\Delta_{\min}^{\max}} \tag{6}$$

2. Compute left normalized value (ls) and right normalized value (rs):

$$x_j^{ls} = \frac{x_{mj}}{1 + x_{mj} - x_{lj}} \tag{7}$$

$$x_j^{rs} = \frac{x_{rj}}{1 + x_{rj} - x_{mj}} \tag{8}$$

3. Calculating total normalized crisp value:

$$x_j^{\text{crisp}} = \frac{\left[x_j^{ls}\left(1 - x_j^{ls}\right) + \left(x_j^{rs} x_j^{rs}\right)\right]}{1 - x_j^{ls} + x_j^{rs}} \tag{9}$$

4. Calculating final crisp values:

$$f_{ij} = l_i^{\min} + \left(x_j^{\text{crisp}} \Delta_{\min}^{\max}\right) \tag{10}$$

The generated crisp matrix is considered as a direct relation matrix (Z).

Step 3—Normalized direct relation matrix (X): Calculated using Eqs. 11 and 12

$$X = \left[x_{ij}\right]_{n*n} = Z * s \tag{11}$$

$$s = \frac{1}{\max\left(\left(\max_{1 \le i \le n} \sum_{j=1}^{n} z_{ij}\right), \left(\max_{1 \le j \le n} \sum_{i=1}^{n} z_{ij}\right)\right)} \tag{12}$$

Step 4—Total relation matrix (T): Total relation matrix (TRM) reflects the summing of all the direct and indirect influences.

$$T = X + X^2 + X^3 + X^4 + \_\_\_\_\_\_\_\_ X^h = X(I - X)^{-1} \tag{13}$$

Step 5—Prominence and relation: Two variables R and C are calculated from TRM

$$R = \sum_{j=1}^{n} t_{ij} \tag{14}$$

$$C = \sum_{i=1}^{n} t_{ij} \tag{15}$$

Sum of a row i in a TRM shows that addition of all the direct and indirect effects is transmitted from factor F_i to the remaining other factors, while sum of a column in a TRM represents that addition of all the direct and indirect influences obtained by factor F_i from all the other factors.

Now $R + C$ called 'prominence' and $R{-}C$ called 'relation' of all the factors are calculated. $R + C$ portrays the strength of the factor F_i based on the given and received influences of the factor. And $R{-}C$ demonstrates the net consequences of the factor on the remaining system. If $R{-}C$ of a factor is positive, it is taken into the 'cause group' depicting that the factor acts an influence on all the factors. If $R{-}C$ of a factor is negative, it shows that the factor is influenced by some other factors of the system and this factor is taken into an 'effect group'. Likewise, some factors come under cause group and some factors come under effect group.

Optimum placing order of the pressure sensors. As explained above, prominence reflects the net influence on the system contributed by a node. So, the node, that is having the highest prominence value will be given the first priority for placing the sensor. In the same way, given available number of sensors can be placed on the nodes according to the decreasing value of the prominence values.

2.2 *Entropy-Based Approach*

Entropy-based approach is applied for optimum placing of acoustic sensors in WDNs. The proposed method uses the Shannon's entropy and the sensing range or sensing radius of the sensor to find the probability of sensing range within the pipe. Using this probability an entropy is assigned to each pipe. And thereby, optimum location of monitoring points is found by maximizing the network's total entropy using subadditivity and maximality properties of entropy.

Overview of Entropy. In physics, entropy is defined as a quantity indicating the unavailability of energy in system to perform a work. Entropy is often referred as an order or disorder level of a system and is a metric of degree of randomness of a system. Its primary concern is to measure the ease of transformation of energy from one state to other state in a system. Entropy was first related to thermodynamics system in which it quantifies the level of disorder in the system and measures the amount of energy wasted at the time of transformation between different states in the thermodynamic system. According to classical definition, entropy is a measure of stability and order level of a system. Mathematically, entropy is calculated using the probability mass function (p_x) of a variable x. It is a product of probability and its inverse in a logarithm.

$$\text{Shannon's entropy } (H_x) = \sum_x p_x \ln \frac{1}{p_x} \tag{16}$$

Properties of entropy. Subadditivity property: It stated that if a function has two variables, then the individual sum of the function values of both the variables is always greater than or equal to the total function value for the addition of both the variables.

$$\forall_{x,y} \in Af(x, y) \leq f(x) + f(y)$$

Maximality property: According to this property, when there is no any information about the distribution function, then considering that one distribution function with highest entropy is having the highest information about the system. When all the possible outcomes are having the same probabilities ($p_1 = p_2 = \ldots = p_n$), then the entropy function $H(p_1, p_2, \ldots p_n)$ takes the maximum value.

Equivocation property: Equivocation property with respect to conditional entropy states that the total amount of information needed to define the possible outcome of a random variable Y, given that another random variable X value is already known. Condition entropy referred to entropy of Y given X is written as $H(Y|X)$.

$$H(Y|X) = \sum_{x \in X, y \in Y} \left[p(x, y) \ln \frac{p(x)}{p(x, y)} \right] \tag{17}$$

All the above three properties along with an illustrative example were clearly explained in [5].

Sensing Range of an acoustic sensor. The sensing range or a sensing radius of an acoustic sensor indicates the sensor to sensor spacing required when sensors are placed in the network for efficient functioning. Also, it is the maximum distance that a sensor can sense and is also called as controllable length of a sensor. Usually, in case of fixed sensors, 200 to 500 m of sensor to sensor spacing is preferred in case of metal pipes and a 100 m of sensor to sensor spacing is preferred in case of plastic pipes.

Entropy formula. Probability mass function (p_x) of an entropy equation was firstly substituted with the ration of sensing range of the sensor to the total network length. Keeping in view that a single sensor cannot cover the entire network and also to reduce the clustering of sensors at least in few parts of the network, the total length of the network in probability mass function is replaced by length of each pipe segment (L_i). Now, total entropy of the network is calculated by adding the entropies of all pipe segments. Total entropy of the network (H_T) is given for two different cases based on the number of types of sensors used.

CASE 1: When a single sensor type is used, which implies the constant sensing radius (r_i)

$$H_T = \sum_{i=1}^{n} \left(\frac{r_i}{L_i} \ln \frac{1}{(r_i/L_i)} \right) \tag{18}$$

CASE 2: When multiple sensor types are used, which implies different sensing radius.

$$H_T = \sum_{j=1}^{n_r} \sum_{i=1}^{n_t} \left(\frac{r_{i,j}}{r_{T,j}} \ln \frac{r_{T,j}}{r_{i,j}} \right) \tag{19}$$

where j is for sensor-type index; n_r is number of different sensor types used; n_t is total number of sensors in the network; $r_{i,j}$ is number of unit types of sensor j used at node i and $r_{T,j}$ is total number of unit types of sensor j used in the entire network.

The objective for the present acoustic sensor placement problem is to maximize the entropy and to reduce the overlap between the sensors using a single-type sensor. Objective function is considered as given in Eq. 20.

$$H_T = \max\left(\sum_{i=1}^{n} \left(\frac{r_i}{L_i} \ln \frac{L_i}{r_i} \right) \right) \tag{20}$$

Again, the value of r_i is changed based on the length of the pipe as explained further (Fig. 1).

In case a, $r_i = x$; case b, $r_i = 2x$; case c, $r_i =$ Minimum of total sensing radius and length of pipe.

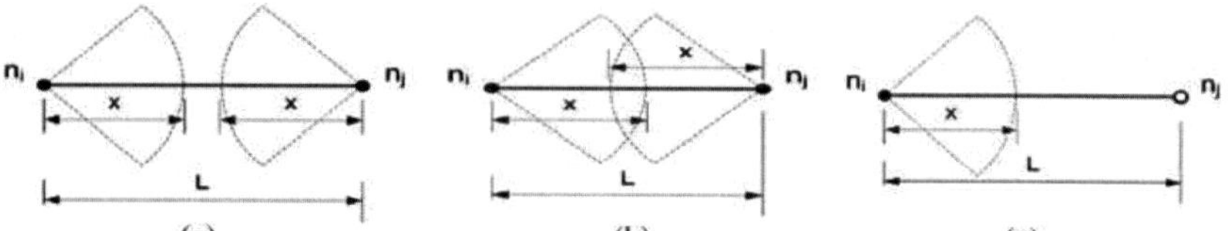

Fig. 1 Sensors placed at **a** one end node of a pipe; **b** both end nodes of a pipe without overlapping of sensing radius; **c** both end nodes of a pipe with overlapping of sensing radius

2.3 Methodology

Step 1: Calculate the entropy of a node by adding the entropy of the pipes connected to that node and assuming that the sensor is placed at that node.
Step 2: Further, select the node that is having the highest entropy. As it reflects that the particular node is contributing more to the total network entropy compared to the other nodes, first sensor is placed at the node having the highest entropy.
Step 3: Calculate the total entropy of the network.
Step 4: Once first sensor is placed, the remaining entropy of the pipes that are connected with the node of the sensor is calculated. Remaining entropy is nothing but, once the sensor is placed at one end of the pipe, what would be the remaining uncertainty. This is calculated by taking the difference of entropies when the sensor is placed at one node of the pipe and when the sensor is placed at both the nodes of the pipe.
Step 5: A new nodal entropy is calculated at the nodes that are connected with the node of the sensor by adjusting the previous nodal entropies using remaining entropy.
Step 6: The node with highest entropy is selected for placing of second sensor.
Step 7: Repeat the same procedure from step 2.
Step 8: After selecting few nodes for placing of sensors, the total entropy of network reaches its peak and starts decreasing. This is because almost the entire network is covered, and if the sensors are increasing, it results in overlapping of the sensors. Also, sensors start placing at both ends of the pipe which results in decreasing of the entropy.
Step 9: Number of sensors to be placed and where to be placed is equal to number of sensors selected until the entropy reached its maximum value and is to be placed at the nodes selected.

3 Study Area

The study area is Ramnagar Ground Storage Reservoir (GSR) hydraulic zone which is located in western part of Nagpur city. The water supply to this zone is from Pench and Gorewada lake. For better operation and maintenance of water supply in Nagpur

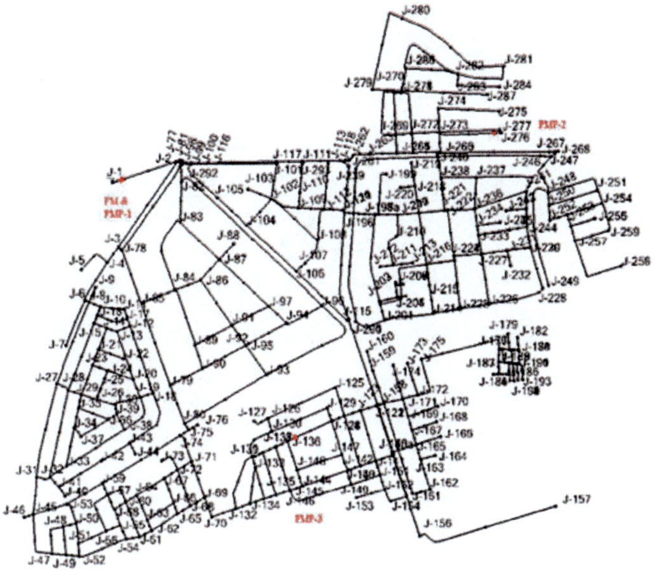

Fig. 2 Layout of Ramnagar GSR zone (water distribution zone under study)

city, water work department, Nagpur Municipal Corporation (NMC), has divided the WDN of Nagpur city into ten zones. Dharampeth zone is one of the zones and is further divided into two subzones for the better organizing. Ramnagar GSR zone is one of the subzones of Dharampeth zone. Network model of this zone consists of 292 junctions and 375 pipes with one ground storage reservoir (Fig. 2).

4 Results and Discussions

4.1 Optimum Placing of Pressure Sensors Using Fuzzy DEMATEL Approach—Results

A detailed analysis of fuzzy DEMATEL approach was carried out for Ramnagar GSR WDN. Leakage rate of 10 LPS is taken in order to derive sensitivity matrix and the leak is modelled using EPANET software. From the sensitivity matrix, $R + C$ (Prominance) and $R-C$ (Relation) values are calculated. The largest value of $R + C$ is obtained at J-2 that indicates pressure sensor at J-2 can inform about likely leakages at several other nodes. Junction locations are arranged with decreasing value of $R + C$ and first 15 sensor locations are shown in Table 2. It can be observed that there are several locations having same $R + C$, for example, first five junctions in Table 2. They have the same priority order (Fig. 3).

Table 2 Optimum placing order of first 15 sensors using fuzzy DEMATEL approach

Priority order	$R + C$ (Prominence)	$R-C$ (Relation)	Priority order	$R + C$ (Prominence)	$R-C$ (Relation)
(1)	(2)	(3)	(1)	(2)	(3)
Junc J-2	2.479356	1.357396	Junc J-120	2.376688	0.551805
Junc J-77	2.479356	1.357396	Junc J-121	2.368665	0.193697
Junc J-81	2.479356	1.357396	Junc J-196	2.313098	0.452035
Junc J-98	2.479356	1.357396	Junc J-122	2.303592	0.128623
Junc J-116	2.479356	1.357396	Junc J-117	2.290155	0.956966
Junc J-1	2.4351	1.365075	Junc J-132	2.28439	0.146591
Junc J-115	2.415595	0.494455	Junc J-128	2.280886	0.097334
Junc J-294	2.398085	0.505808			

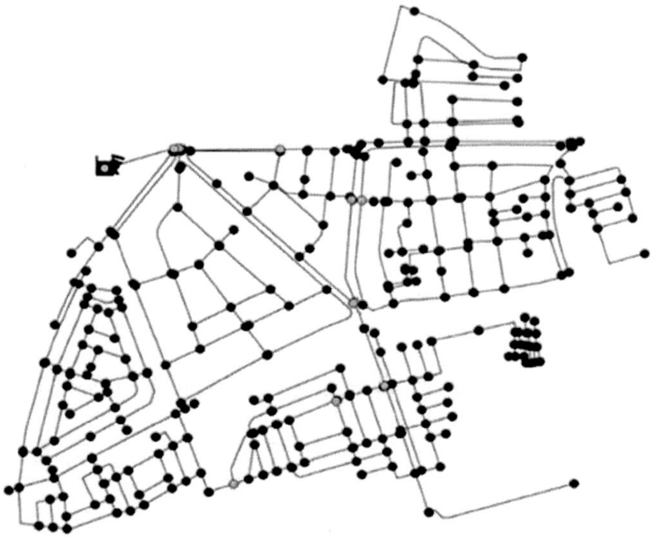

Fig. 3 Image showing selected points of first 15 sensors using fuzzy DEMATEL

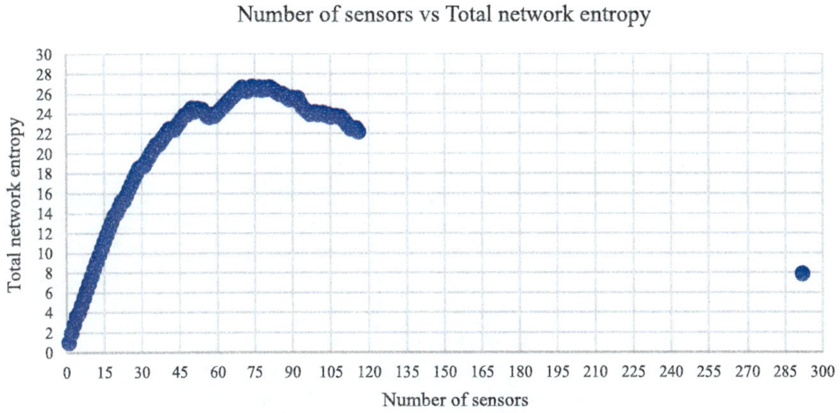

Fig. 4 Graph representing number of sensors vs total entropy of the network

4.2 Optimum Placing of Acoustic Sensors Using Entropy-Based Approach—Results

A detailed analysis of entropy-based approach was carried out for Ramnagar GSR water distribution network. Noise loggers with a sensing range of 100 m were assumed. Small sensing range is considered that helps in finding the leak.

A graph between the number of sensors and network entropy is shown in Fig. 4. It can be observed from the figure that total entropy for this network initially increases rapidly. It reaches to maximum at about 74 sensors. It starts gradually decreasing with further with increase in number of sensors and reaches to a minimum value when sensors are located at all the nodes.

Therefore, a total of 74 nodes are considered optimal to host the acoustic sensors using entropy-based approach. These nodes with their covering range are shown in Fig. 5. From the figure, it can be observed that some of the nodes, especially on the periphery, are not covered by any acoustic sensor.

5 Summary and Conclusions

It is summarized that the present work deals with the application of DEMATEL approach for optimum placement of pressure sensors and application of entropy-based approach for optimum placement of acoustic sensors in Ramnagar GSR WDN of Nagpur city. DEMATEL approach helps in understanding the mutual influences among the pressure sensors placed at different nodes and aids in identifying the nodes which are having more influence over the network. Entropy-based approach maximizes the entropy using the subadditivity and maximality properties of entropy

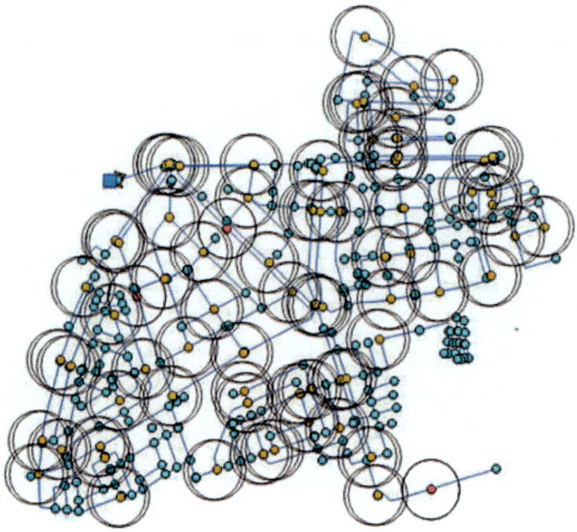

Fig. 5 Buffered area of sensors placed optimally at entropy derived nodes

to increase the sensor coverage area by scattering the sensors over the network and to reduce the overlapping of the sensors.

Acknowledgements The authors sincerely thank Mr. Sanjoy Roy, Chief Officer, Orange City Water Pvt. Ltd, Nagpur, for providing network details of Ramnagar used in this study.

References

1. Blesa J, Nejjari F, Sarrate R (2016) Robust sensor placement for leak location: analysis and design. J Hydroinf 18(1):136–148
2. Boatwright S, Romano M, Mounce S, Woodward K, Boxall J (2018) Optimal sensor placement and leak/burst localisation in a water distribution system using spatially-constrained inverse-distance weighted interpolation. J Hydroinf 3:282–289
3. Boulos PF, Trent S (2007) A GIS-centric approach for optimal sensor placement in large-scale water distribution systems. In: World environmental and water resources congress 2007, ASCE, p 513
4. Casillas MV, Puig V, Garza-Castanon LE, Rosich A (2013) Optimal sensor placement for leak location in water distribution networks using genetic algorithms. Journal Sens MDPI 2013:13
5. Christodoulou SE, Gagatsis A, Xanthos S, Kranioti S (2013) Entropy-based sensor placement optimization for water loss detection in water distribution networks. Eur Water Resour Assoc (EWRA) 27(13):4443–4468
6. Eryiğit M (2019) Water loss detection in water distribution networks by using modified Clone alg. J water Supply Res Technol IWA 68(4):253–263
7. Hamilton S, Charalambous B (2013) Leak detection technology and implementation. IWA Publishing Alliance House
8. Jadhao RD, Gupta R (2018) Calibration of water distribution network of the Ramnagar zone in Nagpur City using online pressure and flow data. Appl Water Sci 8(1):1–10
9. Francés-Chust J, Brentan BM (2020) Optimal placement of pressure sensors using fuzzy DEMATEL-based sensor influence. Water 2020(12):493

10. El-Zahab S, Zayed T (2019) Leak detection in water distribution networks: an introductory overview. Smart Water J 4(5)
11. Serafim O, Tzeng GH (2003) Defuzzification within a multicriteria decision model. Int J Uncertain Fuzziness Knowl-Based Syst 11(5):635–652
12. Shafiqul Islam M, Sadiq R, Rodriguez MJ, Francisque A, Najjaran H, Hoorfar M (2011) Leakage detection and location in water distribution systems using a fuzzy-based methodology. Urban Water J 8(6):351–365
13. Si S-L, You X-Y (2018) DEMATEL technique: a systematic review of the state-of-the-art literature on methodologies and applications. Math Prob Eng 33
14. Tafuri AN (2000) Locating leaks with acoustic technology. Am Water Works Assoc 92(7):57–66
15. Xue Z, Tao L, Fuchun J, Riehle E, Xiang H, Bowen N, Singh RP (2020) Application of acoustic intelligent leak detection in an urban water supply network. J Water Supply Res Technol 69(5):512–520

Least Cost Path Pipeline Routing Using Spatial Multi-criteria Analysis for Vidarbha Region: A Case Study

Abhishek Mhamane and A. B. Mirajkar

Abstract The Western Vidarbha region of Maharashtra state is experiencing drought, whereas the Eastern Vidarbha region is experiencing floods. National Water Development Agency (NWDA) proposed an intra-state link project to divert surplus water from the Wainganga basin (Eastern Vidarbha) to the Nalganga reservoir (Western Vidarbha) via canals. This study investigates possible alternative routes for intra-state link via pipeline, using spatial multi-criteria analysis (SMCA) and least cost path (LCP) approach in the GIS environment. In this study, SMCA implementing analytical hierarchal process (AHP), was used to combine different spatial data coherently. This approach also incorporates the subjective opinion of the decision-maker into the route planning process. Pathfinding was carried out using the LCP approach implementing Dijkstra's Algorithm. The route planning process considered four main factors viz. economy, engineering, environmental, and social. These factors in turn comprised of different criteria and sub-criteria such as slope, soil type, geomorphology, land use/land cover, drainage network, and settlements, which were structured in a hierarchal fashion. Four alternative routes were generated which prioritized one of the four main factors. These routes were then evaluated and ranked using the AHP method, and the optimal route is suggested. It was found that the total length and number of road-rail-river intersections for LCP generated route were, respectively, 10.7% and 5.88% less than NWDA proposed route. The average ground slope for the LCP-generated route was 6.22% and 12.15% for NWDA-proposed route. The results show that incorporating SMCA and LCP for route planning in a GIS environment generates results better than the conventional approach. Further studies are needed to investigate the utility of anisotropic cost surfaces or the use of alternative LCP algorithms such as the A^* algorithm or the development of local route optimization techniques.

A. Mhamane (✉) · A. B. Mirajkar
Department of Civil Engineering, Visvesvaraya National Institute of Technology, Nagpur, Maharashtra 440010, India
e-mail: amhamane22@iitk.ac.in; abhishekmhamane@students.vnit.ac.in

A. B. Mirajkar
e-mail: abmirajkar@civ.vnit.ac.in

P. V. Timbadiya et al. (eds.), *Geospatial and Soft Computing Techniques*, Lecture Notes in Civil Engineering 339, https://doi.org/10.1007/978-981-99-1901-7_30

Keywords Spatial multi-criteria analysis · Least cost path · QGIS · Pipeline routing · Vidarbha region

1 Introduction

Water scarcity is a very real and serious issue faced by countries around the world. Climate change has further aggravated the situation. Water access and availability are essential for agricultural productivity and livelihood sustenance of farmers globally. To mitigate the temporal and spatial water disparity in different regions of the country, governments and researchers have proposed policy and infrastructural interventions for the equitable distribution of water resources. Water transfer is a water management strategy aimed at transferring water from surplus regions to water-deficit regions. The transfer scheme can be either inter-basin–between two basins or intra-basin–from one sub-basin to another in the same basin. Many such schemes have been implemented around the world and in India [1].

Water disparity between the Eastern and Western Vidarbha region of the state of Maharashtra is well documented. The National Water Development Agency (NWDA) proposed a Wainganga (Godavari basin)–Nalganga (Tapi basin) intra-state link project to mitigate this challenge. This project would also address the irrigation backlog faced by this region [2, 3].

This study proposes a GIS-based methodology for efficient and economical route alignment for a piped water transfer system. This methodology provides a synoptic view for solving the problem, in contrast to conventional methods which mostly rely on subjective judgement. This approach incorporates various conflicting planning criteria, expert judgement and geospatial data using the spatial multi-criteria analysis (SMCA) framework. This study provides a case study of applying the proposed methodology to the Wainganga-Nalganga Intra-State Link project.

2 Materials and Method

2.1 Study Area and Data Sources

Vidarbha is the eastern region of Maharashtra, comprising of 11 districts viz Amravati, Akola, Buldhana, Washim, Yeotmal, Nagpur, Wardha, Bhandara, Gondia, Chandrapur, and Gadchiroli. Vidarbha accounts for 31.62% of the area and 21.3% population of Maharashtra state. Vidarbha region has underdeveloped irrigation infrastructure as compared to the rest of the state, due to the absence of irrigation infrastructure agriculture in upland areas of western Vidarbha is mainly dependent on rainfall and

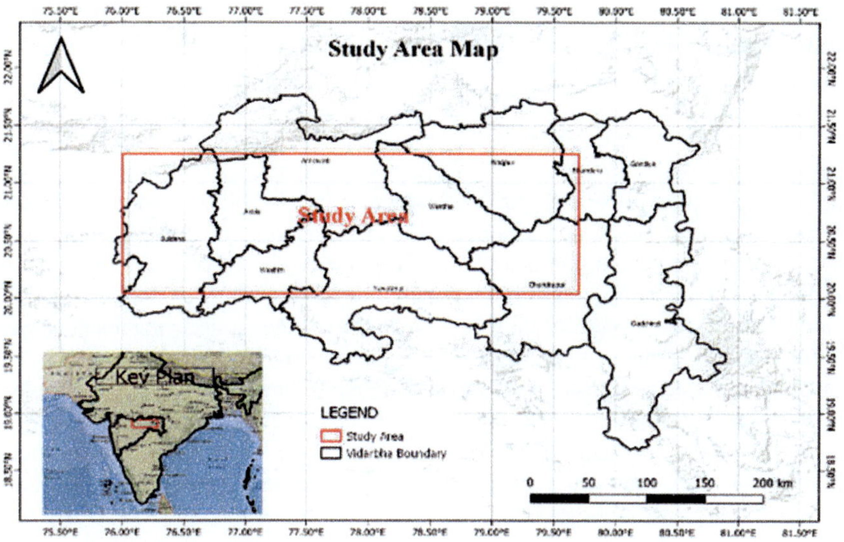

Fig. 1 Study area map

groundwater for irrigation [2, 3]. The NWDA project proposes linking of four reservoirs viz. Gosikhued Reservoir, Lower Wardha Reservoir, Katepurna Reservoir, and Nalganga Reservoir [2]. The study area is shown in Fig. 1.

Data was collected from various government and commercial organizations using their data dissemination platform. The study utilized various remote sensing products such as 30 m spatial resolution Shuttle Radar Topography Mission (SRTM) digital elevation model (DEM) and Landsat 8 satellite imagery. Additional geological data such as geomorphological, geological, soil, and rock maps were obtained in digital vector format from the Geological Survey of India using the Bhukosh portal. Additional geospatial data such as the road network, railway network, stream network, protected areas, settlement locations, and water bodies were obtained from the OpenStreet Map database. QGIS was used as the geographical information system (GIS) software package along with plugins developed for QGIS. Data collected was checked for quality and necessary rectifications were made.

3 Methodology

Pipeline route alignment is about achieving a trade-off between minimum straight-line distance and existing ground conditions both on and below the ground [4]. Pipeline route planning or routing is a spatial decision problem and often spatial decisions involve multiple criteria or factors, which involves evaluation of a large number of alternatives based on suitable criteria. Spatial multi-criteria analysis (SMCA) or

GIS-based multi-criteria analysis pairs the forte of GIS at geographical data storage, acquisition, retrieval, manipulation, and analysis capabilities with the ability of multi-criteria decision-making (MCDM) methods to incorporate decision maker's preference in an objective sense, resulting in data-driven decision making. In short, SMCA is a process that combines and transforms geographical data into a resultant decision [5–7].

SMCA requires the determination of planning and evaluation criteria. These are further used by the decision-maker to measure the performance of alternatives, decision maker's preferences are incorporated into the model by assigning weights to different criteria and scoring different alternatives based on these criteria. A decision rule or aggregation function is required to integrate decision judgement across various criteria. Sensitivity analysis helps understand the robustness of the model, the interaction between data, preferences and their influence on the output. Finally, all alternatives are ranked and optimal is recommended [5, 7].

Incorporating geospatial data such as slope, soil, geology, geomorphology, drainage network, road and rail networks, human settlement, protected and cultural areas, land use and land cover (LULC), allows having a synoptic view. This study incorporates all these different criteria in a hierarchical structure where economy, engineering, environmental, and social being four major factors, which in turn, comprise different criteria and sub-criteria [5, 6, 8, 9].

Data cannot be utilized in its raw form, basic data processing is essential such as DEM sink/void filling, vector geometry rectification [10]. Multiple criteria have varying influences on route planning, these variances can be resolved by the reclassification of these layers on a common standard scale. The scale can be either relative suitability scores ranging from 1–9 to 1–10 [11]. This study adopts the approach of a relative suitability scoring scale ranging from 1–9 with the assumption that a higher score indicates unsuitability. Standardization scores have been organized in the Fig. 2 along with brief justification and reference.

Analytical hierarchal process (AHP) is a multi-criteria decision analysis (MCDA) technique that uses a pairwise comparison matrix to determine the weights of criteria. The hierarchal structure of route planning criteria naturally facilitated the use of AHP for weight determination of criteria and sub-criteria [9, 12]. The weights of criteria and sub-criteria are tabulated in Fig. 3a and b. To mathematically combine data, an aggregation rule is required. The use of the GIS environment further facilitated the use of weighted linear combination (WLC). This is a simple yet powerful method with its limitations. Standardized data layers were used to obtain economical, engineering, environmental, and social criteria cost surface. These cost surfaces were further aggregated using AHP obtained weights and WLC to emulate different decision scenarios. These alternative scenarios and their compositions are presented in Fig. 4.

Cost surfaces can either be isotropic or anisotropic. The distinction is based on the cost being independent or dependent on the direction of movement, respectively. The isotropic approach is simple and idealistic in nature, whereas the anisotropic approach is more realistic and more complex. This study adopts an isotropic approach due to computational simplicity [8, 11, 13]. Least cost path (LCP) studies are traditionally associated with vector networks in GIS analysis. The raster-based implementation

Criteria	Cost Value		Reference/Justification
LULC	Water	9	Modified (Marjuki & Rudiarto, 2020; Saha et al., 2005)
	Built-Up	7	
	Bare Soil/ Rocky	3	
	Agriculture	5	
	Vegetation	9	
Land Value	Water	9	Surrogate data due to lack of field data, (Marjuki & Rudiarto, 2020)
	Built-Up	5	
	Bare Soil/ Rocky	1	
	Agriculture	3	
	Vegetation	9	
Soil	Cambisols	3	Based on agricultural or non-agricultural soils (Yildirim et al., 2012)
	Lithosols	3	
	Luvisols	5	
	Vertisols	7	
Geomorphology	Pediplain	1	Modified Terrain Classes (Durmaz et al., 2019)
	Pediment	5	
	Plateau	∞	
	Alluvial Plain	2	
	Gullied Land	7	
	Hill & Valley	∞	
	Scrap, Mesa. & Custa	∞	
Rock	Igneous Rock (Deccan Trap)	1	Modified (Saha et al., 2005; Yildirim et al., 2012)
	Sedimentary Rock	7	
	Fluvial Sediments	9	
	Metamorphic Rock	3	
Slope (%)	0 - 1.75%	1	Modified (Arabi & Gharehhassanloo, 2018; Marjuki & Rudiarto, 2020; Saha et al., 2005; Yildirim et al., 2012)
	1.75% - 5.25%	2	
	5.25% - 7.00%	3	
	7.00% - 8.75%	4	
	8.75% - 10.50%	5	
	10.50% - 12.30%	6	
	12.30% - 14.00%	7	
	14.00% - 15.85%	8	
	15.85% - 128%	9	
Water Bodies & Fault Proximity Reclass	< 0m	∞	Modified (Marjuki & Rudiarto, 2020; Whanda et al., 2015; Yildirim et al., 2012)
	0 - 500m	9	
	500 - 1000m	8	
	1000 - 3000m	7	
	3000 - 5000m	5	
	5000 - 7000m	3	
	> 7000m	1	
Cities	< 5000m	∞	Modified (Marjuki & Rudiarto, 2020)(Yildirim et al., 2012)
	5000 - 6000m	7	
	6000 - 7000m	5	
	7000 - 10000m	3	
	> 10000m	1	
Towns	< 2000 m	∞	
	2000 - 3000 m	7	
	3000 - 4000 m	5	
	4000 - 6000 m	3	
	> 6000 m	1	
Village	< 500 m	∞	
	500 - 1000 m	7	
	1000 - 2500 m	5	
	2500 - 4000 m	3	
	> 4000 m	1	
Archaeological sites/ Forts/ Monuments etc...	< 500 m	∞	(Effat & Hassan, 2013; Marjuki & Rudiarto, 2020; Singh & Singh, 2017; Yildirim et al., 2012)
	500 - 1000 m	7	
	1000 - 200 m	5	
	2000 - 4000 m	3	
	> 4000 m	1	

Fig. 2 Criteria and sub-criteria

Factor	Criteria	Criteria Weights	Sub-Criteria	Sub-Criteria Wrights
Economical	Physical Barriers/ Extra Infrastructure	0.724	Road Network	0.050
			Railway Network	0.105
			Water Bodies	0.602
			Mines	0.243
	Proximity to Growth Centers	0.193	Proximity to City	0.600
			Proximity to Towns	0.400
	Land Value	0.083	~	~
Engineering	Slope	0.643	~	~
	Geotechnical	0.074	Soil Type	0.500
			Rock Type	0.500
	Geological	0.283	Geomorphology	0.700
			Faults	0.300
Environmental	Protected Area	0.669	~	~
	River	0.267	~	~
	Vegetation	0.064	~	~
Social	Land Use	0.633	~	~
	Archaeological sites/ Forts/ Monuments etc...	0.106	~	~
	Settlements	0.260	Villages	0.074
			Towns	0.283
			Cities	0.643

(a)

Criteria	Sub-Criteria	Sub -Criteria Weight
Road Network	National Highway (NH)	0.602
	State Highway (SH)	0.243
	Major District Roads(MDR)	0.105
	Other District Roads (ODR)	0.05
Water Bodies	Rivers	0.29
	Streams	0.089
	Canals	0.1
	Reservoirs	0.484
	Ponds, lakes and other water bodies	0.036

(b)

Fig. 3 **a** Criteria and sub-criteria weights. **b** Criteria and sub-criteria weights

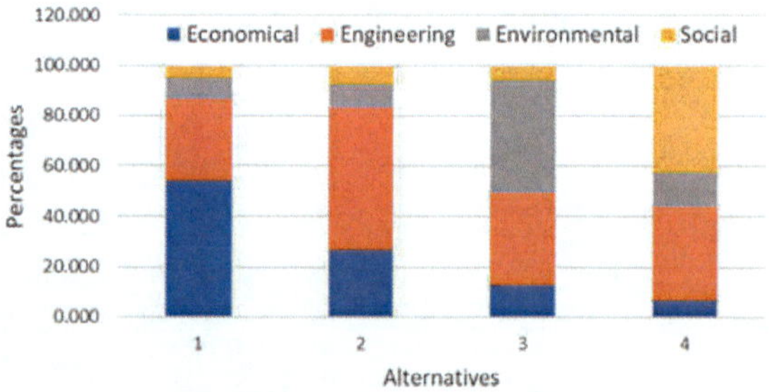

Fig. 4 Composition of different alternative cost surfaces

of LCP utilizes algorithms similar to the vector-network-based approach. A virtual network is established using raster cells as nodes and defining connected neighbours to establish links. The higher the number of connected neighbours, denser the network and has higher computational complexity is involved. Movement cost is assigned to each link using the underlying cost surface, connection pattern, and distance. Queen's pattern is used in this study.

In this study, Dijkstra's algorithm was used as the LCP algorithm [8, 14]. Dijkstra's algorithm was developed by the computer scientist Dijkstra in 1959 [15] based on graph or tree. Dijkstra's algorithm is computationally efficient and thus implemented in many GIS software.

4 Results and Discussion

This study intended to investigate alternative route possibilities for Wainganga–Nalganga Link Project via pipeline using Spatial Analysis in the GIS environment. The methodology adopted for the study has been discussed in the previous section. The following section discusses the major results obtained for major intermediate steps along with LCP evaluation and ranking using the AHP technique.

4.1 Reclassification and WLC

To apply these weights to data layers (Raster data), reclassification has to be done. The results have been summarized in the form of maps in Figs. 5a and b, and 6.

As per the methodology mentioned before, these standardized data layers were combined using WLC to obtain economical, engineering, environmental, and social criteria cost surfaces. The results are shown in Fig. 7a–d.

These cost surfaces were combined to generate alternative cost surfaces and are presented in Fig. 8a–d.

4.2 Evaluation and Comparison of Results

To find an optimal route, route evaluation criteria were decided upon based on a literature study. Weights of these evaluation criteria were obtained using the AHP technique. Evaluation criteria such as land-use length transition, maximum slope, mean slope, rail-road-river intersection and settlement proximity were computed using raster statistics function, profile tool and by overlaying operations. Scores were assigned on a scale of 1–9 based on the relative performance of alternatives on the set evaluation criteria. Ranking of routes within each link was done using the AHP technique and an optimal route was suggested [6, 16]. The data is presented in form of a stacked column chart in Fig. 9a–c.

Alternative routes 3, 2, and 4 were found to be optimal routes for link-1, link-2, and link-3, respectively. It was found that the total length and number of Road-Rail-River intersections for the LCP-generated route were 10.7% and 5.88%, respectively, less than NWDA proposed route. The average ground slope for the LCP-generated route was 6.22% and 12.15% for NWDA proposed route. Other statistics comparison

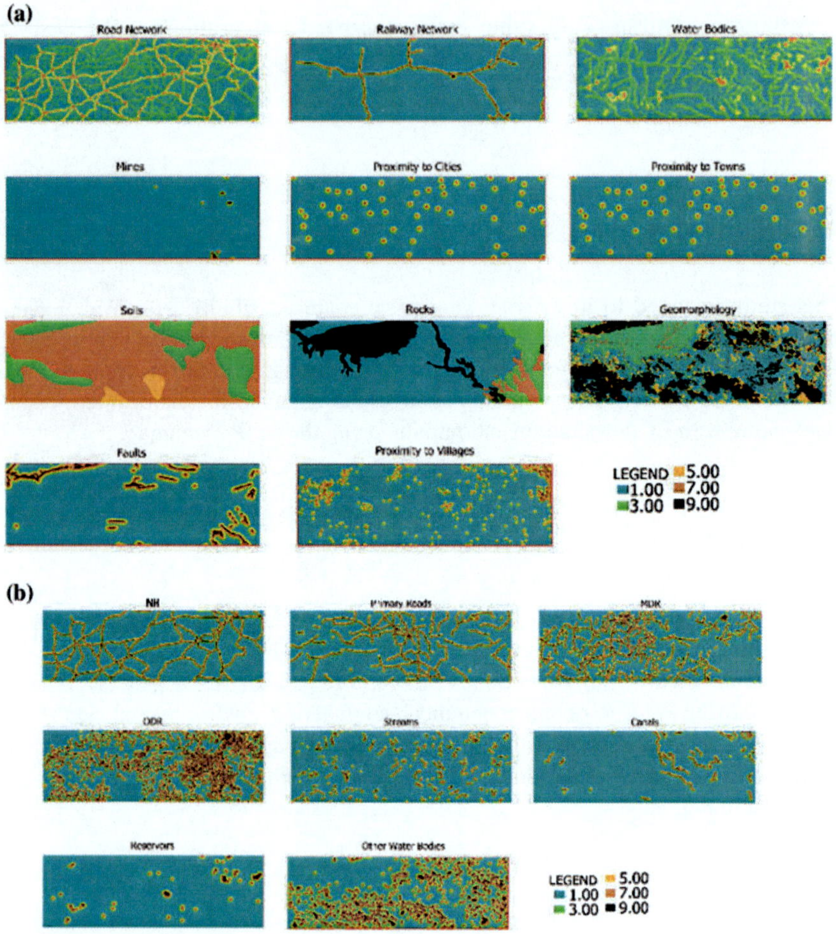

Fig. 5 **a** Reclassified sub-criteria map layers. **b** Reclassified sub-criteria map layers

is tabulated in Fig. 10 along with results in Fig. 10a–c, Alternative routes Figs. 11 and 12.

5 Conclusions

- The adopted methodology incorporates various criteria along with decision-makers judgement, leading to an objective decision based on geospatial data. The results obtained are promising, the length and average slope of the entire route decreased by 17.08% and 48.06%, respectively.

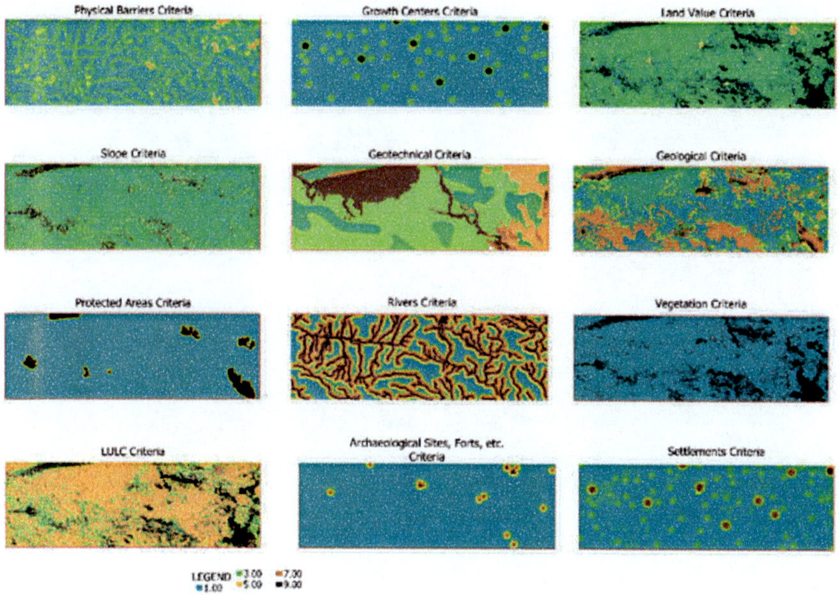

Fig. 6 Criteria layers obtained from sub-criteria layer using AHP and WLC

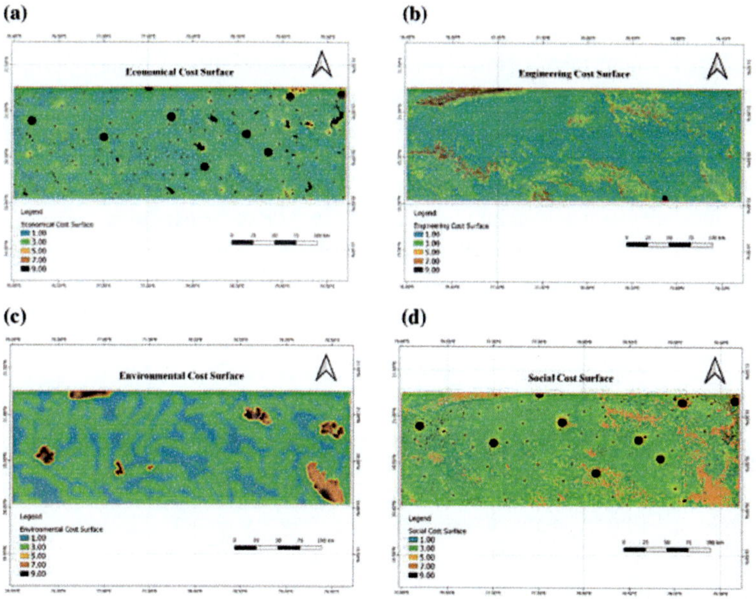

Fig. 7 **a** Economical cost surfaces. **b** Engineering cost surfaces. **c** Environmental cost surfaces. **d** Social cost surfaces

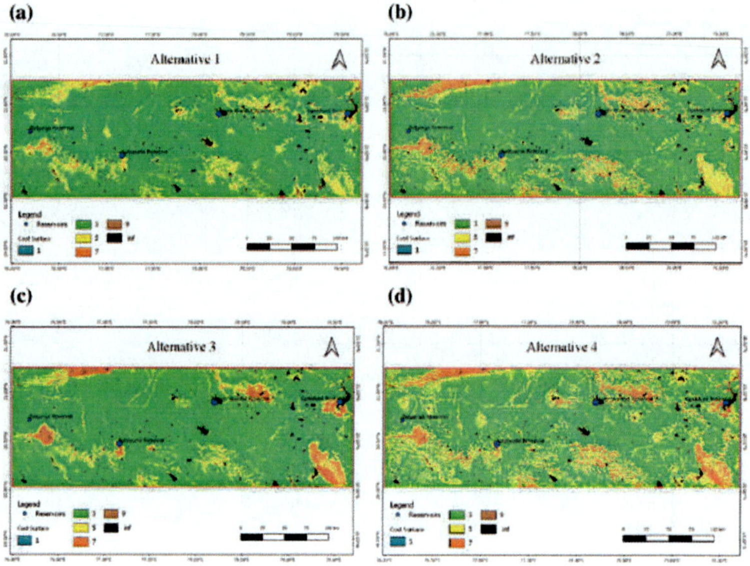

Fig. 8 **a** Alternative 1 cost surface. **b** Alternative 2 cost surface. **c** Alternative 3 cost surface. **d** Alternative 4 cost surface

- The model lacked in some areas such as not being able to avoid settlement and railway line intersections. Which may require special route optimization algorithms or using the cost of construction or traversing for standardization (reclassification) instead of suitability score. Further study is required to study the anisotropic approach and develop an efficient GIS implementation.
- It can be concluded that the GIS-based SMCA approach is useful for routing pipelines, as this is a general approach, it can further be extended for routing of any linear engineering structure such as roadways with necessary modifications. The Vidarbha region has a serious irrigation infrastructure backlog, which is threatening livelihood, water security, land, and groundwater degradation. Serious remedial measures are required to mitigate the worsening situation.

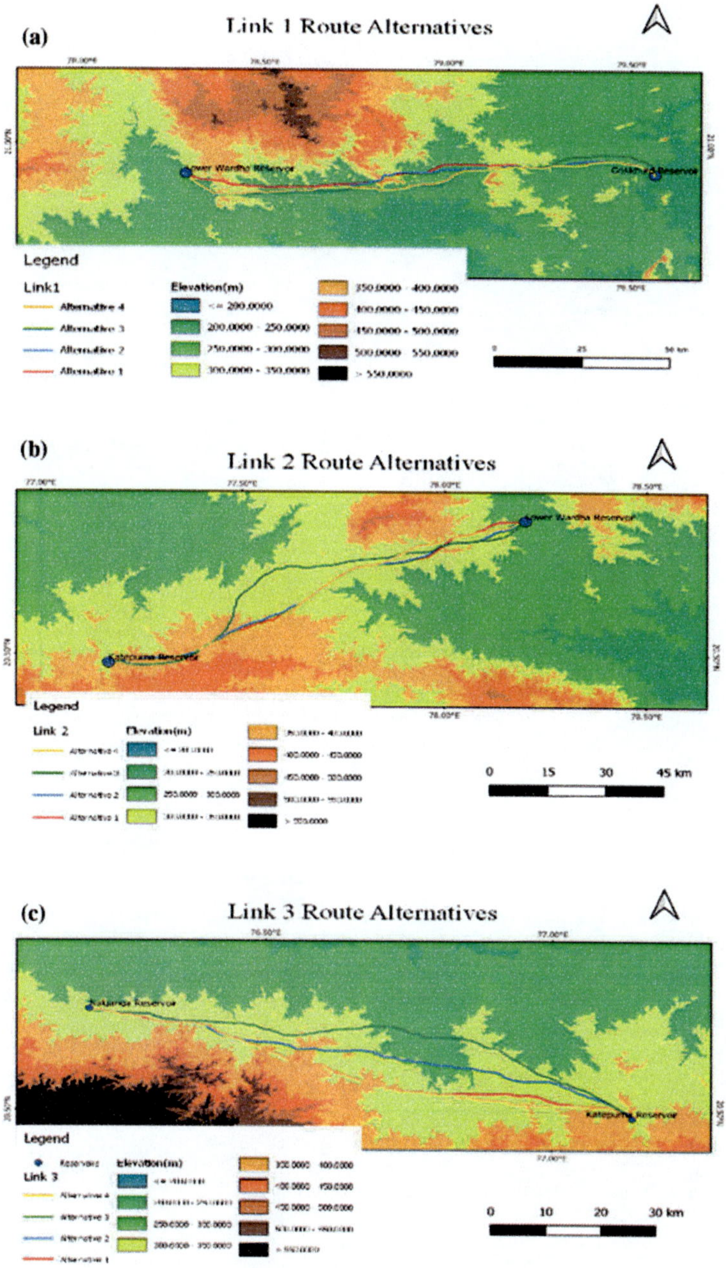

Fig. 9 **a** LCP routes for link 1. **b** LCP routes for link 2. **c** LCP routes for link 3

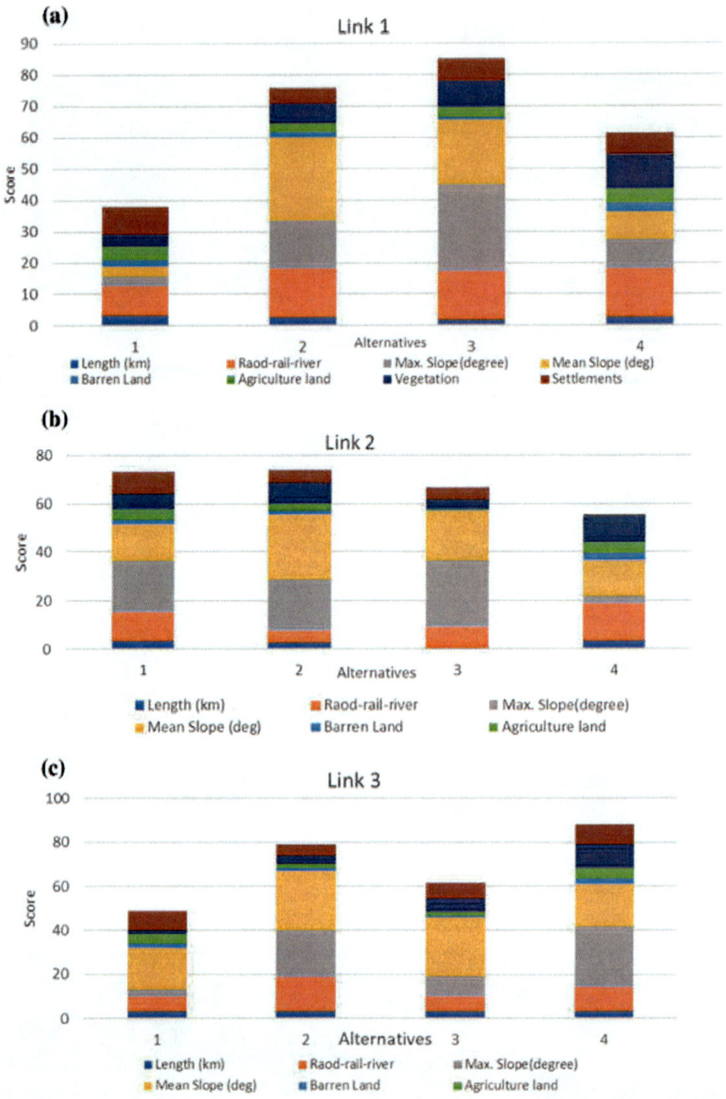

Fig. 10 **a** Ranking of alternative LCPs for link 1. **b** Ranking of alternative LCPs for link. **c** Ranking of alternative LCPs for link 3

	LCP				Proposed				% Change
	Link-1	Link-2	Link-3	Total	Link-1	Link-2	Link-3	Total	
Water (Km)	0.23	0.07	0.00	0.30	1.66	3.08	1.63	6.38	95.29%
Built-up (Km)	3.77	3.38	0.30	7.46	6.62	3.48	6.04	16.13	53.75%
Barren Land (Km)	27.05	24.42	52.68	104.15	22.58	25.97	35.73	84.27	-23.58%
Agriculture land (Km)	115.86	99.35	57.91	273.13	114.15	94.41	82.78	291.34	6.25%
Vegetation (Km)	3.09	0.78	0.10	3.96	31.90	3.80	1.72	37.42	89.41%
Total Length(km)	150.00	128.00	111.00	389.00	176.90	130.73	127.91	435.54	10.69%
Fault Intersection(Nos)	1.00	0.00	0.00	1.00	1.00	1.00	0.00	2.00	50.00%
Road Intersection(Nos)	52.00	38.00	22.00	112.00	50.00	33.00	26.00	109.00	-2.75%
river/canals/stream intersection(Nos)	13.00	6.00	8.00	27.00	22.00	6.00	14.00	42.00	35.71%
Railway intersection	3.00	1.00	1.00	5.00	1.00	1.00	0.00	2.00	-150.00%
Road-Railway-River Intersection (Nos)	68.00	45.00	31.00	144.00	73.00	40.00	40.00	153.00	5.88%
Max. Slope(degree)	6.93	10.37	8.71	8.67	31.54	12.73	24.97	23.08	62.44%
Mean Slope (deg)	1.11	1.10	1.35	1.19	3.11	1.75	2.07	2.31	48.63%
Max. Elevation	377.00	408.00	383.00	389.33	298.86	335.02	395.00	342.96	-13.52%
protected area intersection(Km)	0.00	0.46	0.00	0.46	2.04	0.16	0.27	2.47	81.49%
Villages (Nos)	4.00	1.00	2.00	7.00	2.00	1.00	4.00	7.00	0.00%
Towns(Nos)	0.00	1.00	0.00	1.00	0.00	0.00	0.00	0.00	NA
Cities(Nos)	0.00	0.00	0.00	0.00	0.00	0.00	0.00	0.00	NA
Settlements(Nos)	4.00	2.00	2.00	8.00	2	1	4	7.00	-14.29%

Fig. 11 Optimal LCP statistics and comparison

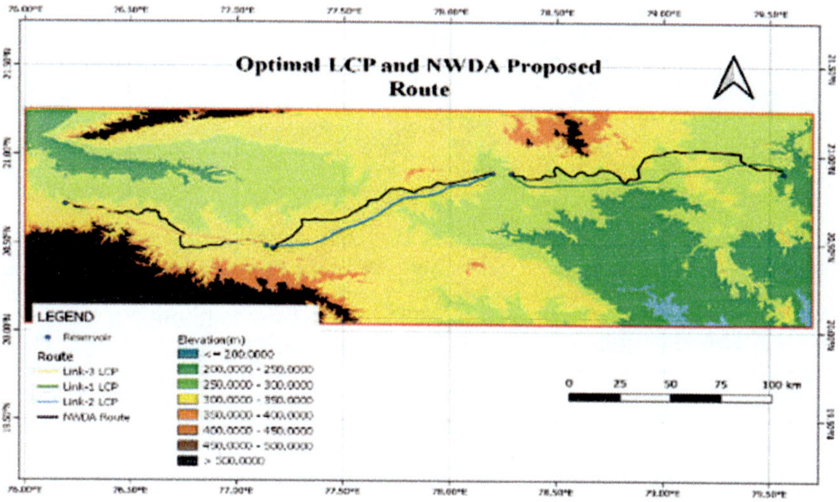

Fig. 12 Optimal route and NWDA route on map

Acknowledgements The authors would like to acknowledge the different agencies/organizations for making geospatial data freely and easily accessible. Authors are also thankful to the various contributors of open-source projects like QGIS and OpenStreetMap.

References

1. Shao X, Wang Z (2003) Interbasin transfer projects and their implications: a china case study. Int J River Basin Manage 1(1):5–14. https://doi.org/10.1080/15715124.2003.9635187
2. NWDA (2015) Detailed Project Report of Wainganga (Gosikhurd)-Nalganga (Purna Tapi) link project reviewed. NWDA
3. Nair S, Mirajkar AB (2021) Spatio-temporal rainfall trend anomalies in Vidarbha region using historic and predicted data: a case study. Model Earth Syst Environ 7(1):503–510. https://doi.org/10.1007/s40808-020-00928-1
4. Dubey R (2003) A remote sensing and GIS based least cost routing of pipelines. Geospatial Today (September–October) 47–50
5. Marjuki B, Rudiarto I (2020) Spatial multi-criteria analysis and least-cost path on the highway route planning: a case study of Bawen—Yogyakarta Highway, Indonesia. Geoplann J Geomatics Plann 7(2):113–130. https://doi.org/10.14710/geoplanning.7.2.113-130
6. Effat HA, Hassan OA (2013) Designing and evaluation of three alternatives highway routes using the analytical hierarchy process and the least-cost path analysis, application in Sinai Peninsula, Egypt. Egypt J Remote Sens Space Sci 16(2):141–151. https://doi.org/10.1016/j.ejrs.2013.08.001
7. Greene R, Devillers R, Luther JE, Eddy BG (2011) GIS-based multiple-criteria decision analysis. Geograph Compass 5(6):412–432
8. Saha AK, Arora MK, Gupta RP, Virdi ML, Csaplovics E (2005) GIS-based route planning in landslide-prone areas. Int J Geogr Inf Sci 19(10):1149–1175. https://doi.org/10.1080/13658810500105887
9. Yildirim V, Yomralioglu T, Nisanci R, Al Y et al (2012) A raster based geospatial model for natural gas transmission line routing. Natural gas—extraction to end use, (October). https://doi.org/10.5772/45814
10. Valencia J, Monserrate F, Casteleyn S, Bax V, Francesconi W, Quintero M (2020) A GIS-based methodological framework to identify superficial water sources and their corresponding conduction paths for gravity-driven irrigation systems in developing countries. Agric Water Manag 232(February):106048. https://doi.org/10.1016/j.agwat.2020.106048
11. Medrano FA (2021) Effects of raster terrain representation on GIS shortest path analysis. PLoS ONE 16(4):e0250106. https://doi.org/10.1371/journal.pone.0250106
12. Drobne S, Lisec A (2009) Multi-attribute decision analysis in GIS: weighted linear combination and ordered weighted averaging. Informatica (Ljubljana) 33(4):459–474
13. Collischonn W, Pilar JV (2000) A direction dependent least-cost-path algorithm for roads and canals. Int J Geogr Inf Sci 14(4):397–406. https://doi.org/10.1080/13658810050024304
14. Yu C, Lee J, Munro-Stasiuk MJ (2003) Extensions to least-cost path algorithms for roadway planning. Int J Geogr Inf Sci 17(4):361–376. https://doi.org/10.1080/1365881031000072645
15. Dijkstra EW (1959) A note on two problems in connexion with graphs. Numerishe Mathematik 1:3
16. Singh MP, Singh P (2017) Multi-criteria GIS modeling for optimum route alignment planning in outer region of Allahabad City, India. Arabian J Geosci 10(13). https://doi.org/10.1007/s12517-017-3076-z

Application of Random Forest and Model Tree for Discharge and Water Level Estimation and Prediction

S. N. Londhe, P. R. Dixit, P. S. Kulkarni, and H. Dhumal

Abstract Data-driven models (DDM) have been emerging as promising tools for estimation and prediction of hydrological parameters. The popularity of DDM is due to the method's competitive performance and relative lack of strict distributional assumptions which prove to be an excellent addition with the traditional methods in the field. In the current study, Model Tree (MT) and Random Forest (RF), a popular DD technique is utilized to estimate and predict the discharge and water level at 1 station which are at downstream with discharge and water level at upstream station/s. For the present study stations: Krishna Bridge, Khodashi, Bahe, Bhilawadi at upstream and Irwin at downstream situated in Krishna Basin are selected. Two set of models were developed with Set 1 as estimation of discharge at time t at Irwin with discharge at Krishna and Kodhshi at time t as input parameters and prediction of discharge at t at Irwin with discharge at Krishna at t−1 and Khodshi at t−1 as input parameters. In set 2 model, separate models to estimate water level Irwin were developed with, water level at Bahe at time t and with water level at Bhilawadi at time t as input. Prediction of water level at t at Irwin was done using water level at t−1 at Bahe and next with water level at t−1 at Bhilwadi. Performance of RF and MT models were evaluated and compared using metrics: Coefficient of correlation(r), Root mean square error (RMSE) and mean absolute error (MAE). RF, characterized by ensemble of decision trees, enable to understand the variable importance and MT are characterized by series of equations. The results display good performing RF and MT models in terms of correlation between observed and predicted values(r) and root mean square error (RMSE). RF displays better performance as compared to MT in both the set of models with higher r and lower RMSE values.

S. N. Londhe · P. R. Dixit · P. S. Kulkarni (✉)
Vishwakarma Institute of Information Technology, Pune 411048, India
e-mail: Preeti.kulkarni@viit.ac.in

S. N. Londhe
e-mail: shreenivas.londhe@viit.ac.in

P. R. Dixit
e-mail: pradnya.dixit@viit.ac.in

H. Dhumal
Water Resource Department, Mumbai, Maharashtra, India

P. V. Timbadiya et al. (eds.), *Geospatial and Soft Computing Techniques*, Lecture Notes in Civil Engineering 339, https://doi.org/10.1007/978-981-99-1901-7_31

Keywords Estimation · Random forest · Model tree · Discharge · Water level

1 Introduction

Precise and accurate measurement of discharge is necessary and vital for almost all the activities of water resource departments including hydrologic and hydraulic modeling. Estimating and prediction of discharge and water level at a certain station will be useful in mitigating the risk of floods. Traditionally, discharge which represents volume flowing per unit time and can be predicted by traditional (theoretical, deterministic and/or empirical) equations of hydrology requires large, complex and consistent parametric data and also consume time which makes it a plausible case modelling using Data driven techniques (DDT). DDTs are found to deal with uncertainty, partial truth and approximation to achieve reasonably accurate results. In the family of DDT, Artificial Neural Networks, Genetic Programming has proved its presence through their performance since last 3 decades or so. However, tools such as Model Tree (MT) and Random Forest (RF) which are paving their way in Discharge and water level estimation and prediction are under explored. Quinlan presented four case studies which compared M5 with other methods showed the better efficiency of M5 model tree [1]. The M5 Model trees (MT) models are advantageous due to the generated tree like structure of linear models, reproducible and easy to understand for decision makers [2]. For instance, it was used in the water level-discharge relationship and it was found that M5 had the same predictive accuracy as an ANN built with the same data. M5 learns efficiently and tackles tasks with very high dimensionality [3, 4]. Zia et al. predicted drainage discharge using an M5 decision tree modelling technique and validated the discharge prediction model for implementation on a system with limited resources [5]. Researchers have also made attempts in utilizing RF for Discharge and Water level estimation and prediction. Examples of RF applications in hydrology include precipitation downscaling [6], flood prediction [7], predicting flow characteristics at ungauged locations [8]. In 33% of the water-related studies random forests were used to complement other modelling approaches to improve inference. This highlights its usefulness in water science [9]. This study is an attempt to explore the performance of M5P Model Tree (MT) and Random Forest (RF) through Estimating and Predicting of discharge and water level at Irwin station on Krishna River in the South-western part of Maharashtra state of India using the discharge and water levels at upstream stations. The variable/input parameter importance analysis for each model helps in increase of confidence in these tools. The paper further is organized as follows: the next section gives brief introduction of MT and RF followed by the description of study area and data characteristics followed by the methodology section with a clear explanation of Input and output parameters, model design and the performance measures to judge the model performance. Finally, the results are presented and discussed along with the hydrographs and scatter plots. Conclusion and takeaways will close the current study.

2 Techniques Utilized

2.1 Model Tree (MT)

The M5 model tree algorithm was originally developed by Quinlan [1]; MT combines a conventional decision tree with the possibility of generating linear regression functions at the leaves. The M5 tree is a piecewise linear model, so it takes an intermediate position between the linear models such as ARIMA and truly nonlinear models such as ANN. Smoothing and pruning is done to the trees to overcome the over fitting problem. For decision tree the splitting criterion is based on the standard deviation error reduction of the values in the subset T of the training data that reaches a particular node (which is an analogue of entropy). For further details of M5 Model Tree readers are directed to [1, 10].

2.2 Random Forest (RF)

Proposed by Breiman, RF is a semi-unsupervised, non-parametric algorithm within the decision tree family that comprises an ensemble of uncorrelated trees to yield prediction for classification and regression tasks [11]. It combines bagging and ensemble learning theory with the random subspace method. RF is a classifier consisting of a collection of tree-structured classifiers: $\{h(x, \Theta k), k = 1 \ldots\}$, where $\{\Theta k\}$ are independent and identically distributed random vectors and each tree casts a unit vote for the most popular class of input x. The method utilizes bootstrap re-sampling technology to sample the original data and generate a number of training samples, each of which randomly selects the feature attributes through random subspace methods to construct the decision tree. Finally, the optimal result is obtained by a voting or averaging method. Estimation of variable importance (i.e., assessing the relative significance of predictor variables in modeling the behavior of response variables is doable with RF, through the use of variable importance metrics in two stages [9].

3 Study Area and Data

For the present study, Krishna Basin in the Deccan Plateau covering large areas in the States of Maharashtra, Karnataka and Andhra Pradesh. The mainstream Krishna River rises from the Western Ghats near Jor village of Satara district of Maharashtra at an altitude of 1337 m just north of Mahabaleshwar. The total length of river from origin to its outfall into the Bay of Bengal is 1400 km. The river Krishna in Krishna Basin consists of 4 main stations gauging stations namely; Krishna Bridge, Khodshi,

Bahe and Bhilawadi at upstream and and Irwin station at downstream. Figure 1 displays the pictorial view of the Krishna River, Maharashtra state, India with stations under consideration: Krishna bridge, Khodshi, Bahe and Bhilawadi at upstream of Irwin and at downstream Irwin [12]. The daily measured Discharge Q (in cumecs) and water level WL (in Meter) values at respective stations: Krishna bridge, Khodshi, Bahe and Bhilawadi and Irwin, was made available from the year 2012 to 2019 by the Water Resource Department (WRD), Sangli Division, Government of Maharashtra, The discharge and water level values at upstream stations: Krishna Bridge, Khodshi, Bahe and Bhilwadi are used for estimating and predicting the discharge and water level at the downstream station of Irwin in the same reach. The characteristics of the data is presented in the Tables 1 and 2 for discharge and water level respectively.

Table 1 displays the range of discharge values between the stations. These range of values are indication of the variation in the data. A larger variation in data can contribute towards a universal model, however it can also be be determinantal towards the model performance. The table also shows positive skewness for all the stations indicating its positively skewness. However, a larger skewness of Krishna and Khodshi mean a longer distribution of the data in the right tail than left thus indicating asymmetry in the data. A large standard deviation in discharge values of Krishna and Khodshi indicate higher dispersion of data from the mean. Table 2 displays the range of water levels at respective stations in meters. A higher range of

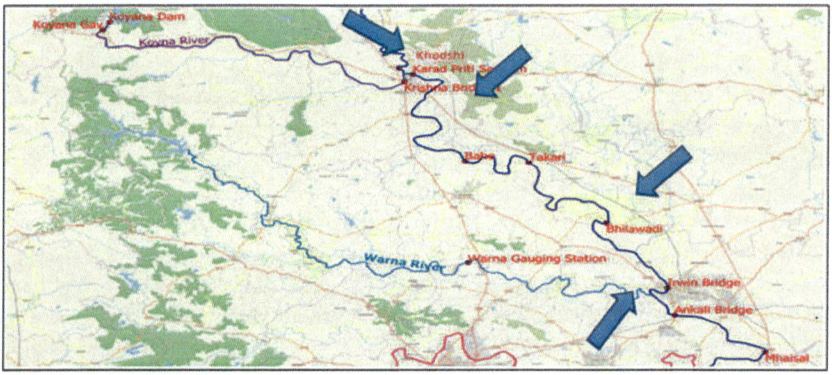

Fig. 1 Study area [13]

Table 1 Statistical analysis of discharge (Q) at stations under consideration in Krishna Basin

Characteristics of data	Krishna (KRI) (Q)	Khodshi (KHO)(Q)	Irwin (IR) (Q)
Minimum Q in Cumecs	533	95	100
Maximum Q in Cumecs	262,967	95,690	223,466
Standard Deviation	57,547.63	16,119.18	57,505.79
Skewness	1.893578	1.79206	1.292693
Mode	593	1007	8100

Table 2 Statistical analysis of water level (WL) at stations under consideration in Krishna Basin

Characteristics of data	Krishna (KRI)	Khodshi (KHO)	Bahe(B)	Bhilawadi (BH)	Irwin (IR)
Minimum WL in meters	549.7	568.38	530.26	534.22	527.06
Maximum WL in meters	565.65	573.5	560.28	571.26	548.48
Standard deviation	1.84	0.39	1.09	2.84	2.26
Skewness	0.75	0.40	−3.44	3.27	3.09
Mode	554.20	570.13	550.42	534.22	530.11

Table 3 Model development for estimation and prediction of discharge at Irwin station

Model No.	Input (Discharge Q, time t)	Output	Remark
Model 1	Krishna (Qt), Khodshi (Qt)	Irwin (Qt)	Estimation of Q
Model 2	Krishna (Qt−1), Khodshi (Qt−1)	Irwin (Qt)	Prediction of Q:24 h ahead
Model 3	Krishna (Qt), Khodshi (Qt), Krishna (Qt−1), Khodshi (Qt−1)	Irwin (Qt)	Prediction of Q:24 h ahead
Model 4	Krishna (Qt), Khodshi (Qt), Krishna (Qt−1), Khodshi (Qt−1)	Irwin (Qt + 1)	Prediction of Q:24 h ahead
Model 5	Krishna (Qt), Khodshi (Qt), Krishna (Qt−1), Khodshi (Qt−1)	Irwin (Qt + 2)	Prediction of Q:72 h ahead

Table 4 Model development for estimation and prediction of water level at Irwin station

Model No.	Input (WL in meters)	Output	Remark
Model 6	Krishna (t), Khodshi (t)	Irwin (t)	Estimation of WL
Model 7	Krishna (t−1), Khodshi (t−1)	Irwin (t)	Prediction of WL:24 h ahead
Model 8	Krishna (t−1), Krishna (t), Khodshi (t−1), Khodshi(t)	Irwin (t)	Prediction of WL:48 h ahead
Model 9	Krishna (t−1), Krishna (t), Khodshi (t−1), khodshi(t)	Irwin (t + 1)	Prediction of WL:48 h ahead
Model 10	Bahe(t−1),Bahe(t)	Irwin (t)	Estimation of WL
Model 11	Bahe(t−1), Bahe(t)	Irwin (t + 1)	Prediction of WL:24 h ahead
Model 12	Bhilawadi(t−1), Bhilawadi(t)	Irwin (t)	Estimation of WL
Model 13	Bhilawadi(t−1), Bhilawadi(t)	Irwin (t + 1)	Prediction of WL:48 h ahead

values (variation in minimum and maximum values) can be seen at Bhilawadi station followed by Bahe and Irwin. Higher standard deviation at Bhilawadi also indicate dispersion of values at grater pace as compared to water levels at other stations.

Bhilawadi also shows higher skewness of 3.27 as compared to other stations indicating its larger distribution of data at right tail and thus higher asymmetry in data. Negative skewness at Bahe indicate a larger weight distribution of data towards the left.

4 Methodology Adopted

The aim of this study is to estimate and predict the discharge and water level at Irwin station which is at downstream using previously measured discharge and water level values at the upstream stations namely—Krishna Bridge, Khodshi, Bahe and Bhilawadi. For this DDT of Random Forest and M5 Model Tree are employed to explore the capacity of both the techniques in estimation and prediction. Two set of models were designed for estimation and prediction of discharge (Q) at Irwin using MT and RF. For the design of Input and output parameters refer Tables 3 and 4. To estimate the discharge at Irwin station at time "t", discharges of the same time "t" at the two upstream stations namely; Krishna and Khodshi were used as input to develop data driven models (Model 1). Also, discharge on the previous day (Qt−1) at Krishna and Kodshi were considered to predict Q at Irwin at time"t" (Model 2). Discharge at Irwin is predicted 24 h and 48 h and 72 h in advance using the previously measured discharges values at the same two upstream station of Krishna and Khodshi with Qt and Qt−1 at respective stations (Model 3, 4 and 5). In Set 2 models i.e. models from Model 6 to Model 13, Water level (WL) estimation (at time "t") and forecast (24 h in advance and 48 h in advance) is done using the previously measured data at Krishna, Khodshi, Bahe and Bhilawadi stations in the same reach of Krishna River. Model 1, Model 5 and Model 7 were designed for estimating WL with water level at Krishna, Khodshi; at Bahe and at Bhilwadi respectively. Prediction of water levels were done in other models with water level at time t and t−1 at respective stations as input parameters. Tables 3 and 4 represents the different models developed for estimation and prediction of discharge in set 1 as well as of water levels at Irwin station in set 2 respectively.

The current study also in a subtle way attempts to select the discharge and water level at appropriate stations and at appropriate times i.e. appropriate input parameters for better estimation and prediction of discharge Q and Water level WL, especially in set 2 models. The aim in set models is to estimate and forecast water level at Irwin. However, selection of Water level at Bahe or Bhilwadi or Krishna Bridge and Khodshi as input parameter can be methodized and thus the input parameter which helps in better performing model can be selected. Model development was done using the techniques: Model Tree and Random Forest. Weka software version 3.9 developed by university of Waikato, Newzeland [14] was used for the same. For MT M5P algorithm developed by Quinlan was used. For Random Forest the Bagging iterations were restricted to 100. Each model was calibrated using approximately 70% of data for training model and the remaining 30% of data for model testing. The results of the developed models were judged using graphically in the form of scatter plots and hydrograph and through measures: Correlation coefficient (r), Root Mean Squared Error (RMSE), one relative error measure Mean Absolute Error (MAE) was considered to judge the accuracy of the models [15]. The results presented in the current study are for testing dataset.

5 Results and Discussion

The current study is an attempt to Estimate and Predict Discharge(Q) and water level (WL) at Irwin station situated on Krishna River, Maharashtra. Table 5 displays the performance of models developed in Set 1.

As seen in Table 5, model 1 with estimation of Qt at Irwin shows a good performance with $r = 0.8$ followed by middle order of RMSE and MAE values with MT. However, for estimation of the same, RF scores over MT with $r = 0.88$ followed by better RMSE ad MAE values. A similar trend can be seen for Prediction of Qt at Irwin with Model 2. An increase of 0.04 can be seen in r as compared to model in MT and a similar increase in RF as well. In both the models, decrease in RMSE and MAE values are also appreciable. Thus model 2 is of special mention with input parameters Q, one day before at respective stations and prediction of Q at time t. Variable/parameter importance plays a major role in model development as it helps in wiping the myth that such types of techniques are grey boxes. The parameters which significant impact on the outcome values in Model 1 and Model 2 for MT and RF are Q at Krishna followed by Q at Khodshi. The parameter influence can be seen through the coefficients of the parameters in equation developed through MT. The variable importance in RF is judged through Gini importance or Attribute importance based on average impurity decrease (and number of nodes using that attribute) i.e. it is a method of computing the feature importance on permuted out-of-bag (OOB) samples based on mean decrease in the accuracy. Feature importance is calculated as the decrease in node impurity weighted by the probability of reaching that node. The node probability can be calculated by the number of samples that reach the node, divided by the total number of samples. The higher the value the more important the feature. In Model 1 and Model 2 the Discharge Qt at Krishna is the most contributing parameter to the output followed by Qt at Khodshi. Model 3 with Qt and Qt−1 at Krishna and Khodshi as input parameters display an increased accuracy compared to Model 1 and 3 for Estimation/prediction of Qt at Irwin for MT. Also, the attribute which contributes significantly in predicting the output in RF is Krishna (Qt−1), Krishna (Qt) for both MT and RF. Refer Fig. 2c. A model tree developed for Model 3 and its equations is shown in Fig. 2a and b. A sample classifier Random trees for Model 3 is also shown in Fig. 2d.

Prediction of Discharge one day ahead and two days ahead in Model 4 and Model 5 also show good results with $r = 0.90$ and $r = 0.83$ with Model Tree and $r = 0.93$ and $r = 0.92$ with RF and accompanying lower RMSE and MAE values. It is worth mentioning that the property of Random tree of bootstrap and splitting the data into number of trees seems to work better than the Model Tree as evident by the results.

Table 5 Performance of models in set 1: estimation and prediction of discharge (Q)

Model No.	Output	Model tree				Random forest		
		r	RMSE (in cusec)	MAE (in cusec)	No. of rules	r	RMSE (in cusec)	MAE (in cusec)
Model 1	Irwin (Qt)	0.80	31,146.9	14,462.82	5	0.88	27,318.71	13,413.46
Model 2	Irwin (Qt)	0.84	29,934.6	13,863.93	8	0.92	25,087.75	12,469.70
Model 3	Irwin (Qt)	0.90	27,264.4	13,166.44	16	0.93	24,323.46	12,221.40
Model 4	Irwin (Q t + 1)	0.90	22,405.7	11,730.42	9	0.93	23,185.71	11,904.86
Model 5	Irwin (Q t + 2)	0.83	27,464.1	13,315.21	8	0.92	23,303.09	12,483.26

(a)

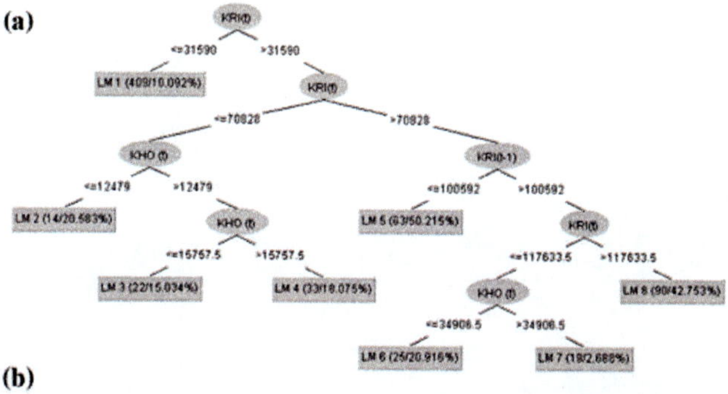

(b)

LM 1: IR (t) = 0.1652 * KHO(t-1) - 0.0433 * KHO (t) + 0.1468 * KRI(t-1) + 0.0152 * KRI(t) + 2766.1857
LM 2: IR (t) = 0.1652 * KHO(t-1) - 0.1333 * KHO (t) + 0.1792 * KRI(t-1) + 0.0152 * KRI(t) + 5583.4099
LM 3: IR (t) = -0.181 * KHO(t-1) - 0.1305 * KHO (t) + 0.3974 * KRI(t-1) - 0.218 * KRI(t) + 6272.0213
LM 4: IR (t) = -0.181 * KHO(t-1) - 0.1305 * KHO (t) + 0.3974 * KRI(t-1) - 0.1555 * KRI(t) + 4603.0812
LM 5: IR (t) = -0.2876 * KHO(t-1) - 0.9655 * KHO (t) + 0.465 * KRI(t-1) - 0.0563 * KRI(t) + 9331.1286
LM 6: IR (t) = 0.1709 * KHO(t-1) + 1.2532 * KHO (t) + 0.4763 * KRI(t-1) + 0.0167 * KRI(t) + 2272.6212
LM 7: IR (t) = -0.488 * KHO(t-1) + 0.0279 * KHO (t) + 0.9512 * KRI(t-1) + 0.0179 * KRI(t) + 27732.6838
LM 8: IR (t) = 0.4471 * KHO(t-1) + 1.5158 * KHO (t) + 1.1899 * KRI(t-1) - 0.7833 * KRI(t) + 31149.5131
LM 9: IR (t) = -0.2834 * KHO(t-1) - 0.1058 * KHO (t) + 0.1594 * KRI(t-1) + 0.0268 * KRI(t) + 124788.6868
LM 10: IR (t) = -2.5882 * KHO(t-1) + 1.2651 * KHO (t) + 0.3842 * KRI(t-1) + 0.0268 * KRI(t) + 144214.0109
LM 11: IR (t) = -3.8858 * KHO(t-1) - 0.1058 * KHO (t) + 0.7555 * KRI(t-1) + 0.0268 * KRI(t) + 162336.8629
LM 12: IR (t) = -0.2966 * KHO(t-1) + 0.1192 * KHO (t) - 0.1496 * KRI(t) + 179072.3289
LM 13: IR (t) = -0.2966 * KHO(t-1) - 0.0865 * KHO (t) - 0.2252 * KRI(t) + 196324.9048
LM 14: IR (t) = -0.9 * KHO(t-1) - 0.8095 * KHO (t) + 0.244 * KRI(t) + 182741.2392
LM 15: IR (t) = -0.017 * KHO(t-1) - 1.6158 * KHO (t) + 0.0778 * KRI(t-1) + 0.1581 * KRI(t) + 170155.279
LM 16: IR (t) = -0.1445 * KHO(t-1) - 2.0351 * KHO (t) + 0.0849 * KRI(t-1) + 0.1581 * KRI(t) + 202495.0043

(c)

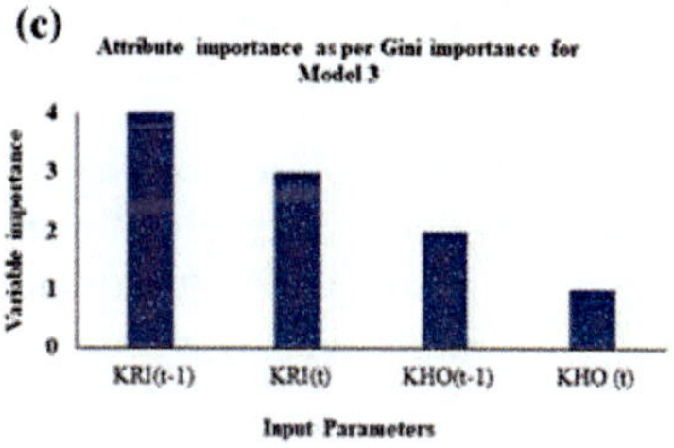

(d)

```
RandomTree
==========

KRI(t-1) < 66278.5
|   KHO(t-1) < 11375
|   |   KRI(t) < 8758.5
|   |   |   KRI(t-1) < 3235.5
|   |   |   |   KRI(t) < 720.5 : 300 (20/9000)
|   |   |   |   KRI(t) >= 720.5
|   |   |   |   |   KHO (t) < 1178.5 : 3528.78 (60/3717169.94)
|   |   |   |   |   KHO (t) >= 1178.5 : 1580.25 (8/3818250.44)
|   |   |   KRI(t-1) >= 3235.5
|   |   |   |   KRI(t-1) < 6212
|   |   |   |   |   KRI(t) < 5015 : 6907.29 (52/2168093.17)
|   |   |   |   |   KRI(t) >= 5015
|   |   |   |   |   |   KRI(t-1) < 5284.5
```

Fig. 2 **a** M5 model tree in model 3 **b** Equations developed for model 3, **c** Attribute importance for model 3, **d** Trees in RF for model 3

The Scatter plot (Fig. 3b) and hydrograph (Fig. 3a) display the testing results for model 4. The hydrograph displays that the observed and predicted values by MT and RF in tune with each other. However, at larger discharges values of MT and RF both need a better accuracy. This can also be seen in scatter plot in which at higher discharge values are slightly underpredicted.

Set 2 models were designed to explore the potential of MT and RF when water level at Irwin is to be estimated and predicted using water level (WL) at Krishna, Khodshi, Bahe and Bhilawadi at different times. Table 6 displays the performances of the models developed in set 2. Table 6 shows that as compared to the estimation and predication of discharge at Irwin, the accuracy of water level estimation and

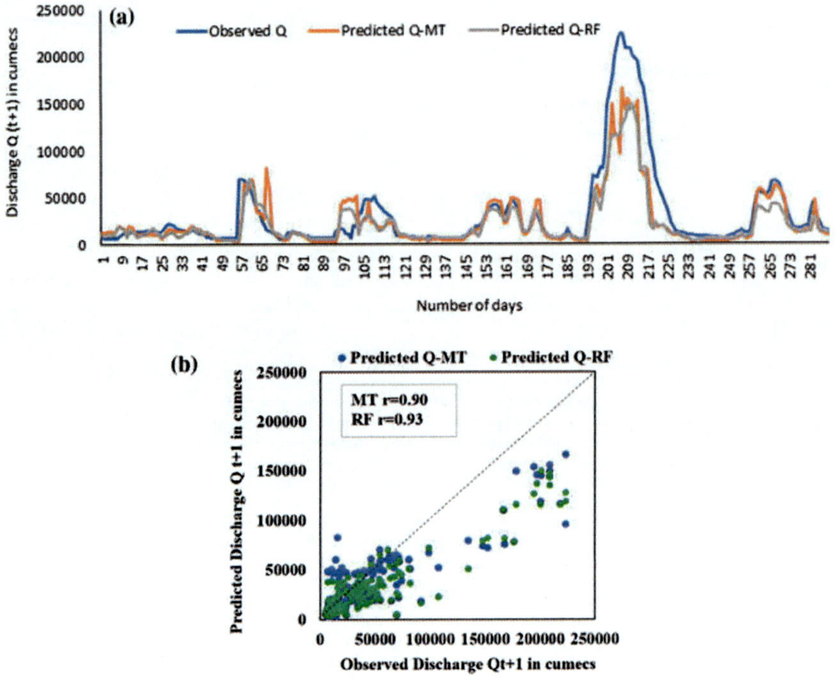

Fig. 3 **a** Observed and predicted values for model 4 using MT and RF, **b** Scatter plot for model 4

prediction decreases as compared to set 1 model, with reduction in R, increase in RMSE and MAE. The performance of model/s developed with RF to predict estimate water level with input as WL at Krishna and Khodshi is less as compared to the model 5 to 6 where WL at Bahe and Bhilawadi at t and t−1 for respective models are the input parameters.

Table 6 Performance of models in set 2: estimation and prediction of water level (WL)

Model No.	Output	Model tree				Random forest		
		r	RMSE	MAE	No. of rules	r	RMSE	MAE
Model 6	Irwin (t)	0.78	2.05	1.15	4	0.725	2.20	1.35
Model 7	Irwin (t)	0.73	2.24	1.24	8	0.701	2.39	1.33
Model 8	Irwin (t)	0.77	2.06	1.15	3	0.753	2.18	1.28
Model 9	Irwin ($t + 1$)	0.79	1.95	1.11	10	0.766	2.14	1.32
Model 10	Irwin (t)	0.82	2.55	0.99	5	0.817	1.74	0.93
Model 11	Irwin ($t + 1$)	0.80	2.86	1.07	3	0.805	1.79	0.95
Model 12	Irwin (t)	0.91	1.23	0.54	10	0.892	1.42	0.65
Model 13	Irwin ($t + 1$)	0.87	1.52	0.77	7	0.852	1.62	0.85

It is also evident from Fig. 1 that the large distance between the Krishna, Khodshi station and Irwin station has been instrumental in decrease of the accuracy of Models 6 to 9 and is also true for set 1 model with respective stations The number of equations(rules) developed through Model Tree for Models 6 to 9 are seen in Table 6. In all the four models, water level at Khodshi at t and $t-1$ is the important variable in estimating and predicting the output at Irwin t and Irwin $t + 1$, and the same trend is seen in both the techniques. Model 10 and Model 12 Estimate the water level at Irwin at time t and the models display a good performance with $r = 0.815$ (MT) and $r =$ 0.817(RF) and $r = 0.915$ (MT) and $r = 0.892$ (RF) followed by lower RMSE and MAE values. The proximity of Irwin station to Bhilwadi contributes in the increased performance of Model 12 over Model 10 with Bahe as input parameter. Model 12 consists of water level at Bhilwadi at $t-1$ and water level at Bhilwadi at t as input parameters. The model tree developed for Model 12 is as shown in Fig. 4a with the 10 equations developed in Fig. 4b.

Figure 4b shows that the slightly higher coefficients are being calculated for the parameter BH($t-1$) than BH(t). However, in RF the attribute performance as per Gini importance shows BH(t) as first influential parameter followed by BH($t-1$). Refer Fig. 4c. A sample of RF classifier is also shown in Fig. 4d. Figure 4e and f display the scatter plot for Model 12. Figure 4e shows a balanced scatter with no obvious under or over prediction. However, a slight underprediction in case of higher water levels by RF can be seen. The same trend can also be viewed in the hydrograph in Fig. 4f. The current study also highlights the fact that in case of absence of the water level data at the nearby station, the water level can be estimated through water level at Krishna bridge and Khodshi stations with satisfactory performance. Equations developed through MT are user-friendly and can be implemented practically. RF uses decision trees as base-learners, so that each tree is dependent on a set of random parameters. RFs can be employed for either a categorical dependent parameter (response variable) for classification purposes or a continuous response variable, for developing regressive models. RFs, similar to other Ensemble Machine learning algorithms combine multiple individual models to make predictions. Therefore, this method can be a better choice than a single decision tree since it decreases the over-fitting by averaging the result. Both these tools are advantageous not only from the performance perspective but also in the aspect that the variable /parameter importance is in tune with the knowledge which helps to build in more confidence in these data driven tools.

6 Inference and Takeaway

The present work judges the performance of models developed using MT and RF for estimation and forecasting of water levels and discharge at Irvin station on Krishna using the same on the upstream side. Set 1 which is dedicated towards estimating and predicting the discharge Q at Irwin at time t, $t + 1$ and $t + 2$ with input parameters as discharge of respective stations at t and $t-1$. Similarly in Set 2 water level was

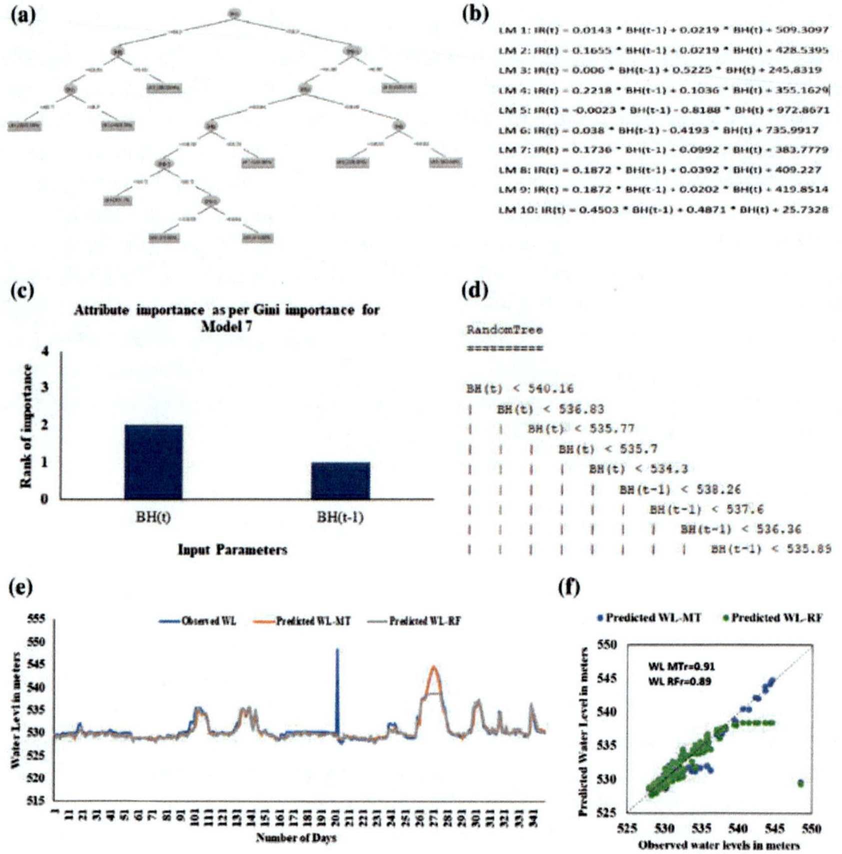

Fig. 4 **a** Model tree for model 12, **b** Equations developed for model 12, **c** Attribute importance for model 12, **d** Trees in RF for model 12, **e** Observed and predicted values for model 12, **f** Scatter plot for model 12

estimated and predicted with water level at previous stations at time t and $t-1$ respectively as input parameters. The study inferred that MT and RF can predict the discharge and water level with a reasonable accuracy ranging from $r = 0.80$ to 0.90 for Model Tree and $r = 0.88$ to 0.93 with Random Forest. Similarly in water level estimation and prediction the MT performance could be viewed from $r = 0.73$ to 0.91 and for RF as $r = 0.70$ to 0.90. In both the sets and for estimation and prediction of respective parameters, Random Forest wins the race in set 1 and shows a slight performance setback in set 2 models. The influential parameters seen in Model Tree through coefficients of parameters and through Gini importance in RF are in sync. Thus, these techniques can no more be labelled as grey boxes as they understand the basic underlying theory. In M5 Model tree the generated tree like structure of the linear model is reproducible and easy to understand for the decision makers. It makes it possible for a hydrologist to have a good overview of the relationships between the hydrological characteristics. RFs can be employed for either a categorical dependent parameter (response variable) for classification purposes or a continuous response variable, for developing regressive models. The advantage of the MT and RF is that the result is more understandable and allows one to build family of models of varying complexity and accuracy.

References

1. Quinlan JR (1992) Learning with continous classes. In: Adams & Sterling (eds) Proceedings AI"92, Singapore, World Scientific, pp 343–348
2. Londhe SN, Dixit PR (2012) Forecasting stream flow using support vector regression and M5 model tree. Int J Eng Res Developm 2(5):1–12
3. Bhattacharya B, Solomatine DP (2005) Neural networks and M5P model trees in modelling water level-discharge relationship. Neurocomputing 63:381–396
4. Solomatine DP (2004) Optimisation of hierarchical modular models and M5 trees. In: Proceedings of international joint conference on neural networks. Budapest, Hungary
5. Zia H, Harris H, Merrett H, Rivers M (2015) Predicting discharge using a low complexity machine learning model. Comput Electron Agr 372(118):350–360
6. Diez-Sierra J, del Jesus M (2019) Subdaily rainfall estimation through daily rainfall downscaling using random forests in Spain. Water 11:1–25
7. Muñoz P, Orellana-Alvear J, Willems P, Célleri R (2018) Flash-flood forecasting in an andean mountain catchment—development of a step-wise methodology based on the random forest algorithm. Water 10:15–19
8. Prieto C, Vine NL, Kavetski D, Garcia E, Medina R (2019) Flow prediction in ungauged catchments using probabilistic random forests regionalization and new statistical adequacy tests. Water Resour Res 55:4364–4392
9. Tyralis H, Georgia P, Langousis A (2019) A brief review of random forests for water scientists and practitioners and their recent history in water resources. Water 11(5):1–37
10. Witten IH, Frank E (2000) Data mining: practical machine learning tools and techniques with java implementations. Morgan Kaufmann, Los Altos
11. Breiman L (2001) Random forests. Machine Learn 45(1):5–32
12. http://www.cwc.gov.in/sites/default/files/admin/Krishna-kgbo-map.pdf
13. Google maps
14. https://waikato.github.io/weka-wiki/downloading_weka/
15. Legates DR, McCabe GJ (2010) Evaluating the use of goodness-of-fit measures in hydrologic and hydroclimatic model validation. Water Resour Res 35(1):233–241

Multi-step Ahead Forecasting of Streamflow Using Deep Learning-Based LSTM Approach

Mohd Imran Khan and Rajib Maity

Abstract With the availability of enormous datasets and computational resources in the present decade and effectiveness of artificial intelligence (AI)-based machine learning (ML)/ deep learning (DL) in understanding nonlinear hidden associations among meteorological precursors, it is worthwhile to explore the streamflow forecasting performance using the same. Among several meteorological variables, forecasting of streamflow, one of the major components in the hydrologic cycle, at basin scale is a vital task for sustainable management of water resources. The present study aims to develop a DL-based long short-term memory (LSTM) neural network approach to forecast multi-step ahead streamflow (1- to 3-month(s) ahead). The developed DL model is applied to a medium sized rainfed river basin, namely Bhadra basin, a part of Tungabhadra River basin located in Karnataka, India. The input dataset comprises of previous three-month lagged values of six meteorological precursors, namely precipitation, temperature, soil moisture, evaporation, relative humidity, streamflow and a month index (i.e. Jan = 1, Feb = 2 and so on), which are fed as inputs to the model. The results obtained from the developed model are assessed using three statistical measures, i.e., correlation coefficient (r), root mean square error (RMSE) and Nash–Sutcliffe Efficiency (NSE) and are found to be promising in capturing the hidden complex associateship between streamflow and meteorological variables. Moreover, the performance of the proposed approach is also compared with three other popular regression approaches viz. multi-layer perceptron (MLP), support vector regression (SVR) and multiple linear regression (MLR). It is noticed that the LSTM exhibits the best performance at all leads (1- to 3-month(s)). The benefit can be attributed to the memory cell structure of LSTM which can store previous time steps information.

Keywords Deep learning · Data-driven approach · Long short-term memory (LSTM) neural network · Streamflow forecasting · Model evaluation

M. I. Khan · R. Maity (✉)
Department of Civil Engineering, Indian Institute of Technology, Kharagpur 721302, India
e-mail: rajib@civil.iitkgp.ac.in

M. I. Khan
e-mail: imran.khan@iitkgp.ac.in

P. V. Timbadiya et al. (eds.), *Geospatial and Soft Computing Techniques*, Lecture Notes in Civil Engineering 339, https://doi.org/10.1007/978-981-99-1901-7_32

1 Introduction

Forecasting streamflow at various spatio-temporal scales is very crucial in planning, operation and management of droughts, floods and numerous water resources systems [1, 2]. In many cases, the streamflow forecast of a river basin is carried out through physically based and conceptual models [2]. However, since the past decade, there has been an unprecedented rise in the application of artificial intelligence (AI) and machine learning (ML) approaches that is found to be an effective alternative over many other approaches [3]. The AI/ML and one of its relatively new subsets, namely deep learning (DL), have a long history in quantitative and qualitative modelling of different water resource elements [2–5]. However, with the increase in computational resources and availability of vast amount of data the subset of AI/ML, i.e., DL has shown benchmark performances in various fields, such as image processing, healthcare, natural language processing and speech recognition, and has gained the attention of researchers to analyze the its potential in different other domains [6–11].

In the domain of hydroclimatology also, the DL has been proven successful to forecast different hydroclimatic events, such as rainfall, streamflow, soil moisture, drought, and many more at different spatio-temporal scales in different parts of the world. For instance, [12] used the potential of one-dimensional convolutional neural network (Conv1D) and multi-layer perceptron (MLP) in predicting multi-step ahead rainfall. The authors also compared the performance with other ML approaches and it was found better. Fu et al. [13] showed the intelligence of LSTM in forecasting streamflow at daily scale. The analysis was carried out over the Kelantan River of Malaysian Peninsula, and the performance was found better than the older ANN approach. Oh et al. [14] exploited the potential of deep neural network (DNN), a DL-based forward propagation technique, to capture the intensity and characteristics of the urban heat (both spatial and temporal) by deploying two separate models. The authors also studied the effect of urban heat on outdoor workers, thus examining productivity at a daily scale. Kaur and Sood [15] also deployed the DNN algorithm to assess the severity level and predict drought accurately at different time frames. The authors established the effectiveness of DNN in capturing drought for three different climatic blocks and comparing the result with optimized hybrid approach of ANN and genetic algorithm (GA) and ANN only. Maity et al. [16] developed a Conv1D model to assess the drought condition of a basin by utilizing several hydrometeorological variables as an input to the model. The performance of the DL-based model was found better than SVR and MLP approaches. Gauch et al. [17] demonstrated the effectiveness of long short-term memory (LSTM) approach in predicting the streamflow of ungauged basins. The developed model was first calibrated on datasets of different basins, and thereafter, its accuracy was tested on several ungauged basins of similar climatic conditions. Fang et al. [18] performed an experiment to model soil moisture and streamflow using a data synergy approach. They demonstrated the effectiveness of DL in training the model on big dataset that was achieved by merging different datasets of smaller regions, also known as the process of unification. Thus, a single DL model (LSTM) was able to learn the common features of the different subregions.

Therefore, from the aforementioned literature, it can be concluded that due to effective performance of DL approaches, they are recurrently applied in many hydro-climatological studies to improve the forecasting/simulation accuracy of different hydrological variables. However, there is a long way to go to fully utilize the huge potential of DL approaches. The present study aims to forecast multi-step ahead (1- to 3-month(s) ahead) streamflow by utilizing the efficacy of DL-based LSTM model viz. capturing nonlinear relationship between meteorological precursors and streamflow, at monthly scale. The inputs fed to the model are previous 3 month lagged values of 6 different meteorological variables and a month index (i.e., Jan = 1, Feb = 2, …, Dec = 12). Furthermore, the performance of the developed DL model is also compared with three popular regression approaches, namely support vector regression (SVR), MLP, and multiple linear regression (MLR) approaches by means of three statistical metrics, i.e., correlation coefficient (r), root mean square error (RMSE) and Nash–Sutcliffe efficiency (NSE).

2 Study Area and Data Source

2.1 *Bhadra River Basin*

A portion of Tungabhadra River basin, i.e., up to Bhadra dam, known as Bhadra River basin (BRB), situated in Karnataka, India, is considered as the study area (Fig. 1). The selected basin is of medium size and has no major man-made structure. According to the Koppen classification system, BRB falls under the tropical wet climate region and therefore, it experiences a high rainfall although it is seasonal in nature. The basin receives an annual rainfall of about 2300 mm, most of which occurs during the monsoon months i.e., from July to September and least during the dry months, i.e., January to May. However, the annual variation in the range of temperature is very less as compared to the rainfall. Generally, the daily average temperature over the basin varies in a very narrow range, i.e., between 3 and 4 °C. Bhadra river is the major river in BRB which originates near Gangamoola, located in western ghat, southern region of India, and flows up to the considered dam, i.e., Bhadra dam. Moreover, due to tropical climate, most of the basin area is covered with dense trees and bushes, and has an undulated ground surface and steep slope. The average slope of BRB is approximately 6% and has an elevation of about 1200 m, with reference as mean sea level.

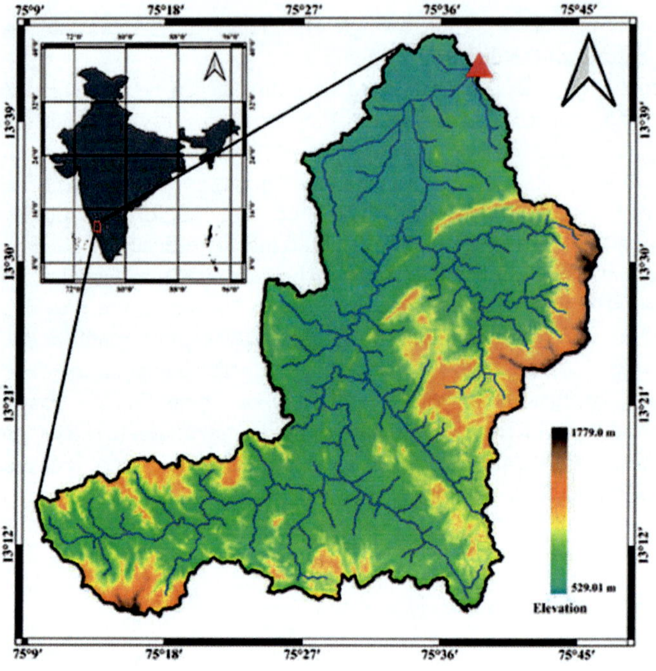

Fig. 1 Study area map showing the outlet as a red triangle where the streamflow is measured

2.2 Data Used

The meteorological variables used in the model development are previous 3-months values of precipitation, air temperature, relative humidity, soil moisture, evaporation and streamflow (Table 1). Out of the following variables, streamflow data is procured from the gauging station site of Karnataka Niravari Nigam Limited (KNNL) located near Bhadra dam and rest of the variables are obtained from ERA5 reanalysis dataset at monthly scale. Precipitation, air temperature, soil moisture, and evaporation are available at 0.1° latitude × 0.1° longitude and relative humidity at 0.25° latitude × 0.25° longitude resolutions https://cds.climate.copernicus.eu/cdsapp#!/dataset/reanalysis-era5-single-levels-monthly-means?tab=form, accessed in September, 2021). The period of dataset is considered from January 1981 to June 2015, based on the availability of the streamflow data.

Table 1 Details of the dataset used in the present analysis

Dataset (1981–2015)	Variables	Spatial resolution	Vertical/ pressure level	Units
ERA5 hourly data on single levels	Soil moisture	0.25° × 0.25°	Surface	W/m^2
	Evaporation	0.25° × 0.25°	Surface	J/m^2
	Rainfall	0.25° × 0.25°	Surface	mm
	Air temperature	0.1° × 0.1°	Surface	°C
ERA5 hourly data on pressure levels	Relative humidity	0.25° × 0.25°	1000 hPa	%
Gauging station data	Streamflow	–	–	m/s

3 Methodology

A schematic presentation of the proposed DL-based LSTM architecture is shown in Fig. 2. The entire process is performed in spyder notebook, an integrated developed environment written in Python, using Keras library, built on the top of tensorflow. The finalized architecture is tuned and optimized using grid search fivefold cross-validation of scikit-learn library.

3.1 *Data Preprocessing*

The process starts with averaging monthly reanalysis values of ERA5 data across the BRB and dividing the dataset into two parts, i.e., training and testing dataset. The training dataset consists of 80% of the total dataset, and the remaining 20% is considered as the testing dataset. Thereafter, the input datasets comprising of

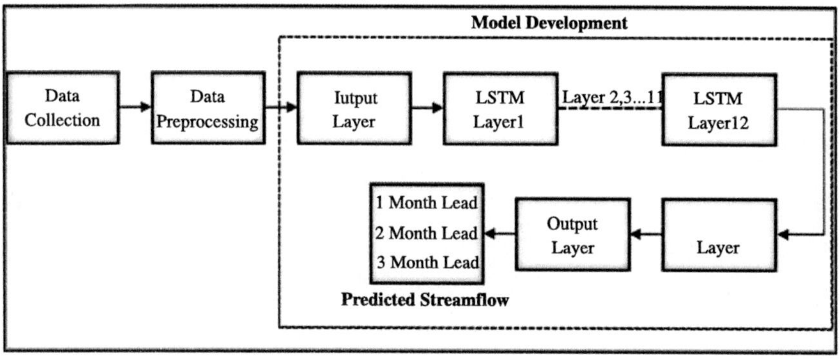

Fig. 2 Schematic presentation of the proposed DL based LSTM methodology

six meteorological values and month index are converted to standard anomalies, excluding the month index. The standard anomaly of the dataset is obtained by subtracting each value of the time series from its respective monthly average value and dividing it by corresponding monthly standard deviation obtained from the training dataset.

3.2 Deep Learning Model

The LSTM, a DL-based time series prediction model, also known as sequential model, is proposed in this study to learn the hidden associations between meteorological forcings and streamflow. It was first proposed by Hochreiter and Schmidhuber in 1996 to overcome the problem of decaying weights during the back propagation of recurrent neural networks (RNN) [19].

A typical structure of LSTM-based DL model generally comprises an input layer, hidden layers (LSTM layers and dropout layer) and an output layer (Dense layer) whose function are as follows: The input layer is responsible for feeding the data, arranged in a three-dimensional form (i.e., number of samples or batch size, time steps, number of input features), to the memory cell of the LSTM layer. The memory cells are the most important component of the LSTM network. They are capable of memorizing the time sequence information of the hydrometeorological forcings over a period of time. Also, the information stored within the memory cells is updated at each time step $\left(\tilde{C}_t\right)$ with the help of three gates viz. forget gate (f_t), input gate (i_t) and an output gate (o_t). The process of updating cell state at each time step can be illustrated with the help of the following six equations:

$$i_t = \sigma\left(W_i.\left[h_{t-1},\ X_t\right] + b_i\right) \tag{1}$$

$$\tilde{C}_t = \tanh\left(W_c\left[h_{t-1}, X_t\right] + b_c\right). \tag{2}$$

$$f_t = \sigma\left(W_f.\left[h_{t-1},\ X_t\right] + b_f\right) \tag{3}$$

$$C_t = i_t * \tilde{C}_t + f_t * C_{t-1} \tag{4}$$

$$o_t = \sigma\left(W_0\left[h_{t-1},\ X_t\right] + b_0\right) \tag{5}$$

$$h_t = o_t * \tanh(C_t) \tag{6}$$

where t represents time step, X_t is the input given to the memory cell at time t, h_{t-1} is the output of the previous cell state, and $W_{i,}\ W_f,\ W_c,\ W_0,\ b_i, b_f, b_c, b_0$ are the corresponding weights and biases. $\tilde{C}_t$ and C_{t-1} are the updated and previous cell state information. h_t is the output of the cell state at any time step t. This process occurs in all the LSTM layers which are connected to the model (if any). Next, a dropout layer (if required) is added either at the end of the LSTM layers or in between depending on the suitability of the model structure achieved during the tuning process. Dropout layer helps in removing the redundant information from the model (if any), thereby reduces the complexity and chances of underfitting/overfitting of the model [20]. At the end, a dense layer, an output layer comprising of a neuron and an activation function, is added to obtain the final result, i.e., monthly streamflow in this study. A more detail about the interpretation of the LSTM network in the field of hydroclimatology can be found in [21].

3.3 Model Evaluation

The performance of the proposed model is cross validated with the help of five-fold cross validation and assessed through r, RMSE and NSE values and their expressions are shown in Eqs. (7), (8) and (9), respectively, as follows:

$$r = \frac{\sum_{t=1}^{n}(Y_t - \overline{Y})(Y_t' - \overline{Y'})}{\sqrt{\sum_{t=1}^{n}(Y_t - \overline{Y})^2 \sum_{t=1}^{n}(Y_t' - \overline{Y'})^2}} \tag{7}$$

$$RMSE = \sqrt{\frac{\sum_{t=1}^{n}(Y_t - Y_t')^2}{n}} \tag{8}$$

$$NSE = 1 - \frac{\sum_{t=1}^{n}(Y_t - Y_t')^2}{\sum_{t=1}^{n}(Y_t - \overline{Y})^2} \tag{9}$$

where Y_t and Y_t' are the observed and the predicted values at any time step t, $\overline{Y}$ and $\overline{Y'}$ are the mean of the observed and the predicted values.

3.4 Comparison with Other Models

Performance of the proposed DL-based LSTM model is compared with three other popular models, namely MLP, SVR and MLR models through the same statistical metrics mentioned before. The models used for comparison are trained, validated, tested and proportions of the training and testing datasets as used in the DL based LSTM model. The detailed methodology of these models can be found in [22–24].

4 Results and Discussions

The finalized architecture of the DL-based LSTM model, shown in Fig. 2, is used for predicting monthly streamflow with a lead time of 1- to 3-month(s) in advance. It can be observed from the figure that DL model comprises of twelve LSTM layers, one dropout layer, and one fully connected output layer which is obtained after an exhaustive grid search optimization. The optimized/tuned parameters in various layers are as follows:

(a) Each LSTM layer comprises of 20 memory cells, he normal kernel initializer, and leaky rectified linear unit (Leaky ReLU) with $\alpha = 0.3$.
(b) Dropout rate of 10% in the dropout layer and
(c) Three neurons along with same activation function as that of LSTM layer in the output layer. These three neurons give the monthly predicted streamflow values at 1, 2 and 3-month lead, respectively.

Having configured the layers with the aforesaid parameters, model is compiled using a batch size of 48, mean squared error loss function, adaptive learning optimization (Adam) technique with learning rate 1×10–4 and 1000 number of epochs. The trained model is also validated using a fivefold cross-validation strategy and thereafter it is used for predicting monthly streamflow.

Table 2 illustrates the three metrics r, RMSE and NSE values across all the five folds obtained from the proposed DL based LSTM model and other three models at all three months lead time. It is observed that during testing period, performance on fold1 of the LSTM model: r, RMSE and NSE ranges between 0.89 to 0.99, 14.64 to 58.13 and 0.80 to 0.98, respectively. This is much better than MLP, SVR and MLR model performances obtained considering the same testing period. A similar observation is also found in case of 2 and 3-month lead performance as reflected through the performance metrics.

Table 2 also contains the average values (across all 5 folds) of the performance metrics of the models. The average prediction efficiency (NSE) during the testing period of the proposed model is found greater than 90% for all the three lead times, whereas it ranges between 81–82, 76–78 and 75–76% for MLP, SVR, and MLR, respectively. Similarly, the lower range of RMSE values, i.e., 33–37 and higher range of r values, i.e., 0.95–0.96 are obtained as compared to MLP (RMSE: 49–52 and r: 0.91–92), SVR (RMSE:55–58 and r: 0.88–0.89), and MLR (RMSE: 56–58 and r: 0.88–0.89). This proves the superiority of the LSTM model in capturing the better relationship between the meteorological forcings and streamflow variation for all three lead times, i.e., 1-month, 2-month and 3-month.

Scatter plots are drawn between the observed and the predicted stream flow at all the lead times to demonstrate the effectiveness of LSTM in capturing monthly variation of streamflow. However, a typical plot illustrating 1-month-ahead prediction performance of the proposed model along with performances of all other three models used for comparison is shown here (Fig. 3). The best performance of the DL-based LSTM model is reflected from the least deviation of the best fit line from the 45° line

Table 2 Performance metrics obtained for 1- to 3-month(s) ahead streamflow forecasting from the proposed DL-based LSTM model and other three models used for comparison

Folds	Models	1 Month lead		2 Month lead		3 Month Lead	
		Training	Testing	Training	Testing	Training	Testing
Fold1	LSTM	0.98	0.89	0.97	0.89	0.97	0.91
		24.51	58.13	26.19	59.08	30.21	55.87
		0.96	0.80	0.95	0.79	0.94	0.82
	MLP	0.94	0.85	0.94	0.87	0.95	0.85
		39.49	69.47	40.27	65.12	38.21	69.57
		0.89	0.71	0.88	0.75	0.89	0.71
	SVR	0.92	0.83	0.91	0.85	0.91	0.86
		45.66	74.74	48.10	71.44	48.22	68.18
		0.85	0.67	0.83	0.70	0.83	0.72
	MLR	0.92	0.84	0.91	0.87	0.91	0.87
		45.98	70.45	47.91	66.89	48.24	67.56
		0.84	0.71	0.83	0.74	0.83	0.73
Fold2	LSTM	0.98	0.97	0.98	0.97	0.96	0.91
		24.88	30.38	26.04	32.02	35.58	54.09
		0.96	0.95	0.95	0.94	0.91	0.83
	MLP	0.93	0.90	0.94	0.92	0.94	0.92
		44.76	58.01	41.91	52.87	40.34	51.46
		0.86	0.80	0.88	0.83	0.89	0.84
	SVR	0.90	0.85	0.90	0.87	0.90	0.89
		52.12	70.58	52.98	66.97	52.48	61.74
		0.82	0.71	0.81	0.74	0.81	0.77
	MLR	0.91	0.86	0.90	0.88	0.90	0.89
		51.76	67.34	51.96	62.46	52.46	61.46
		0.82	0.73	0.82	0.77	0.81	0.78
Fold3	LSTM	0.98	0.99	0.96	0.99	0.96	0.99
		22.64	14.64	34.68	15.01	32.81	11.95
		0.97	0.98	0.92	0.98	0.93	0.99
	MLP	0.93	0.95	0.93	0.93	0.94	0.93
		46.23	33.33	44.06	37.89	41.84	37.57
		0.85	0.89	0.87	0.86	0.88	0.86
	SVR	0.90	0.95	0.90	0.91	0.89	0.91
		54.18	32.39	54.91	41.46	54.61	40.71
		0.80	0.90	0.79	0.83	0.80	0.83
	MLR	0.90	0.93	0.90	0.90	0.89	0.89
		53.31	38.01	53.71	46.21	54.19	45.21
		0.81	0.86	0.80	0.79	0.80	0.79
Fold4	LSTM	0.96	0.94	0.96	0.95	0.96	0.96
		32.84	46.15	33.86	43.17	32.40	35.43
		0.92	0.87	0.92	0.89	0.92	0.93
	MLP	0.93	0.89	0.93	0.91	0.93	0.92
		44.85	60.61	42.55	55.59	42.94	53.14
		0.85	0.78	0.87	0.82	0.87	0.83

(continued)

Table 2 (continued)

Folds	Models	1 Month lead		2 Month lead		3 Month Lead	
		Training	Testing	Training	Testing	Training	Testing
	SVR	0.90 50.63 0.81	0.83 72.97 0.69	0.91 49.66 0.82	0.86 68.49 0.72	0.91 49.31 0.82	0.87 65.38 0.75
	MLR	0.90 50.60 0.81	0.85 69.23 0.72	0.90 50.07 0.82	0.88 64.23 0.76	0.90 50.54 0.81	0.88 63.56 0.76
Fold5	LSTM	0.97 30.25 0.94	0.97 22.93 0.94	0.98 25.43 0.95	0.98 18.57 0.96	0.97 30.71 0.93	0.97 24.32 0.93
	MLP	0.93 45.61 0.85	0.95 36.22 0.86	0.93 44.14 0.86	0.94 35.98 0.86	0.95 38.67 0.89	0.96 35.63 0.86
	SVR	0.89 54.88 0.79	0.94 37.54 0.85	0.90 54.13 0.79	0.92 39.57 0.83	0.91 49.87 0.82	0.91 39.28 0.83
	MLR	0.90 52.12 0.81	0.93 44.38 0.79	0.90 52.10 0.81	0.90 50.54 0.72	0.90 50.79 0.82	0.91 46.25 0.76
Average	LSTM	0.97 27.02 0.95	0.95 34.45 0.91	0.97 29.24 0.94	0.96 33.57 0.91	0.96 32.34 0.93	0.95 36.33 0.90
	MLP	0.93 44.19 0.86	0.91 51.53 0.81	0.94 42.59 0.87	0.91 49.49 0.82	0.94 40.40 0.88	0.92 49.47 0.82
	SVR	0.90 51.49 0.81	0.88 57.64 0.76	0.90 51.96 0.81	0.88 57.59 0.76	0.91 50.90 0.82	0.89 55.06 0.78
	MLR	0.91 50.76 0.82	0.88 57.88 0.76	0.90 51.15 0.82	0.89 58.06 0.75	0.90 51.24 0.82	0.89 56.81 0.76

The three values shown in each cell represent (top to bottom) the r, RMSE, and NSE values, respectively

as compared to the other three models. Moreover, the higher monthly magnitude of streamflow is better captured by the proposed model. Thus, the better efficacy of the DL-based LSTM model is also true for all streamflow regimes.

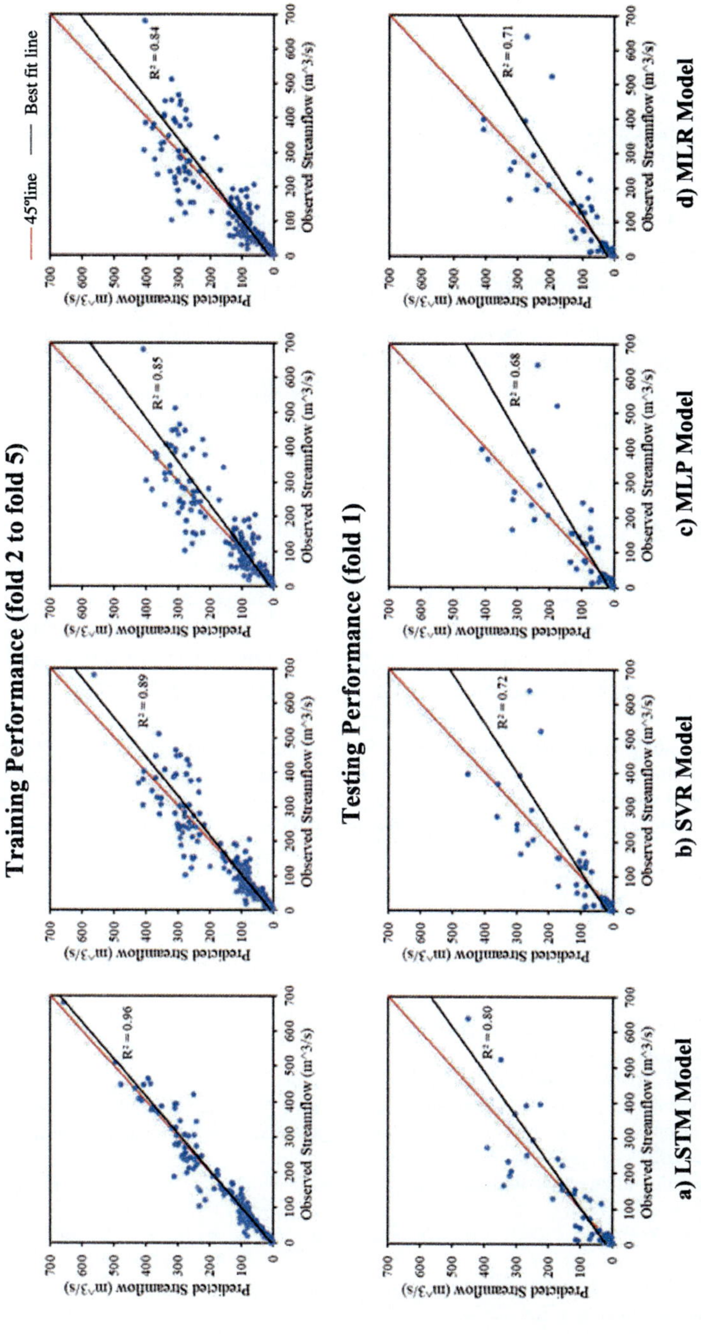

Fig. 3 Scatter plots obtained between observed and actual values of monthly streamflow at 1-month lead from the proposed LSTM model as well as from other three models used for comparison

5 Conclusions

This study portrays the effectiveness of a DL-based LSTM model in capturing the complex relationship between hydrometeorological precursors and monthly streamflow variation with 1- to 3-month(s) lead time for Bhadra river basin, located in Karnataka, India. The performance of the model is assessed using r, RMSE and NSE values for all the three leads. A comparison of the proposed model is also carried out with three other popular approaches, namely MLP, SVR and MLR. Analysis reveals that a fairly high correlation value, least standard deviation of the residuals and high prediction efficiency is obtained in case of the DL-based LSTM model as compared to the other three models. Specific conclusions are as follows:

- Overall, a better correspondence/agreement is established between the observed and predicted values of monthly streamflow.
- It is able to identify and learn the hidden complex association between meteorological forcings and streamflow with high accuracy at monthly scale.
- The DL-based LSTM is better even in capturing the higher magnitude of stream flow which occurs during the monsoon period.

The DL-based LSTM model is expected to be useful in planning and management of different water resource problems, such as irrigation, hydropower generation, flood and droughts. Moreover, the proposed model creates a scope to be applied to other Indian basins, and in this era of data deluge, robustness of the DL-based LST models can be explored to be utilized for forecasting the streamflow in ungauged basins with similar climate characteristics. This is kept as the future scope of this study.

Acknowledgements The authors acknowledge the National Supercomputing Mission (NSM) for providing computing resources of 'PARAM Shakti' at IIT Kharagpur, which is implemented by C-DAC and supported by the Ministry of Electronics and Information Technology (MeitY) and Department of Science and Technology (DST), Government of India.

References

1. Partal T, Kisi O (2007) Wavelet and neuro-fuzzy conjunction model for precipitation forecasting. J Hydrol 342:199–212
2. Muhammad R, Yuan X, Kisi O, Yuan Y (2017) Streamflow forecasting using artificial neural network and support vector machine models. Am Sci Res J Eng Technol Sci 29:286–294
3. Rolnick D, Donti PL, Kaack LH, et al (2019) Tackling climate change with machine learning. arXiv Prepr arXiv190605433
4. ASCE Task Committee (2000) Artificial neural networks in hydrology. I: preliminary concepts. J Hydrol Eng 5:115–123
5. ASCE Task Committee (2000) Artificial neural networks in hydrology. II: hydrologic applications. J Hydrol Eng 5:124–137
6. Collobert R, Weston J, Bottou L et al (2011) Natural language processing (Almost) from scratch. J ofMachine Learn Res 12:2493–2537

7. Ker J, Wang L, Rao J, Lim T (2017) Deep learning applications in medical image analysis. IEEE Access 6:9375–9379
8. Krizhevsky A, Sutskever I, Hinton GE (2012) Imagenet classification with deep convolutional neural networks. In: Proceedings of the 25th international conference on neural information processing systems. pp 1097–1105
9. Liu X, Faes L, Kale AU et al (2019) A comparison of deep learning performance against health-care professionals in detecting diseases from medical imaging: a systematic review and meta-analysis. Lancet Digit Heal 1:e271–e297
10. Khan MI, Maity R (2022) Hybrid deep learning approach for multi-step-ahead prediction for daily maximum temperature and heatwaves. Theor Appl Climatol 149(3-4):945–963. https://doi.org/10.1007/s00704-022-04103-7
11. Khan MI, Sarkar S, Maity R (2023) Artificial intelligence/machine learning techniques in hydroclimatology: a demonstration of deep learning for future assessment of stream flow under climate change. In: Visualization techniques for climate change with machine learning and artificial intelligence, Elsevier Ltd., pp. 247–273. https://doi.org/10.1016/b978-0-323-99714-0.00015-7
12. Khan MI, Maity R (2020) Hybrid deep learning approach for multi-step-ahead daily rainfall prediction using GCM simulations. IEEE Access 8:52774–52784
13. Fu M, Fan T, Ding Z et al (2020) Deep learning data-intelligence model based on adjusted forecasting window scale: application in daily streamflow simulation. IEEE Access 8:32632–32651
14. Oh JW, Ngarambe J, Duhirwe PN et al (2020) Using deep-learning to forecast the magnitude and characteristics of urban heat island in Seoul Korea. Sci Rep 10:1–13
15. Kaur A, Sood SK (2020) Deep learning based drought assessment and prediction framework. Ecol Inform 57:101067
16. Maity R, Khan MI, Sarkar S, et al (2021) Potential of deep learning in drought assessment by extracting information from hydrometeorological precursors. J Water Clim Chang
17. Gauch M, Mai J, Lin J (2021) The proper care and feeding of CAMELS: how limited training data affects streamflow prediction. Environ Model Softw 135:104926
18. Fang K, Kifer D, Lawson K, et al (2021) The data synergy effects of time-series deep learning models in hydrology. arXiv Prepr arXiv210101876
19. Hochreiter S, Schmidhuber J (1997) Long short-term memory. Neural Comput 9:1735–1780
20. Srivastava N, Hinton G, Krizhevsky A et al (2014) Dropout: a simple way to prevent neural networks from overfitting. J Mach Learn Res 15:1929–1958
21. Kratzert F, Klotz D, Brenner C et al (2018) Rainfall—runoff modelling using long short-term memory (LSTM) networks. Hydrol Earth Syst Sci 22:6005–6022
22. Haidar A, Verma B (2018) Monthly rainfall forecasting using one-dimensional deep convolutional neural network. IEEE Access 6:69053–69063
23. Maity R, Bhagwat PP, Bhatnagar A (2010) Potential of support vector regression for prediction of monthly streamflow using endogenous property. Hydrol Process 24:917–923
24. Rasouli K, Hsieh WW, Cannon AJ (2012) Daily streamflow forecasting by machine learning methods with weather and climate inputs. J Hydrol 414–415:284–293

Analysis of Water Resources of Bisalpur Dam Using Time Series Forecasting Models

Shraddha Laxmi and Rohit Goyal

Abstract Water is the source of all life and plays a very essential role in human life and the wellbeing of our nation. Continuous supply of water whether it is for drinking, agriculture and urban use, hydroelectric power, and wildlife management has become one of the greatest challenges of the twenty-first century. Therefore, a technical solution is required for water resource management. The time series forecasting model is one of the ways that have cost-effective procedures for analyzing and forecasting data as well as a reliable method when it comes to water resources management. In this paper, the best fit time series forecasting model is identified using ARIMA for inflow to the reservoir, outflow, rainfall amount, drinking water requirement amount, evaporation amount, irrigation water demand, and water storage of Bisalpur dam, Tonk district, Rajasthan. Monthly data from 2011 to 2019 of these variables are used. Using the Ljung-box method for identification of best fit model is done, together with auto correlation function plot (ACF) and partial auto correlation function plot (PACF). Several models AR, MA, and ARIMA were used in modeling the data in R-studio using R-script. The best fitted seasonal ARIMA model selected for seven parameters and forecasting for the next nine years, i.e., 2020 to 2028, is done.

Keywords ARIMA · Time series analysis · Seasonal ARIMA · ACF · PACF

Disclaimer: The presentation of material and details in maps used in this chapter does not imply the expression of any opinion whatsoever on the part of the Publisher or Author concerning the legal status of any country, area or territory or of its authorities, or concerning the delimitation of its borders. The depiction and use of boundaries, geographic names, and related data shown on maps and included in lists, tables, documents, and databases in this chapter are not warranted to be error free nor do they necessarily imply official endorsement or acceptance by the Publisher or Author.

S. Laxmi · R. Goyal (✉)
Department of Civil Engineering, Malviya National Institute of Technology Jaipur, Jaipur 302017, India
e-mail: rgoyal.ce@mnit.ac.in

P. V. Timbadiya et al. (eds.), *Geospatial and Soft Computing Techniques*, Lecture Notes in Civil Engineering 339, https://doi.org/10.1007/978-981-99-1901-7_33

1 Introduction

The water resource system plays a crucial role and benefited both people's lives and economies for centuries. Many regions of the world are not able to meet even the basic need of sanitation and drinking water demand due to extreme weather events, too polluted water resources, scarcity of water resources, and improper management of reservoirs [9]. Goyal and Arora [6] developed a ground water model to determine the response of aquifer against scenarios used in this study. The result indicates that around 4139.2 ha waterlogged area could be reduced when one of the proposed scenarios is followed. Similarly, modeling has also been applied to other environmental issues. Dadhich et al. [4] performed geostatistical analysis of air quality of Jaipur city, using spatial temporal variation of pollutants SO_2, NO_x, SPM and PM_{10}. The result shows particulate pollutant (PM_{10}) and suspended particulate pollutant (SPM); the seasonal average concentration was higher than the prescribed limits by Central Pollution Control Board (CPCB). Studies have shown that mapping the future trend of any parameter helps experts in better decision-making. Remote sensing together with ArcGIS can be very useful when it comes to land use and land cover. Rajput et al. [11] mapped groundwater quality using to show that the traces of high concentration of metal like Cr, Ni, and Pb are present around different water bodies of Bhiwadi industrial area.

These issues can be resolved using a technical solution, planning, and creating awareness in society. Modeling is one of the solution that offers the analysis of even the complex real-life situation in form of models with more precision in less time than obtained by manual calculation. In this research, analysis of seven different parameters contributing to the water resource system of Bisalpur dam, Tonk district, Rajasthan is analyzed using ARIMA models and forecasting time series of the parameter is done.

Water resource management is a complex problem that involves many variables for analyzing a single variable. These increases cost as well as the time for analysis of variables. The time series forecasting model is one of the ways that have cost-effective procedures for analyzing and forecasting data as well as a reliable method when it comes to water resources management. These models are simple as compared to other methods of forecasting. Time series forecasting models are reasonable as future values are based on analysis of historical data and assume that past patterns in data can be used to forecast future data points [1]. Time series can be continuous or discrete and can be single variate or multivariate. In a continuous-time series, observations are measured at every instance of time, whereas a discrete-time series contains observations measured at discrete points of time [7]. In this way, time series analysis can replicate real-life complex problem of water resource management in from of model.

Researches have shown the different combination of ARIMA and ARMA models have been extensively used in the field of hydrology [2, 3] to get accuracy in forecasting. Wang [13] used an improved ARIMA model which accounts for both inter-annual and inter-monthly variation of any given series for precipitation forecasting.

Analysis was done by using Box-Cox transformation analysis to stabilize the variance of the monthly precipitation data series whereas the use of ACF and PACF plot was done for generating the ARIMA model. According to Machekposhti [10] collected annual mean flow data from 1958 to 2005, then the time series analysis was done by using ARIMA (p, d, q) method also known as the Box-Jenkins model. For comparing different models, Akaike information criterion (AIC) and conditional least square (CLS) parameters were used. Different evaluation techniques can be applied for ARIMA model validation.

In this paper, augmented Dickey-Fuller test (ADF) was used for checking stationarity in data and the Ljung-box with ACF and PACF plot was used for evaluating the presence or absence of serial autocorrelation. Time series analysis is done using nine years monthly data. Seven parameters (variables), i.e., inflow, outflow, evaporation amount, drinking water requirement, irrigation water demand, rainfall amount, and capacity, of Bisalpur dam are used for the analysis with the help of RStudio software. Using seasonal ARIMA model future values of the variables are plotted.

2 Study Area and Data Source

Bisalpur dam, Tonk district, Rajasthan was completed in 1999 on Banas River, constructed to overcome the shortage of water in central-eastern Rajasthan. The height of the Bisalpur dam is 38.50 M and the length is 574 m. The state has extreme climatic and geographical conditions, 2/3rd part of the state is a part of the great Thar Desert thus state faces water scarcity. So, the management of water resources became a crucial task [5] (Fig. 1).

Banas basin location:
Latitude: 24°015′0″ to 27°020′0″ N.
Longitude: 73,025′0″ to 7700′0″ E.
Bisalpur dam location:
Latitude: 25°055′20" N.
Longitude: 75°027′30" E.

2.1 *Methodology*

ARIMA models attempt to identify patterns in the historical data. It aims to identify the process that is generating and influencing the historical pattern hence called the data generating process. ARIMA model is obtained by combining autoregressive (AR) and moving average (MA) models [12]. This model has been widely applied and tested for different types of time series. The 'I' or integrated component defines a trend or other "integrative" process in the data. The AR and MA components have an associated model order indicating how the current value of the data is affected by

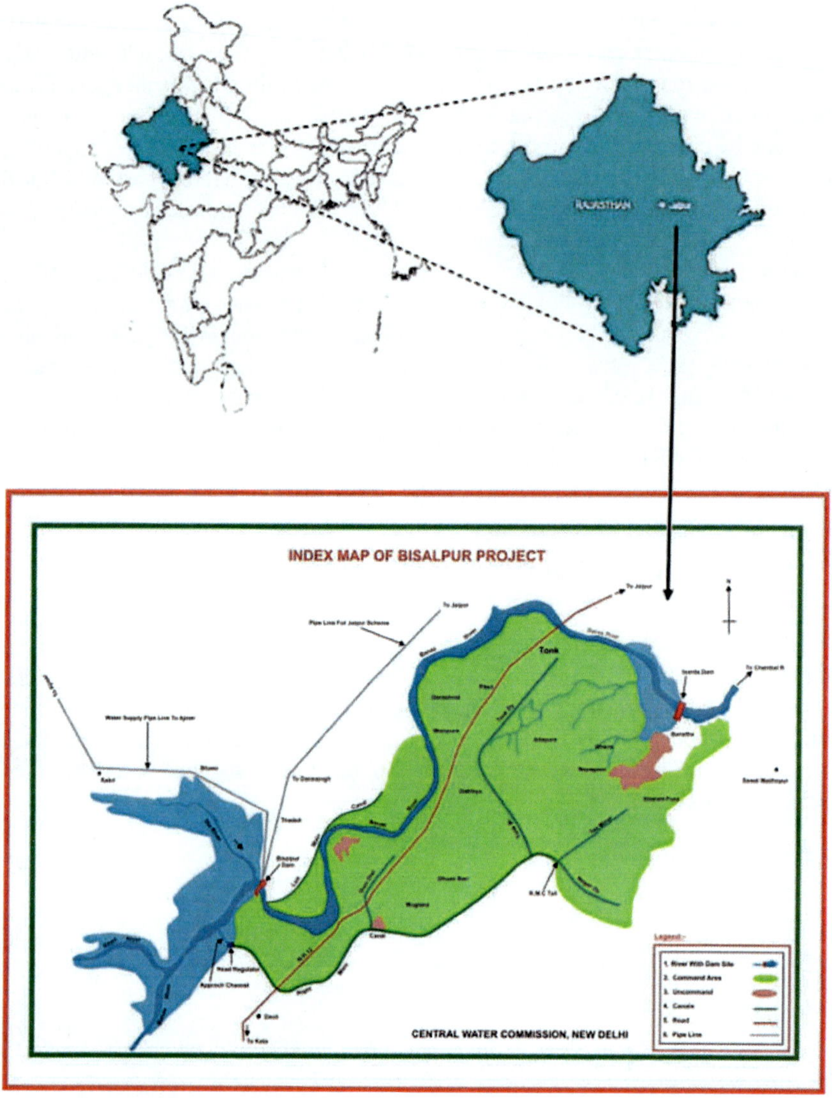

Fig. 1 Map of study location

previous values (lags) of itself [7].

$$y_t = \alpha + \beta_1 y_{t-1} + \beta_2 y_{t-2} + \ldots + \beta_p y_{t-p} \varepsilon_t + \varphi_1 \varepsilon_{t-1} + \varphi_2 \varepsilon_{t-2} + \ldots + \varphi_q \varepsilon_{t-q} \quad (1)$$

α = constant term, ε_t = error term, t = time period. Real-world data mostly consist of seasonal time series. The seasonal time series is known as the multiplicative ARIMA (p, d, q) (P, D, Q)m model. (p, d, q) are non seasonal part and (P, D, Q)m are seasonal part of ARIMA model.

'p' represents the number of lag observations included in the model, also called the lag order. 'd' represents the number of times that the raw observations are differenced, also called the degree of difference. 'q' represents the size of the moving average window, also called the order of moving average, and P, Q, D represents the seasonal part of the model which have the same meaning as non-seasonal part. For example ARIMA(1,1,1)(1,1,1)4 model (without a constant) for quarterly data ($m = 4$) and can be written as

$$(1-\phi 1\mathrm{B})(1-\Phi 1)(1-B)\ (1-B4)\,\mathrm{yt} = (1+\theta 1B)(1+\Theta 1)\,\varepsilon\mathrm{t} \quad (2)$$

$(1-\phi 1B)$ = backshift operator, $(1-\Phi 1\ B^4)$ = seasonal backshift operator, ε_t = error term [8]. For building time series of monthly data evaluation of trend, mean of the data, standard deviation was done by using the "autoplot" function (Fig. 2).

Separation of time series into four different time series components, i.e., trend series, cyclic series, seasonal series, irregular variation series, and further after removing seasonality from the data best fit model is selected. After selecting the best model, the third stage of the diagnostic check is performed. In diagnostic check, the goodness of fit of the selected model is analyzed. Mostly, researchers perform two tests to check the goodness of fit of the selected model. The first test is to check the residuals by using ACF and PACF graphs. If the selected model is appropriate, the residuals graphs of both correlation functions should be white noise that indicates residuals have no remaining correlation with each other. The second test is the

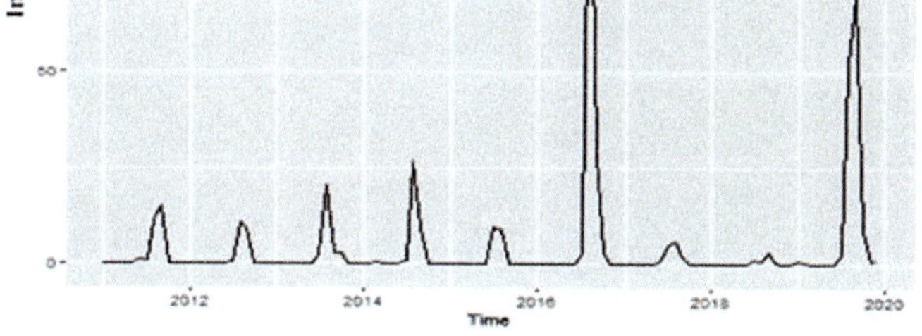

Fig. 2 Autoplot of inflow (2011–2019)

Ljung-Box test (1978). If the values of p-value in this test exceed 5%, it indicates that residuals do not have a significant departure from white noise. If the selected model residuals are very large or a model fails to pass the Ljung-Box test, then the model return to the selected alternative model and follows the same procedure until satisfactory model results are obtained [3].

3 Results and Discussions

Figure 3 shows the ACF graph, i.e., the correlation of the time series with itself at a specified lag, and the PACF graph of inflow shows that the maximum lag is at 1 (12 months), which indicates a positive relationship with the 12-month cycle. This shows it has seasonality and all the lag values are inside the error band for each year (12 months period), this result confirms the good performance of the ARIMA model and suggests further applications for which a seasonal model could be tested. The best model for forecasting the inflow series was ARIMA (0,0,0) (2,1,1)[12]. Model validation with Ljung-Box test gives p-value = 0.9976. Hence, it can be concluded that the predicted model's residue has no autocorrelation and hence shows a good fit for time series.

The prediction by ARIMA and the confidence intervals are illustration of the best model in Fig. 4b. The use of these confidence intervals is used for anomaly detection. If a reading lies outside these intervals, then it is said to the value is anomalous with 95% confidence. The residue of ARIMA follows a bell-shaped curve with high mean and outlier, showing increasing inflow trend in the future.

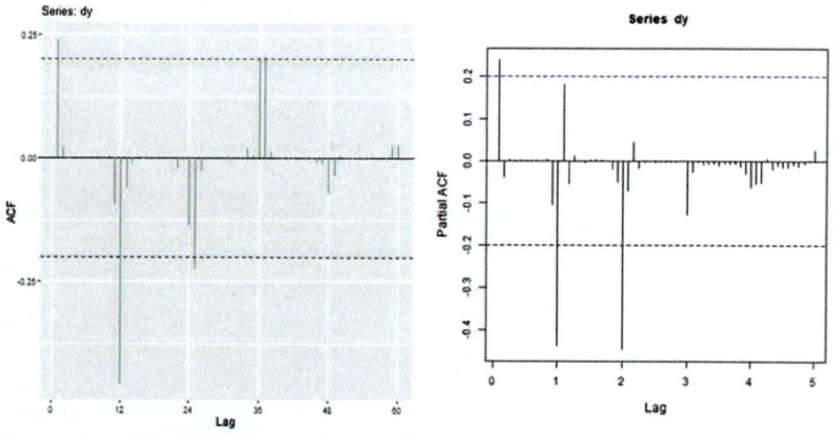

(a) (b)

Fig. 3 **a** ACF graph (left); **b** PACF graph (right) of inflow parameter

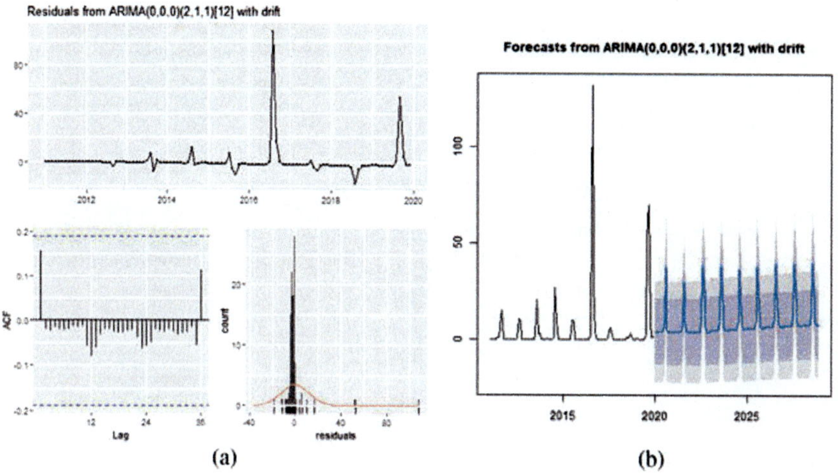

Fig. 4 **a** ARIMA statistics graph of residuals (left), **b** forecasted series (right) of inflow parameter

Illustration of the best model for forecasting the outflow series is shown in Fig. 5. The Ljung-Box test has a p-value = 0.9997 which greater than 0.05; hence, it can be concluded that residues are independent.

Using the ARIMA model it was possible to forecast the evaporation time series from 2020 to 2028. Figure 6 shows best ARIMA model for the series is ARIMA (1,0,0) (0,1,0)[12] with a sigma square estimated as 18.04 which tells that there is not much deviation of data from the mean and the value of the Ljung-Box test is 0.9879, hence the null hypothesis is rejected and it was concluded residues which are independent.

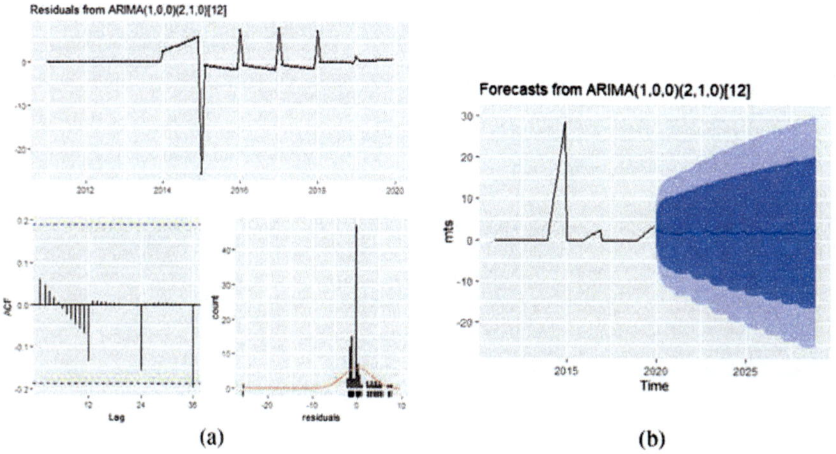

Fig. 5 **a** ARIMA statistics graph of residuals (left), **b** forecasted series (right) of outflow parameter

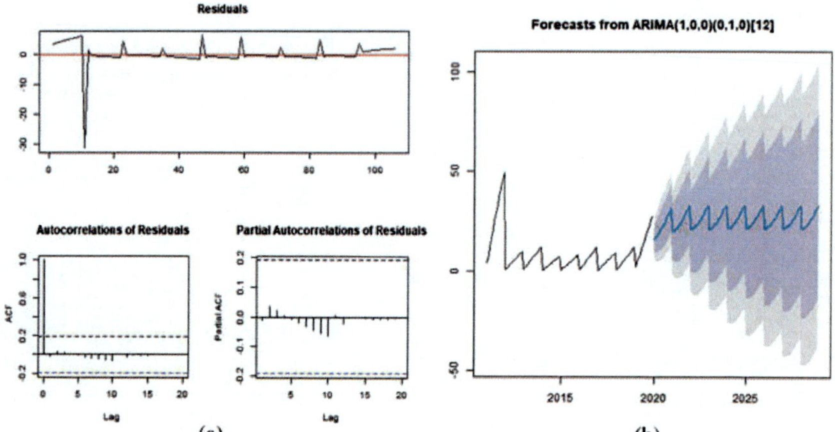

Fig. 6 **a** ARIMA statistics graph of residuals (left), **b** forecasted series (right) of evaporation parameter

The autocorrelation of the time series is shown in Fig. 7. Lag at 12 is the maximum that confirms data has seasonality. The best ARIMA model for the "drinking" series is ARIMA (1,0,0) (1,1,0)[12] with a sigma square estimated as 1.08.

The autocorrelation of the residuals is shown in Fig. 8. The values are very low except maxima for lag = 12. This result confirms the good performance of the ARIMA model. ADF test statistics of irrigation water supply time series shows standard deviation exist in the residue plot, and test statistics is smaller than critical values so the null hypothesis is not rejected. So, time series is further differenced for removing seasonality.

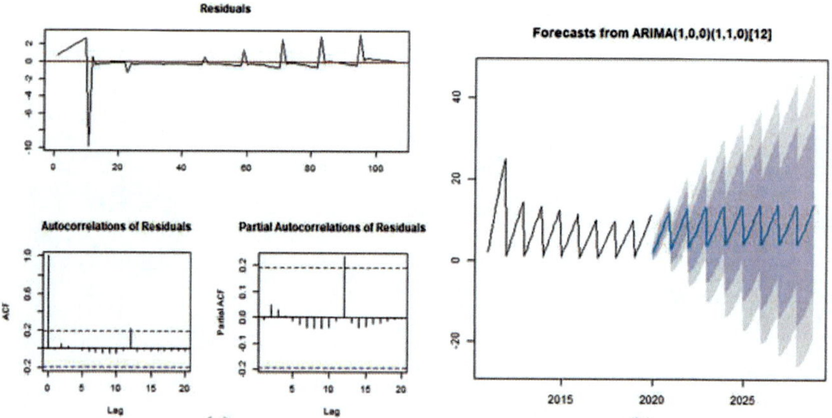

Fig. 7 **a** ARIMA statistics graph of residuals (left), **b** forecasted series (right) of drinking parameter

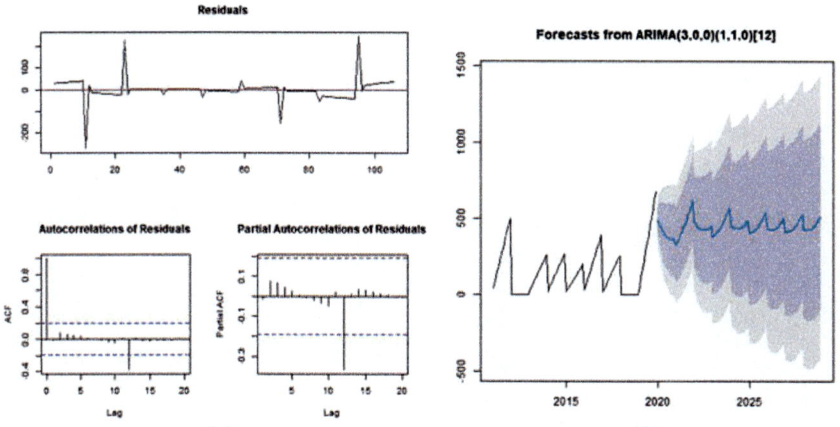

(a) (b)

Fig. 8 **a** ARIMA statistics graph of residuals (left), **b** forecasted series (right) of irrigation parameter

The autocorrelation of the residuals is shown in Fig. 9 which has maxima for lag = 24. From the analysis, it was concluded that the best model for forecasting time series is ARIMA(1,0,1)(2,1,0)[12] and model validation was done using the Ljung-Box test with a p-value of 0.9961. Hence, it can be concluded that residues are independent. It is confirmed that the model is good for forecasting (Fig. 10).

The value of the test statistic of the ADF test is −2.4939 which is greater than the critical value. Hence, series can be used for further analysis of predicting best fitted ARIMA model. From the analysis, it was concluded that the best model for

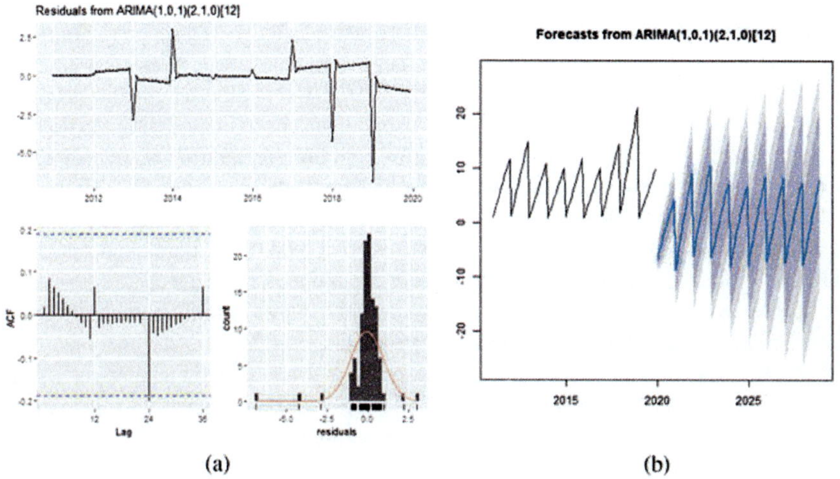

(a) (b)

Fig. 9 **a** ARIMA statistics graph of residuals (left), **b** forecasted series (right) of rainfall parameter

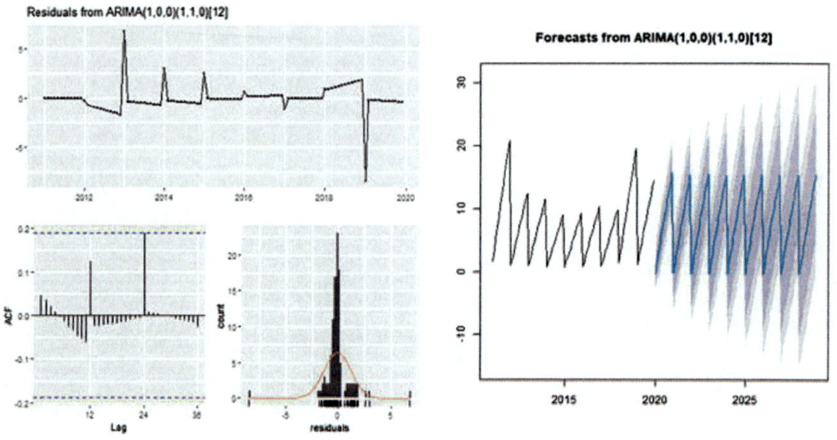

(a) (b)

Fig. 10 **a** ARIMA statistics graph of residuals (left), **b** forecasted series (right) of capacity parameter

forecasting rainfall time series is ARIMA(1,0,0)(1,1,0)[12]. Ljung-Box test p-value is 0.9999 which is greater than 0.05; hence, it was concluded residues of the ARIMA model do not have autocorrelation. Hence, it can be concluded model has good fitness and can be used for further forecasting.

4 Conclusions

In the context of time series analysis and forecasting model, the seven parameters contributing to the water resource system of the Bisalpur dam, Rajasthan are predicted and forecasted for the next 9 years, i.e., 2020 to 2028. Using time series plots original data were analyzed, and a seasonal ARIMA model was used in modeling the data in R-studio using R-script, using autocorrelation function, and partial autocorrelation function analysis of the time series was done. The best fitted seasonal ARIMA model selected for seven parameters was selected after validating it with the "Ljung-Box test" using statistics criteria of "p-value greater than 0.05". The best fitted seasonal ARIMA model for the different parameters is mentioned (Table 1).

Forecasted series of inflow shows increasing trend, while rainfall forecasted series shows declining trend this is because nine years data was used for the analysis.

The forecasted time series can be used for further analysis for managing water resources of Bisalpur dam for drinking water supply and water supply for irrigation together with other models of hydrology for accurate forecasting.

Table 1 Variables with best fitted ARIMA model

Variable	Model
Inflow	ARIMA(0,0,0)(2,1,1)[12]
Outflow	ARIMA(1,0,0)(2,1,0)[12]
Evaporation	ARIMA(1,0,0)(0,1,0)[12]
Drinking	ARIMA(1,0,0)(1,1,0)[12]
Irrigation	ARIMA(3,0,0)(1,1,0)[12]
Capacity	ARIMA(1,0,0)(1,1,0)[12]
Rainfall	ARIMA(1,0,1)(2,1,0)[12]

Acknowledgements Authors would like to thank the Water Resource Department Jaipur, Rajasthan for providing data of the Bisalpur dam used in this paper.

References

1. Adamowski (2010) Comparison of multivariate regression and artificial neural networks for peak urban water-demand forecasting: evaluation of different ANN learning algorithms. ASCE J Hydrol Eng 15(10):729–743
2. Adhikary (2012) A stochastic modeling technique for predicting groundwater table fluctuations with time series analysis. Int J Appl Sci Eng Res 1:238-249
3. Adnan (2017) Application of time series models for streamflow forecasting. Civil and Environ Res 9:2224–5790
4. Dadhich AP, Goyal R, Dadhich PN (2018) Assessment of spatio-temporal variations in air quality of Jaipur city, Rajasthan, India. Egypt J Remote Sens Space Sci 21(2):173–181
5. Devi (2018) An assessment of Bisalpur dam: a major water project of Rajasthan. Int J Soc Sci Econ Res ISSN: 2455–8834
6. Goyal R, Arora AN (2012) Predictive modelling of groundwater flow of indira gandhi nahar pariyojna, Stage I. Ish J Hydraul Eng 18(2):119–128
7. Gupta (2020) Two-step daily reservoir inflow prediction using ARIMA-machine learning and ensemble models. 10.002/essoar.10502185.1
8. Hyndman (eds) (2018) Forecasting: principle and practices. Monash University, Australia
9. Loucks DP, Van Beek E (2017) Water resource systems planning and management: an introduction to methods, models, and applications. Springer
10. Machekposhti (2017) Prediction of annual inflow to Karkheh Dam reservoir using time series models. Civil Eng J 3. https://doi.org/10.28991/cej-2017-00000095
11. Rajput H, Goyal R, Brighu U (2020) Modification and optimization of DRASTIC model for groundwater vulnerability and contamination risk assessment for Bhiwadi region of Rajasthan, India. Environ Earth Sci 79(6):1–15
12. Stellwagen (2013) ARIMA: the models of box and jenkins. Foresight: The Int J Appl Forecast 28–33
13. Wang (2014) An improved ARIMA model for hydrological simulations. Nonlinear Process Geophys 21:1159-1168

Comparison of Multiple Linear Regression and Artificial Neural Network for Inflow Prediction of Ukai Reservoir

Ayushi Panchal and Sanjaykumar M. Yadav

Abstract In recent years, the soft computing techniques have arisen as an alternative for overcoming the limitations of traditional methods. Artificial neural networks (ANNs) can effectively approximate the nonlinear relationship between input and target parameters. Multiple linear regression (MLR) is also used widely to find the relationship between dependent and independent parameters. In the present study, the monthly inflows are predicted to compare the performance and reliability of ANN and MLR models for the Ukai reservoir. The performance measures have been computed to evaluate the model performance. The models based on ANNs with lesser values of RMSE and higher values of co-efficient of determination proved to be more accurate. The results demonstrate that ANN is reliable and effective tool as compared to multiple linear regression (MLR) and can be adopt as a better alternative to make predictions.

Keywords Multiple linear regression · Artificial neural network · Simulation · Inflow prediction

1 Introduction

The prediction of streamflow is the most important issue for the water resources engineers and hydrologists. The estimation of stream flows helps in the design and operations of hydraulic structures, water infrastructures, flood warnings, water distribution networks, river transports as well as hydropower generation. Therefore, the prediction of streamflow is one of the most important research topics in hydrology. Frequently, the difficulties are faced by the engineers while predicting and estimating the water resources parameters. The majority of the variables such as rainfall, runoff,

A. Panchal · S. M. Yadav (✉)
Department of Civil Engineering, Sardar Vallabhbhai National Institute of Technology, Surat, India
e-mail: smy@ced.svnit.ac.in

A. Panchal
e-mail: ayushipanchal44@gmail.com

P. V. Timbadiya et al. (eds.), *Geospatial and Soft Computing Techniques*, Lecture Notes in Civil Engineering 339, https://doi.org/10.1007/978-981-99-1901-7_34

water discharge, sediment discharge reveal highly nonlinear behaviour due to spatial and temporal variations. There are two methods generally for building a streamflow prediction model, physically based methods and data driven methods [1]. The hydrological processes are aimed to reproduce in the realistic manner based on the physical laws in the physically based modelling. In the data driven modelling, the relationship between the stream flows and other variables is modelled directly based on the statistical correlations.

Although the data driven methods require the datasets as similar as the physically based models, they take considerable lesser development time and suitable for the field applications. The data driven models are proven capable of predicting the stream flows accurately [2]. In last decades, the computational methods such as ANNs, fuzzy inference systems (FIS), neuro-fuzzy inference systems (NFIS) have been demonstrated in the fields for simulating the complex, nonlinear systems as these techniques have the capability of "learning" from the data. Aichouri et al. [3] developed the river flow model with the use of ANN and showed that neural network model provides superior predictions and appears to be the promising approach that can be applied for various hydrological systems.

Hamzah et al. [4] compared three different imputation methods including NN and MLR for the prediction of streamflow datasets. Also, Ghourdoyee et al. [5] compared ANN with MLR for the simulation of groundwater head for Imam-Zadeh Jafar aquifer located at Iran. Turhan [6] carried out the comparative evaluation for use of ANN and MLR for rainfall-runoff modelling for water resources management and concluded that ANN models were showing statistically good results for rainfall-runoff modelling and the model developed using ANN can be adopted successfully for estimating the average monthly flows. Mohammadi et al. [7] carried out simulation of streamflow with combined hydrologic process as well as artificial intelligence-based models and concluded that the hybrid models on the basis of AI are the suitable alternatives to the hydrological models where there is lack of measured data. Patle et al. [8] modelled the monthly pan evaporation with the use of MLR and ANN and concluded that ANN models showed better performance in comparison with MLR models in predicting the monthly pan evaporation in the study area. Gaya et al. [9] estimated the water quality index with the use of AI approaches as well as multi-linear regression and after evaluation of the models with MSE, RMSE, DC, etc. It has been concluded that for reliable estimation of Water Quality Index, the artificial intelligence techniques are much useful. Poul et al. [10] carried out the comparative study using MLR, ANN as well as ANFIS models for streamflow prediction and the results revealed that the coupled models were showing improved Nash-Sutcliff co-efficient values.

Noor et al. [11] carried out the comparative study for prediction of marine diesel engine performance using ANN and MLR and concluded that ANN is more reliable and accurate model as compared to the mathematical model. Nathan et al. [12] used ANN and MLR on ground water quality by applying Canadian Water Quality Index (CWQI) and revealed that ANN model did fairly well performance. The rainfall-runoff modelling has been done using ANN and MLR techniques by Singh et al. [13] and the results showed that ANN approach shows more reliable results. Mustafa et al. [14] showed various applications of ANNs for water resources engineering. Alp and Cigizoglu [15] developed different two ANN models for simulating the relation of suspended sediment load with rainfall and river flow with the use of hydro meteorological data. From the results, it has been seen that the ANN models produced significantly better results than multiple linear regression (MLR). Melesse et al. [16] compared the results of artificial neural network model with the results from multiple linear regressions (MLRs), multiple nonlinear regression (MNLR) as well as the autoregressive integrated moving average (ARIMA) for daily and weekly prediction of the suspended sediment at different period length of training as well as the testing data. The ANN predictions were found superior compare to the results of MLR, MNLR and ARIMA [16]. Ghorbani et al. [17] carried out study on estimation of sediment load by using ANN, MLR, NF as well as sediment rating curve and concluded that the best results were achieved by the models of neuro-fuzzy and artificial neural networks.

There has been a rapidly growing interest in hydrologists and water resources engineers to apply the soft computing techniques in the area of water resources engineering and management. During the past two decades, the researchers have used the ANNs for predicting the streamflow. Raman and Chandramouli [18] applied the ANNs for the derivation of better operating policy for the Aliyar Dam located in Tamil Nadu. Using the dynamic programming model, the general operation policies were being derived using the ANNs. The results of dynamic programming algorithm with artificial neural networks given the better performance comparing to the other models. The artificial neural network was used by Jain et al. [19] for the inflow prediction as well as the operation for the upper Indravati Multipurpose Project located in Orissa, India. Two ANN models were developed for modelling the reservoir inflows and mapping the operation policy. The ANNs were found suitable for predicting the high flows. Kang et al. [20] used three layered ANN to predict the daily as well as hourly stream flow for the Chang River basin in Korea. It has been concluded by Kang et al. that ANNs are promising for predicting the streamflow compared to the autoregressive moving average model. Adamowski et al. [21] have carried out the comparison of MLR, MNLR, AIMA, ANN as well as wavelet ANN for forecasting of urban water demand. Adamowski et al. [22] carried out demand forecast modelling with ANN at daily basis and concluded that ANN provides better prediction for peak daily water demand as compared to MLR approach. Riad et al. [23] carried out the rainfall-runoff modelling with the use of ANN, and the results indicated that the ANNs are more suitable for prediction of the river runoff in comparison with the regression model.

The multiple linear regression (MLR) is simple tool for bringing down the efforts in modelling. MLR models are suitable for the various filed applications. The effectiveness of linear models like auto regression (AR), MLR, etc., has been established by previous studies. Sahay and Sehgal [24] used the wavelet-based auto regression models (WR) for the river stage prediction for Kosi River of North Bihar, India and found that the WR models outperformed ANN as well as the simple AR models for river stage predicting with one day lead time. Rezaeianzadeh et al. [25] used ANN, ANFIS as well as regression models for flood forecasting and showed that superior results were found in AI approaches as compared to the regression models. The multiple linear regression approach was used by Magar and Jothiprakash [26] the result were compared to the ARIMA models, and they concluded that both the models were performing equally well.

Based on the various literature, many studies are being carried out for the comparison of ANN and MLR for various aspects such as simulation of suspended sediments, prediction of sediments, river stage prediction and streamflow predictions. However, investigation for the reservoir inflows is still limited. Thus, in the view of the superior performance of ANN and MLR in different applications of water resources management, both the approaches are adopted in the present study to compare the model performances for the prediction of monthly inflows for the Ukai reservoir located at Tapi River, Surat.

2 Study Area and Data Used

The Ukai reservoir has been selected as the study area for the present study, which is constructed across the river Tapi. This is the second largest reservoir in Gujarat state. The reservoir is being utilised for various purposes like irrigation hydropower generation, flood control, etc. The catchment area of Ukai reservoir is 62,255 km^2. The Ukai reservoir is located 94 km away from Surat city. The left bank canal of the dam feeds water to the area of 1522 km^2, and the bank canal feeds water to 2275 km^2 of the land. The map of study area is as shown in Fig. 1.

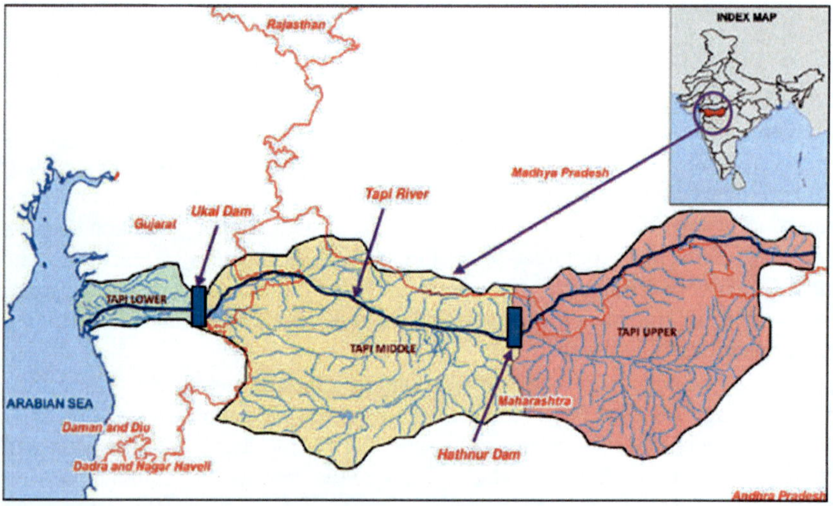

Fig. 1 Map of study area (Tapi basin) (*Source* iomenvis.nic.in)

The monthly inflows (MCM), monthly evaporation data (in MCM), the releases from the dam (MCM) as well as the data of storage capacity of the reservoir (MCM) at the beginning and at the end of the month of past 45 years were collected from the Irrigation department, Ukai as well as the Irrigation circle, Surat.

3 Methodology

3.1 Multiple Linear Regression

To find out the appropriate relationship between a dependent variable and several independent variables is of the classical problems in the statistical analysis [27]. The MLR is an approach that is applied to establish the relationship between a dependent variable and one or more independent variables in the linearly and it is based on method of least squares [12]. The multiple linear regression is analogous to single regression that permits the researchers to have multiple predictor variables.

In the present study, the model is developed using software SPSS. SPSS is the software which is generally used for statistical analysis. Its name was standing for Statistical Package for the Social Sciences (SPSS), then later changed and known as Statistical Product and Service Solutions (SPSS). Various market researchers, health researchers, survey companies and education researchers found this software helpful for statistical analysis. The descriptive statistics, i.e. mean and standard deviation, for the observed data of sample size 312 is shown in Table 1.

The frequency histogram for the regression residuals is plotted and shown in Fig. 2. The plot shown that it follows the normal distribution.

The relation obtained by developing the models in SPSS is as below.

$$S = 335.867 + 0.713(I) - 5.237\,(E) + 0.862\,(C) - 0.687\,(O)$$

where

S = Storage capacity at the end of the month (MCM),
I = Inflows (MCM),
E = Evaporation (MCM),

Table 1 Descriptive statistics for the observed data

Parameters	Mean	Std. deviation	N
Storage capacity at the end of the month (MCM)	4484.9535	1938.11887	312
Inflows (MCM)	1315.8735	1798.61575	312
Storage capacity at the beginning of the month (MCM)	4494.2290	1936.87987	312
Evaporation (MCM)	27.9519	18.81331	312
Outflow (MCM)	753.2132	888.72156	312

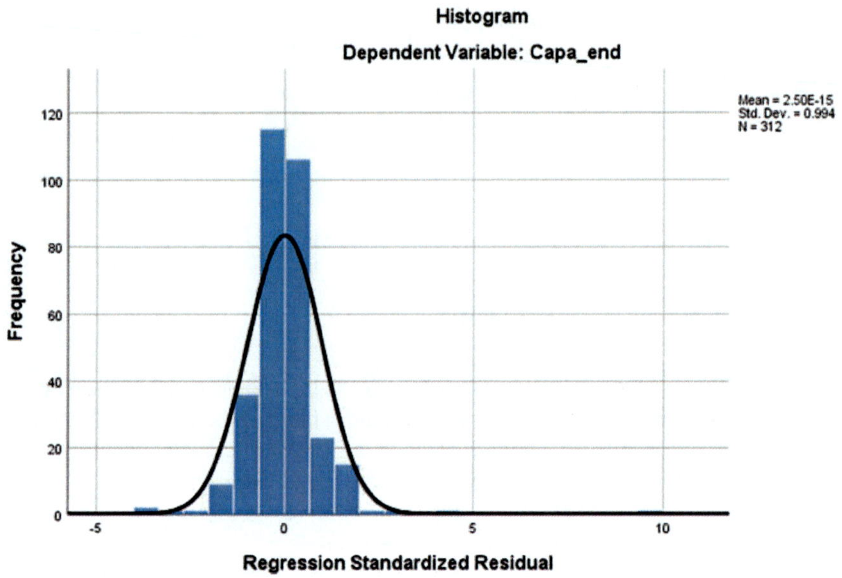

Fig. 2 Frequency histogram of regression residual (SPSS software)

C = Capacity (volume stored) at the beginning of the month (MCM),
O = Outflows (MCM).

From the predicted storage capacity (MCM) by MLR, the inflows are computed by the following equation.

Predicted inflows (MCM) = capacity at the end of the month (MCM)—capacity at the beginning of the month (MCM) + evaporation (MCM) + releases (MCM).

3.2 *Artificial Neural Networks (ANN)*

In this study, the inflow prediction is carried out using the artificial neural network (ANN) model. The performance of developed model has been tested and validated using MATLAB software. Using the observed data, the model is prepared on MATLAB using "nnstart" tool.

The model has been developed using monthly data. The monthly inflows (MCM), monthly evaporation (MCM), storage capacity at beginning of the month (MCM) as well as monthly releases (MCM) from year 1975 up to the year 2000 were given as inputs to the model. The inflows are predicted from the year 2001 up to the year 2020. The storage capacity at the end of the month was given as target to the network. The network architecture is as shown in Table 2. From the predicted storage capacity (MCM) by ANN, the inflows were computed by the following equation.

Table 2 Network architecture of developed model

Network architecture	
Number of inputs	4
Number of targets	1
Number of hidden neurons	10
Network type	Feedforward back propagation
Learning rate	0.001

Predicted inflows (MCM) = capacity at the end of the month (MCM)—capacity at the beginning of the month (MCM) + evaporation (MCM) + releases (MCM).

3.3 Evaluation of the Model Using Performance Measures

The performance measures calculated for evaluating the model are obtained as described below.

The root mean square error can be measured by

$$RMSE = \sqrt{\frac{\sum_{i=1}^{n}\left(X_{obs,i} - X_{\text{model},,i}\right)^2}{n}} \tag{1}$$

where

$X_{\text{obs},i}$ = the observed data,
$X_{\text{model},i}$ = the modelled data,
n = total number of observations.

Co-efficient of determination can be obtained by

$$R^2 = \left[\frac{\sum_{i=1}^{N}\left(O_i - O_{\text{avg}}\right)\left(P_i - P_{\text{avg}}\right)}{\sqrt{\sum_{i=1}^{N}\left(O_i - O_{\text{avg}}\right)^2}\sqrt{\sum_{i=1}^{N}\left(P_i - P_{\text{avg}}\right)^2}}\right]^2 \tag{2}$$

where

O_{avg} = mean observed discharge,
P_{avg} = mean modelled discharge,
P_i = computed discharge,
O_i = computed discharge.

Mean absolute error can be defined as.

$$MAE = \frac{1}{n} \times \sum_{i=1}^{n} |O_i - P_i| \tag{3}$$

where

P_i = Computed discharge,
O_i = Computed discharge,
n = Number of observations.

The normalised root means square error can be defined as.

$$NRMSE = \frac{RMSE}{\overline{O}} \tag{4}$$

where

$\overline{O}$ = average of observation value.

The Pearson's correlation co-efficient measures the strength of linear relationship between two variables, which is computed using the following equation.

$$r = \frac{\sum_{i=1}^{n} (O_i - \overline{O})(P_i - \overline{P})}{\sqrt{\sum_{i=1}^{n} (O_i - \overline{O})^2}\sqrt{\sum_{i=1}^{n} (P_i - \overline{P})^2}} \tag{5}$$

where

O_i = observed value,
$\bar{O}$ = average of observation value,
P_i = simulated value,
$\overline{P}$= average of simulated value.

The Nash-Sutcliff efficiency (NSE) is computed to know how well the plot of simulated and observed data fits 1:1 line. NSE can be computed using the following equation:

$$NSE = 1 - \frac{\sum_{i=1}^{n} (OBS_i - SIM_i)^2}{\sum_{i=1}^{n} (OBS_i - \overline{OBS})^2} \tag{6}$$

where

OBS_i = the observed value,
SIM_i = the simulated value,
$\overline{OBS}$ = the average of observation values.

From the observed and simulated data, various performance measures are computed using the formulas above, to evaluate the model performance.

4 Results and Discussion

The model has been trained using 70% of the total data using nntool in MATLAB. Remaining 15% data were used for testing the model, and another 15% data were being used to validate the network. The regression plot obtained from MATLAB for the developed model is as shown in Fig. 3. In Fig. 3, the R-values that are computed using ANN, are the R-values for the data up to the year 2000 that were used for training, testing as well as validation of the model. The regression plots give the idea about how close the outputs are from the actual target values. From Fig. 3, it can be seen that R-values are higher for training, testing as well as validation of the neural network which shows the good correlation between the target data and the output data.

Figures 4 and 5 show the comparison between observed vs predicted inflows by MLR and ANNs, respectively. The performance measures are computed to evaluate the model performances using the Formulas (1) to (6), and the values are as shown

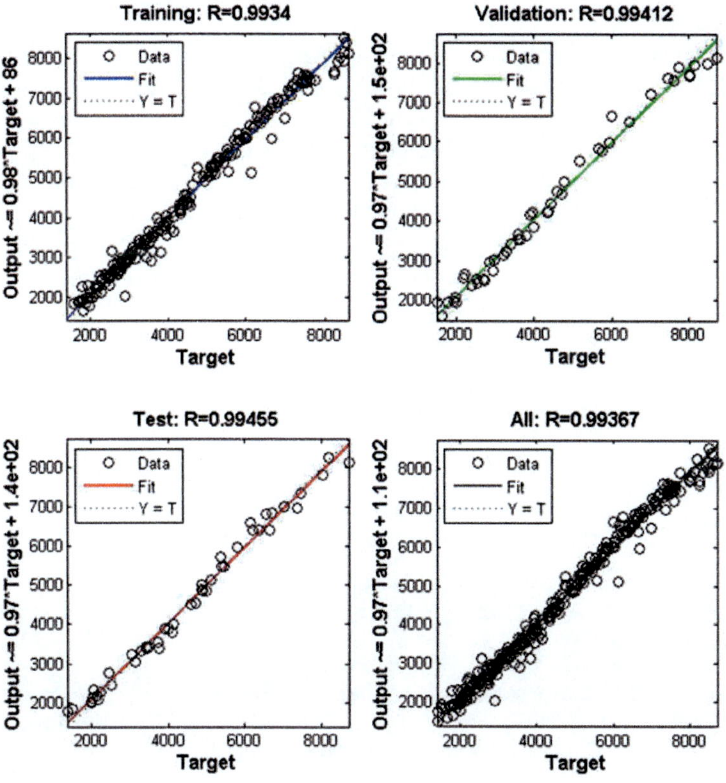

Fig. 3 Regression plot for training, testing and validation of neural network

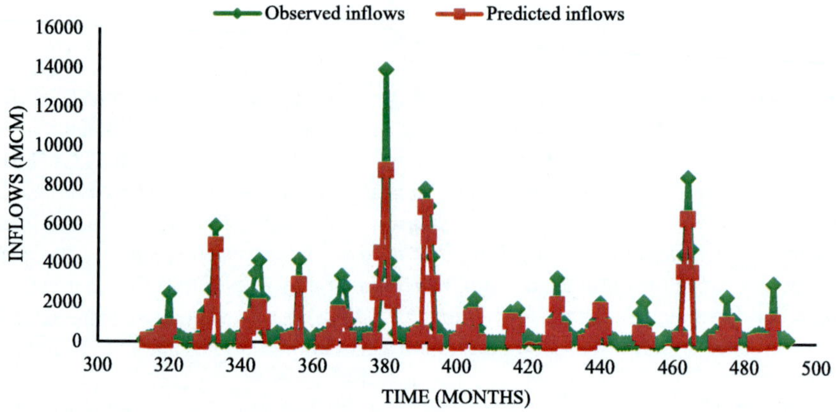

Fig. 4 Observed versus predicted inflows by MLR

in Table 3. The performance measures are computed for the predicted data from the year 2001–2020.

The observed and predicted inflows by MLR as well as ANN approach are compared with the scatter plot as shown in Fig. 6. From Table 3, it is seen that the co-efficient of determination is higher for the ANN approach comparing to MLR approach of inflow prediction. The other performance measures such as RMSE, MAE, NRMSE values are lesser in ANN approach. Also, the Nash-Sutcliff efficiency (NSE) co-efficient of the variables should be nearer to one to be the good correlated variables and that is also observed in good correlation with ANN approach rather than MLR approach for inflow prediction.

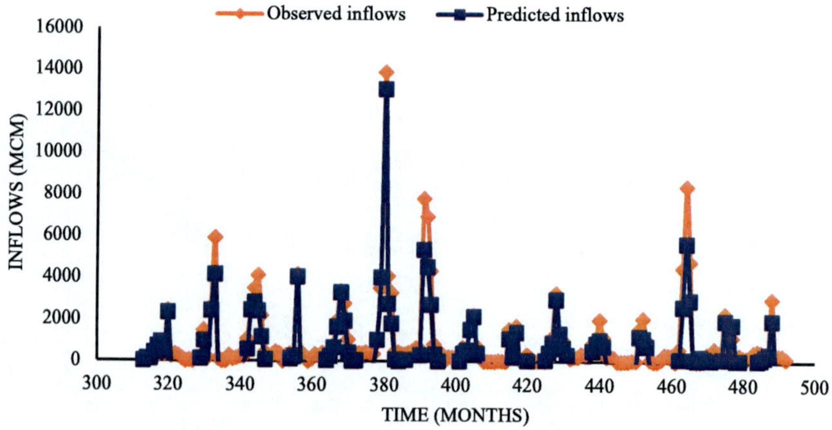

Fig. 5 Observed versus predicted inflows by ANN

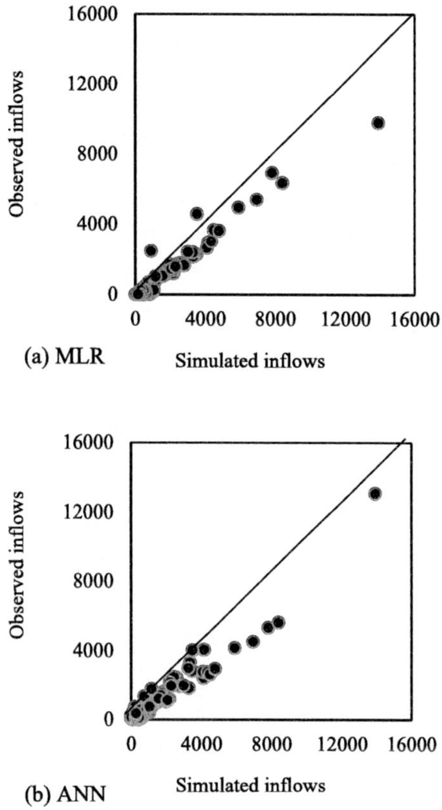

Fig. 6 Observed versus predicted inflows (MCM) by MLR and ANN

Table 3 Statistical performance measures

Performance measure	MLR	ANN
Co-efficient of determination	0.89	0.94
RMSE	801.39	615.06
MAE	585.40	451.91
NSE	0.78	0.87
NRMSE	0.87	0.66
Pearson's correlation co-efficient	0.95	0.97

5 Conclusions

The following conclusions are derived from the foregoing study.

- The artificial neural networks and multiple linear regression approaches are used for predicting the inflows. From the results, the errors were found lesser in the ANN approach comparing to the MLR approach.

- The co-efficient of determination as well as the Pearson's correlation co-efficient are found higher in ANN approach than MLR approach.
- The Nash-Sutcliff efficiency co-efficient is computed to see how well the plot of modelled vs observed data fits 1:1 line. The value of NSE = 1 shows the perfect match of the model to observed data, which is observed nearer to one for ANN approach comparatively to MLR approach.
- From the present study, it can be seen that the ANN is powerful tool for reservoir inflow prediction from the historical observed data.

References

1. Nguyen TT, Baxter H, Barber ME, Hossain A, Orr CH, Adam JC (2013) Impacts of future changes on groundwater recharge and flow in highly-connected river-aquifer systems: A case study of the Spokane Valley-Rathdrum Prairie Aquifer. In AGU Fall Meeting Abstracts (Vol. 2013, pp. H23P–07)
2. Awchi TA, Srivastava DK (2009) Analysis of drought and storage for mula project using ANN and stochastic generation models. Hydrol Res 40(1):79–91
3. Aichouri I, Hani A, Bougherira N, Djabri L, Chaffai H, Lallahem S (2015) River flow model using artificial neural networks. Energy Proc 74:1007–1014
4. Hamzah FB, Hamzah FM, Razali SFM, Samad H (2021) A comparison of multiple imputation methods for recovering missing data in hydrological studies. Civil Eng J 7(9):1608–1619
5. Ghourdoyee Milan S, Aryaazar N, Javadi S, Razdar B (2020) Simulation of groundwater head using LS-SVM and comparison with ANN & MLR. Hydrogeology 5(1):118–133
6. Turhan E (2021) A comparative evaluation of the use of artificial neural networks for modeling the rainfall-runoff relationship in water resources management. J Ecol Eng 22(5):166–178
7. Mohammadi B, Moazenzadeh R, Christian K, Duan Z (2021) Improving streamflow simulation by combining hydrological process-driven and artificial intelligence-based models. Environ Sci Pollut Res 1–17
8. Patle GT, Chettri M, Jhajharia D (2020) Monthly pan evaporation modelling using multiple linear regression and artificial neural network techniques. Water Supply 20(3):800–808
9. Gaya MS, Abba SI, Abdu AM, Tukur AI, Saleh MA, Esmaili P, Wahab NA (2020) Estimation of water quality index using artificial intelligence approaches and multi-linear regression. Int J Artif Intell 2252:8938
10. Poul AK, Shourian M, Ebrahimi H (2019) A comparative study of MLR, KNN, ANN and ANFIS models with wavelet transform in monthly stream flow prediction. Water Resour Manage 33(8):2907–2923
11. Noor CWM, Mamat R, Ahmed AN (2018) Comparative study of artificial neural network and mathematical model on marine diesel engine performance prediction. Int J Innov Comput Inform Control 14(3):959–969
12. Nathan NS, Saravanane R, Sundararajan T (2017) Application of ANN and MLR models on groundwater quality using CWQI at lawspet, Puducherry in India. J Geosci Environ Protect 5(03):99
13. Singh VK, Kumar P, Singh BP (2016) Rainfall-runoff modeling using artificial neural networks (ANNs) and multiple linear regression (MLR) techniques. Indian J Ecol 43(2):436–442
14. Mustafa MR, Isa MH, Rezaur RB (2012) Artificial neural networks modeling in water resources engineering: infrastructure and applications. In: Proceedings of World academy of science, engineering and technology (No. 62), February, World Academy of Science, Engineering and Technology

15. Alp M, Cigizoglu HK (2007) Suspended sediment load simulation by two artificial neural network methods using hydrometeorological data. Environ Model Softw 22(1):2–13
16. Melesse AM, Ahmad S, McClain ME, Wang X, Lim YH (2011) Suspended sediment load prediction of river systems: An artificial neural network approach. Agric Water Manage 98(5):855–866
17. Ghorbani MA, Hosseini SH, Fazelifard MH, Abbasi H (2013) Sediment load estimation by MLR, ANN, NF and sediment rating curve (SRC) in Rio Chama river. J Civil Eng Urbanism 3(4):136–141
18. Raman H, Chandramouli V (1996) Deriving a general operating policy for reservoirs using neural network. J water resour plann manage 122(5):342–347
19. Jain SK, Das A, Srivastava DK (1999) Application of ANN for reservoir inflow prediction and operation. J water resour plann manage 125(5):263–271
20. Kang KW, Kim JH, Park CY, Ham KJ (1993) Evaluation of hydrologic forecasting system based on neural network model. In: proceedings of the congress-international association for hydraulic research,1:(257–257). local organizing committee of the xxv congress
21. Adamowski J, Fung Chan H, Prasher SO, Ozga-Zielinski B, Sliusarieva A (2012) Comparison of multiple linear and nonlinear regression, autoregressive integrated moving average, artificial neural network, and wavelet artificial neural network methods for urban water demand forecasting in Montreal, Canada. Water Resour Res 48(1)
22. Adamowski JF (2008) Peak daily water demand forecast modeling using artificial neural networks. J Water Resour Plan Manag 134(2):119–128
23. Riad S, Mania J, Bouchaou L, Najjar Y (2004) Rainfall-runoff model using an artificial neural network approach. Math Comput Model 40(7–8):839–846
24. Sahay RR, Sehgal V (2013) Wavelet regression models for predicting flood stages in rivers: a case study in Eastern India. J Flood Risk Manage 6(2):146–155
25. Rezaeianzadeh M, Tabari H, Yazdi AA, Isik S, Kalin L (2014) Flood flow forecasting using ANN, ANFIS and regression models. Neural Comput Appl 25(1):25–37
26. Magar RB, Jothiprakash V (2011) Intermittent reservoir daily-inflow prediction using lumped and distributed data multi-linear regression models. J earth syst science 120:1067–1084
27. Tabari H, Sabziparvar AA, Ahmadi M (2011) Comparison of artificial neural network and multivariate linear regression methods for estimation of daily soil temperature in an arid region. Meteorol Atmos Phys 110:135–142
28. Awchi TA (2014) River discharges forecasting in northern Iraq using different ANN techniques. Water Resour Manage 28(3):801–814
29. Cigizoglu HK (2008) Artificial neural networks in water resources. In: Integration of information for environmental security, Springer, Dordrecht, pp 115–148
30. Sehgal V, Tiwari MK, Chatterjee C (2014) Wavelet bootstrap multiple linear regression-based hybrid modeling for daily river discharge forecasting. Water Resour Manage 28(10):2793–2811

Rainfall-Runoff Modelling Using Artificial Neural Networks (ANNs) for Upper Krishna Basin, Maharashtra, India

Aparna M. Deulkar, S. N. Londhe, R. K. Jain, and P. R. Dixit

Abstract Rainfall-runoff (R-R) modelling is a process in which rainfall gets converted into runoff after some losses like infiltration, evaporation, transpiration, etc. The rainfall-runoff process is highly complex nonlinear in nature owing mostly to the random nature of its most important parameter, the "rainfall". Accurate estimation and prediction of runoff at a location in a basin helps in flood control operations, reservoir operations, and design of water impounding structures and optimizing different water resource systems. The rainfall-runoff (R-R) modelling can be exercised using various methods like physics-based model, conceptual and mathematical model, and empirical equations. The major limitation of such types of models lies in the exogenous requirement of data in the form of basin parameters like basin slope, soil type, and other basin characteristics. The unavailability of such a data becomes a major impediment in application of these models in any basin. Therefore, soft computing techniques like artificial neural networks (ANNs) have been extensively applied to model the R-R modelling owing to their adaptive, model free, ease of operation, and less time-consuming nature. In the present work, R-R modelling is done using the artificial neural networks (ANNs) in Upper Krishna Basin of Maharashtra, India. For this, previously measured rainfall and runoff values from 1997–2013 were used. ANN is applied to predict the runoff at Shivade station of Krishna River basin using the previously measured rainfall values at nearby 7 stations in the same basin. The results of developed models were compared with actual observations made by Water Resources Department, Government of Maharashtra. Though all the models shown reasonable performance it has been noticed that there is a need of improvising the

A. M. Deulkar (✉) · R. K. Jain
Department of Civil Engineering, JSPM Rajashri Shahu College of Engineering, Tathwade, Pune 411033, India
e-mail: apama.deulkar2015@gmail.com

S. N. Londhe · P. R. Dixit
Department of Civil Engineering, Vishwakarma Institute of Information Technology, Pune 411048, India
e-mail: shreenivas.londhe@viit.ac.in

P. R. Dixit
e-mail: pradnya.dixit@viit.ac.in

P. V. Timbadiya et al. (eds.), *Geospatial and Soft Computing Techniques*, Lecture Notes in Civil Engineering 339, https://doi.org/10.1007/978-981-99-1901-7_35

accuracy in prediction and it can be made possible by applying a makeshift technique of deep neural network instead of shallow neural network.

Keywords Rainfall-runoff · Upper Krishna Basin · Artificial neural network · Deep neural network

1 Introduction

Precise rainfall-runoff modelling is one of the important topics for researchers in hydrology, as it has a critical role in water resources management, hydropower development, urban planning, irrigation, and other agro-hydrological and meteorological activities. Accurate runoff estimation becomes important and dynamic as global water demand rises. Thus, more precise methodology is needed to predict rainfall-runoff process. Accurate estimation of runoff at basin level is a challenging since rainfall-runoff have complicated relationship. Nevertheless, transforming rainfall into runoff is a time-consuming and complicated process which depends upon large number of characteristics parameters of basin and meteorological parameters are associated with it. Thus, assessment of this process with accuracy plays an important role in designing, operating, and maintaining various hydraulic structures as well as for many water resource systems. Record of rainfall and runoff helps to fix the design parameter like storage capacity, water demand for power generation and irrigation. Accurate forecasting also assists farmers who majorly depends on rainfall to decide crop pattern. Rainfall-runoff process helps to understand, control, and monitor the quality and quantity of water resources [24].

The R-R modelling can be classified using various methods like physics-based model, conceptual, mathematical model, and empirical equations. Even though these kinds of models give good results, some challenges are still there to explore the research further. The major challenges and limitation of such types of models are data availability in the form of basin parameters like basin slope, soil type, land cover, and land use along with other meteorological characteristics. Additionally, rainfall varies in space and time relationship which leads to nonlinearity, non-stationarity, and complexity in the R-R process. Considering this, there is always a need for a technique which will be able to map rainfall-runoff process accurately and in the last two decades machine learning models, such as neural network (ANN)-based, fuzzy-based, and regression-based machine learning models have been applied successfully by the researcher's community [21]. Out of these, ANN-based models have been used extensively due to their capability of mapping the random input with output very precisely.

In recent past years, machine learning models based on ANN technique have been applied for modelling rainfall-runoff process. [5, 23] used two techniques namely radial basis function (RBF) neural network model and MLP neural network model for rainfall-runoff modelling and RBF outperformed in comparison with MLP. Jain et al. [11] used the back propagation (BP) neural network model for modelling daily runoff for Kentucky River basin. Then, [3, 9] applied ANN for R-R modelling. [7, 8] used the ANN model for predicting runoff. Kisi [14] applied ANN model in comparison with LM and BP algorithm and found that the LM was winner. Kisi [15] used generalized regression neural network (GRNN) model, RBF model, and feedforward neural network (FFNN) model; conclusion was GRNN performed better compared to others. Furthermore, [12, 13, 16, 22] have used ANN to model the rainfall-runoff process in various areas and found the applicability of ANN over the traditional methods.

Oyebode and Stretch [21] have reviewed the application of ANNs. Their study shows difficulties in finding causative input parameters, model complexity, and training data sets and generalization issues. Asadi et al. [2] suggested input parameter selection is important in soft computing tool like ANNs. Chen et al. [7] applied ANNs to forecast flow using monsoon flood events. Then, [4, 27, 29] have successfully applied ANN for runoff modelling using deep neural network that is LSTM models and received an interest from researchers. Additionally, [1, 6] used rainfall data to develop ANN models to estimate and forecast the runoff events. [10, 28] studied R-R process using ANNs models and concluded that performance of the model depends on accuracy of input data. From the above literature, it can be said that though ANN has been applied widely for modelling R-R process by many researchers since last many years and considering the importance of rainfall in the water cycle and its random nature; location specific R-R model is still need for today's era. Considering this as a motivation to work further in this filed, authors have decided to model the R-R process using ANN in the present study for forecasting the runoff at the downstream stations of the Upper Krishna Basin along the Krishna River reach of Maharashtra, India and detailed work of which is presented in the subsequent sections of this paper.

Present paper includes total 6 sections out of which introduction of the study is presented in the first part, study area and data are explained in Sect. 2. Brief information about ANN is given in the Sect. 2.2, and in Sect. 2.3 model development is explained in detail. Result and discussions are presented in Sect. 3, and concluding remarks are given in the last Sect. 4.

2 Study Area and Data Source

2.1 Shivade Basin

Present study is done in the Shivade catchment of Upper Krishna Basin which is situated in the Satara district of Maharashtra, India. This Shivade catchment consists of seven rain gauge stations namely Upshinge, Thoseghar, Targaon, Sandavali, Nagthane, Marali, Jawalwadi, and Shivade (73°17' to 81°9' East and 13°10' to 19°22' North) as the last station which is the runoff measuring station.

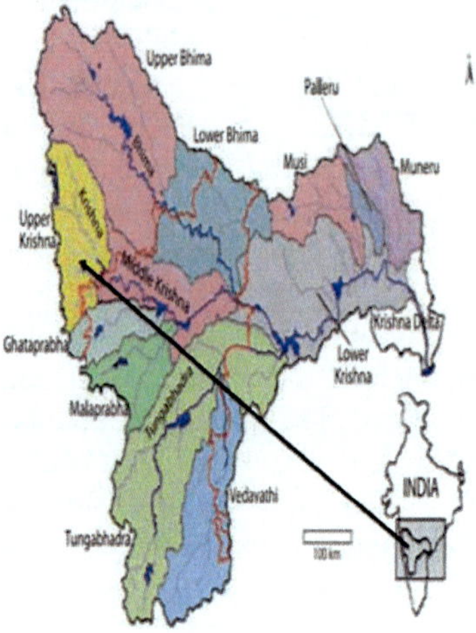

(Location of Maharashtra)
(Source: www.in-
diawris.gov.in)

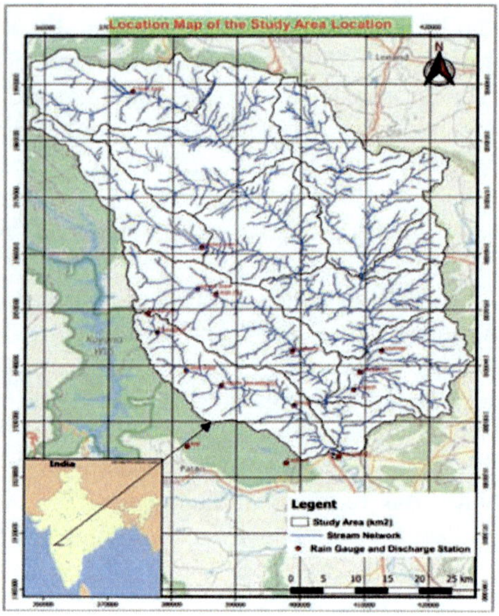

(Source: map developed using QGIS
software. Area:3149.122 sq.km)

Fig. 1 Location map of study area

As the objective of this study is to explore R-R modelling process in the Shivade catchment of Krishna River, daily measured rainfall values of the seven rain gauge stations and daily measured runoff values at Shivade station are considered. After the data received from hydrology project, Nashik through Hydrology Data Users Group (HDUG; https: www.mahahp.gove.in) it was noticed that for some year data is missing. Therefore, it was decided to omit missing year. Thus, 17 years of data from 1997–2013 considered for present work. According to the statistical analysis of the data, the maximum rainfall for monsoon (June–September) was 338.46 mm; average rainfall for all considered monsoon months was 908.8 mm in this catchment. Readers are directed to "https: www.cwc.gov.in" for more details of the catchment area. Figure 1 showcases the location map of study area.

2.2 Artificial Neural Networks (ANNs)

In the present study, artificial neural network (ANN) is applied as a tool to model the rainfall-runoff process in the Shivade catchment. Consequently, ANN is explained briefly in this section.

Artificial neural network (ANN) is basically designed to mimic the cognition process followed by biological neurons in human brains. The feedforward type of

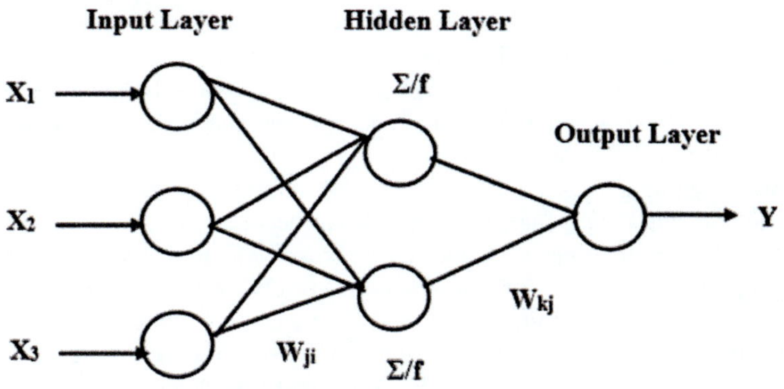

Fig. 2 Typical 3-layered feedforward neural network

neural networks (FFNNs) is the most used to solve engineering problem. An FFNN is made up with three layers: input layer, one or more hidden layers, and an output layer. The input signal passes through the network on a layer-to-layer in the forward direction, with a connection strength called as "weight". Each neuron uses a nonlinear activation function to initiate a response. Figure 2 depicts a feedforward ANN structure. ANN is one of the useful driven techniques to model nonlinear systems which has been used extensively in hydraulic and hydrologic modelling since last two decades. Readers are directed to [17, 18, 25, 26] for further details.

2.3 *Model Development*

As mentioned earlier, present study aims if forecasting the runoff at Shivade station one day ahead in time using previously measured (monsoon data) rainfall values of nearby seven rain gauge stations. To model this rainfall-runoff process, ANN is used as technique to map the "inputs; rainfall of seven stations" to the "output; runoff" of Shivade station at one day ahead in time step. For these 17 years of daily runoff and rainfall values were made available from HDUG, Nashik. But authors want to humbly note here that as the rainfall of Shivade station is not available in consistent records, here in the present study, it was constrain of unavailability of it. And thus to forecast the runoff at Shivade at time "t + 1", antecedent and current rainfall values of nearby stations (excluding the rainfall at Shivade station itself) were used as inputs with different lag time intervals. It was also thought to explore the capacity of ANN as a soft computing technique to model the rainfall-runoff process for Shivade station without using the rainfall of Shivade itself and understand whether there is need of any advance technique to model this R-R process accurately or ANN as a shallow network is sufficient to do it well? It was decided to explore ANN; shallow neural network; in the present study as a soft computing technique which do not assume any mathematical model a priori and hence are more flexible in data mining. Soft computing techniques treat human brain as their role model and mimic the ability of the human mind to effectively employ modes of reasoning that are approximate rather than exact. The conventional hard computing techniques require a precisely stated analytical model and often a lot of computation time whereas soft computing techniques requires very less time.

While developing the R-R models, input selection was done on the basis of cross-correlation analysis. The degree of dependency between past and future rainfall values on the runoff values was assessed by cross-correlation analysis for all the rain gauge stations. Figure 3 shows the cross-correlation for Targaon rain gauge station as an example wherein it is clear that the past three-time step values have greater influence on the current value and hence past three values, i.e. rainfall at "t−1", "t−2", and "t−3", were considered for model development along with the current time step value of rainfall, i.e. at time "t". Like this, number of input values of each station to forecast the runoff at Shivade was decided depending upon the cross-correlation analysis. In the model development, short notifications were used instead of using the full names of these stations: these abbreviations are in the form of numbers and as follows: Upshinge is abbreviated as R1, Thoseghar as R2, Targaon as R3, Sandavali as R4, Nagthane as R5, Marali as R6, and Jawalwadi as R7. Thus, Model I comprises of total 21 input parameters wherein 3 antecedent values of R1 and R2 stations (last 3 days values), 2 antecedent values of R3 station, 4 antecedent values at R4, 2 antecedent values of R5 and R6, and 3 antecedent values of R7 station. Likewise Model II consists of 7 inputs, Model III consists of 21 inputs, Model IV includes 14, and Model V includes 7 inputs which were developed using different inputs–output combinations. Table 1 represents these models in depth.

To get the runoff forecast at Shivade, ANN models were trained with 70% of the total data set and the remaining 30% of data was used for testing (validation:15%; testing 15%). Levernberg-Marquardt algorithm was used to train these models with

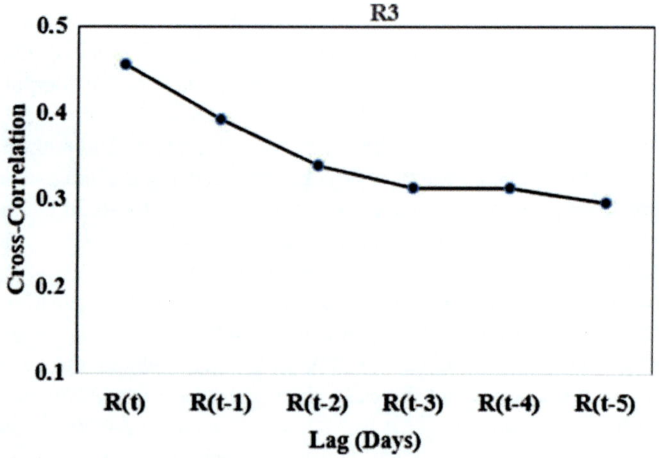

Fig. 3 Cross-correlation graph for targaon station

Table 1 Model development

Model No	Inputs (rainfall at input stations); "R" in mm	Output: Runoff at Shivade; "Q" in m^3/s	ANN architecture
I	R1(t), R1(t−1), R1(t−2), R2(t), R2(t−1), R2(t−2) R2(t−3), R3(t), R3(t−1), R4(t), R4(t−1), R4(t−2), R4(t−3), R5(t), R5(t−1), R5(t−2), R6(t), R6(t−1), R7(t)R7(t−1), R7(t−2)	Q t + 1	21:1:1
II	R1(t−2), R2(t−3), R3(t−1) R4(t−3), R5(t−2), R6(t−1) R7(t−2)	Q t + 1	7:44:1
III	R1(t), R1(t−1), R1(t−3), R2(t), R2(t−1), R2(t−3), R3(t), R3(t−1), R3(t−3), R4(t), R4(t−1), R4(t−3), R5(t), R5(t−1), R5(t−3), R6(t), R6(t−1), R6(t−3), R7(t), R7(t−1), R7(t−3)	Q t + 1	21:16:1
IV	R1, R1(t−2), R2(t), R2(t−3), R3(t), R3(t−1), R4(t), R4(t−3), R5, R5(t−1), R6(t), R6(t−1), R7(t), R7(t−2)	Q t + 1	14:2:1
V	R1(t), R2(t), R3(t), R4 (t), R5(t), R6 (t), R7 (t)	Q t + 1	7:2:1

"log-sigmoidal" and "Plurilinear" as transfer functions in the first and second layer of the network and the data was normalized between 0 to 1 and −1 to 1 in the two layers, respectively.

3 Results and Discussions

As mentioned earlier, total 5 models were developed to forecast the runoff at Shivade station 24 h ahead in time, i.e. at "t + 1" time step, using the previously measured rainfall values of seven rain gauge stations in the same catchment. All the developed models were tested with testing data set and the performance of these models was judged by the error measures like root mean squared error (RMSE), correlation coefficient (r), mean absolute error (MAE) along with the hydrographs and scatter plots. Table 2 represents results of all the five developed models.

From these results, it is can be said that one day ahead prediction of runoff at Shivade station is done by the ANN successfully. Results of correlation coefficient of Model I (0.66), III(0.68), and V(0.61) are better than Model II and IV. It is evident from these results that ANN has overpredicted all the results of runoff by all the models against the observed runoff and thus predicted runoff by ANN models is very high as compared to observed values. For example, in case of Model-1: observed-732.58 m3/s and predicted-1443.08 m3/s. Figure 4 depicts the same for Model-1.

Table 2 Results of developed ANN models

Model	Architecture	(r)	(RMSE) (m3/s)	(MAE) (m3/s)
I	21:1:1	**0.6680**	147.76	90.29
II	7:44:1	0.4621	145.54	89.83
III	21:16:1	**0.6811**	114.65	73.33
IV	14:2:1	0.5398	283.92	264.43
V	7:2:1	**0.6169**	176.96	96.35

Model II and Model IV results are too lower side which indeed needs to be improvised. Model-3 results show reasonable well compared to Model-1 with correlation coefficient (r) 0.68111 between observed and predicted runoff. Predicted runoff by the model shows that, Model-1 and Model-3 are better models than 2, 4, and 5. Thus, hydrograph and scatter plot for Models 1 and 3 are presented here. Overall, all model's prediction for extreme runoff values is poor compared to observed record, which is clearly seen into Figs. 4 and 6.

The lower performance of these models is itself a topic of further interest which is attributed to the complexity of rainfall to runoff process, poor quality of the rainfall data. Additionally, readers are requested to note that while selecting the inputs of all these R-R models, "rainfall" at Shivade which is one of the most important parameters, is not considered purposely as it was not available in consistent form. Still these models have captured the rainfall-runoff process reasonably well and predicted the runoff at Shivade with correlation coefficient more than 0.65. This depicts the capacity of ANN as a soft computing technique to map the random input with the output. ANN models are more flexible towards data, since they do not assume any mathematical model, thus it stands true with the guiding principle of any soft computing technique which is to exploit the tolerance for imprecision, uncertainty, partial truth, and approximation to achieve tractability, robustness while giving reasonable solutions to the problems. Definitely these results indicated need of

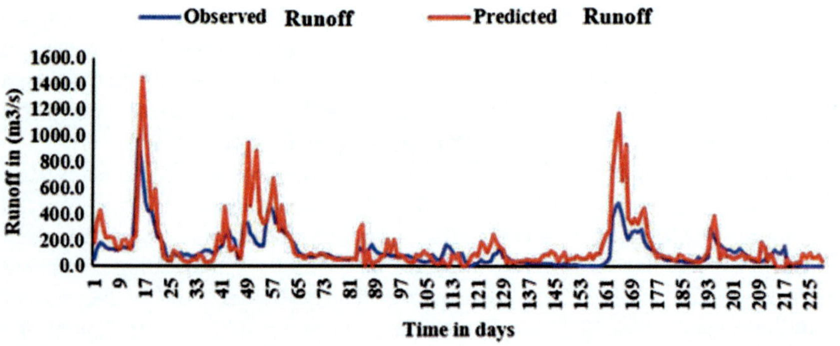

Fig. 4 Observed and predicted runoff for model-1

the advance technique/method to improvise the accuracy of the prediction perhaps which may be satisfied by the use of deep neural network. Deep neural network (DNN) facilitates the increased number of hidden layers which in turn suitable to capture the random and complex nature of the problems and thus helps to improvise the prediction accuracy by providing more degrees of freedom. Thus, from the results, it can be concluded that to improvise the results there is need of advance technique like deep neural network or so and authors will continue this study using DNN and will present the improvised results of runoff prediction at Shivade station (Figs. 5 and 7).

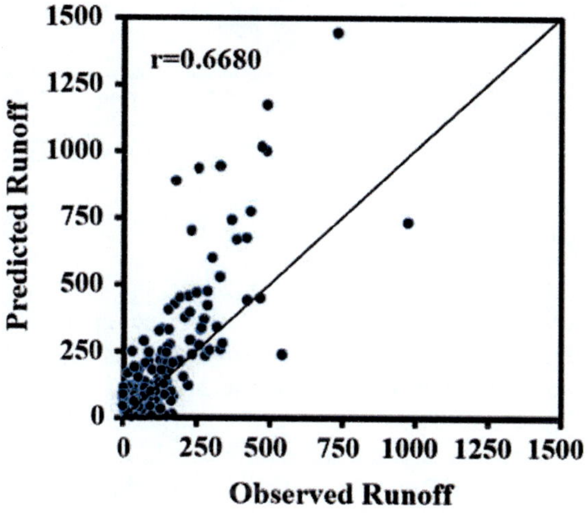

Fig. 5 Scatter plot for model-1

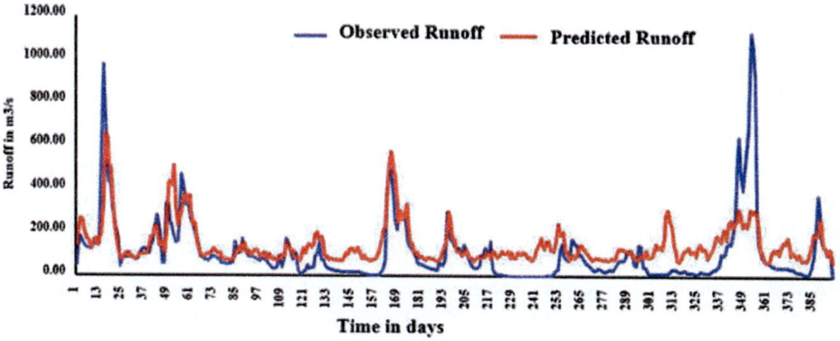

Fig. 6 Time series plot of observed and predicted runoff for model-3

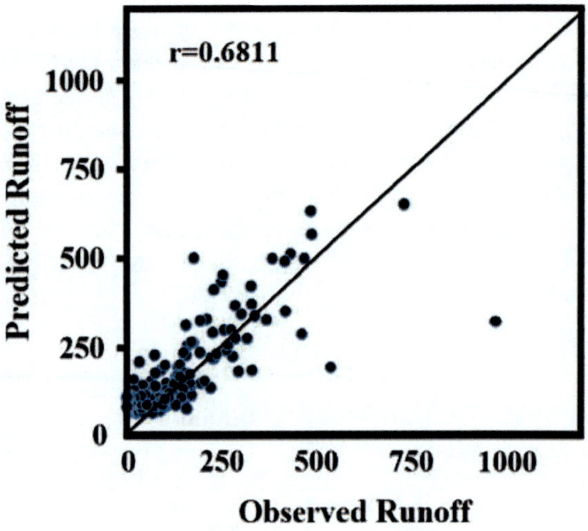

Fig. 7 Scatter plot for model-3

4 Conclusions

In the present work, runoff forecasting one day in advance at Shivade which lies on Upper Krishna Basin, Maharashtra, India was carried out using ANN. Models were developed with different combination of inputs, and its applicability was investigated for selected area. All the developed models performed reasonably in testing with few exceptions. It was observed that trial and error method for hidden neuron selection performed well in this work. Results of the present work show a no peak is predicted well by developed model. Model-3 performed better compared to other models. The results of the models largely affected by irregularity in measured rainfall data. Quality of the measured data impacts on the ANN model results. The all models were developed using past rainfall values of each selected stations. By adding one more layer to this, three-layered feedforward network may improve the model results and its potential. That means by using multiple hidden layers in the simple neural network forecasting may improve.

By moving from simple neural network (three-layered feedforward network) to deep neural network (addition of more hidden layer) could improve the accuracy and need to verify this reason in future work. Finally, it is concluded that ANN model is an important alternative to the conceptual model for the rainfall-runoff analysis.

Acknowledgements The authors are also grateful to Hydrology Project User Group (HDUG) Nashik, India for supplying the information needed to conduct the current study.

References

1. Abdalhi MA, Jingyi Z, Ali O (2020) Application of artificial neural networks (ANNs) based rainfall-runoff model for flood forecasting. J Agricul Sci Eng 6(2):17–25
2. Asadi H, Shahedi K, Jarihani B, Roy CS (2019) Rainfall-runoff modelling using hydrological connectivity index and artificial neural network approach. J mdpi Water 11(2):212. https://doi.org/10.3390/w11020212
3. Antar MA, Elassiouti I, Alam MN (2006) Rainfall–runoff modeling using artificial neural networks technique: a blue Nile catchment case study. Hydrol Process 20(5):1201–1216
4. Aghelpour P, Varshavian V (2020) Evaluation of stochastic and artificial intelligence models in modeling and predicting of river daily flow time series. Stoch Env Res Risk Assess 34(1):33–50
5. Birikundavyi S, Labib R, Trung HT, Rousselle J (2002) Performance of neural networks in daily streamflow forecasting. J Hydrol Eng 7(5):392–398
6. Caihong H, Wu Q, Li H, Jian S, Li N, Lou Z (2018) Deep learning with a long short-term memory networks approach for rainfall-runoff simulation. J mdpi Water 10(1):15–43. https://doi.org/10.3390/w1011154
7. Chen SM, Wang YM, Tsou I (2013) Using artificial neural network approach for modelling rainfall-runoff due to typhoon. J Earth Syst Sci 122(2):399–405
8. Chen L, Singh VP, Guo S, Zhou J, Ye L (2014) Copula entropy coupled with artificial neural network for rainfall–runoff simulation. Stoch Env Res Risk Assess 28(7):1755–1767
9. de Vos NJ, Rientjes THM (2005) Constraints of artificial neural networks for rainfall–runoff odeling: trade-offs in hydrological state representation and model evaluation. Hydrol Earth Syst Sci 9:111–126
10. Gholami V, Khaleghi M (2021) A simulation of the rainfall-runoff process using artificial neural network and HEC-HMS model in forest lands. J Forest Sci 67(4):165–174. https://doi.org/10.17221/90/2020-JFS
11. Jain A, Sudheer KP, Srinivasulu S (2004) Identification of physical processes inherent in artificial neural network rainfall–runoff models. Hydrol Process 18:571–581
12. Kasiviswanathan KS, Sudheer KP (2013) Quantification of the predictive uncertainty of artificial neural network-based river flow forecast models. Stoch Env Res Risk Assess 27(1):137–146
13. Kasiviswanathan KS, Sudheer KP (2017) Methods used for quantifying the prediction uncertainty of artificial neural network based hydrologic models. Stoch Env Res Risk Assess 31(7):1659–1670
14. Kisi O (2007) Streamflow forecasting using different artificial neural network algorithms. J Hydrol Eng 12(5):532–539
15. Kisi O (2008) River flow forecasting and estimation using different artificial neural network techniques. Hydrol Res 39(1):27–40
16. Londhe SN, Dixit PR (2012) Forecasting streamflow using support vector regression and M5 model trees. Int J Eng Res Developm 2:1–12
17. Londhe SN, Narkhede S (2017) Forecasting streamflow using hybrid neuro-wavelet technique. J of ISH Hydraulic Eng 24:1–10
18. Londhe SN, Shah S (2017) A novel approach for knowledge extraction from artificial neural networks. J ISH Hydraulic Eng 1–13. https://doi.org/10.1080/09715010.2017.1409667
19. Londhe SN (2008) Soft computing approach for real-time estimation of missing wave heights. J Ocean Eng 35:1080–1089
20. Le X, Ho HV, Lee G, Jung S (2019) Application of long short-term memory (LSTM) neural network for flood forecasting. J mdpi Water 11(1):1387
21. Oyebode O, Stretch D (2018) Neural network modeling of hydrological systems: a review of implementation techniques. J Natural Resour Model 32(1):1–14. https://doi.org/10.1002/nrm.12189
22. Partal T, Cigizoglu HK, Kahya E (2015) Daily precipitation predictions using three different wavelet neural network algorithms by meteorological data. Stoch Env Res Risk Assess 29(5):1317–1329

23. Senthil Kumar AR, Sudheer KP, Jain SK, Agarwal PK (2004) Rainfall–runoff modeling using artificial neural network: comparison of networks types. Hydrol Process 19(6):1277–1291
24. Sitterson J, Knightes R, Wolfe K, Muche M, Avant B (2017) An overview of rainfall runoff model types. Environmental Protection Agency United States. EPA/600/R-14/152
25. The ASCE Task Committee (2000) Artificial neural networks in hydrology. II: hydrologic applications. J Hydrol Eng 5(2):124–137
26. The ASCE Task Committee (2000) Artificial neural networks in hydrology I: preliminary concepts. J Hydrol Eng 5(2):115–123
27. Yuan X, Chen C, Lei X, Yuan Y, Adnan RM (2018) Monthly runoff forecasting based on LSTM–ALO model. Stoch Env Res Risk Assess 32(8):2199–2212
28. Zhihua L, Zuo J, Rodriguez D (2020) Predicting of runoff using an optimized SWAT—ANN: a case study. J Hydrol-Reg Stud 29:100–688
29. Zhu S, Luo X, Yuan X, Xu Z (2020) An improved long short-term memory network for streamflow forecasting in the upper Yangtze River. Stoch Environ Res Assess 34:1–17

Prediction of Seasonal and Annual Rainfall of Pune and Mahabaleshwar Regions Using ANN and Regression Approaches

N. Vivekanandan, Aayushi Ghule, and Vaishnavi Darade

Abstract Prediction of rainfall has always been one of the most important issues in hydrological cycle and also essential in water resource development, planning and management of flood and drought. With the development of Artificial Intelligence (AI), a number of AI methods such as Artificial Neural Network (ANN), Adaptive Neuro-Fuzzy Inference System, Fuzzy Logic, Support Vector Machine and Evolutionary Optimization Algorithm are widely applied for rainfall prediction. In this paper, the ANN is considered because it represents a complex nonlinear relationship and extracts the dependence between the variables through training process. In ANN, Multi-Layer Perceptron (MLP) and Radial Basis Function (RBF) algorithms are used for training the network data. In addition to MLP and RBF, the Multiple Linear Regression (MLR) model based on regression (REG) approach is applied in rainfall prediction. This paper illustrates a study on prediction of seasonal daily and annual daily rainfalls by adopting ANN (viz., MLP and RBF) and REG (viz., MLR) approaches for Pune and Mahabaleshwar regions, and the results are obtained thereof. The performance of the MLP, RBF and MLR models applied in rainfall prediction is evaluated by using Model Performance Indicators (MPIs) such as correlation coefficient, Nash–Sutcliffe model efficiency and root mean squared error. On the basis of the MPI's values, it is found that the RBF is better suited model for prediction of seasonal daily rainfall while MLP for annual daily rainfall for Pune and Mahabaleshwar regions.

N. Vivekanandan (✉)
Central Water and Power Research Station, Pune, Maharashtra, India
e-mail: anandaan@rediffmail.com

A. Ghule · V. Darade
Department of Civil Engineering, Sinhgad College of Engineering, Pune, Maharashtra, India

P. V. Timbadiya et al. (eds.), *Geospatial and Soft Computing Techniques*, Lecture Notes in Civil Engineering 339, https://doi.org/10.1007/978-981-99-1901-7_36

Keywords Correlation coefficient · Multi-layer perceptron · Radial basis function · Nash–Sutcliffe model efficiency · Rainfall · Regression · Root mean squared error

1 Introduction

Knowledge of rainfall characteristics plays an important role in understanding hydrology of a region as well as for planning and management of water resources. Rainfall is one of the key natural resources that have a varying impact on human society such as agricultural activities, hydropower generation, flood control and sustainability of biodiversity. Apart from this, rainfall prediction is needed for estimating the water requirement in a particular area or a region. Since the distribution of rainfall varies over space and time, it is required to analyze the data covering long periods and recorded at various locations to arrive at a reliable information for decision support.

A number of approaches based on numerical, statistical, machine learning and empirical [1, 6] are generally applied for rainfall prediction. Due to nonlinear nature of Indian rainfall, machine learning-based models are gaining more popularity over empirical, numerical and statistical methods for accurate prediction of rainfall [18]. With more focus on Artificial Intelligence (AI) and availability of high computational devices, a number of various AI methods, viz., Artificial Neural Network (ANN), Adaptive Neuro-Fuzzy Inference System (ANFIS), Fuzzy Logic, Support Vector Machine and Evolutionary Optimization Algorithm, have gained a lot amount of attention in predicting the meteorological data [11]. Out of these methods, ANN can represent a complex nonlinear relationship and extract the dependence between the variables through training process and hence used. In ANN, the training algorithms such as Bayesian, cascade correlation, conjugate gradient, Multi-Layer Perceptron (MLP) and Radial Basis Function (RBF) networks are generally applied for training the network data.

Chattopadhyay [3] also indicated that the feedforward ANN has less error than Multiple Linear Regression (MLR) in predicting the average summer monsoon rainfall over India. Dahamsheh and Aksoy [7] suggested that the ANNs were slightly better than MLR in forecasting the monthly total precipitation of arid regions. Azadi and Sepaskhah [2] concluded that ANNs did not significantly increase prediction accuracy compared with the MLR. Study by Nayak et al. [14] revealed that the rainfall prediction using ANN is more suitable than traditional statistical and numerical methods. Mislan et al. [13] applied the ANN with Backpropagation Neural Network (BPNN) algorithm for monthly rainfall prediction for Tenggarong station of East Kalimantan, Indonesia. Choubin et al. [5] compared the performance of MLR, MLP and ANFIS used in forecasting precipitation based on large-scale climate signals. Lee et al. [12] implemented the spatial prediction of flood susceptibility by using random forest and boosted-tree models in Seoul metropolitan city, Korea.

Satish et al. [17] applied the ANN-based hybrid genetic algorithm to train the network data for rainfall prediction. Sofian et al. [19] carried out the study on monthly rainfall prediction using BPNN and RBF networks. Zhang et al. [22] applied Support Vector Regression (SVR)–MLP method for prediction of annual and non-monsoon rainfall for Odisha. By considering the research works carried out by various researchers on rainfall prediction using ANN, it is found that the MLP and RBF algorithms are widely applied for training the network data and hence used in the study. In addition to ANN, many of the researchers have also applied regression approach for rainfall prediction. Swain et al. [20] developed a MLR model to reckon annual precipitation over Cuttack district, Odisha. India. Study by Refona et al. [16] applied the linear regression model for prediction of rainfall in Chennai district. Gnanasankaran and Ramaraj [9] applied the machine learning (ML) algorithm and MLR model for rainfall forecasting by using a set of meteorological data including the monthly wise rainfall in India. They have found that the MLR gave better results than those values of ML. Patil et al. [15] applied MLR, neural networks and decision trees algorithm to predict the rainfall by using the Austin weather dataset that was collected from Kaggle.

In the present study, both ANN and REG approaches are considered in prediction of rainfall. The performance of the ANN (viz., MLP and RBF) and REG (viz., MLR) approaches adopted in rainfall prediction is evaluated by using Model Performance Indicators (MPIs) such as correlation coefficient (CC), Nash–Sutcliffe model efficiency (NSE) and root mean squared error (RMSE). This paper illustrates a study on prediction of seasonal daily (herein after called as seasonal) rainfall and annual daily (herein after called as annual) rainfall using MLP, RBF and MLR models for Mahabaleshwar and Pune regions, and the results obtained thereof.

2 Methodology

The concept of ANN and theoretical descriptions of MLP and RBF models applied in training the network data are briefly described in the following sections.

2.1 Artificial Neural Network

ANN modeling procedures adapt to complexity of input–output patterns and accuracy goes on increasing as more and more data become available. The ANN architecture (Fig. 1) consists of input layer, hidden layer and output layer [21]. From ANN structure, it can be easily understood that input units receive the data from external sources to the network and send to the hidden units, in turn, the hidden units send and receive the data only from other units in the network and output units receive and produce the data generated by the network, which goes out of the system. In this process, a typical problem is to estimate the output as a function of the input.

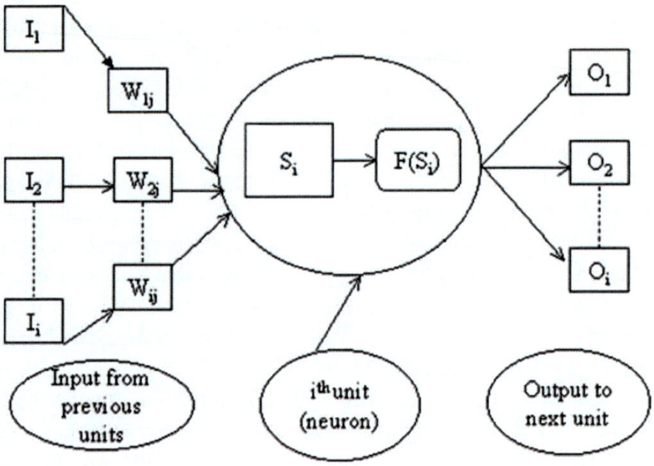

Fig. 1 Architecture of ANN

This unknown function may be approximated by a superposition of certain activation function such as tangent, sigmoid, polynomial and sinusoid in ANN [8]. A common threshold function used in ANN is the sigmoid function ($f(S)$) expressed by Eq. (1), which provides an output in the range of $0 \leq f(S) \leq 1$.

$$f(S) = \left[1 + \exp(-S_i)\right]^{-1} \text{ wherein } S_i = \sum_{i=1}^{N} I_i W_{ij} + O_i,\ j = 1, 2, 3, \ldots, M, \tag{1}$$

where $f(S)$ is the characteristic function of S, S_i is the characteristic function of ith layer, I_i is the input (I) unit of ith layer, O_i is the output (O) unit of ith layer, W_{ij} is the synaptic weights between input (i) and hidden (j) layers, N is the number of observations and M is the number of neurons (or units) of hidden layer. A network output is compared with target output and output error (E) is given by Eq. (2).

$$E = \frac{1}{2}\sum_{i=1}^{N} \left(X_i - X_i^*\right)^2, \tag{2}$$

where X_i is the observed value of ith sample and X_i^* is the predicted value for ith sample.

2.2 Theoretical Description of MLP

MLP network [8] is based on architecture with single hidden layer as shown in Fig. 1. Gradient descent method is commonly used in MLP in which each input unit of the training dataset is passed through the network from the input layer to output layer.

$$\Delta W_{ij}(M) = -\varepsilon \frac{\partial E}{\partial W_{ij}} + \alpha \Delta W_{ij}(M-1), \tag{3}$$

where $\Delta W_{ij}(M)$ is the weight increments between ith and jth layers during M neurons (units) and $\Delta W_{ij}(M-1)$ is the weight increments between ith and jth layers during $M{-}1$ neurons. In MLP, momentum factor (α) is used to speed up training in very flat region of the error surface to prevent oscillation in the weight and learning rate (ε) is used to increase the chance of avoiding the training process being trapped in local minima instead of global minima.

2.3 Theoretical Description of RBF

RBF network is supervised and three-layered feedforward neural network. The hidden layer of RBF network consists of a number of nodes and a parameter vector called a 'centre', which can be considered the weight vector. In RBF, the standard Euclidean distance is used to measure the distance of an input vector from the center. The design of neural networks is a curve-fitting problem in a high dimensional space in RBF [10]. Training the RBF network implies finding the set of basis nodes and weights. Therefore, the learning process is to find the best fit to the training data. The transfer function of the nodes is governed by nonlinear function that is assumed to be an approximation of the influence that data points have at the center. The transfer function of a RBF is mostly built up of Gaussian rather than sigmoid.

The transfer function of the nodes is governed by nonlinear function that is assumed to be an approximation of the influence that data points have at the center. The Euclidean length is represented by r_j that measures the radial distance between the datum vector $\underline{X}(X_1, X_2, \ldots X_M)$ and the radial center $\underline{X}^{(j)} = (W_{1j}, W_{2j}, \ldots W_{Mj})$ which can be written as:

$$r_j = \left\| \underline{X} - \underline{X}^{(j)} \right\| = \left[\sum_{i=1}^{M} (X_i - W_{ij})^2 \right]^{1/2}, \tag{4}$$

where $r_j = \| \|$ is the Euclidean norm and $\Phi(\ldots)$ is the activation function. A suitable transfer function is then applied to r_j to give $\Phi(r_j) = \Phi \left\| \underline{X} - \underline{X}^{(k)} \right\|$. Finally, the output layer ($k{-}1$) receives a weighted linear combination of $\Phi(r_j)$.

$$X^{(k)} = W_0 + \sum_{j=1}^{N} c_j^{(k)} \Phi(r_j) = W_0 + \sum_{j=1}^{N} c_j^{(k)} \Phi(\|\underline{X} - \underline{X^{(j)}}\|), \quad (5)$$

where $c_j^{(k)}$ is the center (c) of the neuron (j) in the hidden layer (k), $\Phi(r_j)$ is the response of r_j in the jth hidden unit and W_0 is the bias term.

2.4 Theoretical Description of Regression

Regression is a statistical technique of data mining that has wide range of application in various fields like rainfall–runoff modeling, prediction of meteorological events, stream flow forecasting, etc. Also, regression in simple term is defined as the prediction of one variable from another variable that can be easily obtained by using simple linear regression. Thereafter, the MLR is used to describe the process by which several variables that are used to predict the desired variable. A general form of the MLR model is given as below:

$$Y = a_0 + a_1 X_1 + a_2 X_2 + a_3 X_3 + \ldots + a_n X_n, \quad (6)$$

where Y is the predicted value, a_i's ($i = 1$ to n) are the predictor coefficients and X_i's ($i = 1$ to n) are the predictors.

2.5 Model Performance Analysis

The performance of ANN (viz., MLP and RBF) and REG (viz. MLR) approaches adopted in prediction of seasonal and annual rainfalls has been evaluated by using MPIs such as CC, NSE and RMSE. The theoretical descriptions of MPIs are given as below:

$$\begin{aligned} CC &= \frac{\sum_{i=1}^{N} (X_i - \overline{X})(X_i^* - \overline{X^*})}{\sqrt{\left(\sum_{i=1}^{N} (X_i - \overline{X})^2\right)\left(\sum_{i=1}^{N} (X_i^* - \overline{X^*})^2\right)}} \\ NSE(\%) &= \left(1 - \frac{\sum_{i=1}^{N} (X_i - X_i^*)^2}{\sum_{i=1}^{N} (X_i - \overline{X})^2}\right) * 100, \\ RMSE &= \left(\frac{1}{N} \sum_{i=1}^{N} (X_i - X_i^*)^2\right)^{1/2} \end{aligned} \quad (7)$$

where $\overline{X}$ is the average of observed data and $\overline{X^*}$ is the average of predicted data [4]. The model with high *CC*, better *NSE* and minimum *RMSE* is considered as better suited for prediction of seasonal and annual rainfalls.

3 Application

In this paper, a study on prediction of seasonal and annual daily rainfall for Pune and Mahabaleshwar regions using ANN and REG approaches is carried out. Mahabaleshwar is a vast plateau bounded by valley from all sides. Mahabaleshwar region is located at approximately 17° 55′ 18″ N latitude and 73° 39′ 20″ E longitude. Mahabaleshwar receives heavy rainfall during monsoon and is cold enough in winter. Likewise, Pune lies on the western side of Deccan Plateau and is on leeward side of Sahyadri mountain range which forms a barrier from Arabian Sea. Pune region is located at approximately 18° 31′ 00″ N latitude and 73° 51′ 22″ E longitude. Pune has hot semi-arid climate and receives moderate rainfall. The index map of the study area with locations of Pune and Mahabaleshwar regions is shown in Fig. 2.

In the present study, the daily series of the meteorological data, viz., rainfall (RFL), maximum temperature (T_{max}) and minimum temperature (T_{min}), average wind speed (AWS) and evaporation (EPR) (for Pune only) observed at Pune for the period 1997–2016 and Mahabaleshwar for the period 1997–2014 is collected from India Meteorological Department and also used. From the scrutiny of the data, it is found that the observed data for few days in a month of a year are not available, and hence, those values are not considered in data analysis. The seasonal (monsoon and post-monsoon) and annual rainfall data series is extracted from the daily rainfall data series and used in rainfall prediction by applying MLP, RBF and MLR models. In the present study, 80% of the observed data is used for training (TRG) and the remaining 20% is used for testing (TES). As the units of the meteorological data, viz., RFL (in mm), T_{max} and T_{min} (in °C), AWS (in km/hour) and EVA (in mm/day) considered in the study are in different units, these values are normalized through Eq. (8) and used in rainfall prediction. After completion of the training and testing processes, the output data are again denormalized through Eq. (8) to achieve the results in original domain.

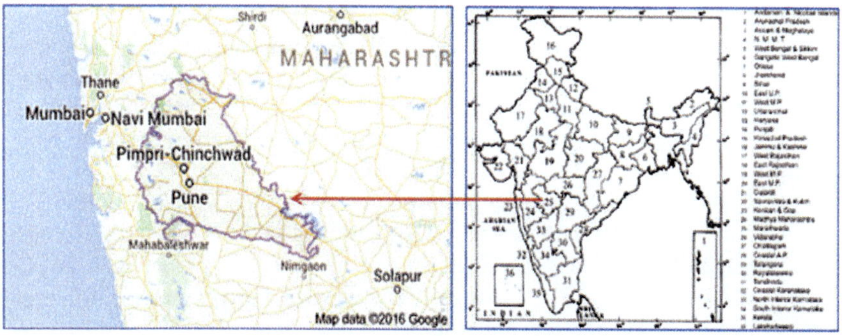

Fig. 2 Index map of the study area with locations of Pune and Mahabaleshwar regions

$$\text{Nor}(X_i) = \frac{X_i - \text{Min}\,(X_i)}{\text{Max}\,(X_i) - \text{Min}\,(X_i)}, \quad (8)$$

where Nor (X_i) is the normalized value of X_i, Min (X_i) is the series minimum value of X_i and Max (X_i) is the series maximum value of X_i.

4 Results and Discussion

By applying the procedures of MLP and RBF models, as described above, prediction of seasonal and annual rainfall for Pune and Mahabaleshwar regions was carried out. In ANN approach, the meteorological data series was trained with MLP and RBF. Table 1 presents the details on Optimum Network Architecture (ONA), number of data points used in data analysis, parameters of MLP and RBF models used in training the network data and the input and output data (i.e., units) considered in seasonal and annual rainfall prediction. In the present study, Statistical Package for the Social Sciences (SPSS) was used to train the network data with MLP and RBF models.

4.1 Prediction of Seasonal Rainfall Using ANN and REG Approaches

By using the parameters, as given in Table 1, the network data were trained with MLP and RBF models for prediction of seasonal (monsoon and post-monsoon) rainfall. In regression approach, the normalized values of meteorological data were used to develop a MLR model for rainfall prediction. The MLR models used in prediction of seasonal rainfall for Pune and Mahabaleshwar regions are given in Table 2. By using the MLR models, the normalized values of rainfall were estimated and thereafter denormalized through Eq. (8) to get the predicted rainfall in original domain.

The descriptive statistics of the observed and predicted values of seasonal rainfall using MLP, RBF and MLR models for Pune and Mahabaleshwar are given in Tables 3 and 4. The plots of predicted rainfall using MLP, RBF and MLR models with observed rainfall for monsoon season of Pune and Mahabaleshwar are given in Fig. 3, while the plots of post-monsoon season are shown in Fig. 4.

By using the descriptive statistics, as given in Table 3, the percentage of variation in the average of predicted rainfall using MLP, RBF and MLR models with reference to the average of observed rainfall during testing period is computed as 20.0%, 18.8% and 6.3%, respectively, for monsoon season of Pune. Likewise, for post-monsoon season of Pune, these values are computed as 13.1%, 11.5% and 10.7%, respectively. Similarly, from the values of descriptive statistics as presented in Table 4, the percentage of variation in the average of predicted rainfall using MLP, RBF and MLR models with reference to the average of observed rainfall during testing period is computed as 9.7%, 4.4% and 6.0%, respectively for monsoon season of

Table 1 Parameters considered in prediction of seasonal and annual rainfall using MLP and RBF models for Pune and Mahabaleshwar

Parameters	Pune				Mahabaleshwar			
	MLP		RBF		MLP		RBF	
	TRG	TES	TRG	TES	TRG	TES	TRG	TES
Input variables	T_{max}, T_{min}, AWS, EPR				T_{max}, T_{min} and AWS			
Output variable	Rainfall				Rainfall			
Activation function	Sigmoid				Sigmoid			
Monsoon (June–September)								
ONA	4–8–1		4–6–1		3–6–1		3–5–1	
Learning rate (ε)	0.7		–		0.6		–	
Momentum factor (α)	0.6		–		0.5		–	
Number of data samples	1254	314	1254	314	1284	321	1284	321
Post-monsoon (October–November)								
ONA	4–5–1		4–4–1		3–5–1		3–4–1	
Learning rate (ε)	0.6		–		0.7		–	
Momentum factor (α)	0.7		–		0.6		–	
Number of data samples	174	43	174	43	178	45	178	45
Annual (January–December)								
ONA	4–9–1		4–6–1		3–8–1		3–6–1	
Learning rate (ε)	0.5		–		0.7		–	
Momentum factor (α)	0.7		–		0.6		–	
Number of data samples	1512	378	1512	378	1512	378	1512	378

For example, the architecture 4–9–1 indicates that the network consists one input layer with four input units, one hidden layer with nine hidden units and one output layer with one output unit

Mahabaleshwar. For post-monsoon of Mahabaleshwar, these values are computed as 37.5%, 23.6% and 18.1%, respectively. However, from Figs. 3 and 4, it can be seen that the predicted rainfall using RBF is closer to the observed rainfall for monsoon and post-monsoon seasons of Pune and Mahabaleshwar regions.

4.2 Prediction of Annual Rainfall Using ANN and REG Approaches

By using the parameters, as given in Table 1, the network data were trained with MLP and RBF models for prediction of annual rainfall. In regression approach, the normalized values of meteorological data were used to develop a MLR model for rainfall prediction. The MLR models used in prediction of annual rainfall for Pune

Table 2 MLR models used in prediction of seasonal rainfall for Pune and Mahabaleshwar

Region	MLRs using normalized values of predictors
Pune	
Monsoon	$NRFL = (1.641)NT_{max} - (0.260)NT_{min} + (2.237)NAWS + (0.367)NEPR - 1.143$
Post-monsoon	$NRFL = (0.840)NT_{max} - (0.348)NT_{min} + (1.807)NAWS + (0.317)NEPR - 0.695$
Mahabaleshwar	
Monsoon	$NRFL = (0.328)NT_{max} - (0.242)NT_{min} + (0.129)NAWS + 0.014$
Post-monsoon	$NRFL = (0.009)NT_{max} - (0.008)NT_{min} + (0.019)NAWS + 0.099$
NT_{max}:	Normalized value of maximum temperature
NT_{min}:	Normalized value of minimum temperature
NAWS:	Normalized value of average wind speed
NEPR:	Normalized value of evaporation
NRFL:	Normalized value of rainfall

Table 3 Descriptive statistics of observed and predicted seasonal rainfall using MLP, RBF and MLR models for Pune

Descriptive statistics	Observed rainfall		Predicted rainfall					
			MLP		RBF		MLR	
	Training	Testing	Training	Testing	Training	Testing	Training	Testing
Monsoon								
Average (mm)	8.4	8.0	9.3	9.6	9.6	9.5	9.0	8.5
SD (mm)	14.5	14.1	14.4	14.2	14.1	13.9	14.6	14.2
CS	2.998	3.253	2.942	3.169	3.032	3.316	2.959	3.165
CK	10.054	11.463	9.827	11.395	10.535	12.047	9.669	10.897
Post-monsoon								
Average (mm)	10.0	12.2	10.9	13.8	11.5	13.6	11.7	13.5
SD (mm)	13.9	18.1	13.5	18.6	14.6	18.8	15.4	19.3
CS	2.671	2.789	2.696	3.246	2.613	2.832	2.248	2.645
CK	9.819	9.951	9.547	13.688	9.725	10.139	6.254	8.283

SD: Standard Deviation; CS: Coefficient of Skewness; CK: Coefficient of Kurtosis

and Mahabaleshwar regions are given in Table 5. By using the MLR models, the normalized values of rainfall were computed and thereafter denormalized through Eq. (8) to get the predicted rainfall in original domain.

The descriptive statistics of the observed and predicted values of annual rainfall using MLP, RBF and MLR models for Pune and Mahabaleshwar are given in Table 6. The time series plots of predicted annual rainfall using MLP, RBF and MLR models together with observed annual rainfall for Pune and Mahabaleshwar are presented in

Table 4 Descriptive statistics of observed and predicted seasonal rainfall using MLP, RBF and MLR models for Mahabaleshwar

Descriptive statistics	Observed rainfall		Predicted rainfall					
			MLP		RBF		MLR	
	Training	Testing	Training	Testing	Training	Testing	Training	Testing
Monsoon								
Average (mm)	50.7	54.6	54.1	59.9	52.3	57.0	53.2	57.9
SD (mm)	62.9	62.3	60.7	59.4	61.9	59.9	62.0	63.4
CS	2.460	2.099	2.612	2.289	2.468	2.145	2.393	1.985
CK	8.228	6.065	9.576	7.672	8.203	6.458	8.106	5.188
Post-monsoon								
Average (mm)	11.5	7.2	14.2	9.9	13.6	8.9	13.4	8.5
SD (mm)	14.7	9.3	15.7	9.1	14.4	9.1	15.6	10.2
CS	2.415	1.898	3.052	1.237	2.496	1.817	2.117	1.828
CK	8.165	3.795	14.667	1.417	8.312	3.627	5.551	3.404

Fig. 5. By using the values of descriptive statistics, as given in Table 6, the percentage of variation in the average of predicted rainfall using MLP, RBF and MLR models with reference to the average of observed rainfall during testing period is computed as 18.6%, 19.8% and 5.8%, respectively, for Pune. Likewise, for Mahabaleshwar, these values are computed as 5.4%, 14.9% and 4.5%, respectively, for Mahabaleshwar. However, from Fig. 5, it can be seen that the predicted annual rainfall using MLP is closer to the observed annual rainfall for Pune and Mahabaleshwar regions.

4.3 Analysis of Results Based on MPIs

The performance of MLP, RBF and MLR models applied in prediction of seasonal and annual rainfall for Pune and Mahabaleshwar was evaluated by using MPIs and the results are presented in Tables 7 and 8.

Based on the MPIs' values, some of the observations drawn from the study are summarized and given as below:

(i) The RMSE on the predicted rainfall using RBF is found as minimum for monsoon and post-monsoon seasons when compared with those values of MLP and MLR models. For annual rainfall, the RMSE given by MLP is lesser than those values of RBF and MLR models for Pune and Mahabaleshwar.
(ii) From the CC values, it is noted that there is generally good correlation between the observed and predicted values using MLP, RBF and MLR models for seasonal and annual rainfalls. The CC values in the rainfall prediction using MLP, RBF and MLR models varied from 0.902 to 0.994 for Pune while 0.930 to 0.993 for Mahabaleshwar.

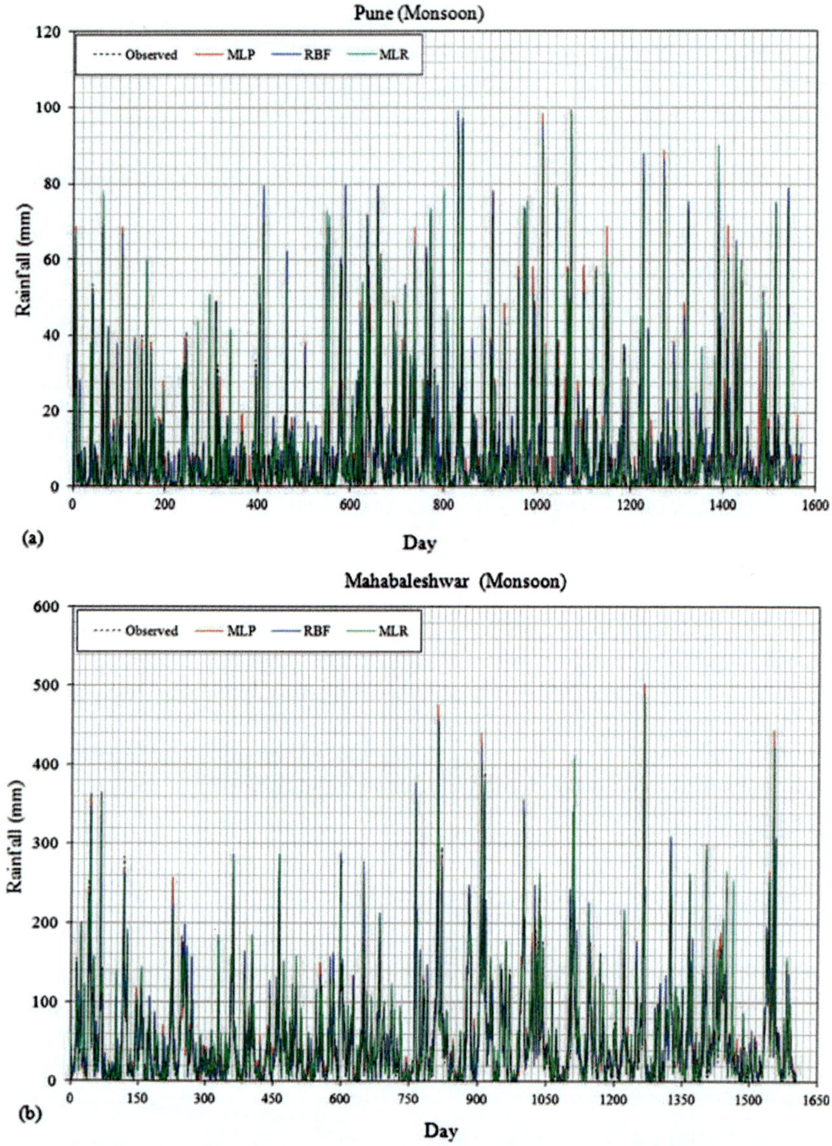

Fig. 3 Plots of predicted rainfall using MLP, RBF and MLR models with observed rainfall for monsoon season of Pune and Mahabaleshwar

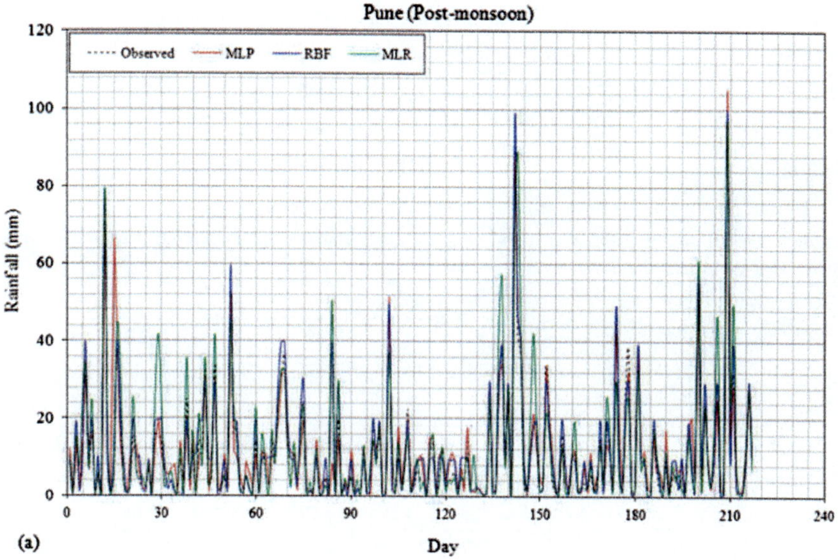

(a)

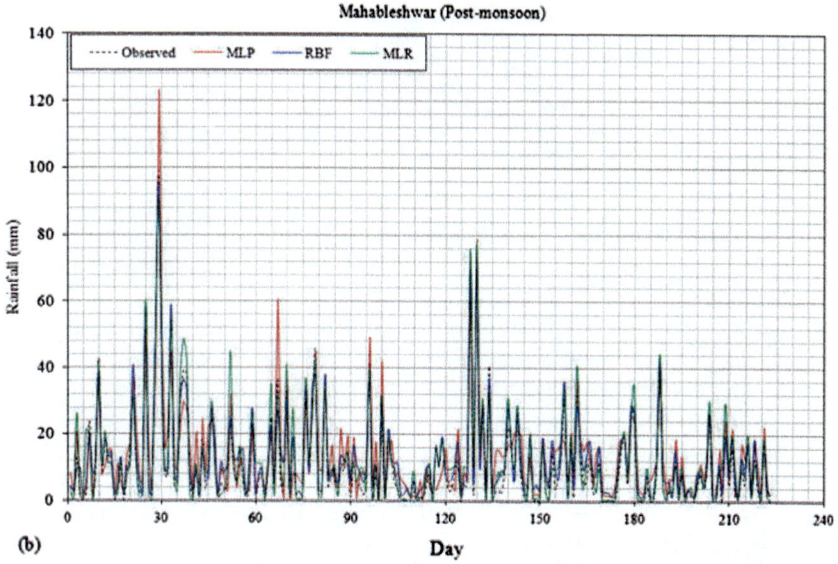

(b)

Fig. 4 Plots of predicted rainfall using MLP, RBF and MLR models with observed rainfall for post-monsoon season of Pune and Mahabaleshwar

Table 5 MLR models used in prediction of annual rainfall for Pune and Mahabaleshwar

Region	MLRs using normalized values of predictors
Pune	NRFL = (0.146)NT_{max}−(0.085)NT_{min} + (0.223)NAWS + (0.236)NEPR-0.243
Mahabaleshwar	NRFL = (0.032)NT_{max}−(0.024)NT_{min} + (0.014)NAWS + 0.013

Table 6 Descriptive statistics of observed and predicted annual rainfall using MLP, RBF and MLR models for Pune and Mahabaleshwar

Descriptive statistics	Observed rainfall		Predicted rainfall					
			MLP		RBF		MLR	
	Training	Testing	Training	Testing	Training	Testing	Training	Testing
Pune								
Average (mm)	8.5	8.6	9.5	10.2	9.7	10.3	10.2	9.1
SD (mm)	14.1	15.3	14.2	15.5	13.8	15.1	14.7	14.7
CS	2.988	3.298	2.908	3.295	3.012	3.415	2.643	2.907
CK	10.310	12.348	9.959	12.862	10.670	13.484	8.147	9.249
Mahabaleshwar								
Average (mm)	43.7	44.2	45.4	46.6	48.5	50.8	43.9	46.2
SD (mm)	59.2	59.2	58.4	56.9	56.5	57.2	58.8	58.9
CS	2.692	2.399	2.681	2.409	2.805	2.510	2.751	2.464
CK	9.892	7.622	9.741	7.968	11.410	9.271	10.478	8.278

(iii) The NSE given by RBF for seasonal rainfall while MLP for annual rainfall is comparatively better than those values of other models adopted in rainfall prediction.

(iv) For monsoon and post-monsoon seasons of Pune, the NSE given by RBF in testing period is computed as 98.6 and 96.8%. For Mahabaleshwar, the NSE given by RBF is computed as 97.6% for monsoon season, while 92.6% for post-monsoon.

(v) For annual rainfall, the NSE given by MLP during testing period is computed as 94.5% for Pune and 98.3% for Mahabaleshwar.

(vi) Based on the analysis of the results using MPIs, for Pune and Mahabaleshwar, it is identified that the RBF is better suited among three models (viz., MLP, RBF and MLR) applied in predicting the seasonal rainfall, whereas MLP for annual rainfall.

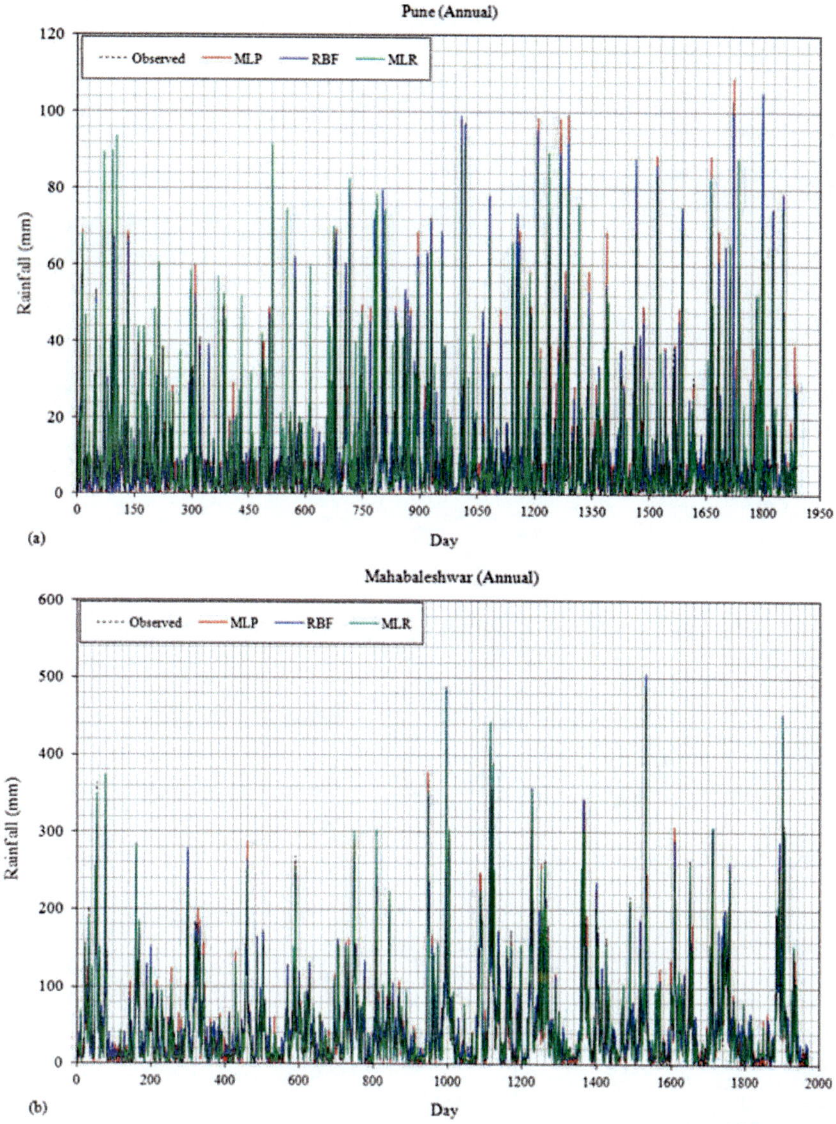

Fig. 5 Plots of predicted annual rainfall using MLP, RBF and MLR models with observed annual rainfall for Pune and Mahabaleshwar

Table 7 MPIs values given by MLP, RBF and MLR models for seasonal and annual rainfall of Pune

MPIs	MLP		RBF		MLR	
	Training	Testing	Training	Testing	Training	Testing
Monsoon						
CC	0.974	0.968	0.990	0.994	0.977	0.973
NSE (%)	94.4	92.4	97.9	98.6	94.8	93.2
RMSE (mm)	3.427	3.894	2.119	1.649	3.307	3.662
Post-monsoon						
CC	0.902	0.984	0.976	0.988	0.912	0.962
NSE (%)	89.4	95.8	93.6	96.8	78.2	90.9
RMSE (mm)	6.341	3.648	3.520	3.181	6.512	5.385
Annual						
CC	0.973	0.975	0.967	0.978	0.906	0.949
NSE (%)	94.1	94.5	92.9	94.3	80.4	89.6
RMSE (mm)	3.422	3.650	3.770	3.800	6.508	4.729

Table 8 MPIs values given by MLP, RBF and MLR models for seasonal and annual rainfall of Mahabaleshwar

MPIs	MLP		RBF		MLR	
	Training	Testing	Training	Testing	Training	Testing
Monsoon						
CC	0.990	0.985	0.988	0.989	0.987	0.973
NSE (%)	97.6	96.1	97.5	97.6	97.3	94.4
RMSE (mm)	9.673	12.243	9.895	9.632	10.402	15.013
Post-monsoon						
CC	0.930	0.961	0.974	0.980	0.963	0.967
NSE (%)	81.4	83.3	92.9	92.6	90.4	86.3
RMSE (mm)	6.314	3.774	3.916	2.506	4.580	3.303
Annual						
CC	0.993	0.992	0.972	0.970	0.986	0.987
NSE (%)	98.6	98.3	93.3	92.8	97.1	97.2
RMSE (mm)	7.093	7.615	14.699	15.839	10.127	9.848

5 Conclusions

The paper presented a study on prediction of seasonal daily and annual daily rainfall using ANN (viz., MLP and RBF) and regression (viz., MLR) for Pune and Mahabaleshwar regions. The performance of the models applied in rainfall prediction was evaluated by using MPIs such as CC, NSE and RMSE. On the basis of the evaluation of the results through MPIs, some of the conclusions drawn from the study are summarized and presented as below:

- The Optimum Network Architectures with parameters of MLP and RBF, as given in Table 1, were used for training the network data.
- The time series plots showed that the predicted values by RBF for seasonal rainfall and MLP for annual rainfall are comparatively better than those values of other models used in the study.
- The CC values indicated that there was generally a good correlation between the observed and predicted rainfall using MLP, RBF and MLR models, and these values vary from 0.902 to 0.994 for Pune, while 0.930 to 0.993 for Mahabaleshwar.
- For Pune, the NSE in predicting the rainfall using RBF for monsoon and post-monsoon seasons during testing period was computed as 98.6 and 96.8%. For Mahabaleshwar, the NSE for monsoon and post-monsoon seasons in testing period was computed as 97.6 and 92.6%.
- The NSE in predicting the annual rainfall using MLP models for Pune and Mahabaleshwar during testing period was computed as 94.5% and 98.3%.
- For monsoon and post-monsoon seasons of Pune, the percentage of variation in the average of predicted rainfall using RBF with reference to the average of observed rainfall was computed as 18.8 and 11.5% in testing period. For Mahabaleshwar, these values were computed as 4.4% for monsoon season and 23.6% for post-monsoon season.
- The percentage of variation in the average of predicted annual rainfall using MLP with reference to the average of observed rainfall during testing period was computed as 18.6% for Pune and 5.4% for Mahabaleshwar.

In light of the above, it is suggested that the predicted seasonal daily rainfall using RBF and annual daily rainfall using MLP could be used for design purposes. The outcomes of the study would also be useful for stakeholders for planning, design and management of water resources projects in Pune and Mahabaleshwar regions.

Acknowledgements The authors are thankful to Shri A.K. Agrawal, Director, Central Water and Power Research Station, Pune, for providing the research facilities to carry out the study. The contents and views expressed in this research paper are the views of the authors and do not necessarily reflect the view of the organization/institution they belong to.

References

1. Al Mamun A, Bin Salleh MN, Noor HM (2018) Estimation of short-duration rainfall intensity from daily rainfall values in Klang valley, Malaysia. Appl Water Sci 8(7):1–10. https://doi.org/10.1007/s13201-018-0854-z
2. Azadi S, Sepaskhah AR (2012) Annual precipitation forecast for west, southwest, and south provinces of Iran using artificial neural networks. Theoret Appl Climatol 109(1–2):175–189. https://doi.org/10.1007/s00704-011-0575-9

3. Chattopadhyay S (2007) Feed forward artificial neural network model to predict the average summer monsoon rainfall in India. Acta Geophys 55(3):369–382. https://doi.org/10.2478/s11600-007-0020-8
4. Chen J, Adams BJ (2006) Integration of artificial neural networks with conceptual models in rainfall-runoff modelling. J Hydrol 318(1–4):232–249. https://doi.org/10.1016/j.jhydrol.2005.06.017
5. Choubin B, Malekian S, Golshan M (2016) Application of several data-driven techniques to predict a standardized precipitation index. Atmósfera 29(2):121–128. https://doi.org/10.20937/ATM.2016.29.02.02
6. Cramer S, Kampouridis M, Freitas AA, Alexandridis AK (2017) An extensive evaluation of seven machine learning methods for rainfall prediction in weather derivatives. Expert Syst with Appl 85(November issue):169–181. https://doi.org/10.1016/j.eswa.2017.05.029
7. Dahamsheh A, Aksoy H (2009) Artificial neural network models for forecasting intermittent monthly precipitation in arid regions. Meteorol Appl 16(3):325–337. https://doi.org/10.1002/met.127
8. Dubey A (2015) Artificial neural network models for rainfall prediction in Pondicherry. Int J Computat Appl 120(3):30–35. https://doi.org/10.5120/21210-3910
9. Gnanasankaran N, Ramaraj E (2020) A multiple linear regression model to predict rainfall using Indian meteorological data. Int J Adv Sci Technol 29(8):746–758. Google Scholar
10. Kaltech M (2008) Rainfall-runoff modelling using artificial neural networks: modelling and understanding. Caspian J Environ Sci 6(1):53–58. Google Scholar
11. Ko CM, Jeong YY, Lee YM, Kim BS (2020) The development of a quantitative precipitation forecast correction technique based on machine learning for hydro-logical applications. Atmosphere 11(1):1–17. https://doi.org/10.3390/atmos11010111
12. Lee S, Kim JC, Jung HS, Lee MJ, Lee S (2017) Spatial prediction of flood susceptibility using random-forest and boosted-tree models in Seoul metropolitan city, Korea. Geomatics. Natural Hazards and Risk 8(2):1185–1203. https://doi.org/10.1080/19475705.2017.1308971
13. Mislan H, Hardwinarto S, Sumaryono MA (2015) Rainfall monthly prediction based on artificial neural network: a case study in Tenggarong Station. East Kalimantan-Indonesia. Proc Comput Sci 59:142–151. https://doi.org/10.1016/j.procs.2015.07.528
14. Nayak DR, Mahapatra A, Mishra P (2013) A survey on rainfall prediction using artificial neural network. Int J Comput Appl 72(16):32–40. https://doi.org/10.5120/12580-9217
15. Patil D, Badarpura S, Jain A, Aniket Gupta A (2020) Rainfall prediction using linear approach and neural networks and crop recommendation based on decision tree. Int J Eng Res Technol 9(4):394–399. Google Scholar
16. Refonaa J, Lakshmi M, Abbas R, Raziullha M (2019) Rainfall prediction using regression model. Int J Recent Technol Eng 8(2S3):543–546. Google Scholar
17. Satish P, Srinivasulu S, Swathi R (2019) A hybrid genetic algorithm based rainfall prediction model using deep neural network. Int J Innov Technol Explor Eng 8(12):5370–5373, Google Scholar
18. Singh P (2017) Indian summer monsoon rainfall forecasting using time series data: a fuzzy-entropy-neuro based expert system. Geosci Front 9(4):1243–1257. https://doi.org/10.1016/j.gsf.2017.07.011
19. Sofian IM, Affandi AK, Iskandar I, Apriani Y (2018) Monthly rainfall prediction based on artificial neural networks with back propagation and radial basis function. Int J Adv Intell Inform 4(2):154–166. https://doi.org/10.26555/ijain.v4i2.208
20. Swain S, Patel P, Nandi S (2017) A multiple linear regression model for preci-pitation forecasting over Cuttack district, Odisha, India. In: 2nd international conference for convergence in technology (I2CT). https://doi.org/10.1109/i2ct.2017.8226150
21. Tokar S, Markus M (2000) Precipitation runoff modeling using artificial neural network and conceptual models. J Hydrol Eng 5(2):156–161. https://doi.org/10.1061/(ASCE)1084-0699(2000)5:2(156)
22. Zhang X, Mohanty SN, Parida AK, Pani SK, Dong B, Cheng X (2020) Annual and non-monsoon rainfall prediction modelling using SVR-MLP: an empirical study from Odisha. IEEE Access 8:30223–30233. https://doi.org/10.1109/ACCESS.2020.2972435

A Review on the Techniques Employed in Prediction of Northeast Monsoon Rainfall over Peninsular India

H. R. Pawar, S. S. Kashid, and S. D. Jagdale

Abstract The agricultural sector in India depends heavily on monsoon rainfall. The monsoon rainfall and the resulting stream flow in rivers have a tremendous impact on the socio-economic elements of India. Given that the Indian economy is mostly based on agriculture, it is necessary to comprehend rainfall variability, its relationship to other meteorological indicators, and how to predict it. Scarce rainfall in monsoon causes drought situations. Shortage of food and fodder has devastating effects on human beings and animals. To mitigate these effects, government can plan if it receives good rainfall and drought forecasts before the onset of the monsoon. 'Northeast Monsoon' (NEM) is one more form of monsoon activity experienced by the Peninsular Indian region. The NEM provides between 30 and 60 percent of the annual mean rainfall in the meteorological sub-divisions of Tamil Nadu, Kerala, Coastal Andhra Pradesh, Rayalseema, and South Interior Karnataka. Compared to the summer monsoon, the NEM is significantly understudied even though having considerable agricultural and economic significance. It shows substantial variations at intraseasonal and interannual. NEM season is vital for stream flow, reservoir yields, and water resource availability for agricultural and other purposes. Therefore, predicting NEM rainfall and corresponding stream flow is quite essential. Regression, Multi-target Regression, Artificial Neural Network, Recurrent Neural Network, Fuzzy Logic, Genetic Programming, and Deep Learning are the most often used techniques for climate prediction. In the present work, Discipulus software is used to implement Genetic Programming for the prediction of NEM rainfall over peninsular India. For the prediction of NEM rainfall, global climate indices like ENSO, EQUINOO and OLR were considered.

Disclaimer: The presentation of material and details in maps used in this chapter does not imply the expression of any opinion whatsoever on the part of the Publisher or Author concerning the legal status of any country, area or territory or of its authorities, or concerning the delimitation of its borders. The depiction and use of boundaries, geographic names, and related data shown on maps and included in lists, tables, documents, and databases in this chapter are not warranted to be error free nor do they necessarily imply official endorsement or acceptance by the Publisher or Author.

H. R. Pawar (✉) · S. S. Kashid · S. D. Jagdale
Department of Civil Engineering, Walchand Institute of Technology, Solapur 413006, India
e-mail: hemchandra.pawar@gmail.com

P. V. Timbadiya et al. (eds.), *Geospatial and Soft Computing Techniques*, Lecture Notes in Civil Engineering 339, https://doi.org/10.1007/978-981-99-1901-7_37

Keywords Northeast Monsoon · Rainfall prediction · Peninsular Indian region · Stream flow

1 Introduction

The process of forecasting future rainfall in a specific region is known as rainfall prediction. It collects, analyzes, verifies, models, simulates, and conducts research on different meteorological parameters that influence rainfall.

The Southwest Monsoon period is the primary rainy season for the Indian subcontinent. India receives nearly 75% of its rainfall in the Southwest Monsoon season. Typically, the Southwest Monsoon starts in early June and ends by the end of September.

'Northeast monsoon' is one more form of monsoon experienced by the Peninsular Indian region. Around September, the sun starts retreating toward the south; due to this, the Indian subcontinent's northern landmass rapidly cools. As a result, air pressure starts to increase across northern India. Because the Indian Ocean and its surroundings still retain their heat, the cold wind is forced to blow down from the Himalayas and Indo-Gangetic Plain toward the Indian Ocean, i.e., south of the Deccan peninsula. The peninsular Indian region receives moisture from the Bay of Bengal when a cold, dry wind blows that ways toward the Indian Ocean.

2 Reviews on Prediction Techniques

The most commonly used and studied rainfall forecasting methods, especially for the NEM rainfall prediction, are reviewed in this paper.

2.1 Multiple Linear Regressions

Regression is a statistical technique that looks for patterns in the connection between a single dependent variable, typically represented by the letter Y, and a number of independent variables. The term 'Multiple Regression Model' refers to regression models with more than two predictor variables.

Selvaraj and Aditya [1] adopted the Multiple Linear Regression Technique to predict the northeast rainfall for Tamil Nadu. They verified their results for 2007, 2008, and 2009 and concluded that the Multiple Linear Regression model could be used to predict the Northeast Monsoon of Tamil Nadu. Ahmed et al. [2] employed Multiple Linear Regression techniques to predict long-term rainfall across the targeted area in the Nilgiris District of Tamil Nadu. They built the model using

Visual Studio 2012 and Python. By using six years, climatological data prediction is made, and they got satisfactory results.

2.2 *Autoregressive Integrated Moving Average (ARIMA) Model*

A value in a response time series is predicted using ARIMA as a linear combination of its past values, past errors, and the present and past values of other time series.

KhadarBabu et al. [3] used the ARIMA model for the prediction of the NEM rainfall of Vellore in Tamil Nadu. They compared the observed rainfall flow and the synthetically generated data. They observed that the ARIMA approach is a more appropriate prediction for the future meteorological parameters compared with the probability MARKOV chain models.

Sundaram and Lakshmi [4] used the classical statistical Auto-Regressive Integrated Moving Average (ARIMA) model for monthly rainfall prediction of Northeast Monsoon Rainfall. It is suggested that the seasonal autoregressive integrated moving average process (SARIMA) time series model is a useful tool for rainfall forecasting.

2.3 *Support Vector Machine (SVM)*

Zhang et al. [5] using the Multiple Regression Analysis SVM (PUK Kernel) model predicted the yearly rainfall for Odisha. They found that the Multiple Regression Analysis SVM (PUK Kernel) model performs better than Multilayer Perceptron in their research.

2.4 *Artificial Neural Network (ANN)*

The ANN is a particular kind of machine learning technique that has been extensively used in the prediction of rainfall given its ability to identify highly complex nonlinear relationships between input and output variables without requiring an understanding of the nature of the underlying physical processes.

Chattopadhyay et al. [6] for determining the typical NEM rainfall for a given year used an Artificial Neural Network (ANN) and an exponential regression equation. In the months of the previous year's winter monsoon, they employed rainfall totals and SST anomalies as predictors. They found that the Prediction Error for ANN was lower (27%) than that for exponential regression (30%) and suggested this method above other nonlinear approaches to predicting rainfall.

Dash et al. [7] for the first time assessed the implication of machine learning techniques such as Linear Regression (LR), Artificial Neural Network (ANN), and Extreme learning machine (ELM) with Principal Component Analysis (PCA) for the forecasting of NEM rainfall utilizing global SST anomaly as a predictor.

Aarthi et al. [8] made an effort to predict the NEM rainfall for the regions of Coimbatore, Erode, Tiruppur, Dindigul, and Theni using an Artificial Neural Network with Multilayer Perceptron. They developed a model based on actual and anticipated rainfall levels during the training and testing phases of the execution, and the result was satisfactory.

Nirmala and Sundaram [9] utilized an Integrated Hybrid Approach for Rainfall Prediction Modeling in Tamil Nadu. Moving average and the traditional time series model, Box Jenkins ARIMA, were integrated with Artificial Neural Networks such as MA-ANN and ARIMA-ANN models to improve rainfall forecasting in Tamil Nadu. Four new hybrid prediction models, including the original ANN model, MA-ANN1, MA-ANN2, and ARIMA-ANN, have been presented to estimate the annual rainfall in Tamil Nadu. The ARIMA-ANN model, however, was the most effective of the connected models in terms of model performance and modeling complexity.

2.5 Genetic Programming (GP)

Kashid and Maity [11] used Genetic Programming for forecasting rainfall. They found that ENSO and EQUINOO have the highest correlation with rainfall during the Indian Summer Monsson in peninsular India. The steps involved in GP are shown in Fig. 1.

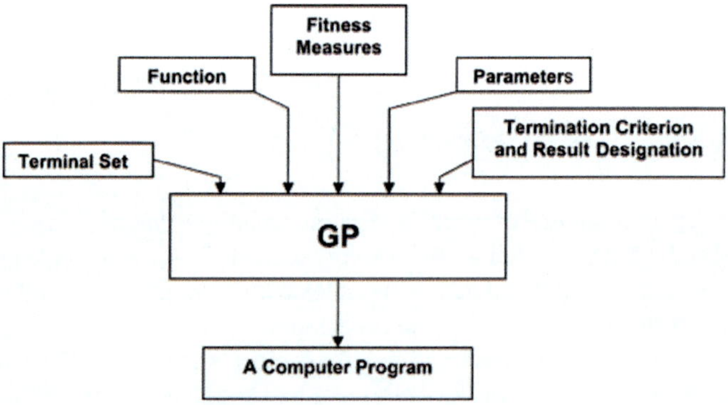

Fig.1 Schematic illustration of the preparations of GP [10]

3 Reviews on Parameters Influencing NEM Rainfall

Tamil Selvi [12] has mentioned that global climate variables including the Indian Ocean Dipole, Madden–Julian Oscillation, and El Nino/La Nina and Southern Oscillation Index (ENSO) affect rainfall in the NEM (MJO). Dimri et al. [13] have demonstrated that anticyclonic circulations in the Northern Hemisphere and cyclonic circulations in the Southern Hemisphere, which are symmetric over the equatorial Pacific Ocean, start to build up in December and peak in February during an El Nino phase. Dash et al. [7] have demonstrated a direct correlation between NEM rainfall and the Indian Ocean dipole mode (IODM), indicating that the positive (negative) phase influences the Northeast Monsoon activity. Sengupta and Nigam [14] showed that the NEM rainfall is significantly impacted by ENSO. Over southeastern peninsular India, the NEM rainfall intensifies during the El Nino phase. Saroja et al. [15] used Ground-based Microwave Radiation (MWR) for researching NEM onset features across the Sriharikota Range (SHAR) region in advance of the years 2014, 2016, and 2017. They evaluated basic measurement metrics from MWR such as temperature and moisture profiles and thermodynamic indicators before one week of the monsoon's commencement date. They arrived at the conclusion that a network of MWRs throughout the coastal regions may strengthen the observational network for the monsoon onset study. Misra [16] has shown that over India, summer monsoon rainfall has been linked to El Nio episodes that are below average. However, El Nio has the opposite effect on the NEM, causing rainfall that is above average.

4 Methodology

4.1 *Forecasting Technique*

Monthly NEM rainfall in Tamil Nadu is predicted using the Genetic Programming (GP) approach. 'Discipulus' software is used for the implementation of Genetic Programming for this purpose.

4.2 *Study Area*

The meteorological divisions of Rayalaseema, Coastal Andhra Pradesh, Puducherry, and Karaikal, as well as Tamil Nadu, Puducherry, and Karaikal (RYS) receive a substantial amount of rainfall during the Northeast Monsoon. The NEM is responsible for roughly 50–60% of the rain that Tamil Nadu receives. Therefore, Tamil Nadu state is considered for the study area. Figure 2 illustrates the map of the meteorological sub-divisions of India receiving significant rainfall during the NEM season.

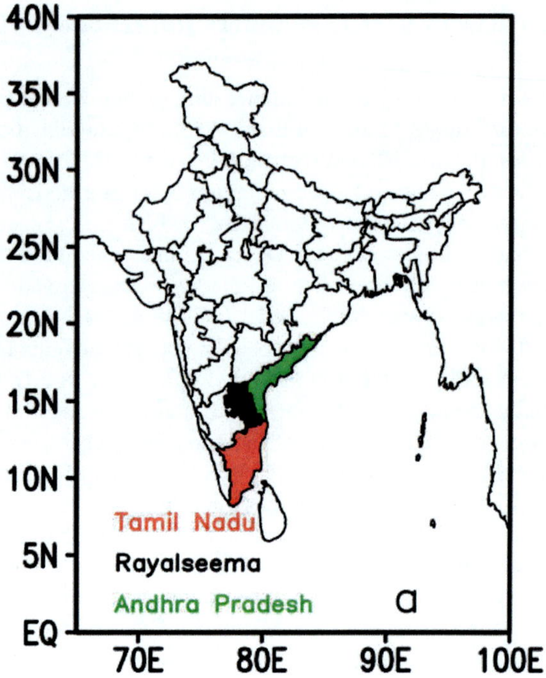

Fig. 2 Meteorological sub-divisions of India receiving significant rainfall during the NEM season [15]

4.3 *Data Source*

The monthly rainfall data have been collected from India Meteorological Department (IMD), Pune, and other weather parameters, viz. ENSO, EQUINOO, OLR data, have been collected from National Oceanic and Atmospheric Administration (NOAA) website.

4.4 *Selection of Input Parameters*

The input parameters were selected based on the observations and works carried out on Northeast Monsoon in the past. Historical Average Monthly Rainfall (HAMR), ENSO, *EQUINOO, and OLR* are taken as input parameters for the prediction of the NEM monthly rainfall with one month lead time. The monthly mean OLR from 1979 to 2016 has been used. In this work as the ENSO index from 1972 to 2016, sea surface temperature anomaly (SSTA) data have been used from the Nino-3.4 region (58S–58N, 1208–1708W).

5 Results and Discussions

Seven models with various variable pairings were created, tested, and used for analysis. The HAMR of the months October–December, ENSO, EQUINOO, and OLR data of the months October–December with a time lag up to 3 months from the year 1979 to 2016 is used for the development of the models in GP. Out of these seven different models, Model Number Five has shown the best results for the training phase as well as the testing phase as shown in Table 1. For the training of a model in GP, 50% (from 1979 to 1998) of the total data (1979–2016) is used, whereas for validation 25% (from 1999 to 2007) and for testing 25% (from 2008 to 2016) of remaining data are used. The result of the best combination during the training phase is shown in Fig. 3 and the result of the best combination during the testing phase is shown in Fig. 4. It is evident from Figs. 3 and 4 that the GP has captured both the highest and lowest values of rainfall. The HAMR for that month has the biggest impact (1.00), which is not surprising considering that it represents the climatological average for that month. While OLR (t−2) shows relatively much smaller consequences, EN (t−2), EQ (t−1), and OLR (t−1) all exhibit higher impacts. The best combination gave the highest value of 'r' during training as 0.92 and testing as 0.64. Along with this, the combination has provided the highest Nash–Sutcliffe model efficiency coefficient and the lowest root mean square error (RMSE) value (NSE) (Table 1).

6 Conclusions

The following conclusions are derived from the foregoing study:

- Forecasting techniques like Multiple Linear Regression, SVM, ARIMA, Genetic Programming, ANN have been used by many researchers. Most of these techniques are suitable for prediction and give satisfactory results.
- Most of the researchers have used limited parameters for the prediction of NEM Rainfall. So, there is a vast scope to explore other climatological parameters and new forecasting techniques for the prediction of the NEM rainfall.
- Genetic Programming has given satisfactory results. The value of Pearson's Correlation Coefficient (r) for the training phase ($r = 0.92$), testing phase ($r = 0.64$), NSE for the training phase (0.85), testing phase (0.37), and RMSE for the training phase (79.50), testing phase (132.80) are satisfactory.

Table 1 Results of the best combination tested for the monthly rainfall prediction using the coefficient of determination (r2), Pearson's C. C. (r), NSE, and RMSE values

Sr. no	Name of variable	Combinations of input variable
1	HAMR	√
2	EN (*t*−1)	√
3	EN (*t*−2)	√
4	EN (*t*−3)	
5	EQ (*t*−1)	√
6	EQ (*t*−2)	√
7	EQ (*t*−3)	
8	OLR (*t*−1)	√
9	OLR (*t*−2)	√
10	OLR(*t*−3)	
r^2 (training)		0.85
r^2 (testing)		0.42
r (training)		0.92
r (testing)		0.64
NSE (training)		0.85
NSE(testing)		0.37
RMSE (training)		79.50
RMSE (testing)		132.80

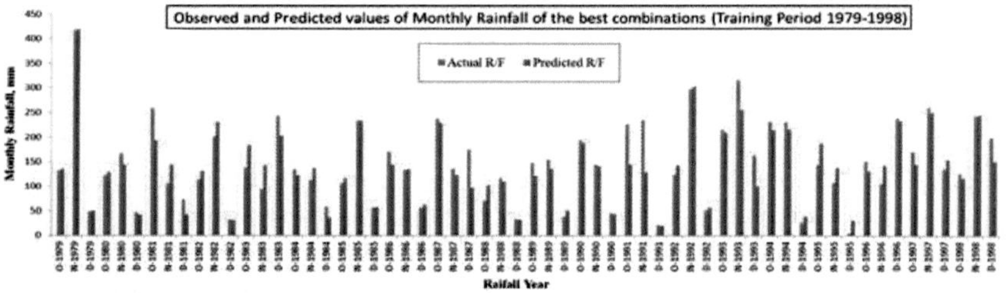

Fig. 3 Observed and predicted values of monthly rainfall of the best combinations (training period 1979–1998)

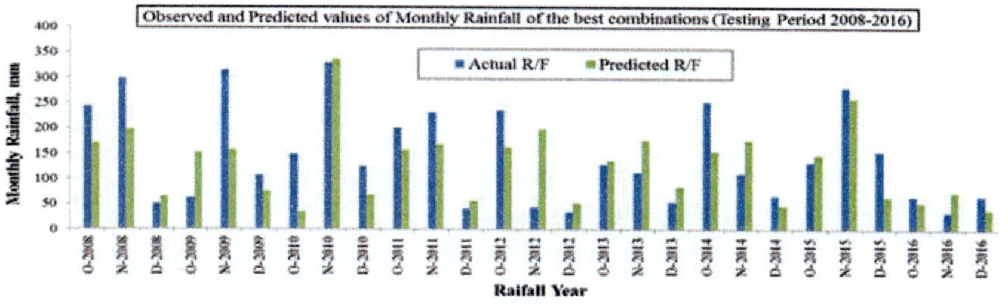

Fig. 4 Observed and predicted values of monthly rainfall of the best combinations (testing period 2008–2016)

Acknowledgements The authors would like to acknowledge the India Meteorological Department Pune, the Government of India (GoI), National Oceanic and Atmospheric Administration (NOAA) for providing data for this research.

References

1. Samuel RS, Raajalakshmi A (2011) Statistical method of predicting the Northeast rainfall of Tamil Nadu. Universal J Environ Res Technol 1(4):557–559
2. Imran A, Shruti M, Nikitha KB (2013) Rainfall prediction using multiple regression technique. Int J Appl Eng Res
3. Khadar SKB, Karthikeyan K (2011) Prediction of rain-fall flow time series using auto-regressive models. Adv Appl Sci Res 128–133
4. Meenakshi S, Lakshmi M (2014) Rainfall prediction using seasonal auto-regressive integrated moving average model. Paripex-Indian J Res 2250:1991
5. Xiaobo Z, SachiNandan M (2020) Annual and non-Monsoon rainfall prediction modelling using SVR-MLP: an empirical study from Odisha. IEEE Access 30223–30233
6. Goutami C, Surajit C, Rajni J (2010) Multivariate forecast of winter monsoon rainfall in India using SST anomaly as a predictor. In: Neurocomputing and statistical approaches. Elsevier Masson SAS https://doi.org/10.1016/j.crte.2010.06.004
7. Yajnaseni D, Mishra SK, Bijaya K et al (2018) Predictability assessment of northeast monsoon rainfall in India using sea surface temperature anomaly through statistical and machine learning techniques. Environmetrics 2018:e2533
8. Aarthi R, Shanmugavadivu R (2019) An efficient model for seasonal predictability on North-East Monsoon using multilayer perceptron. Int J Comput Sci Mobile Comput 8(12):101—108
9. Nirmala M, Sundaram SM (2010) Integrated hybrid approach for modeling rainfall prediction in Tamil Nadu. In: International conference on advances and emerging trends in computing technologies, Chennai, India
10. Maity R, Kashid SS (2009) Short-term basin-scale stream flow forecasting using large-scale coupled atmospheric–oceanic circulation and local outgoing longwave radiation. J Hydrometeorol 11(2):370–387
11. Kashid SS, Maity R (2012) Prediction of monthly rainfall on homogeneous monsoon regions of India based on large-scale circulation patterns using genetic programming. J Hydrol 26–41
12. Tamil Selvi S, Samuel S (2012) Trend analysis of Northeast Monsoon rainfall of Tamil Nadu. IJCRR 04(03)
13. Dimri AP, Yasunari T (2016) Indian Winter Monsoon: present and past. Earth-Sci Rev Earth 2334

14. Sengupta A, Nigam S (2019) The Northeast Winter Monsoon over the Indian Subcontinent and Southeast Asia: evolution, interannual variability, and model simulations. J Climate of American Meteorol Soc 32:1
15. Pushpa RS, Rajasekhar M (2019) Study on the North-East monsoon onset features using a ground-based microwave radiometer over SHAR. 128:208
16. Vasubandhu M, Amit B (2019) Defining the Northeast Monsoon of India. American Meteorological Society.https://doi.org/10.1175/MWR-D-18-0287.1

Sustainable Multiobjective Reservoir Optimization Considering Environmental Flow Using Python

Pushpak D. Dabhade and D. G. Regulwar

Abstract Rising human water demands from all sectors are challenging for policymakers to improve water allocation policies for storage reservoirs. In addition, there are so many other living beings and species which present in river water which also require water for their survival. Most of the time, the quantity and quality of water required for ecosystem conservation are not taken into consideration while optimizing the reservoir for different releases. A sustainable approach is necessary to conserve the river ecosystem with environmental flow while meeting the different demands. This paper presents a multiobjective reservoir operation model for maximization of releases for irrigation, maximization of releases for hydropower by considering environmental flow demand. The environmental flow requirements are considered as per the recommendations of the Central Water Commission (CWC) studies from a report submitted by a working group to advise Water Quality Assessment Authority on the minimum flows in the rivers. CWC report recommends two different minimum environmental flow criteria for the Himalayan rivers and other rivers from India. The optimal reservoir operation policy is presented considering constraints including irrigation release, turbine release, reservoir storage and hydrologic continuity. A monthly multiobjective reservoir optimization model is developed using Python programming. Model applied for Jayakwadi Stage-I reservoir is constructed across river Godavari in Maharashtra State, India.

P. D. Dabhade (✉)
Department of Civil Engineering, CSMSS Chh. Shahu College of Engineering, Kanchanwadi, Aurangabad, Maharashtra, India
e-mail: pushpak3dabhade@gmail.com

D. G. Regulwar
Department of Civil Engineering, Government Engineering College, Aurangabad, India
e-mail: regulwar@gmail.com

P. V. Timbadiya et al. (eds.), *Geospatial and Soft Computing Techniques*, Lecture Notes in Civil Engineering 339, https://doi.org/10.1007/978-981-99-1901-7_38

Keywords Optimization · Multiobjective analysis · Reservoir operation

1 Introduction

The whole world is facing constant climate change, erratic rainfall, global warming and other natural disasters every year. On the other hand, the global population has reached the threshold of 9 billion and the goal of meeting their growing needs is before the whole world. The most basic of these needs is an adequate supply of food and abundant electricity for industrialization, and these needs can be fulfilled by availability of water. Uncertain rainfall and the constant increasing need for water have made it necessary to build big reservoirs on rivers and manage the water in the reservoir for different purposes. The tendency to store more water often leads to a lack of water flow in the river at downstream of the dams, and this makes very serious impacts on the biodiversity that is dependent on the river water and grows in it. While achieving the targets of human water needs, environmental flow needs to maintain the river ecosystem should be taken into consideration. Thus, reservoir optimization techniques are helpful to implement optimal policy for multipurpose reservoirs. To achieve sustainable development, a complex problem of deriving the optimal operational policy for multiobjective reservoir is presented in this study considering the environmental flow releases.

Labadie [1] have presented a review on optimal operation model for multireservoir systems by all different methods of optimization for water resources' operations and management. Wurbs [2] have discussed different reservoir system simulation and optimization models to give guideline about different methods and its usefulness in various types of decision-support situations. Kim et al. [3] have developed a monthly operating rule for single-reservoir operation for Soyanggang dam basin in Korea Peninsula, with objectives of minimization of shortage and maximization of sum of hydropower production. The simulation results are obtained by using the developed piecewise linear operating rule. Chen et al. [4] have suggested an interactive dynamic programming model for optimize reservoir operation to support the policymakers for balancing humans and environmental water requirement. Mousavi et al. [5] have presented reservoir operation model using a dynamic programming fuzzy rule-based approach. This model was applied to the reservoirs system in Iran. Developed model gives better values of the simulated objective function with the higher reliability of meeting the demand. Kumar et al. [6] have developed a reservoir operation model for flood control purpose with the help of folded dynamic programming. Regulwar and Anand Raj [7] have proposed a multireservoir operation model using genetic algorithm under fuzzy environment considering multiobjectives. Regulwar and Kamodkar [8] derived a fuzzy-constrained reservoir operation model for multipurpose reservoir. Model is useful in dealing with imprecise constraints.

Environmental Flow: Brisbane declaration [9] provides a clear idea of Environmental Flow (E-Flow), and it can be defined as "The quantity, timing, and quality of water flows required to sustain freshwater and estuarine ecosystems and the human livelihoods and wellbeing that depend on these ecosystems." Jain and Kumar [10] have highlighted Indian scenario of environmental flow assessment with different Indian river case studies. Lokgariwar et al. [11] have discussed a case study of Upper Ganga River cultural water requirements for environmental flow assessment. Jain [12] have suggested adaptive management of environmental flow, which involves estimation of environmental flow requirements, consideration of those requirements in reservoir operation policies and monitoring the results after implementation to revise the estimation. Recommendation by Central Water Commission (CWC) [13]: To address the issue of Environmental flow in India a report is submitted by CWC in year 2007 about study of different Indian Rivers and their minimum environmental flow requirements. Study is carried out for different rivers like Krishna, Bhagirathi, Godavari, Alaknanda, Tapi and Yamuna. Report shows that, in the case of Himalayan rivers due to snow melting from different Himalayan glaciers, availability of natural flows is very high. Despite of this, it is not possible to preserve such high availability of water at lower reaches due to high amount of water requirements. CWC recommends environmental flow conditions for Himalayan rivers as: The minimum environmental flow must not be less than 2.5% of the 75% dependable annual flow. Also, a flushing flow with a peak flow of not less than 250% of the 75% dependable annual flow is required during the monsoon period. For other than Himalayan rivers, recommendations are: The minimum environmental flow must not be less than 0.5% of the 75% dependable annual flow. Also, a flushing flow with a peak flow of not less than 600% of the 75% dependable annual flow is required during the monsoon period.

In October 2018 Central Government circulated a notification regarding minimum environmental flow's requirements in Ganga River [14], which provides guide lines to maintain the minimum environmental flow conditions at the downstream of different water resources projects. The order of e-flows applicable to the Ganga River basin stretches from originating glaciers to Unnao district. The minimum environmental flow's conditions are to be maintained in all the existing, under-construction and future water resources' projects. For the existing projects, a period of three years is given to ensure minimum environmental flow which are unable to meet the CWC norms at present.

Government of Maharashtra Water Resources Department published a report on Integrated State Water Plan for Maharashtra [15] about importance of water for ecosystem. It describes river, water sources, aquifers and wetlands will be regarded as natural systems and must be protected from overuse, extinction, pollutants or pollution. In consultation with experts, stake holders and NGO essential steps shall be planed like preparation of annual action plan by incorporating necessary environment flow for restoration and maintenance of river health. From the literature study, it is observed that necessary environmental flow releases should be estimated and incorporated in reservoir operation to achieve a sustainable development policy in reservoir planning. In present study it is observed that availability of inflow in Godavari basin is insufficient, and also no flow condition has also been observed in river during many days of year. It shows inability to satisfy both minimum required environmental flow release conditions recommended by CWC, so the condition of a flushing flow of not less than 600% of the 75% dependable annual flow is not feasible to adopt. In present model, condition of 0.5% of the 75% dependable annual flow is considered as release for environmental flow and considered constant for all months.

2 System Description

This study involves the operations' planning of the Jayakwadi Stage-I reservoir. Jayakwadi project is constructed across Godavari River situated at Paithan in Aurangabad district. Jayakwadi Stage-I project is constructed considering different purposes including irrigation supply, water supply, hydropower generation, etc. The Jayakwadi project is a very important project for Maharashtra and especially for Marathwada region. Irrigation needs are met through Jayakwadi project for drought-stricken Marathwada region. The Jayakwadi project supplies drinking water to Aurangabad, the district headquarters and many other residential colonies. One of the major reasons behind the development of industrial area in Aurangabad is the supply of water through Jayakwadi (Fig. 1).

Fig. 1 Location map of Jayakwadi reservoir

3 Model Development

This study aims to maximize the two objective functions applied on multipurpose Jayakwadi reservoir. Objective functions are as follows:

1. Maximization of release for irrigation (RE_IRR)

$$\text{Max } Z_1 = \sum_{t=1}^{12} (RE\_IRR)_t. \tag{1}$$

2. Maximization of release for hydropower generation (RE_POW)

$$\text{Max } Z_2 = \sum_{t=1}^{12} (RE\_POW)_t. \tag{2}$$

As monthly model is developed, t indicates the number of monthly time steps. Above objective functions are subjected to various system constrains written as follows:

Irrigation Release

Release for irrigation releases should be less than or equal to irrigation demand (IRRD) for all the months and should be greater than or equal to the minimum irrigation demand (IRRDmin). Mathematically, this constraint is given as:

$$\text{RE\_IRR} \leq \text{IRRDt} \quad \forall \text{t} = 1, 2, \ldots, 12, \tag{3}$$

$$\text{RE\_IRR} \geq \text{(IRRDmin)t} \quad \forall \text{t} = 1, 2, \ldots, 12. \tag{4}$$

Turbine Capacity Constraints

Constraints for turbine release is considered according to capacity of turbine installed in power house for hydropower production. The maximum release for power production should be less than or equal to maximum capacity of flow through turbine for all months. Also, the hydropower production for all months should be greater than or equal to the firm power. Mathematically, constraints can be written as:

$$\text{RE\_POWt} \leq \text{TC} \quad \forall \text{t} = 1, 2, \ldots, 12, \tag{5}$$

$$\text{RE\_POWt} \geq \text{FPt} \quad \forall \text{t} = 1, 2, \ldots, 12. \tag{6}$$

Storage Capacity of Reservoir

Reservoir storage (St) for all months should be less than or equal to the maximum reservoir storage capacity (SCmax) of the reservoir and should be greater than or equal to minimum storage capacity (SCmin) of the reservoir. It can be given as:

$$\mathrm{St} \leq \mathrm{SCmax} \quad \forall \mathrm{t} = 1, 2, \ldots, 12, \tag{7}$$

$$\mathrm{St} \geq \mathrm{SCmin} \quad \forall \mathrm{t} = 1, 2, \ldots, 12. \tag{8}$$

Environmental Flow Release

Environmental flow is essential to preserve the river health and river ecosystem and must be considered in reservoir planning. Environmental flow (RE_EF) required for Godavari sub-basin is also considered. As per the CWC study report guidelines, it is suggested that the quantity of minimum environmental flow should be taken as 0.5 of 75% dependable annual flow. The quantity of environmental flow is estimated as per the recommendations of CWC and written as:

$$\begin{aligned} &\mathrm{RE\_EF} = 0.5\% \text{ of } 75\% \text{ dependable annual flow} \\ &\forall \quad \mathrm{t} = 1, 2, \ldots, 12. \end{aligned} \tag{9}$$

Hydrologic Continuity Constraints

Hydrologic continuity constraints deals with release from turbine (RP), release for irrigation purpose (RI), a constant value for drinking water supply release (RWS) and release of environmental flow (RE_EF) is taken, storage of reservoir (SC), inflows (IN), different Losses from the reservoirs like loss due to evaporation for all months. The individual hydrologic continuity constraint for each reservoir is written as:

Jayakwadi Stage-I (R1)

$$\begin{aligned} &(1 + \mathrm{a\_t}\,(1, \mathrm{t}))\,\mathrm{S}(1, \mathrm{t} + 1) = (1 - \mathrm{a\_t}\,(1, \mathrm{t}))\,\mathrm{S}(1, \mathrm{t}) \\ &\quad + \mathrm{IN}(1, \mathrm{t}) - \mathrm{RI}(1, \mathrm{t}) - \mathrm{RP}(1, \mathrm{t}) - \mathrm{OVF}(1, \mathrm{t}) - \mathrm{RWS}(1, \mathrm{t}) \\ &\quad - \mathrm{FCR}(1, \mathrm{t}) + \alpha\_1\ \mathrm{RP}(1, \mathrm{t}) - \mathrm{A\_(0\,)}\ \mathrm{e\_t}\,(1, \mathrm{t}) \\ &\forall \quad \mathrm{t} = 1, 2, \ldots, 12. \end{aligned} \tag{10}$$

Jayakwadi project is a multipurpose reservoir, and water is mainly used for irrigation of agricultural land in drought-prone areas of Marathwada in the state. It supplies water for drinking and industrial use to nearby towns and villages and to municipalities and industrial areas in Aurangabad and Jalna districts. Another feature of this project is the bird sanctuary. Many species of birds from all over the country visit the Nathsagar reservoir created by the dam. The abundance of waterfowl-loving swamps, algae, aquatic plants, small fish, insects and a wide variety of aquatic habitats attract domestic and exotic birds to reservoir. About 200 endemic and about 70

Table 1 Inflow and maximum irrigation demand

Month	Jayakwadi reservoir	
	Inflow	Irrigation demand
June	148.7620	18.55
July	408.2500	26.7
August	610.6600	25.43
September	600.0000	85.79
October	287.7500	267.86
November	196.4600	228.74
December	125.5300	210.88
January	37.6500	230.34
February	21.4620	85.23
March	19.5620	70.06
April	25.5000	85.49
May	46.5870	58.2

migratory birds of different species are found in the vicinity of this reservoir. Also, more than 50 species of fish are found in Jayakwadi reservoir. Total 183,322 Ha. of command area receives irrigation supply from Jayakwadi Stage-I project through left and right bank canals. A pump storage-type hydropower plant of 12 MW capacity is also installed at Jayakwadi Stage-I project. Water released through power plant is stored in a pool at downstream and pumped back to reservoir. This volume of water is also considered in model with 10% of transition loss. Jayakwadi Stage-II is constructed on Sindphana River which also receives water from Jayakwadi Stage-I reservoir through feeder canal. The feeder canal release (FCR) from Jayakwadi Stage-I is added in model with 10% transition loss (Table 1).

4 Results and Discussions

Optimal reservoir operation policies were developed for multiobjective reservoir system by developing a multiobjective fuzzy linear programming model (MOFLP) to operate on monthly time interval. Developed MOFLP model is performed to maximize release for irrigation and maximize the release for hydropower production. Model is developed with Python programming. Latest version of Python 3 is used with an open-source linear programming package PuLP library. At first phase model was run by considering objective of maximization of irrigation release, i.e., (Z1). As the priority is given for maximization of irrigation release, model generates best value of irrigation releases (Z1 +) and worst value of releases for hydropower (Z2−). In second phase model is run for objective of maximization of releases for hydropower, i.e., (Z2). Here, priority is given for maximization of hydropower release, model generates best value of irrigation releases (Z2 +) and worst value of releases for hydropower (Z1−) (Table 2).

Further linear membership functions are developed for fuzzification of objective functions.

Membership Function of Irrigation Release

$$\mu_{Z_1}(x) = \begin{cases} 0 & \text{if } Z1 \leq 417.94 \\ \frac{(Z1-417.94)}{(1065.06-417.94)} & \text{if } 417.94 \leq Z1 \leq 1065.06 \\ 1 & \text{if } Z1 \geq 1065.06 \end{cases} \quad (11)$$

Membership Function of Hydropower Release

$$\mu_{Z_2}(x) = \begin{cases} 0 & \text{if } Z2 \leq 272.04 \\ \frac{(Z2-272.04)}{(407.52-272.04)} & \text{if } 272.04 \leq Z2 \leq 407.52 \\ 1 & \text{if } Z2 \geq 407.52 \end{cases} \quad (12)$$

From Eqs. (11) and (12), a new model is developed by introducing λ as indicator for satisfaction level, and model is solved for objective of maximization of satisfaction level indicator (λ) subjected to new constraints as per Eqs. (11) and (12). Because of these newly added constraints and all the previous constrains, model maximized λ as objective and also maximize the both fuzzified objectives simultaneously. In developed model, maximum value of λ is obtained as 0.97, which indicates that the 97% of satisfaction is achieved in reservoir operation policy for both objectives with corresponding maximum total annual releases for irrigation of 1052.0 Mm3 and maximum total annual releases for hydropower production equal to 403.9 Mm3 (Figs. 2 and 3).

As per the recommendations of CWC, releases for environmental flow are also estimated and made available from reservoir. Total annual environmental flow releases maintained are 12.64 Mm3. Further, 100% maximum irrigation demand from the month of June to November has been fulfilled successfully. For the month of December and January, model provides releases for irrigation of 95% and 47%

Table 2 Best and worst values of objective functions

Objective function (maximization)	Best value Z +	Worst value Z−
Releases for irrigation (Z1)	1065.06 Mm3	417.94 Mm3
Releases for hydropower (Z2)	407.52 Mm3	272.04 Mm3

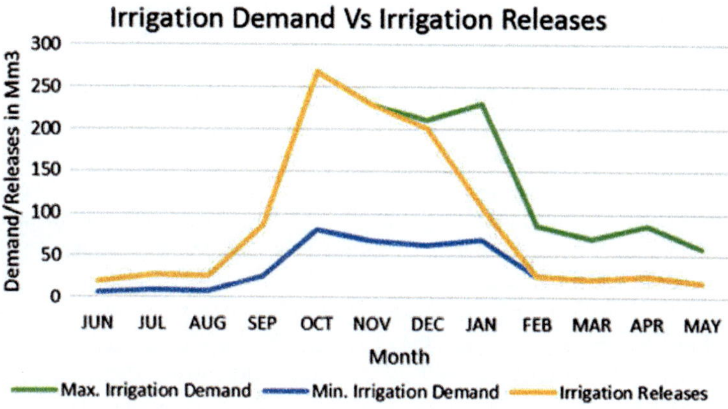

Fig. 2 Irrigation demand and releases for irrigation

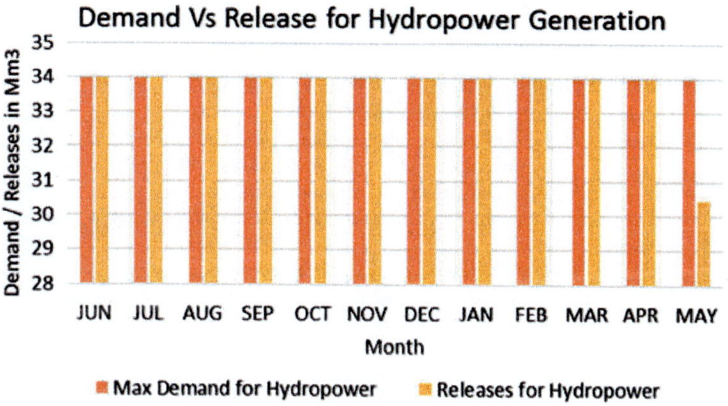

Fig. 3 Demand and releases for hydropower generation

of maximum demand, respectively. Due to the less inflow in reservoir, minimum required irrigation releases are made available for remaining four months, i.e., February–May which is considered as 30% of maximum irrigation demand. For months June to April, model have been able to meet 100% of the demand for releases for hydropower generation. In the month of May, releases for hydropower were 89.67% of the maximum demand.

5 Conclusions

The availability of water has always been a pressing issue for drought-hit Marathwada region of Maharashtra. This situation may worsen in the future, so it is imperative that proper planning of water use from Godavari basin and its reservoirs which are socially and economically important for the region is essential. Developed monthly operated model presents a holistic, sustainable approach for reservoir operation management in line with future water demands of humans as well as environment. Multiobjective fuzzy linear programming model achieves the 97% of level of satisfaction in releases for irrigation and hydropower generation simultaneously. According to sustainable development approach, the supply newly considered environmental flow for the aquatic environment as well as for various organism's dependent on river water has also been maintained uninterrupted through the reservoir.

References

1. Labadie JW (2004) Optimal operation of multireservoir systems: state-of-the-art review. 130(2):93–0. https://doi.org/10.1061/(asce)0733-9496
2. Wurbs RA (1993) Reservoir-system simulation and optimization models. J Water Resour Plan Manag 119(4):455–472. https://doi.org/10.1061/(ASCE)0733-9496(1993)119:4(455)
3. Kim T, Heo J-H, Bae D-H, Kim J-H (2008) Single-reservoir operating rules for a year using multiobjective genetic algorithm. J Hydro Inform 10(2):163. https://doi.org/10.2166/hydro.2008.019
4. Chen L, Yang ZF, Yang SWY (2012) Sustainable reservoir operations to balance upstream human needs and downstream lake ecosystem targets. Proc Environ Sci 13:1444–1457. ISSN 1878-0296. https://doi.org/10.1016/j.proenv
5. Mousavi SJ, Ponnambalam K, Karray F (2005) Reservoir operation using a dynamic programming fuzzy rule–based approach. 19(5):655–672. https://doi.org/10.1007/s11269-005-3275-3
6. Nagesh Kumar D, Baliarsingh F, Srinivasa Raju K (2010) Optimal reservoir operation for flood control using folded dynamic programming. 24(6):1045–1064. https://doi.org/10.1007/s11269-009-9485-3
7. Regulwar, Raj P (2009) Multi objective multireservoir optimization in fuzzy environment for river sub basin development and management. J Water Resour Protect 1(4):271–280. https://doi.org/10.4236/jwarp.2009.14033
8. Regulwar, Kamodkar R (2010) Derivation of multipurpose single reservoir release policies with fuzzy constraints. J Water Resour Protect 2(12):1030–1041. https://doi.org/10.4236/jwarp.2010.212123
9. Declaration B (2007) In: 10th international river symposium and international environmental flows conference, Brisbane 3–6 September 2007
10. Jain SK, Kumar P (2014) Environmental flows in India: towards sustainable water management. Hydrol Sci J 59(3–4):751–769. https://doi.org/10.1080/02626667.2014.896996
11. Lokgariwar, Ravi C, Vladimir S, Luna B, Jay O (2014) Including cultural water requirements in environmental flow assessment: an example from the upper Ganga River, India. Water Int 39(1):81–96. https://doi.org/10.1080/02508060.2013.863684
12. Jain SK (2020) Assessment and implementation of environmental flows. Guest Editorial, Current Sci 118(11)

13. Report of Central Water Commission Working Group (2007) CWC, Ministry of Water Resources, Government of India
14. Jalansh (2018) Monthly Newsletter of Central Water Commission, Issue No.4 (2018) Published by: Water Systems Engineering Directorate Central Water Commission, New Delhi
15. Integrated State Water Plan for Godavari Basin in Maharashtra, Government of Maharashtra Water Resources Department volume I (2017)

Optimization of an Irrigation Reservoir Using Dynamic Programming Model

Nidhi Khare and V. L. Manekar

Abstract In present study, Dynamic Programming optimization model is developed to optimize the water use in an irrigation system of JOBAT PROJECT supported by a reservoir using Dynamic Programming model. It is concluded from the reviewed literature that various investigators have worked for optimization of water management in a reservoir for irrigation. A number of models have been attempted to investigate the problem of optimum management of water resources. A few of them belong to the category of traditional optimization models, and others belong to the new models belonging to evolutionary model's category. The evolutionary algorithms have certain limitations as regards the convergence to an optimum is concerned. The traditional optimization techniques are better in this sense. In a paper, Paul and Panda (J Irrigation and Drainage Eng 126(3):2000 [17]) have demonstrated the use of Stochastic Dynamic Programming (SDP) in water distribution among different crops. The stochasticity of demands is considered. The Stochastic Dynamic Programming (SDP) has serious limitations of the dimensional curse. Moreover, in Indian context, the reference evapotranspiration is almost constant in each of the seasons. Hence, one can take an average value for the reference evapotranspiration. This will save the efforts of bringing the demands in stochastic area. Owing to the facts stated above, present study envisages a deterministic Dynamic Programming for water distribution among the various crops.

Keywords Irrigation reservoir · Cropping pattern · Water management · Reservoir optimization · Dynamic programming

N. Khare (✉) · V. L. Manekar
Department of Civil Engineering, Sardar Vallabhbhai National Institute of Technology, Surat 395007, India
e-mail: nidhi18288@gmail.com

V. L. Manekar
e-mail: vlm@ced.svnit.ac.in

P. V. Timbadiya et al. (eds.), *Geospatial and Soft Computing Techniques*, Lecture Notes in Civil Engineering 339, https://doi.org/10.1007/978-981-99-1901-7_39

1 Introduction

Water management is a subject of increasing concern as demand grows; the major consumer of fresh water resources of the world is irrigation. In India about 90% of water is used for irrigation. Because of rise in demand of water resources, the necessity of water management is becoming the need of the hour. Less irrigation affects the growth of the plant and reduces the productivity of the crops to a great extent. Hence, under irrigation and over irrigation both are harmful and require proper attention of water resource's manager.

Irrigation reservoir operation problems are characterized by uncertainties due to randomness of inflows and demands. This situation calls for optimum utilization of water in irrigation. The objective for operation of an irrigation system should be the optimum use of all available water in the sense of maximization of the crop production for the whole irrigation area. At present in India, none of the reservoir systems for irrigation are operated for optimum irrigation supplies. This calls for a study of improved water management in irrigation systems. Owing to the need, the present study is aimed at optimization of water resources in an irrigation system.

2 Materials and Methods

2.1 *Dynamic Programming Technique*

Dynamic Programming is ideally suited for sequential decision problems in which decisions are made sequentially one after other based on the state of system. It is a mathematical technique which can be used for single-stage decision problems as well as serial multi-stage decision problems. Reservoir operation problems are a type of sequential decision problem; hence, Dynamic Programming is suitable for such problems. Single-stage decision problems can be illustrated using state transformation function $T = t\ (S, X)$ in which S is input and X is decision.

2.1.1 Jobat Project

Jobat irrigation project is one of the 29 major projects identified in the Narmada Master Plan for implementation. The project site is near village Waskel and is about 24 km from Kukshi. The project on completion shall provide irrigation in an area of 9848 hectares of 24 villages of Kukshi Tehsil of Dhar district. The Jobat irrigation project is planned to benefit the drought-prone area declared by Government of India and tribal areas of Kukshi tehsil in Dhar district of M.P. as shown in Fig. 1 *Jobat project map.*

The total geographical area of the state is 308 lakh hectares. Annual average rainfall in the state varies from 60 cm in Northeast past to 100–120 cm in Southeastern

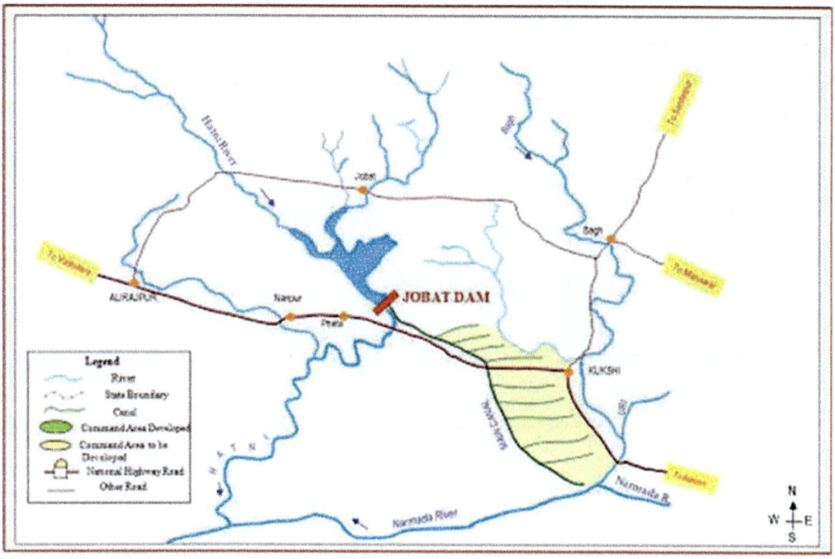

Fig. 1 Jobat project map

region. Total population of the state as per 2011 census is 726.27 lakh. About 69% population of the state is dependent on agriculture.

2.1.2 Data Collection

The catchment area at the proposed dam site is 792 km^2. The construction of project was completed in 2004–05, and first time, the water was released in the main canal in year 2007–08. First time, the reservoir has filled with 67.57 million cubic meter water in 2007–08. Indeed, in most of subsequent years, it has been overflowing due to good rains in the catchments area. Cumulatively, about 71.801 million cubic meter water has been released from the head regulator to the main canals in last three years including 25.858 M.cum in 2010 and 20.085 M.cum in each 2009 and 2008. The topography of the command area is such that the seepage water from the main, distributaries, and minor canal helps in recharging the sub-surface and ground water tables. Hence, it is estimated that an additional 2000-hectare area could be irrigated by arresting 20% of 30% efficiency losses through ground water recharge attributing lift irrigation from open wells and tube wells totaling to 17,000 hectares.

The construction of project was completed in 2004–05, and first time, the water was released in the main canal in year 2007–08. First time, the reservoir has filled with 67.57 million cubic meter water in 2007–08. Indeed, in most of subsequent years, it has been overflowing due to good rains in the catchments area. Cumulatively, about 71.801 million cubic meter water has been released from the head regulator to the main canals in last three years including 25.858 M.cum in 2010 and 20.085 M.cum in each 2009 and 2008. The topography of the command area is such that the seepage water from the main, distributaries, and minor canal helps in recharging the subsurface and ground water tables. Hence, it is estimated that an additional 2,000-hectare area could be irrigated by arresting 20% of 30% efficiency losses through ground water recharge attributing lift irrigation from open wells and tube wells totaling to 17,000 hectares.

2.2 *Selection of Input Parameters*

2.2.1 Irrigation Reservoir Operation

The objective function for maximizing the total net benefits from the irrigation system is expressed as:

$$f1 = \sum \text{Ac}\,(\text{YBc} * \text{RYc} - \text{PCc}),$$

$$R_t^*(x_t, \text{AET}_t) = 1 - k_t\left(1 - \frac{\text{AET}}{\text{PET}}\right)_t$$

2.2.2 Crop Area Constraints

The constraints of minimum and maximum area restrictions are expressed as:
Ac min$\leq$ Ac $\leq$ Ac max ¥ C

2.2.3 Relative Yield Constraint

The process of growing a crop necessarily involves production costs for seeds, fertilizers, labor, cultivation, etc.

2.2.4 Soil Moisture Balance

The rainfall in the study area is scanty. In addition, there are bunds around the field to prevent runoff from this scanty rainfall, and the soil moisture balance equation is stated as follows (Tables 1, 2, 3, 4, 5, 6, 7 and 8).

$$\begin{aligned} \text{SMt} + 1.\text{Zt}11 &= \text{SMt Zt} + \text{Rt} + \text{Xt}1 \\ &+ \text{S0}(\text{Zt}11 - \text{Zt}) - \text{AETt}{-}\text{Pt} \end{aligned}$$

Table 1 Computation of depth of supply for various crops for storage 70.04

N1	C.A	492	Ha		
St	It	Et	Rt	St + 1	Ds (MCM)
70.04	0	0	0.23	68.73259	38.95664
70.04	0	0	0.255	68.70775	43.19106
70.04	0	0	0.26	68.70279	44.03794
70.04	0	0	0.265	68.69782	44.88482
70.04	0	0	0.27	68.69285	45.73171

Table 2 Computation of depth of supply for various crops for storage 68.73

N2	C.A	1672	Ha		
St	It	Et	Rt	St + 1	Ds (MCM)
68.73259	0	0	0.81	66.88109	40.37081
68.73259	0	0	1	66.69231	49.84051
68.73259	0	0	1.2	66.49359	59.80861
68.73259	0	0	1.4	66.29488	69.77671
68.73259	0	0	1.6	66.09617	79.74482

Table 3 Computation of depth of supply for various crops for storage 69.91

N3	C.A	7189	Ha		
St	It	Et	Rt	St + 1	Ds (MCM)
69.91051	0	0	10.1	58.99391	117.077
69.91051	0	0	9.1	59.98864	105.4852
69.91051	0	0	9.3	59.7897	107.8036
69.91051	0	0	9.5	59.59075	110.1219
69.91051	0	0	9.7	59.39181	112.4403

Table 4 Computation of depth of supply for various crops for storage 58.99

N4	C.A	7189	Ha		
St	It	Et	Rt	St + 1	Ds (MCM)
58.99391	0	0	12.3	46.05992	142.5789
58.99391	0	0	11.3	47.05508	130.9872
58.99391	0	0	11.5	46.85604	133.3055
58.99391	0	0	11.7	46.65701	135.6239
58.99391	0	0	11.9	46.45798	137.9422

Table 5 Computation of depth of supply for various crops for storage 46.05

N5	C.A	7189	Ha		
St	It	Et	Rt	St + 1	Ds (MCM)
46.05992	0	0	15.7	29.86767	181.991
46.05992	0	0	13.2	32.35556	153.0115
46.05992	0	0	13.4	32.15653	155.3299
46.05992	0	0	13.6	31.9575	157.6483
46.05992	0	0	13.8	31.75847	159.9666

Table 6 Computation of depth of supply for various crops for storage 29.86

N6	C.A	7189	Ha		
St	It	Et	Rt	St + 1	Ds (MCM)
29.86767	0	0	13.7	15.73681	158.8074
29.86767	0	0	15.8	13.64951	183.1502
29.86767	0	0	16	13.45072	185.4685
29.86767	0	0	16.2	13.25192	187.7869
29.86767	0	0	16.4	13.05313	190.1053

Table 7 Computation of depth of supply for various crops for storage 15.73

N7	C.A	6599	Ha		
St	It	Et	Rt	St + 1	Ds (MCM)
15.73681	0	0	11.4	4.062919	143.9612
15.73681	0	0	10.2	5.255663	128.8074
15.73681	0	0	10.4	5.056873	131.333
15.73681	0	0	10.6	4.858082	133.8587
15.73681	0	0	10.8	4.659291	136.3843

Table 8 Computation of depth of supply for various crops for storage 4.06

N8	C.A	6107	Ha		
St	It	Et	Rt	St + 1	Ds (MCM)
4.062919	0	0	0	3.849891	0
4.062919	0	0	0	3.849891	0
4.062919	0	0	0	3.849891	0
4.062919	0	0	0	3.849891	0
4.062919	0	0	0	3.849891	0

2.2.5 Actual Evapotranspiration

The actual evapotranspiration in each fortnight is found as follows"

$$\mathrm{AET}_t = \begin{cases} 0; & \mathrm{SM}_t \le \mathrm{WP}_t \\ \frac{\mathrm{PET}_t(\mathrm{SM}_t - \mathrm{WP})}{(1-p)(\mathrm{FC}-\mathrm{WP})}; & \mathrm{WP} < \mathrm{SM}_t \le (1-p)(\mathrm{FC}-\mathrm{WP}) \\ \mathrm{PET}_t; & \mathrm{SM} \ge (1-p)(\mathrm{FC}-\mathrm{WP}) \end{cases}$$

The FAO Penman–Monteith equation is:

$$\mathrm{ET_O} = \frac{0.408\,\Delta(R_n - G) + \gamma \frac{900}{T+273} u_2 (e_S - e_a)}{\Delta + \gamma(1 + 0.34\,u_2)}$$

2.2.6 Potential Evapotranspiration

The potential evapotranspiration is given by

$$\mathrm{PET} = \mathrm{Kc}^{*}\mathrm{Eto}.$$

The actual evapotranspiration in relation to its potential rate is determined by considering whether the available water in the root zone is adequate or whether the crop will suffer from stress induced by water deficit.

2.2.7 Root Growth Model

A root growth model (Borg and Grimes 1986) is used in the present study:

$$Z_t = Z_{\max}\left(0.5 + 0.5\,\sin\left[3.03\left(\frac{t}{t_{\max}}\right) - 1.47\right]\right)$$

3 Results and Discussions

The multi-objective model for reservoir operation has maximizing total crop area and maximizing total net benefits, as two competing objectives, and is modeled as relative yield, actual evapotranspiration, potential evapotranspiration, and soil moisture which have been calculated from the data collected from Jobat project including meteorological data, crop coefficient values, yield response factor, evaporation data,

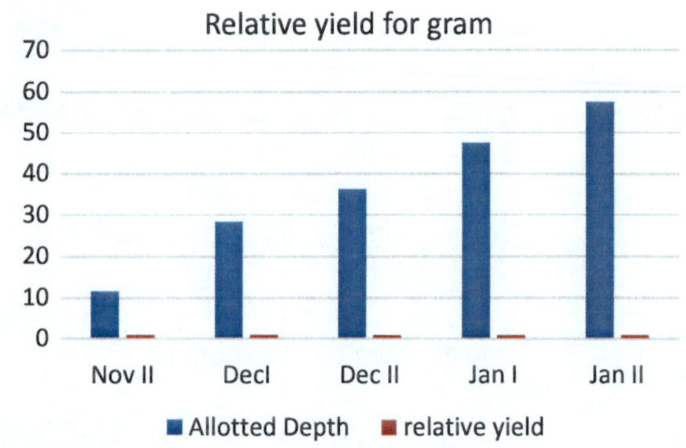

Fig. 2 Relative yield for gram

area of crop and crop water requirement (Figs. 2, 3, 4, 5 and 6; Tables 9, 10, 11, 12 and 13).

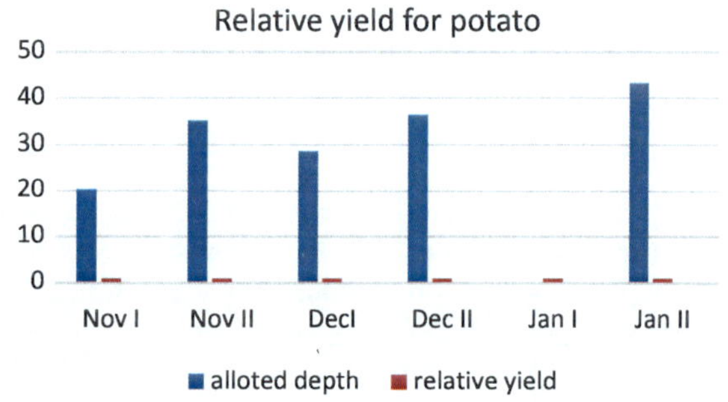

Fig. 3 Relative yield for potato

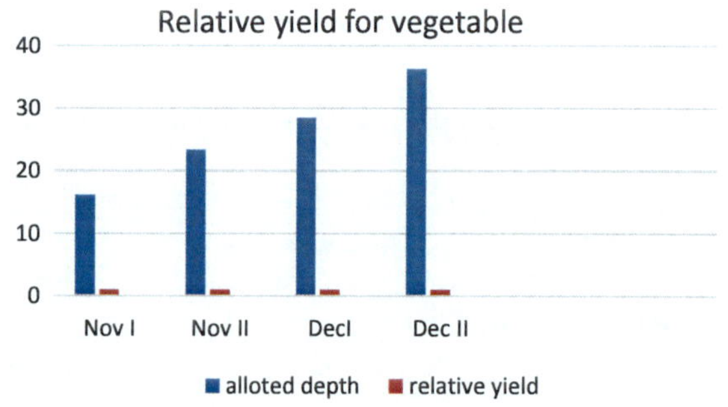

Fig. 4 Relative yield for vegetable

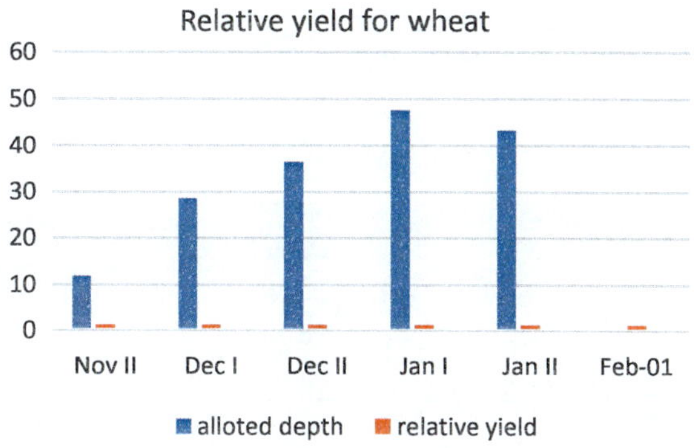

Fig. 5 Relative yield for wheat

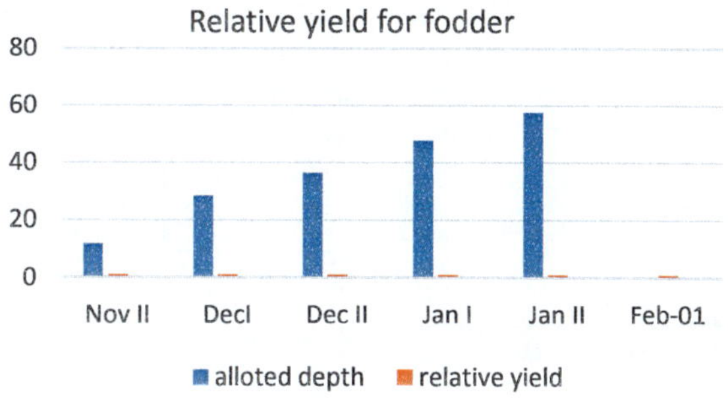

Fig. 6 Relative yield for fodder

Table 9 Computation of final relative yield for gram

Fortnight	Value of Ky	Allotted depth (MCM)	Relative yield
November II	0.2	11.70770158	1.00
December I	0.6	28.51578801	1.00
December II	0.6	36.39820096	1.00
January I	0.5	47.64223119	1.00
January II	0.5	57.5844825	1.00
February 1	0.5	0	1.00
	Final relative yield		1.00

4 Conclusions

The following conclusions are derived from the foregoing study:

(1) The main objective of the present study was to develop a procedure in Dynamic Programming environment for optimization of water distribution over different crops in various periods of time.

Table 10 Computation of final relative yield for potato

Fortnight	Value of Ky	Allotted depth (MCM)	Relative yield
November I	0.45	20.19	1.00
November II	0.45	35.12310474	1.00
December I	0.45	28.51578801	1.00
December II	0.33	36.39820096	1.00
January I	0.7	0.00	1.00
January II	0.7	43.18836187	1.00
Final relative yield			1.00

Table 11 Computation of final relative yield for vegetable

Fortnight	Value of Ky	Allotted depth (MCM)	Relative yield
November I	1.1	16.14832536	1.00
November II	1.1	23.41540316	1.00
December I	1.1	28.52	1.00
December II	0.8	36.40	1.00
	Final relative yield		1.00

Table 12 Computation of final relative yield for wheat

Fortnight	Value of Ky	Allotted depth (MCM)	Relative yield
November II	0.2	11.70770158	1.00
December I	0.6	28.51578801	1.00
December II	0.6	36.39820096	1.00
January I	0.5	47.64223119	1.00
January II	**0.5**	43.18836187	1.00
February 1	0.5	0	1.00
	Final relative yield		1.00

Table 13 Computation of final relative yield for fodder

Fortnight	Value of Ky	Allotted depth (MCM)	Relative yield
November II	0.2	11.70770158	1.00
December I	0.6	28.51578801	1.00
December II	0.6	36.39820096	1.00
January I	0.5	47.64223119	1.00
January II	0.5	57.5844825	1.00
February 1	0.5	0	1.00
	Final relative yield		1.00

(2) The DP model developed in the present study distributes the water optimally among various crops in different time periods. This also accounts for the deficits in the supply and the distribution of deficits to maximize the possible relative yield.
(3) The deficient storage conditions in the beginning of the Rabi season have a sure shot impact on the water supplies and the crops suffer deficits. However, in the present study the analysis was carried out to maximize the yield from the crops by distributing the deficient supplies during the periods of lower relative yield factor.

Acknowledgements The authors acknowledge the support from the Irrigation and water resources department, Indore Government of Madhya Pradesh for present work. The authors are also thankful to Irrigation department of Dhar district of Madhya Pradesh for providing necessary data to conduct the present study.

References

1. Adeyemo J, Stretch D (2018) Review of hybrid evolutionary algorithms for optimizing a reservoir. South African J Chem Eng 25:22–31
2. Adhikary (2011) Simulating impacts of EFR consideration on reservoir operation policy and irrigation management in the Hari Rod River Basin, Afghanistan. In: 19th international congress on modelling and simulation, Perth, Australia
3. Adujiya et al. (2019) Deriving operation policy for multiple reservoir system under irrigation using LPM model. Int J Recent Technol Eng (IJRTE) 8:08–17
4. Asadieh B, Afshar A (2019) Optimization of water-supply and hydropower reservoir operation using the charged system search algorithm. J Hydrol 6:1–16
5. Akbarifard et al. (2020) Data on optimization of the Karun-4 hydropower reservoir operation using evolutionary algorithms. Adv Water Resourc 124
6. Dobson et al. (2019) An argument driven classification and comparison of reservoir operation optimization methods. Adv Water Resour 128:74–8
7. Elmahdi et al. (2005) System dynamics optimization approach to irrigation demand management. Environ Res J 4:1–18
8. Jalali et al (2006) Reservoir operation by ant colony optimization algorithms. Iranian J Sci Technol
9. Kumar DN, Reddy MJ (2007) Multipurpose reservoir operation using particle swarm optimization. J Water Resour Plann Managem 123
10. Kumar et al. (2010) Optimal reservoir operation for irrigation of multiple crops using genetic genetic algorithms. J Irrigation and Drainage Eng ASCE 123
11. Kumar Y (2019) Emerging research and innovations in civil engineering. Global Res Developm J Eng. e-ISSN: 2455-5703
12. Kuo SF, Liu CW (2002) Simulation and optimization model for irrigation planning and management. In: Hydrological processes, Wiley Inter Science, pp 1269
13. Maliwal et al. (2019) Multi-reservoir flood control operation by optimization technique: a review. Int J Eng ResTechnol (IJERT)
14. Mujumdar (2002) Mathematical tools for irrigation water management-an overview. IWRA, Int Water Resour Assoc 27(1):47–57
15. Mujumdar PP, Ramesh TSV (1997) Real-time reservoir operation for irrigation. Water Resour Res 33(5):1157–1164

16. Mathur YP, Nikam SJ (2009) Optimal reservoir operation policies using genetic algorithm. Int J Eng Technol 123
17. Paul S, Panda SN (2000) Optimal irrigation allocation: a multilevel approach. J Irrigation and Drainage Eng 126(3)
18. Ponnambalam K, Adams BJ (1987) Experiences with integrated irrigation system optimization analysis. In: Irrigation and water allocation (Proceedings of the Vancouver Symposium, August 1987). IAHS Publ. no. 16
19. Reddy MJ, Kumar DN (2007) Evolving strategies for crop planning and operation of irrigation reservoir system using multi-objective differential evolution. Irrigation Sci 26:177–190
20. Rao RV (2015) Teaching learning-based optimization algorithm and its engineering applications. ISBN-978-3-319-22732-0
21. Tran et al. (2011) Managing multiple-use resources: optimizing reservoir water use for irrigation and fisheries. In: 55th Annual AARES national conference Melbourne, Victoria
22. Zhihao et al. (2019) Optimization method for joint operation of a double-reservoir-and-double-pumping-station system: a case study of Nanjing, China. J Water Supply: Res Technol

Development of Multipurpose Single Reservoir Release Policy with Fuzzy constraints—A Case Study

S. V. Pawar, P. L. Patel, and A. B. Mirajkar

Abstract The study aims to present a fuzzy linear programming approach using fuzzy technological coefficients for optimal function of a reservoir system. This comprises a model that takes into account the water resource system's inherent uncertainty, which includes lack of adequate data, subjectivity, imprecision, and fuzziness. A single objective function, i.e., maximization of irrigation releases, is considered to solve the problem. The model is run for 75% dependable inflow into the reservoir while using fuzzy linear programming (FLP) approach. Here, uncertainty in reservoir operation parameters, including irrigation demand, reservoir storages, and irrigation releases, is taken into consideration by considering them as fuzzy sets. Construction of membership functions for the objective function and the constraints are included in the model development processes. The Khadakwasala reservoir in Maharashtra State, India, has been used as a case study to demonstrate the methodology. Optimizing the fuzzy objectives and constraints leads to a compromised solution for the suggested FLP model. The resulting degree of truth (λ) for the chosen objective function is 0.48. Also, the monthly release pattern has been obtained in the command area which can be used by the users in the command area.

Keywords Fuzzy linear programming problem · Fuzzy technological coefficient · Khadakwasala reservoir

Disclaimer: The presentation of material and details in maps used in this chapter does not imply the expression of any opinion whatsoever on the part of the Publisher or Author concerning the legal status of any country, area or territory or of its authorities, or concerning the delimitation of its borders. The depiction and use of boundaries, geographic names, and related data shown on maps and included in lists, tables, documents, and databases in this chapter are not warranted to be error free nor do they necessarily imply official endorsement or acceptance by the Publisher or Author.

S. V. Pawar (✉) · P. L. Patel
Sardar Vallabhbhai National Institute of Technology, Surat 395007, India
e-mail: sangita.pawar23@gmail.com

A. B. Mirajkar
Visvesvaraya National Institute of Technology, Nagpur 440010, India

P. V. Timbadiya et al. (eds.), *Geospatial and Soft Computing Techniques*, Lecture Notes in Civil Engineering 339, https://doi.org/10.1007/978-981-99-1901-7_40

1 Introduction

The fuzzy logic method can offer a potential option to the techniques considered for reservoir operation modeling in fuzzy decision-making problems, according to Russell et al. [1]. This approach is further adaptable and allows for the inclusion of expert opinions, which may make it more suitable to operators. To determine the best reservoir operating strategies for the Karanjwan reservoir in Maharashtra State, India, Balve and Patel [2] used fuzzy linear programming (FLP). The obtained truth level is 0.5058, and the related optimal irrigation releases are 88.51 Mm^3. Comparisons between FLP's reservoir release policy and linear programmings (LPs) are presented. Regulwar et al. [3] used three different models to determine the best operating strategy. By creating three models, i.e., when the resources and technological coefficients are fuzzy and when both technological coefficients and resources are fuzzy, then fuzzy set theory was utilized to explain ambiguity in a variety of parameters. Yeh [4] studied models for reservoir operation and management. Computer modeling techniques are necessary for the synchronization of reservoir systems to offer information for intelligent management and operational decisions. The most recent research on multireservoir system optimization was reviewed by Labadie [5]. Panigrahi and Mujumdar [6] advanced a fuzzy rule-based model for the management of a single-purpose reservoir. The "if–then" assumption governs how the model behaves, with "if" denoting a vector of fuzzy premises and "then" denoting a vector of fuzzy consequences. Using genetic algorithms, Oliveira and Loucks [7] established operating principles for multireservoir systems. By utilizing fuzzy membership functions, Fontane et al. [8] gave the vague and non-commensurable objectives for planning reservoir operation and investigated the applicability of the method in dynamic programming. According to Shreshtha et al. [9], fuzzy relations could be used to represent both the inputs (such as storage, inflows, and demands) and the outputs (historical releases) of reservoir operating principles. To get crisp output, these fuzzy inputs were combined and defuzzified. The approach for solving the FLP problem by utilizing a linear membership function was described by Gasimov et al. [10]. With the use of FLP with technology coefficients, FLP with technological coefficient, and fuzzy right-side numbers, a FLP problem has been resolved and methodology is illustrated through a numerical example. The idea of maximizing and minimizing sets was used to propose a novel fuzzy ranking algorithm by Anand Raj et al. [11]. The Ranking Fuzzy Weight's (RANFUWs) approach is easy analytically and simple to use. The suggested approach (RANFUW method) was used to solve a planning and management issue for a river basin. To determine the most appropriate planning for the reservoirs and their related purposes, the technique was used to the Krishna River basin. Using three conflicting objectives, Raju and Kumar [12] used the Multi-Objective Fuzzy Linear Programming (MOFLP) technique in irrigation planning. For the Sri Ram Irrigation Project in India, the study was carried out. The degree of truth (λ) of the compromised solution for these three objective functions was calculated to be 0.69. With the use of a multi-objective constrained linear programming problem, Thakre et al. [13] presented the solution to a FLP problem where the cost coefficients and constraint matrix were both fuzzy. Additionally, they demonstrated that the solutions proposed by them were independent of weights. Singh et al. [14] developed a LP model to suggest the best cropping strategy for the maximum net return at various water availability levels, including 100, 70 and 50% dependability levels. It was discovered that the water in the command area may support the best cropping strategy for the maximum net return.

In the current study, the formulation of a FLP model is used to apply the fuzzy set theory to a water resources system in order to maximize reservoir release rates. Here, the LP model's technological coefficients are considered as having a fuzzy nature. The Khadakwasala reservoir in Pune, Maharashtra State, India, is used as a case study to demonstrate the methodology. LINGO 19 is used to develop and solve the FLP model.

2 Methodology

In this model, the technological coefficients are fuzzy numbers and resources are crisp in nature. A membership function represents the level of truth (λ) of a certain value of the parameter within the fuzzy set. Figure 1 shows the flowchart for solving LP problem with fuzzy technological coefficients. Formulation of FLP model is explained as discussed in following sub-sections:

2.1 LP Problems with Fuzzy Technological Coefficients

A LP problem with fuzzy technological coefficients is given as [3]

$$\max \sum_{j=1}^{n} c_j x_j$$

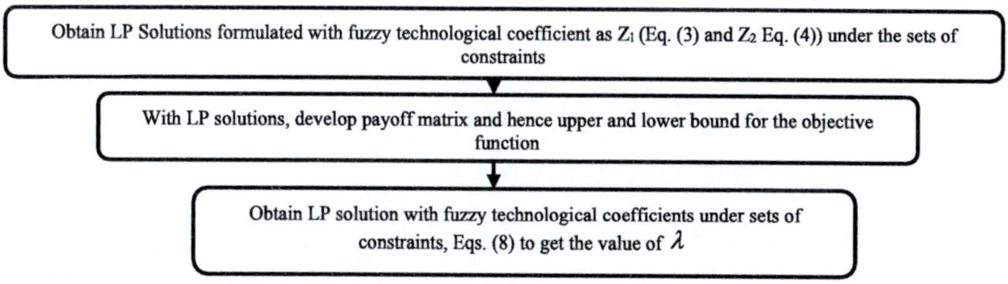

Fig. 1 Flowchart for development of LP solution with fuzzy technological coefficient

$$\text{s.t.} \sum_{j=1}^{n} a_{ij}\, x_j \le b_i \quad 1 \le i \le m$$
$$\text{where,} \quad x_j \ge 0, \quad 1 \le i \le n \tag{1}$$

2.2 Assumptions

Assumption 1. a_{ij} is a fuzzy number and consider the following linear membership function:

$$\mu_{a_{ij}}(x) = \begin{cases} 1 & \text{if}\, x < a_{ij} \\ (a_{ij} + d_{ij} - x)/d_{ij} & \text{if}\, a_{ij} \le x \le a_{ij} + d_{ij}, \\ 0 & \text{if}\, x \ge a_{ij} + d_{ij}, \end{cases} \tag{2}$$

where $x \, \varepsilon \, R$ and $d_{ij} > 0$. Initially, fuzzify the objective function in order to defuzzify this problem. This is accomplished by first determining the optimal values' minimum and maximum bounds. Solving the basic LP problem gives the best values, Z_l and Z_u.

$$z_1 = \max \sum_{j=1}^{n} c_j\, x_j$$
$$\text{s.t.} \sum_{j=1}^{n} a_{ij}\, x_j \le b_i, \quad i = 1, \ldots, m,$$
$$x_j \ge 0, \quad j = 1, \ldots, n, \tag{3}$$

and

$$z_2 = \max \sum_{j=1}^{n} c_j\, x_j$$
$$s.t. \sum_{j=1}^{n} (a_{ij} + d_{ij})\, x_j \le b_i, \quad i = 1 \ldots m,$$
$$x_j \ge 0, \quad j = 1 \ldots n, \tag{4}$$

The objective function takes values between Z_1 and Z_2' while technological coefficients vary between a_{ij} and $a_{ij} + d_{ij}$, where $Z_l = \min\,(Z_1, Z_2)$ and $Z_u = \max\,(Z_1, Z_2)$. Then, Z_l and Z_u are called the minimum and maximum bounds of the optimal values, respectively.

Assumption 2 The solutions to the linear crisp problems are finite. The fuzzy set of optimum values in this situation, G, which is a subset of R^n, is defined as.

$$\mu_G(x) = \begin{cases} 0 & \text{if } \sum_{j=1}^{n} c_i x_j < z_1 \\ (\sum_{j=1}^{n} c_i x_j z_l)/(z_u - z_l) & \text{if } z_l \le \sum_{j=1}^{n} c_i x_j < z_u, \\ 1 & \text{if } \sum_{j=1}^{n} c_i x_j \ge z_u \end{cases} \tag{5}$$

The fuzzy set of the ith constraint, C_i, which is a subset of R^m, is defined by,

$$\mu_{c_i}(x) = \begin{cases} 0, & b_i < \sum_{j=1}^{n} a_{ij} x_j \\ (b_i - \sum_{j=1}^{n} a_{ij} x_j)/\sum_{j=1}^{n} d_{ij} x_j, & \sum_{j=1}^{n} a_{ij} x_j \le b_i < \sum_{j=1}^{n} (a_{ij} + d_{ij}) x_j \\ 1, & b_i \ge \sum_{j=1}^{n} (a_{ij} + d_{ij}) x_j \end{cases} \tag{6}$$

Accordingly, the problem (1) becomes to the subsequent optimization problem

$$\begin{aligned} &\max \lambda \\ &\mu_G(x) \ge \lambda \\ &\mu_{C_i}(x) \ge \lambda, \quad 1 \le i \le m \\ &x \ge 0, \quad 0 \le \lambda \le 1 \end{aligned} \tag{7}$$

By using (5) and (6), the problem (7) can be written as

$$\begin{aligned} &\max \lambda \\ &\lambda(z_1 - z_2) - \sum_{j=1}^{n} c_j x_j + z_2 \le 0, \\ &\sum_{j=1}^{n} (a_{ij} + \lambda d_{ij}) x_j - b_i \le 0, \ 1 \le i \le m \\ &x_j \ge 0, \quad 1, \ldots n, \quad 0 \le \lambda \le 1. \end{aligned} \tag{8}$$

3 Case Study

In the Pune region of Maharashtra State, India, a dam was built across the Mula-Mutha River to form the Khadakwasala reservoir, a multipurpose project considered as a study area. The reservoir has an 86 Mm^3 total storage capacity, 56 Mm^3 of live storage, and a 62,146 Ha irrigable command area. The index map of the Khadakwasala command area is shown in Fig. 2. Table 1 presents the reservoir's monthly 75% dependable inflows.

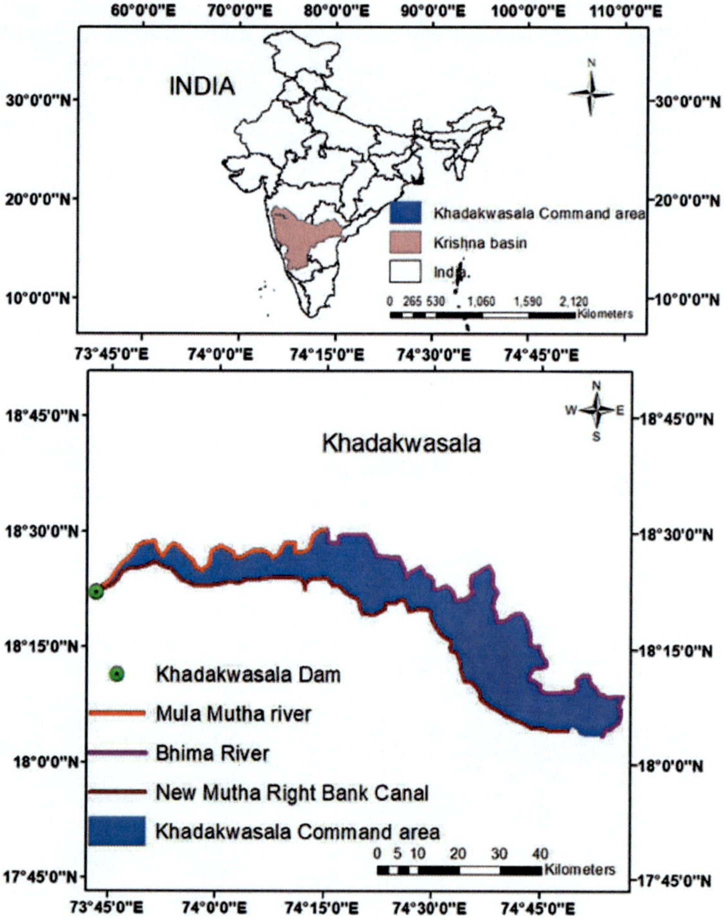

Fig. 2 Index map of Khadakwasala command area

Table 1 Monthly inflows of Khadakwasala reservoir

Month	Inflows in Mm^3	Month	Inflows in Mm^3
June	82.48	December	56.82
July	156.25	January	145.51
August	190.48	February	63.47
September	178.52	March	123.87
October	65.82	April	131.17
November	143.03	May	106.95

4 Fuzzy Linear Programming Model

The FLP model is developed for the reservoir's monthly operation. Here, the constraints are considered as crisp, or non-fuzzy, and the objective functions as fuzzy. Assuming stationary inflows throughout the course of a water year, the following generalized LP model is developed for monthly operation of the reservoir. The following generalized LP model incorporates FLP formulations as described in the methodology.

4.1 *Objective Function*

Maximization of irrigation releases has been considered as an objective function with a relevant set of constraints.

The objective considered in the model is

$$\text{Max } Z = \sum_{t=1}^{12} RIr \quad (9)$$

4.2 *Constraints*

Industrial release constraint
Releases for industry in each month (RIn_t) should be less than the maximum industrial releases ($RIn_{t(\max)}$). Releases in each month also should be greater than the minimum industrial releases ($RIn_{t(\min)}$).

$$RIn_t < RIn_{t(\max)} \quad \forall t = 1 \text{ to } 12, \quad (10)$$

$$RIn_t > RIn_{t(\min)} \quad \forall\, t = 1 \text{ to } 12. \tag{11}$$

Irrigation release constraint

Irrigation releases in each month (RIr_t) should be less than maximum irrigation demand ($IrD_{t(\max)}$). Irrigation releases also should be greater than minimum irrigation demand ($IrD_{t(\min)}$).

$$RIr_t < IrD_{t(\max)} \quad \forall t = 1 \text{ to } 12, \tag{12}$$

$$RIr_t > IrD_{t(\min)} \quad \forall t = 1 \text{ to } 12. \tag{13}$$

Domestic water supply constraint

Releases for domestic in each month (RDw_t) should be less than maximum domestic water demand ($RDw_{t(\max)}$). Releases for domestic also should be greater than minimum domestic water demand ($RDw_{t(\min)}$).

$$RDw_t < RDw_{t(\max)} \quad \forall t = 1 \text{ to } 12, \tag{14}$$

$$RDw_t > RDw_{t(\min)} \quad \forall t = 1 \text{ to } 12. \tag{15}$$

Reservoir storage constraints

Reservoir storage in each month should be less than maximum reservoir storage. Reservoir storage also should be greater than minimum reservoir storage.

$$S_t < S_{t(\max)} \quad \forall t = 1 \text{ to } 12, \tag{16}$$

$$S_t > S_{t(\min)} \quad \forall t = 1 \text{ to } 12. \tag{17}$$

Reservoir storage continuity constraint

Reservoir storage, inflows, irrigation releases, industrial releases, domestic water supply, evaporation losses from the reservoir during the time period t for all months are considered in volume units, and overflows are all subjected to this constraint.

$$S_t + I_t - RIr_t - RIn_t - RDw_t - E_t - O_t = S_{t+1} \quad \forall t = 1 \text{ to } 12. \tag{18}$$

Overflow Constraint

$$O_t > S_t + I_t - RIr_t - RIn_t - RDw_t - E_t - S_{t(\max)} \quad \forall t = 1 \text{ to } 12. \tag{19}$$

5 Results and Discussions

In the present study, model of FLP has studied and it has applied to Khadakwasala reservoir. Data used (inflow, irrigation, domestic and industrial water requirement, evaporation) to formulate aforementioned methodology have been procured form the Sinchan Bhavan, Pune. The objective considered in FLP model is maximization of irrigation releases (*RIr*). As stated in the methodology, the model initially takes into account the uncertainty associated with resources (b_i), i.e., irrigation demands and reservoir storage in any time period t are considered to be fuzzy resources, while technological coefficients are considered to be crisp in nature. For fuzzy technological coefficients, such as irrigation releases, the FLP model is solved (RIr). In this model using Eqs. (3) and (4), the model is solved for upper and lower bounds of irrigation releases, and maximum value of the objective is considered as upper bound (Z_u) and minimum value is considered as lower bound (Z_l) for the objective function. These values are given in Table 2. Equations (5) and (6) have been used to establish a linear membership function for the objective and constraints. Finally, a model is solved using Eq. (8) to maximize the truth level (λ). Table 3 displays the release policy for the maximized value of degree of truth (λ).

Table 3 presents optimal operating policies for the FLP model as described in the methodology section. When the uncertainty in the technological coefficients of the model is taken into account, the annual release for irrigation obtained is 515.63 Mm^3, and the degree of truth (λ) is 0.48.

By this model, in the month of June—90.84%, July—87.76%, October—38.54%, November—62.31%, January—4.41%, February—84.92%, March—40.15%, April—43.14%, May—44.85%, the releases in terms of percentage demand are satisfied, and in the months of August, September, and December, the releases

Table 2 Maximum and minimum bounds of the objective function for b_i

Objective function	
Bounds	Irrigation releases in Mm^3
Upper	593.495
Lower	578.42

Table 3 Release strategy for fuzzy technological coefficients (a_{ij})

Months	Irrigation releases (Mm^3)	Months	Irrigation releases (Mm^3)
June	3.51	December	73.17
July	11.91	January	45.21
August	12.57	February	77.86
September	66.59	March	38.83
October	54.72	April	43.96
November	48.69	May	38.61

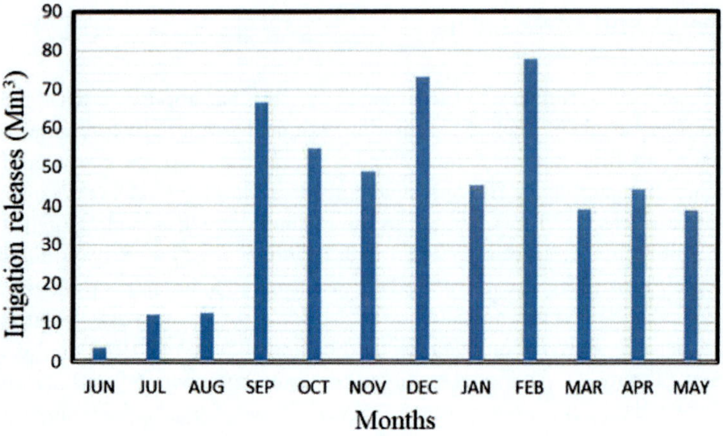

Fig. 3 Irrigation releases

in terms of percentage demand satisfied are 91.29%. As a result, it is observed that the best operating strategy derived by the current methodology taking into account the fuzziness involved in technological coefficients provides more precise results (Fig. 3).

6 Conclusions

The optimal reservoir operation with fuzzy technological coefficients is described as a fuzzy linear programming problem with a single objective. With respect to determine the optimal monthly operation strategy, this model is used for the case study of the Khadakwasala reservoir, which is located on the Mutha River in Maharashtra State, India. Maximizing irrigation releases is the objective function taken into consideration. Within a context of linear modeling, this methodology addresses uncertainty in demands, storages, and releases. The modeling process verified that how uncertainty in different parameters of reservoir operation model can be included progressively in resources, in technological coefficients of the model with fuzzy objective. The key findings of the present study are:

(i) The fuzzy logic model has the benefit that its computations are simple and its structure makes it simple for the operator to understand.
(ii) The model achieves an overall truth level (λ) of 0.48; however, it is up to the operations' manager to understand the sensitivity of the optimal results.

Acknowledgements The authors are thankful to Pune Irrigation Department, Pune, Maharashtra State, India, for giving necessary data for the analysis.

References

1. Russell SO, Campbell PF (1996) Reservoir operating rules with fuzzy programming. J Water Resour Planning and Managem ASCE 122(7):165–170
2. Balve PN, Patel JN (2016) Optimal reservoir operation policy using fuzzy linear programming. Int J Adv Mechan Civil Eng 3(4):77–80
3. Regulwar DG, Kamodkar RU (2010) Derivation of multipurpose single reservoir release policies with fuzzy constraints. J Water Resour Protect 2:1030–1041
4. Yeh WG (1985) Reservoir management and operations models: a state-of-the-art review. Water Resour Res 21(12):1797–1818
5. Labadie JW (2004) Optimal operation of multireservoir system: state of the art review. J Water Resour Plan Manag 130(2):93–111
6. Panigrahi DP, Mujumdar PP (2000) Reservoir operation modelling with fuzzy logic. J Water Resour Manag 14(2):89–109
7. Oliveira R, Loucks DP (1997) Operating rules for multi-reservoir systems. J Water Resour Managem 33(4):839–852
8. Fontane DG, Gates TG, Moncada E (1997) Planning reservoir operations with imprecise objectives. J Water Resour Plan Manag 123(3):154–162
9. Shrestha BP, Duckstein L, Stakhiv EZ (1996) A fuzzy rule based reservoir operation. J Water Resour Manag 122(3):262–269
10. Gasimov RN, Yenilmenz K (2002) Solving fuzzy linear programming problem with linear membership function. Turk J Mathem 26(4):375–396
11. Anand Raj P, Nagesh Kumar D (1999) Ranking alternatives with fuzzy weights using maximizing set and minimizing set. J Fuzzy Sets Syst 105(4):365–375
12. Raju KS, Kumar DN (2012) Irrigation planning of Sri Ram Sagar project using multi objective fuzzy linear programming. J Hydraulic Eng 6(1):55–63
13. Thakre PA, Shelar DS, Thakre SP (2009) Solving fuzzy linear programming problem as multi objective linear programming problem. In: Proceedings of the World congress on engineering (WCE-2009), London, U.K.
14. Singh DK, Jaiswal CS, Reddy KS, Singh RM, Bhandarkar DM (2001) Optimal cropping pattern in a canal command area. Agric Water Manag 50:1–8

A Bayesian Approach to Evaluate Surface Water Quality in the Upper Krishna Basin, India

Chanapathi Tirupathi, Thatikonda Shashidhar, and K. N. Murali Krishna

Abstract The Krishna river is the major source for the domestic, industrial, and agricultural needs of many cities, towns, and villages in its passage. However, the water in the basin was getting contaminated over the last two decades due to point and non-point source pollutants in the basin and it affects the river ecosystem. Thus, understanding the spatial variation of water quality across a river plays a vital role in controlling water pollution and public health safety. Therefore, the present study aims to analyze the spatial distribution of water quality across the Krishna river basin (KRB), India. This study developed a water quality model with a Bayesian approach to assess water quality at selected locations in the basin with seven water quality parameters, namely pH, Dissolved Oxygen (DO), Electrical Conductivity (EC), Biological oxygen demand (BOD), Nitrates, Total Suspended Solids (TSS), and Fecal Coliforms (FCs). The water quality parameters are selected on basis of their impact on water quality evaluation and availability. The water quality data at ten gauge stations were used, which were collected from the report prepared by the Mass Initiative for Truth Research and Action (MITRA). Out of ten-gauge stations, excellent water quality was observed at Panchganga river water at Balinga U/S of Kolhapur Town with a score of 78.6, whereas poor water quality was observed at Mahuli with a score of 69.6. Spatial distribution maps of water quality across the

Disclaimer The presentation of material and details in maps used in this chapter does not imply the expression of any opinion whatsoever on the part of the Publisher or Author concerning the legal status of any country, area or territory or of its authorities, or concerning the delimitation of its borders. The depiction and use of boundaries, geographic names and related data shown on maps and included in lists, tables, documents, and databases in this chapter are not warranted to be error free nor do they necessarily imply official endorsement or acceptance by the Publisher or Author.

C. Tirupathi (✉) · K. N. Murali Krishna
Department of Civil Engineering, Sasi Institute of Technology and Engineering, Tadepalligudem 534101, India
e-mail: ch.tirupathi1@sasi.ac.in

T. Shashidhar
Department of Civil Engineering, Indian Institute of Technology Hyderabad, Hyderabad 502285, India
e-mail: shashidhar@ce.iith.ac.in

P. V. Timbadiya et al. (eds.), *Geospatial and Soft Computing Techniques*, Lecture Notes in Civil Engineering 339, https://doi.org/10.1007/978-981-99-1901-7_41

Upper Krishna basin for various seasons summer, monsoon and winter seasons were prepared with Inverse Distance Weighting (IDW) multivariate interpolation method in ArcGIS. The spatial distribution maps of water quality are useful in identifying the regions that need to improve in terms of water quality. Finally, results from this study can be utilized for the decision-making and management of water quality. The proposed model is helpful in self-assessment of local water qualities and initiating further improvements to it.

Keywords Bayesian approach · Geospatial technologies · Krishna river basin · Spatial distribution mapping · Surface water quality

1 Introduction

The surface water bodies play an important role in human life by providing water for domestic, industrial and agricultural needs. However, the quality of water bodies is degrading day by day due to the release of untreated waste into the water bodies [1]. The majority of the river basins were highly contaminated because of anthropogenic activities like the release of domestic and industrial sewage waste into the nearby rivers [2, 3]. For example, Kengal et al. [1] report that Krishna basin water quality degraded drastically due to the point and non-point source pollutants in the basin along with economic and industrial development of regions in the KRB resulted in a rise in water demand and contamination of water bodies. The water quality of the river basins is extremely volatile with time due to anthropogenic activities in the premises of river networks [4, 5]. Therefore, assessment of spatiotemporal variations of water quality caused by natural or anthropogenic activities plays a vital role in planning effective water quality management and decision-making processes especially when there were no significant data [6].

However, it is an arduous task to evaluate water quality due to mixed interactions and interdependencies among input parameters involved in water quality modeling. The computer-based environmental models, such as Bayesian networks, can support the policy and decision-makers in assessment under uncertainty [7]. The Bayesian network simply represents the real world and can be used for assessment [8]. Various studies in the literature have used Bayesian networks for ecological and environmental modeling such as ecological water quality assessment [9]; groundwater quality [10]; water quality [11–13]; water resources vulnerability (WRV) assessment [14] and river basin sustainability [15]. Bayesian network models have the capability in handling ecological modeling as they can deal with the inherent uncertainties in the system explicitly by considering the interrelationship between parameters in a probabilistic way. It can consider the data from various sources such as outputs from models; expert knowledge and can use these available data sets in the best possible way [16]. Therefore, Bayesian networks (BNs) were adopted to assess water quality in the Upper Krishna basin (UKB).

2 Methodology

2.1 Study Area

The Upper Krishna basin is part of Krishna river basin as shown in Fig. 1, and the study region involves some portion of the Upper Krishna basin from Dhoom Dam, Mahabaleswar to Panchganga River water at Balinga U/S of Kolhapur Town. The study region covers 17 talukas (Wai, Medha, Mahabaleshwar, Islampur, Satara, Koregaon, Vaduj, Karad, Patan, Vite, Shirala, Panhala, Shahuwadi, Kohlhapur, Shirol, Tasgon, and Hatkalangada) in three districts (Satara, Sangli, and Kohlapur). It belongs to a tropical climate and the mean annual rainfall in this region is about 1300 mm. The majority of the basin consists of Deccan basaltic rocks and is dominated by vertisols [17]. The Upper Krishna basin is subjected to frequent droughts while 80% of the agriculture in this basin is rain-fed [18].

2.2 Bayesian Networks (BNs)

BNs are frequently used models for modeling the ecological and environmental aspects under uncertainty as these models can utilize Bayesian inference in estimating corresponding conditional probabilities of the input parameters [8]. In the present study, if–then rules have been utilized in framing the Conditional Probabilities Tables (CPTs); however, it is a time-consuming and cognitively challenging task, which was based on available data, literature, and expert knowledge. It can consider the data from various sources such as outputs from models; expert knowledge and can use the available data sets in the best possible way [7, 8, 15, 16]. In the present study, *Netica* software has been used in developing the water quality model which was a widely used model in the fields of environmental modeling [19].

2.3 Methodology

This study developed a water quality model with a Bayesian approach (Fig. 2) to assess the water quality at selected locations in Upper Krishna basin with seven water quality parameters, namely pH, Dissolved Oxygen (DO), Electrical Conductivity (EC), Biological oxygen demand (BOD), Nitrates, Total Suspended Solids (TSS), and Fecal Coliforms (FC). The selection of water quality parameters is on the basis of their influence on water quality evaluation and availability [20]. In this study, water quality data at ten gauge stations were used, which were collected from the report prepared by the Mass Initiative for Truth Research and Action (MITRA) [21].

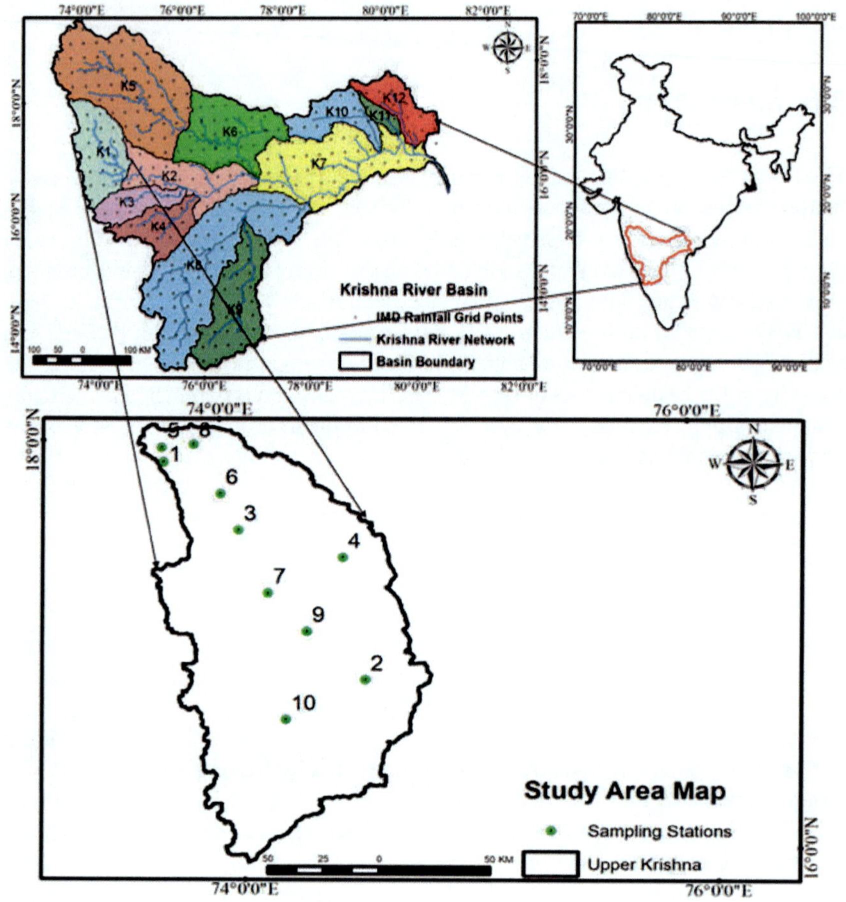

Fig. 1 Study region along with gauge stations

2.4 *Bayesian Water Quality Model (BWQM)*

In this study, assessment of water quality of UKB is represented as a function of seven water quality parameters, which account for physio-chemical and biological characteristics of the water. The classes and ranges of water quality parameters were determined on the basis of Indian water quality criteria as shown in Table 1 [20]. The present study proposes a Bayesian model, namely the Bayesian Water Quality model (BWQM), which attempts to estimate the river water quality for a given DO, BOD, TSS, Nitrates, FC, pH, and EC. Output classes of the proposed model were categorized into four classes such as excellent, good, poor, and extremely poor. The assessment of overall water quality depends on the weight and ratings allocated to each input parameter, and it is based on their relative significance in the assessment

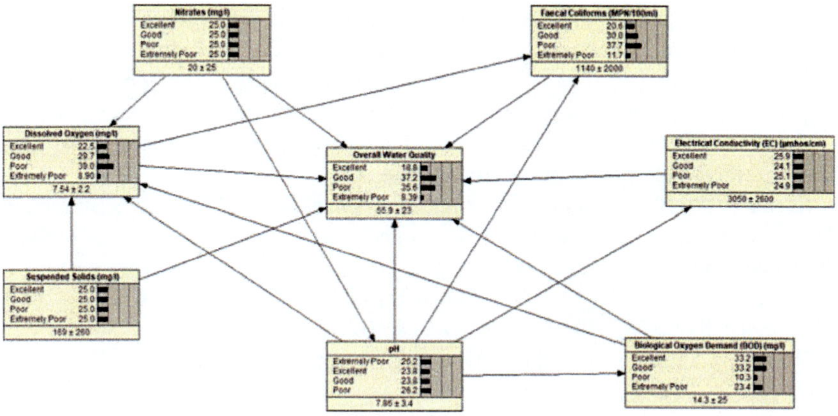

Fig. 2 Bayesian water quality model (BWQM)

of water quality [22]. Higher weight represents higher water quality and these are assigned based on their actual value.

Table 1 Classification of water quality parameters

Parameter	Weight	Range	Excellent (10)	Good (7)	Poor (4)	Extremely poor (1)
1. Dissolved Oxygen (mg/l)	0.18	[0 14]	> 8	6–8	4–6	< 4
2.Fecal Coliforms(MPN/100 ml)	0.16	[0 10000]	< 20	20–150	151–2000	> 2000
3. Biological oxygen demand (BOD) (mg/l)	0.16	[0 100]	< 2	2–3	3–8	> 8
4. Electrical Conductivity (EC) (µmhos/cm)	0.15	[0 10000]	< 1000	1000–2250	2251–4000	> 4000
5. pH	0.11	[1 14]	6.5–8.5	8.5–9.5	> 9.5	< 6.5
6. Nitrates (mg/l)	0.15	[0 100]	< 5	6–10	11–15	> 15
7. Total Suspended Solids (mg/l)	0.09	[0 1000]	< 25	26–50	51–100	> 100
Output class		[0 100]	> 75	55–75	25–55	< 25

Note Class A: water is useful for drinking water source without traditional treatment however after disinfection and can be used for all beneficial uses; Class B: Outdoor bathing and can be used for all beneficial uses but not as well as class A; Class C: Ingesting water source best after traditional remedy and disinfection and it cannot be used for domestic water supply

Seven water quality parameters are considered in this study and each parameter was classified into four classes; thus, the possible number of scenarios will be 16,384 (4^7), which were used in determining the Conditional Probability Table (CPT) for output class (BWQM). The overall score of every scenario is estimated based on allocated weight and rating to each scenario; it was utilized in defining the final state of the overall water quality of the basin. For example:

Scenario 80: If (pH is Excellent) and (DO is Excellent) and (Nitrate is Extremely Poor) and (BOD is Excellent) and (FC is Excellent) and (EC is Good) and (TSS is Extremely Poor), then water quality will be Good.

$= ((10 * 0.11) + (10 * 0.18) + (1 * 0.15) + (10 * 0.16) + (10 * 0.16) + (7 * 0.15) + (1 * 0.09)) = 7.39$

Normalized score is $= ((7.39 * 10)) = 73.9$; therefore, water quality will be Good.

3 Results and Discussions

3.1 Evaluation of Water Quality in Upper Krishna Basin

In this study, water quality state of the Upper Krishna basin is assessed at ten gauge stations for the overall, summer, winter, and monsoon seasons. The selection of gauge stations was based on data availability and was collected from the report prepared by the MITRA [21]. Out of ten-gauge stations, excellent water quality was observed at Panchganga river water at Balinga U/S of Kolhapur Town with a score of 78.6, whereas poor water quality was observed at Mahuli with a score of 69.6 (Fig. 3 and Table 2). The seasonal variations in water quality values of Upper Krishna basin were also evaluated with the BWQM. The model shows better water quality at Panchganga River water at Balinga U/S of Kolhapur Town in all seasons, i.e., summer, monsoon, and winter compared to other stations with a score of 76.8, 78.5, and 76.4, respectively, indicating that the water quality slightly decreases from the overall values in the respective seasons (Figs. 3, 4, 5, and 6). The model showed the lowest water quality at the Venna River, A/p- Mahuli, Satara in the summer season while in monsoon and winter seasons the lowest water quality was observed at Krishna River, Wai, Tal-Wai, Dist-Satara. Overall, the water quality of this basin was good, the model shows improved water quality values in monsoon season compared to overall values, the water quality was slightly reduced in summer season (Fig. 3 and Table 2), and the results were similar to MITRA [21].

Spatial distribution maps of water quality across Upper Krishna basin for various seasons summer, monsoon, and winter seasons were prepared with Inverse Distance Weighting (IDW) multivariate interpolation method in ArcGIS. The results showed excellent water quality in the majority of the Mahabaleshwar, Islampur, Hatkalangada, Panhala, and Tasgon regions, while it shows relatively low water quality in Wai, Vaduj, Karad, and Vite regions (Fig. 4) and it may be due to higher BOD concentration in these regions compared with other stations. In the summer season, a

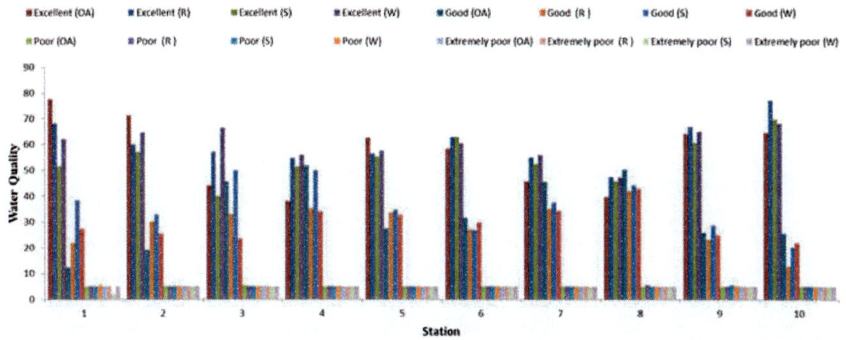

Fig. 3 Seasonal variation of water quality of the Upper Krishna basin. *Note* OA, R, W, and S indicates overall, rainy (monsoon), winter, and summer seasons' water qualities, respectively

few portions of Hatkalangada and Panhala showed excellent water quality, while the water quality of the remaining regions of upper Krishna showed good water quality. The water quality improved in monsoon and winter seasons; the majority of the basin shows excellent water quality except for the Wai region. These spatial distribution maps of water quality are useful in identifying the regions that need to improve in terms of water quality. Finally, results from this study can be utilized for the decision-making and management of water quality. The proposed model is helpful in the self-assessment of local water qualities and to initiate further improvements to it.

4 Conclusions

In the present study, a novel water quality model known as the "Bayesian Water Quality model" was developed to evaluate the river quality with the help of seven water quality parameters, i.e., pH, BOD, Nitrates, FC, DO, TSS, and EC. Overall, the water quality of Upper Krishna basin was good and the model shows improved water quality values in the monsoon season compared with the mean and the water quality was slightly reduced in the summer season. The spatial distribution maps of water quality across the Upper Krishna basin for various seasons summer, monsoon, and winter seasons were prepared and results showed excellent water quality in the majority of the Mahabaleshwar, Islampur, Hatkalangada, Panhala, and Tasgon regions, while it shows relatively low water quality in Wai, Vaduj, Karad, and Vite regions. These spatial distribution maps of water quality are useful in identifying the regions that need to improve in terms of water quality. It helps to predict and understand the influence of natural processes on water quality and determine its impact on the ecosystem. Consequently, it can assist in safeguarding environment by meeting the environmental standards. Finally, results from this study can be utilized for the decision-making and management of water quality. The proposed model is helpful

Table 2 Water quality values of the Upper Krishna basin at various locations

Station no	Station name	Summer	Monsoon	Winter	Average
1	Venna River, Mahabaleshwar	Excellent = 64.6% Good = 25.4% Poor = 5.03% Extremely poor = 5.00% (75.6 ± 21)	Excellent = 57.1% Good = 32.9% Poor = 5.01% Extremely poor = 5.0% (74 ± 21)	Excellent = 59.9% Good = 30.0% Poor = 5.03% Extremely poor = 5.00% (74.6 ± 21)	Excellent = 71.1% Good = 18.9% Poor = 5.00% Extremely poor = 5.00% (77.1 ± 21)
2	Krishna river water at Mai Ghat, Sangli	Excellent = 62.2% Good = 27.2% Poor = 5.54% Extremely poor = 5.00% (75 ± 21)	Excellent = 51.6% Good = 38.3% Poor = 5.13% Extremely poor = 5.01% (72.1 ± 21)	Excellent = 68.2% Good = 21.8% Poor = 5.01% Extremely poor = 5.00% (76.5 ± 21)	Excellent = 77.7% Good = 12.3% Poor = 5.00% Extremely poor = 5.00% (78.6 ± 21)
3	Urmodi River, A/p-Nagthane, Satara	Excellent = 66.4% Good = 23.5% Poor = 5.06% Extremely poor = 5.00% (76.1 ± 21)	Excellent = 40% Good = 49.9% Poor = 5.15% Extremely poor = 5.01% (70.1 ± 21)	Excellent = 57.1% Good = 32.8% Poor = 5.09% Extremely poor = 5.01% (73.9 ± 21)	Excellent = 44.0% Good = 45.5% Poor = 5.43% Extremely poor = 5.03% (70.9 ± 21)
4	Venna River, A/p-Mahuli, Satara	Excellent = 55.9% Good = 34.0% Poor = 5.07% Extremely poor = 5.00% (73.7 ± 21)	Excellent = 51.5% Good = 38.2% Poor = 5.2% Extremely poor = 5.01% (72.7 ± 21)	Excellent = 54.7% Good = 35.2% Poor = 5.04% Extremely poor = 5.00% (73.4 ± 21)	Excellent = 38% Good = 51.9% Poor = 5.11% Extremely poor = 5.01% (69.6 ± 21)

(continued)

Table 2 (continued)

Station no	Station name	Summer	Monsoon	Winter	Average
5	Krishna river (Dhom Dam), Mahabaleshwar	Excellent = 57.4% Good = 32.6% Poor = 5.01% Extremely poor = 5.00% (74 ± 21)	Excellent = 55.2% Good = 34.8% Poor = 5.00% Extremely poor = 5.01% (73.5 ± 21)	Excellent = 56.5% Good = 33.5% Poor = 5.03% Extremely poor = 5.01% (73.8 ± 21)	Excellent = 62.6% Good = 27.4% Poor = 5.01% Extremely poor = 5.00% (75.2 ± 21)
6	Venna River, Varye, Satara	Excellent = 60.3% Good = 29.6% Poor = 5.17% Extremely poor = 5.01% (74.6 ± 21)	Excellent = 62.9% Good = 26.9% Poor = 5.19% Extremely poor = 5.01% (75.2 ± 21)	Excellent = 62.9% Good = 27% Poor = 5.07% Extremely poor = 5.01% (75.3 ± 21)	Excellent = 58.3% Good = 31.7% Poor = 5.01% Extremely poor = 5.00% (74.2 ± 21)
7	Koyna River At-Karad, Tal.—Karad, Dist.—Satara	Excellent = 55.8% Good = 34.2% Poor = 5.04% Extremely poor = 5.00% (73.7 ± 21)	Excellent = 52.5% Good = 37.5% Poor = 5.05% Extremely poor = 5.00% (72.9 ± 21)	Excellent = 54.9% Good = 35.0% Poor = 5.06% Extremely poor = 5.01% (73.5 ± 21)	Excellent = 45.5% Good = 44.3% Poor = 5.15% Extremely poor = 5.00% (71.3 ± 21)
8	Krishna river, Wai, Tal.—Wai, Dist.—Satara	Excellent = 47.2% Good = 42.7% Poor = 5.07% Extremely poor = 5.00% (71.7 ± 21)	Excellent = 45.7% Good = 44.2% Poor = 5.06% Extremely poor = 5.00% (71.4 ± 21)	Excellent = 47.4% Good = 41.9% Poor = 5.66% Extremely poor = 5.04% (71.6 ± 21)	Excellent = 39.6% Good = 50.3% Poor = 5.09% Extremely poor = 5.00% (70 ± 21)

(continued)

Table 2 (continued)

Station no	Station name	Summer	Monsoon	Winter	Average
9	Krishna river water at Walwa D/S of Islampur	Excellent = 64.9% Good = 24.9% Poor = 5.23% Extremely poor = 5.02% (75.7 ± 21)	Excellent = 60.6% Good = 28.6% Poor = 5.76% Extremely poor = 5.05% (74.5 ± 21)	Excellent = 66.9% Good = 23.1% Poor = 5.0% Extremely poor = 5.01% (76.2 ± 21)	Excellent = 64.1% Good = 25.8% Poor = 5.09% Extremely poor = 5.00% (75.5 ± 21)
10	Panchganga River water at Balinga U/S of Kolhapur Town	Excellent = 68.1% Good = 21.9% Poor = 5.0% Extremely poor = 5.00% (76.4 ± 21)	Excellent = 69.7% Good = 20.3% Poor = 5.0% Extremely poor = 5.0% (76.8 ± 21)	Excellent = 77.2% Good = 12.8% Poor = 5.0% Extremely poor = 5.0% (78.5 ± 21)	Excellent = 64.6% Good = 25.4% Poor = 5.00% Extremely poor = 5.00% (78.6 ± 21)

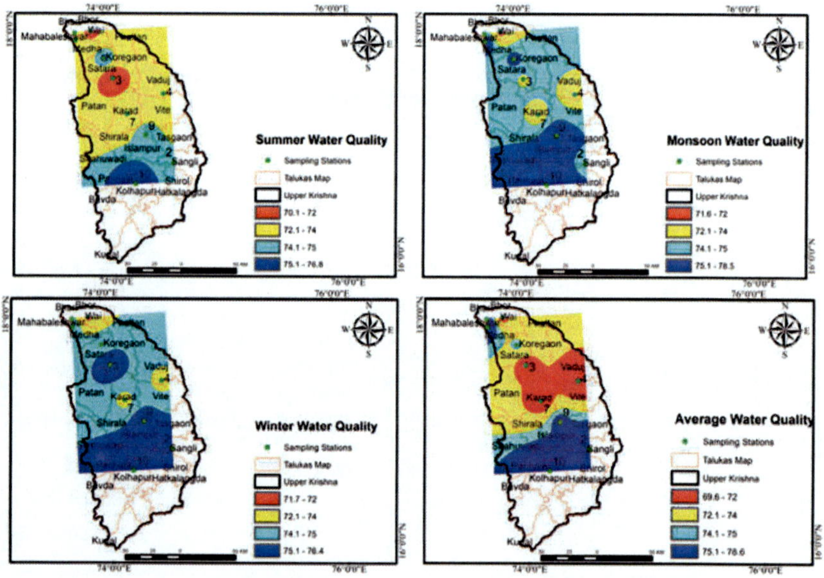

Fig. 4 Spatial distribution of seasonal water quality across the Upper Krishna basin, India

in the self-assessment of local water qualities and to initiate further improvements toward sustainable water resources management.

Acknowledgements The authors like to acknowledge the support from Central Pollution Control Board (CPCB)), Maharashtra Pollution Control Board (MPCB) and Mass Initiative for Truth Research & Action (MITRA) for their support in sharing data.

References

1. Kengnal P, Megeri MN, Giriyappanavar BS, Patil RR (2015) Multivariate analysis for the water quality assessment in rural and urban vicinity of Krishna River (India). Asian J Water Environ Pollut 12(2):73–80
2. Koukal B, Dominik J, Vignati D, Arpagaus P, Santiago S, Ouddane B, Benaabidate L (2004) Assessment of water quality and toxicity of polluted Rivers Fez and Sebou in the region of Fez (Morocco). Environ Pollut 131(1):163–172. https://doi.org/10.1016/j.envpol.2004.01.014
3. Vasistha P, Ganguly R (2020) Water quality assessment of natural lakes and its importance: an overview. Mater Today Proc 32:544–552. https://doi.org/10.1016/j.matpr.2020.02.092
4. Yang S, Liang M, Qin Z, Qian Y, Li M, Cao Y (2021) A novel assessment considering spatial and temporal variations of water quality to identify pollution sources in urban rivers. Sci Rep 11(1):1–11. https://doi.org/10.1038/s41598-021-87671-4
5. Yu R, Zhang C (2021) Early warning of water quality degradation: a copula-based Bayesian network model for highly efficient water quality risk assessment. J Environ Manage 292:112749. https://doi.org/10.1016/j.jenvman.2021.112749
6. Tlili-Zrelli B, Gueddari M, Bouhlila R (2018) Spatial and temporal variations of water quality of Mateur aquifer (northeastern Tunisia): suitability for irrigation and drinking purposes. J Chem. https://doi.org/10.1155/2018/2408632
7. Carmona G, Varela-Ortega C, Bromley J (2013) Participatory modelling to support decision making in water management under uncertainty: two comparative case studies in the Guadiana river basin, Spain. J Environ Manag 128:400–412. https://doi.org/10.1016/j.jenvman.2013.05.019
8. Koski T, Noble J (2011) Bayesian networks: an introduction, vol 924. Wiley
9. Kotta J, Aps R, Orav-Kotta H (2009) Bayesian inference for predicting ecological water quality under different climate change scenarios. WIT Trans Ecol Environ 127:173–184. https://doi.org/10.2495/RAV090151
10. Shihab K, Al-Chalabi N (2014) Bayesian methods for assessing water quality. CCSIT, SIPP, ISC, PDCTA, NLP:397–407. https://doi.org/10.5121/csit.2014.4234
11. Li RA, McDonald JA, Sathasivan A, Khan SJ (2021) A multivariate Bayesian network analysis of water quality factors influencing trihalomethanes formation in drinking water distribution systems. Water Res 190:116712. https://doi.org/10.1016/j.watres.2020.116712
12. Rode M, Arhonditsis G, Balin D, Kebede T, Krysanova V, van Griensven A, van der Zee S (2010) New challenges in integrated water quality modelling. Hydrol Process 24(24):3447–3461. https://doi.org/10.1002/hyp.7766
13. Sperotto A, Molina JL, Torresan S, Critto A, Pulido-Velazquez M, Marcomini A (2019) Water quality sustainability evaluation under uncertainty: a multi-scenario analysis based on Bayesian networks. Sustainability 11(17):4764. https://doi.org/10.3390/su11174764
14. Wang X, Ma F, Li C, Zhu J (2015) A Bayesian method for water resources vulnerability assessment: a case study of the Zhangjiakou Region, North China. Math Probl Eng. https://doi.org/10.1155/2015/120873
15. Chanapathi T, Thatikonda S (2020) Evaluation of sustainability of river Krishna under present and future climate scenarios. Sci Total Environ 738:140322. https://doi.org/10.1016/j.scitotenv.2020.140322

16. Uusitalo L (2007) Advantages and challenges of Bayesian networks in environmental modelling. Ecol Model 203(3–4):312–318. https://doi.org/10.1016/j.ecolmodel.2006.11.033
17. Ashtekar AS, Mohammed-Aslam MA, Moosvi AR (2019) Utility of normalized difference water index and GIS for mapping surface water dynamics in sub-upper Krishna Basin. J Indian Soc Remote Sens 47(8):1431–1442. https://doi.org/10.1007/s12524-019-01013-6
18. Mahajan DR, Dodamani BM (2015) Trend analysis of drought events over upper Krishna basin in Maharashtra. Aquatic Procedia 4:1250–1257. https://doi.org/10.1016/j.watres.2020.116712
19. Frank SK (2015) Expert-based Bayesian Network modeling for environmental management (Doctoral dissertation, Frankfurt am Main, Johann Wolfgang Goethe-Univ., Diss.
20. CPCB (Central Pollution Control Board) (2005) Water quality criteria. Accessed 17 Mar 2020. http://cpcb.nic.in/water-quality-criteria/
21. MITRA (Mass Initiative for Truth Research & Action) (2014) Comprehensive study report on Krishna River Stretch. Accessed 17 Mar 2021. https://mpcb.gov.in/sites/default/files/focus-area-reports-documents/KrishnaRiverReport.pdf
22. Chanapathi T, Thatikonda S (2019) Fuzzy-based regional water quality index for surface water quality assessment. J Hazard Toxic Radioactive Waste 23(4):04019010. https://doi.org/10.1061/(ASCE)HZ.2153-5515.0000443

Fuzzy Optimization Framework for Facilitating Best Management Practices in the Context of Urban Floods

Rohit Dwivedula, Rampalli Madhuri, K. Srinivasa Raju, and A. Vasan

Abstract Placement of best management practices (BMPs) is a constructive approach to control surface runoff. However, deciding where these BMPs need to be placed in practice remains a complex question, often requiring practitioners in the field to analyze trade-offs between the financial capital available and physical goals such as runoff reduction and pollutant reduction. This work describes a multiobjective optimization framework applied to Greater Hyderabad Municipal Corporation (GHMC), India, for the 2016 flooding event. The fuzzy approach converts a multi-objective optimization problem to a single objective problem through a membership function. Three membership functions, namely, nonlinear, exponential, and hyperbolic, were employed. Single Objective Genetic Algorithm (SOGA) is used for performing the optimization. Performing the optimization procedure with hyperbolic membership function yielded a degree of satisfaction, $\lambda = 0.8796$, corresponding to a BMP configuration spanning 61.98 km^2 of the urban case study area. This configuration would have reduced surface runoff by 1.02×10^7 m^3 while removing 73.87 tons of pollutants during this historic extreme rainfall event and arrived at a monetary cost of Rs. 1.16×10^{10}. Using the exponential membership function with 125 different sets of parameters yielded solutions with λ ranging from 0.5479 to 0.6432, and the average value of λ is 0.5950. Similar experiments with a nonlinear membership function yielded λ varying from 0.1307 to 0.9601 with an average λ of 0.5454.

Keywords Best management practices (BMPs) · Fuzzy optimization · SOGA

R. Dwivedula
Department of Computer Science & Information Systems, BITS Pilani, Hyderabad Campus, Hyderabad 500078, India
e-mail: f20170029h@alumni.bits-pilani.ac.in

R. Madhuri (✉) · K. Srinivasa Raju · A. Vasan
Department of Civil Engineering, BITS Pilani, Hyderabad Campus, Hyderabad 500078, India
e-mail: p20140406@hyderabad.bits-pilani.ac.in

K. Srinivasa Raju
e-mail: ksraju@hyderabad.bits-pilani.ac.in

P. V. Timbadiya et al. (eds.), *Geospatial and Soft Computing Techniques*, Lecture Notes in Civil Engineering 339, https://doi.org/10.1007/978-981-99-1901-7_42

1 Introduction

Best management practices (BMPs) have been widely used for over two decades globally to manage flood risks, remove pollutants, or reduce sediment in water bodies [1–3]. To optimize the placement of BMPs, the present approaches use a range of mathematical optimization techniques such as integer programming, nonlinear programming, and evolutionary algorithms [4]. Behroozi et al. [5] used multiobjective particle swarm optimization (PSO) in District 10, Tehran, Iran, to minimize peak water flow rate and pollutant concentration. Singh et al. [6] describe a case study from Heredia, Costa Rica, where bioretention areas, green roofs, and infiltration trenches are placed in an urban setting to control flood risks. They use a nonlinear programming technique to optimize and formulate the trade-offs between land use and cost. Foomani and Malekmohammadi [7] proposed fuzzy logic and analytic hierarchy process for identifying optimum locations of BMPs in the northern region of Tehran, Iran. Li [8] developed SWMM_FLC, a combination of SWMM, fuzzy logic control, and GA, to reduce downstream flooding volume. Zhang et al. [9] applied Storm Water Management Model (SWMM) and System for Urban Stormwater Treatment and Analysis Integration (SUSTAIN) to conduct watershed-level optimization for Sponge City, China. Annual average runoff volume and total pollutants reduced workout to 87.61% and 85%. Dwivedula et al. [10] employed an ensemble of (1) non-dominated sorting genetic algorithm-III and (2) constrained two-archive evolutionary algorithm for optimizing zone-wise BMP placement in GHMC. Studies presented here, including that of [10] and elsewhere, have not reported any applications of fuzzy optimization in the placement of BMPs.

2 Study Area and Data Source

2.1 Greater Hyderabad Municipal Corporation

In this study, a fuzzy multiobjective approach is used to optimize the placement of BMPs for GHMC as a whole (not zone-wise). This section briefly describes the following:

- Case study and process(es) used to identify potential BMP sites.
- Multiobjective optimization problem, i.e., the decision variables, objectives, and constraints.
- Fuzzy optimization framework and membership functions used.
- Single Objective Genetic Algorithm (SOGA).

The fuzzy optimization process allows converting a multiobjective problem to a single objective problem, enabling us to use a Single Objective Genetic Algorithm (SOGA). Figure 1 presents the study area.

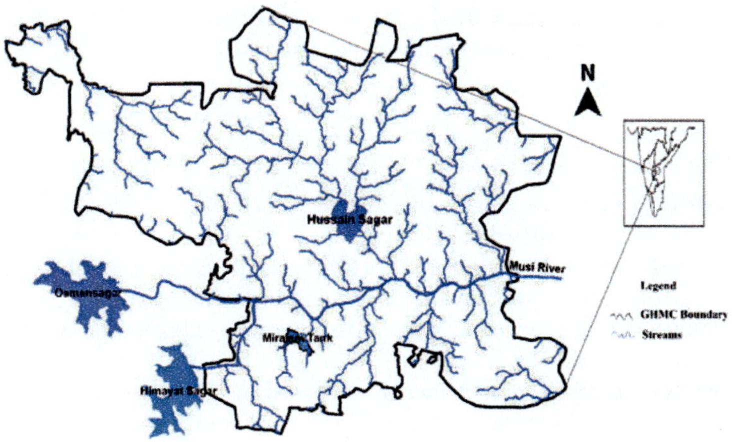

Fig. 1 Study area of GHMC

2.2 *Data Used*

The present study examines a historic extreme rainfall event of 237.5 mm during September 20–28, 2016. We attempt to analyze the impact of BMP placement if a similar extreme event was to happen again. Hydrologic Engineering Centre's-Hydrologic Modeling System (HEC-HMS) was employed to simulate surface runoff [11] and SUSTAIN [12] was used for identifying the potential BMP sites. EPA-SUSTAIN siting tool identified a total of 5,45,895 possible sites. Nine types of BMPs are being considered for placement in the GHMC.

2.2.1 Multiobjective Optimization Problem

The three objectives we wish to optimize are maximizing runoff reduction volume Z_1 (in m^3) and pollutant load reduction Z_2 (in tons) while minimizing the cost of construction Z_3 (in Indian Rupees). For an individual BMP for total area A_k, we can select (all the areas/a fractional part of an area/none of an area. This choice is encapsulated as a *decision variable*, denoted by X_k ($0 \leq X_k \leq 1$). There is a total of K decision variables. The decision variables (X) are related to the objectives (Z) as:

$$\begin{bmatrix} Z_1 \\ Z_2 \\ Z_3 \end{bmatrix} = \sum_k \left\{ \begin{bmatrix} R_k * \rho_k \\ S_k * \eta_k \\ -1 * d_k * c_k \end{bmatrix} * A_k * X_k \right\} \forall k \in [0, k] \tag{1}$$

where R is the rainfall, S is the runoff, ρ is the runoff reduction efficiency of the BMP, η is the pollutant reduction efficiency, d is the depth of the BMP, and c is the construction cost of the BMP per unit volume. More details about case studies, modeling, and data requirements are available from [10].

2.2.2 Fuzzy Optimization and Membership Functions

Our problem in optimization is maximizing the objectives (Z). Lower Z_L and upper limits Z_U for goals are shown in Table 1.

Each objective Z can be represented as a function of the decision variables (X), i.e., $Z_i = f_i(X)$ (Eq. 1). In this work, we define a membership function denoted by $\mu_z(X)$ for each objective. We studied hyperbolic, exponential, and nonlinear membership functions (refer to Table 2).

For all three membership functions, $Z \leq Z_L$ is 0 and $Z \geq Z_U$ is 1.

We notice that the hyperbolic membership function does not have any parameters that the decision-maker must choose, unlike nonlinear or exponential membership functions. S is a non-zero parameter $0 < S \leq 1$ [13]. β determines the shape of the membership function.

The fuzzy optimization problem (with N objectives) is as follows:

Maximize λ, subject to constraints:

$\mu_z(X) \geq \lambda \ \forall i \in \{1, 2, \ldots N\}$
$0 \leq \lambda \leq 1$
$Z_1 \geq 3.5 \times 10^6 \text{ m}^3$ and $Z_2 \geq 25$ tons

along with other existing constraints and bounds.

Table 1 Lower and upper limits of the objective functions

Objective	Units	Z_L	Z_U
Runoff reduction (Z_1)	10^7 m^3	0	1.547
Pollutant load reduction (Z_2)	10^{11} mg	0	1.109
Monetary cost (Z_3)	10^{10} Rs	– 3.497	0

Table 2 Types of membership functions and corresponding equations for $Z_L < Z < Z_U$

Hyperbolic	$\frac{1}{2}\tanh\left[\left(Z - \frac{Z_U+Z_L}{2}\right)\frac{6}{Z_U-Z_L}\right] + \frac{1}{2}$
Exponential	$\left[\frac{e^{-S\left(\frac{Z_U-Z}{Z_U-Z_L}\right)} - e^{-S}}{1-e^{-S}}\right]$
Nonlinear	$\left[\frac{Z-Z_L}{Z_U-Z_L}\right]^{\beta}$

2.2.3 Single Objective Genetic Algorithms

SOGA with a population size of 1000, simulated binary cross-over probability of 0.9 [14], polynomial mutation probability of 0.1 [15], and tournament selection are used for optimization. The PyMoo library [16] is employed for implementing the optimization functions.

3 Results and Discussions

The results of optimization with three membership functions are as follows. All source codes used to run these experiments have been open sourced under the MIT license and are available online.[1]

3.1 Hyperbolic Membership Function

Performing the optimization procedure with hyperbolic membership function for all three objectives yielded a solution of satisfaction $\lambda = 0.8796$, corresponding to a real-world configuration of BMPs spanning 61.98 km^2 of area, which reduce surface runoff by $1.02 \times 10^7 m^3$, while removing 73.87 tons of pollutant at a monetary cost of Rs. 1.16×10^{10}. The progress of SOGA can be visualized by plotting the best-discovered value of λ against the number of function evaluations as depicted in Fig. 2.

3.2 Exponential Membership Function

Next, we present the results of the exponential membership function. Optimization procedure was run with 125 different configurations of the parameter s such that: s_1, $s_2, s_3 \in \{0.2, 0.4, 0.6, 0.8, 1\}$.

Here, s_1, s_2, s_3 are the parameters for runoff reduction, pollutant load reduction, and cost (Z_3). Use of exponential membership function with these 125 different sets of parameters yielded solutions with λ ranging from 0.5479 to 0.6432. The average value of λ is 0.5950. Optimization convergence of all these 125 different sets of parameters can be visualized in Fig. 3. Each line in Fig. 3 represents a different set of parameters. It is noticed that all the lines follow similar trends, suggesting that the optimization process is not very sensitive to changes in parameters s_1, s_2,

[1] Link to code repository: https://github.com/rohitdwivedula/bmp-multiobjective-optimisation (https://doi.org/10.5281/zenodo.6676306).

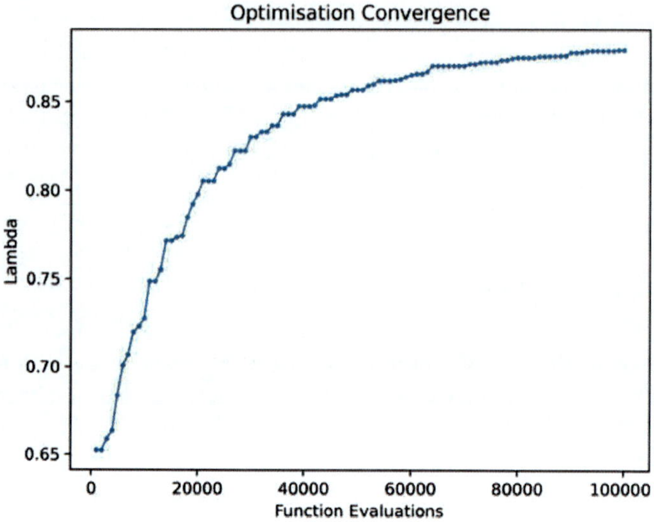

Fig. 2 Optimization process with hyperbolic membership function for all three objectives

s_3. We also notice that the value of λ begins to plateau for most lines after the 60th generation (or 60,000 function evaluations), indicating that the optimization approach has converged.

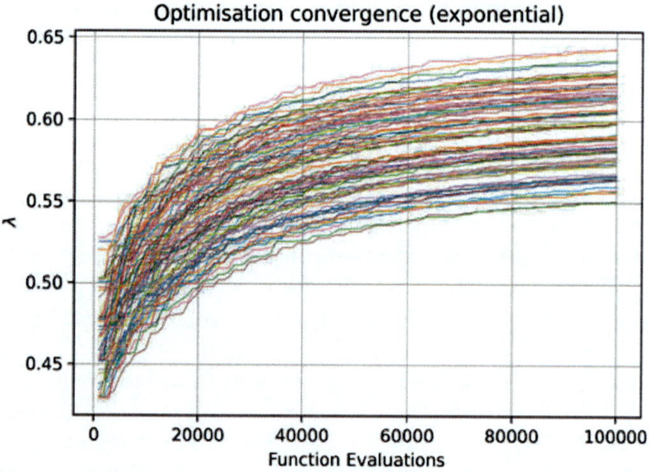

Fig. 3 Optimization process with exponential membership function for all three objectives and various values of the parameter

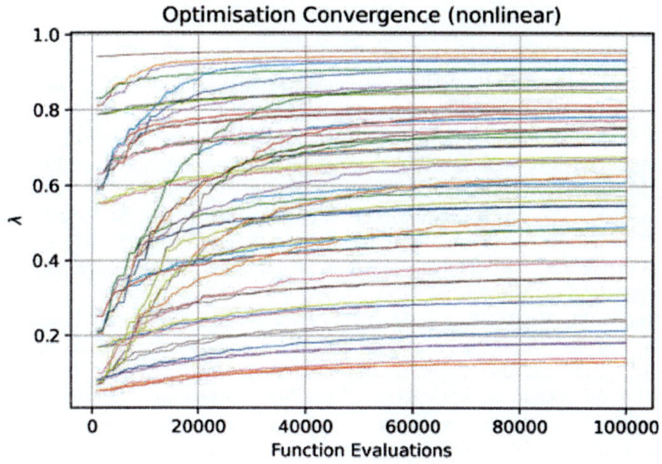

Fig. 4 Optimization process with nonlinear membership function with varying values of membership function parameter β

3.3 *Nonlinear Membership Function*

Use of nonlinear membership function with 125 different sets of parameters yielded solutions with λ ranging from 0.1307 to 0.9601 with a moderate satisfiability of $\lambda = 0.5454$. The parameters of β_i used were: $\beta_1, \beta_2, \beta_3 \in \{0.1, 0.4, 1.0, 3.0, 5.0\}$.

Here, β_1, β_2, and β_3 are the parameters for Z_1, Z_2, and Z_3, respectively. Optimization convergence for nonlinear membership functions is plotted in Fig. 4, similar to the plots in previous sections. One key difference noticed is that solution is susceptible to changes in the parameter β. For example, using $(\beta_1, \beta_2, \beta_3) = (5, 5, 5)$ yields the least value of $\lambda = 0.1307$, while $(\beta_1, \beta_2, \beta_3) = (0.1, 0.1, 0.1)$ yields the highest value of $\lambda = 0.9601$. Values of β_i will have to be decided based on the relative importance of each objective; that is, objectives that are relatively more important must have a larger β relative to others.

4 Conclusions

A fuzzy optimization approach was applied to optimize BMPs in the GHMC. Experimentation was done with a wide range of parameters to analyze the sensitivity of each membership function with its parameters. It is observed that the nonlinear membership function is relatively more sensitive to changes in parameters when compared to the exponential membership function. Future work could include experimentation with more optimization algorithms, extending the analysis for potential future rainfall events, and applying this framework to other case studies.

References

1. Urbonas B (1994) Assessment of stormwater BMPs and their technology. Water Sci Technol 29:347–353. https://doi.org/10.2166/wst.1994.0682
2. Koc CB, Osmond P, Peters A (2017) Towards a comprehensive green infrastructure typology: a systematic review of approaches, methods and typologies. Urban Ecosyst 20:15–35. https://doi.org/10.1007/s11252-016-0578-5
3. Venkataramanan V, Lopez D, McCuskey DJ, Kiefus D, McDonald RI, Miller WM, Packman AI, Young SL (2020) Knowledge, attitudes, intentions, and behavior related to green infrastructure for flood management: a systematic literature review. Sci Total Environ 720:137606. https://doi.org/10.1016/j.scitotenv.2020.137606
4. Janga Reddy M, Nagesh Kumar D (2021) Evolutionary algorithms, swarm intelligence methods, and their applications in water resources engineering: a state-of-the-art review. H2Open Journal 3:135–188. https://doi.org/10.2166/h2oj.2020.128
5. Behroozi A, Niksokhan M, Nazariha M (2018) Developing a simulation-optimization model for quantitative and qualitative control of urban runoff using best management practices. J Flood Risk Manage 11:S340–S351. https://doi.org/10.1111/jfr3.12210
6. Singh A, Sarma AK, Hack J (2020) Cost-effective optimization of nature-based solutions for reducing urban floods considering limited space availability. Environ Process 7:297–319. https://doi.org/10.1007/s40710-019-00420-8
7. Foomani MS, Malekmohammadi B (2020) Site selection of sustainable urban drainage systems using fuzzy logic and multi-criteria decision-making. Water Environ J 34(4):584–599. https://doi.org/10.1111/wej.12487
8. Li J (2020) A data-driven improved fuzzy logic control optimization-simulation tool for reducing flooding volume at downstream urban drainage systems. Sci Total Environ 732(25):138931. https://doi.org/10.1016/j.scitotenv.2020.138931
9. Zhang Z, Gu J, Zhang G, Ma W, Zhao L, Ning P, Shen J (2021) Design of urban runoff pollution control based on the Sponge City concept in a large-scale high-plateau mountainous watershed: a case study in Yunnan China. J Water Clim Change 12(1):201–222. https://doi.org/10.2166/wcc.2019.120
10. Dwivedula R, Madhuri R, Srinivasa Raju K, Vasan A (2021) Multiobjective optimisation and cluster analysis in placement of best management practices in an urban flooding scenario. Water Sci Technol 84(4):966–984. https://doi.org/10.2166/wst.2021.283
11. Feldman AD (2000) Hydrologic modeling system HEC-HMS technical reference manual. US Army Corps of Engineers Hydrologic Engineering Center
12. EPA-SUSTAIN (2014) EPA system for urban stormwater treatment and analysis integration (sustain). https://www.epa.gov/water-research/system-urban-stormwater-treatment-and-analysis-integration-sustain
13. Morankar D, Raju KS, Kumar DN (2013) Integrated sustainable irrigation planning with multi-objective fuzzy optimization approach. Water Resour Manage 27:3981–4004. https://doi.org/10.1007/s11269-013-0391-3
14. Deb K, Sindhya K, Okabe T (2007) Self-adaptive simulated binary crossover for real-parameter optimization. In: Proceedings of the 9th annual conference on genetic and evolutionary computation GECCO '07. Association for Computing Machinery, New York, USA, p 11871194. https://doi.org/10.1145/1276958.1277190
15. Deb K, Deb D (2014) Analysing mutation schemes for real-parameter genetic algorithms. Int J Artif Intell Soft Comput 4:1–28. https://doi.org/10.1504/IJAISC.2014.059280
16. Blank J, Deb K (2020) Pymoo: multiobjective optimization in python. IEEE Access 8:89497–89509. https://doi.org/10.1109/ACCESS.2020.2990567

Machine Learning Framework for Flood Susceptibility Modeling in a Fast-Growing Urban City of Southern India

A. L. Achu, Girish Gopinath, and U. Surendran

Abstract Flooding in urban areas often results severe loss of life and property and has many negative socio-economic impacts. Therefore, identifying the flood prone areas is necessary for future flood hazard mitigation, early warning, and land use planning for infrastructure developments in urban areas. In this study, flood susceptibility modeling is carried out for Kozhikode urban and per-urban area, which is severely affected by 2018 Kerala flood. To begin with, a flood inventory map is prepared with 307 flood location points marked immediately after 2018 flood. Thereafter, the inventory is randomly classified into 70% for model training and remaining 30% for model testing. In addition, twelve independent variables such as land use/land cover, soil texture, lithology, elevation, slope angle, slope aspect, valley depth, topographical wetness index, profile curvature, plan curvature, convergence index, and channel network base level were prepared and used. Subsequently, final modeling is carried out using these flood conditioning factors and flood inventory locations using machine learning random forest method. The result shows that ~ 13.78% of the study area is very highly susceptible to the occurrence of flood. The predicted model shows 85.2% accuracy (ROC-AUC) in training phase and 78.5% in testing phase. Therefore, the model is trustworthy and can be used for future hazard mitigation and land use planning in Kozhikode urban and per-urban area.

Keywords Flood · Machine learning · Kozhikode · Kerala

A. L. Achu (✉) · G. Gopinath
Department of Climate Variability and Aquatic Ecosystems, Kerala University of Fisheries and Ocean Studies (KUFOS), Kochi, Kerala 682508, India
e-mail: achu.geomatics@gmail.com

G. Gopinath
e-mail: ggkufos@ac.in

U. Surendran
Land and Water Management Research Group, Centre for Water Resources Development and Management (CWRDM), Kozhikode, Kerala 673571, India
e-mail: suren@cwrdm.org

P. V. Timbadiya et al. (eds.), *Geospatial and Soft Computing Techniques*, Lecture Notes in Civil Engineering 339, https://doi.org/10.1007/978-981-99-1901-7_43

1 Introduction

Among the different natural calamities, floods are most frequent and affecting millions of peoples across the globe. Urban flooding is a global concern, and it does not just mean "the flooding that happens in an urbanized area." The Federal Emergency Management Agency (FEMA) report 2016 defines urban flooding as: the inundation of property in a built environment, particularly in more densely populated areas, caused by rain falling on increased amounts of impervious surfaces and overwhelming the capacity of drainage systems. Flooding causes huge loss to life and property across the world. Between 2011 and 2012 alone, floods affected around 200 million people and caused economic losses of about $95 billion. Hence, it is of paramount importance to manage floods and reduce their risk, which requires flood prediction and computation of inundation areas [6]. Flood is a complex phenomenon, and hence, predicting the same is difficult [13]. For predicting the probability of flood and for mitigating and managing future flood hazard, modeling flood susceptibility is an essential procedure [10]. To model the flood susceptibility, multi-sourced dataset is required.

With the development of remote sensing techniques, multi-temporal and multi-sourced data have been widely used to predict flood susceptibility with GIS techniques [6]. However, in recent times, with the introduction of the concept of big data analytics and machine learning, accuracy and reliability of flood susceptibility mapping are improved significantly. Many researchers used different machine learning techniques to assess the flood [7, 11, 18]. Methods including random forest [1], support vector machine [17], artificial neural network [3], logistic regression [14, 15] have been widely used for flood prediction. In 2018, Kerala witnessed extreme rainfall event caused huge flooding and numeroous landslides across the state. Kozhikode was one of the coastal cities which was severely affected by flooding during 2018. Absence of flood susceptibility map was one the reason for extended causality, and hence in this study, random forest method is used to model flood susceptibility in Kozhikode urban cluster area. The proposed study will be useful for future hazard mitigation.

2 Materials and Methods

2.1 Study Area

Urban clusters in the Kozhikode District on the southwest coast of India were chosen for the study. Kozhikode urban cluster (KUC) is the largest urban agglomeration in Malabar region (northern Kerala) with an area of 197 km^2. As per the census report 2011 [5], KUC is the second-order urban zone with a population density of 3746 persons/km^2 (State Urbanization Report—Kerala 2012), and it is projected to increase population density and infrastructure development in near future and [9].

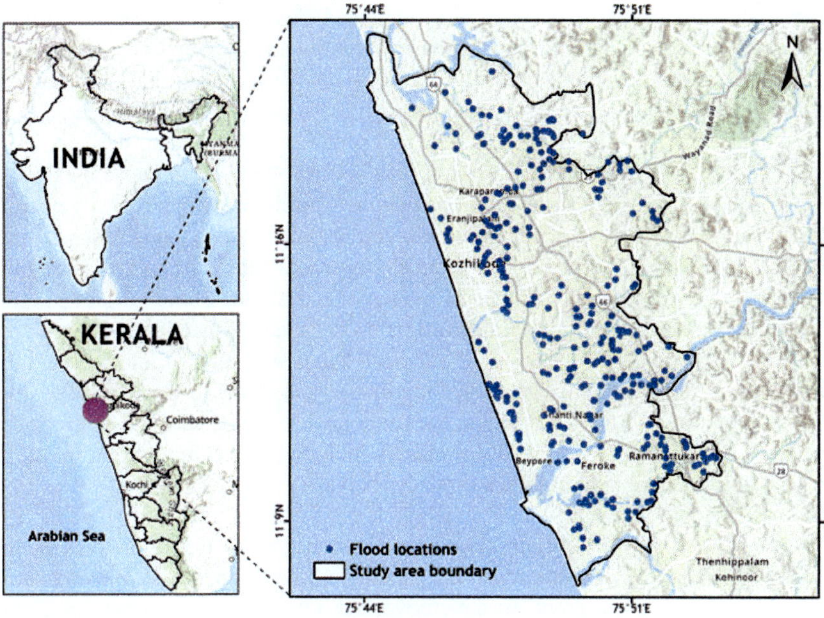

Fig. 1 Location map of the KUC with flood locations

Physiographically KUC is a part of Western coastal plains of Kerala with undulating topography lies between 11° 7′ 27.46″ N to 11° 21′ 17.91″ N latitudes and 75° 44′ 13.09″ E to 75° 52′ 9.11″ E longitudes (Fig. 1). During 2018 Kerala flood, KUC was affected severely, displacing millions. During June–August 2018, KUC received 2898 mm rainfall against its normal average of 2250 mm which caused intense flooding in river valleys and low-lying areas of KUC [16].

2.2 *Spatial Database*

Accurate and reliable flood inventory datasets are essential for flood susceptibility modeling [8]. In this study, flood inventory marking was carried out with intense field visits immediately after 2018 flood. Flooded areas were marked using handheld GPS and flooding height was also measured and attributed to the locations. A total of 307 flood locations were marked which range 0.11 m–2.48 m and used in this study. The locations were randomly divided into 70–30% for model building and model testing. Besides, a ten-fold cross-validation is implemented to avoid over fitting.

The construction of flood susceptibility modeling is a complex decision-making process which involves many geo-environmental variables [8]. In this study, twelve geo-environmental variables including elevation, slope angle, lithology, soil texture, land use/land cover, slope aspect, Topographic Wetness Index (TWI), Valley depth,

Channel network base level (CNBL), Convergence Index, Plan curvature, and Profile curvature were selected on the basis of expert opinion and literature [12, 18, 19].

SRTM Digital Elevation Model (DEM, 30 M) is used to represent elevation of the study area, and other DEM derivatives such as slope angle, slope aspect, TWI, valley depth, CNBL, CI, plan curvature, and profile curvature were derived. The KUC is a low-lying area where elevation ranges from ~ 0 to 90 m above mean sea level (Fig. 2a). Slope angle is an important parameter which determines flow velocity and concentration. KUC is a gently sloping terrain where slope angle ranges from 0 to 20.75° (Fig. 2b). Conventional parameters such as lithology, soil texture, and land use/land cover data are gathered from Geological Survey of India, Kerala State Soil Survey Organization, and Kerala State land Use Board, respectively. Charnokite group of rocks and tertiary deposits of Sand and Silt is the major lithology found in the study area followed minor patches of migmatite complex (Fig. 2c). Gravelly clay is the dominating soil texture found in the study area followed clay and sandy soils (Fig. 2d). KUC has different land use classes including agricultural area, built-up land, waste lands, wetlands, and water bodies (Fig. 2e).

Slope aspect of the study area is shown in Fig. 2f which shows nine slope directions; however, flat and northern slopes are the major slope directions present in KUC. TWI is another important terrain parameter which represents soil moisture concentration at a given point. In the study area, TWI values range from 4.25 to 20.88 (unitless) (Fig. 2g). The KUC has a valley depth which ranges from 0.05 to 56.48 m (Fig. 2h), and in general, valleys having higher depth are considered as higher flood susceptible area. Channel network base level (CNBL) is another important parameter used for flood susceptibility modeling. In the study area, CNBL values range from 0 to 29.55 (Fig. 1f). Convergence index (CI) is a terrain parameter which shows the structure of the relief as a set of convergent areas (channels) and divergent areas (ridges). It represents the convergence or divergence of overland flow. In the study area, CI values range from −93.58 to 96.99 (Fig. 2j). The present study used plan and profile curvatures for modeling (Fig. 3k and l). In general, profile curvature is defined as curvature parallel to the direction of the maximum slope, whereas plan curvature is perpendicular to the direction of the maximum slope.

2.3 Random Forest Method (RF)

RF is a powerful machine learning method proposed by Breiman [4]. RF is a decision tree-based (DT) model that can be used for both classification and regression problems. RF is an ensemble DT model which operates by constructing a multitude of decision trees at the training time and outputting the class that is the mode of the classes (classification) or the mean prediction (regression) of the individual trees. Random-decision forests correct for the decision tree habit of overfitting to a training set [2, 15]. In this study, flood susceptibility is treated as a binary classification, i.e., flood occurrences (1) and non-occurrences (0). Consider training set $D = ((A1, B1),\ldots\ldots,(An, Bn))$ that consists of n vectors, $A = \ \in X$ where X is a set

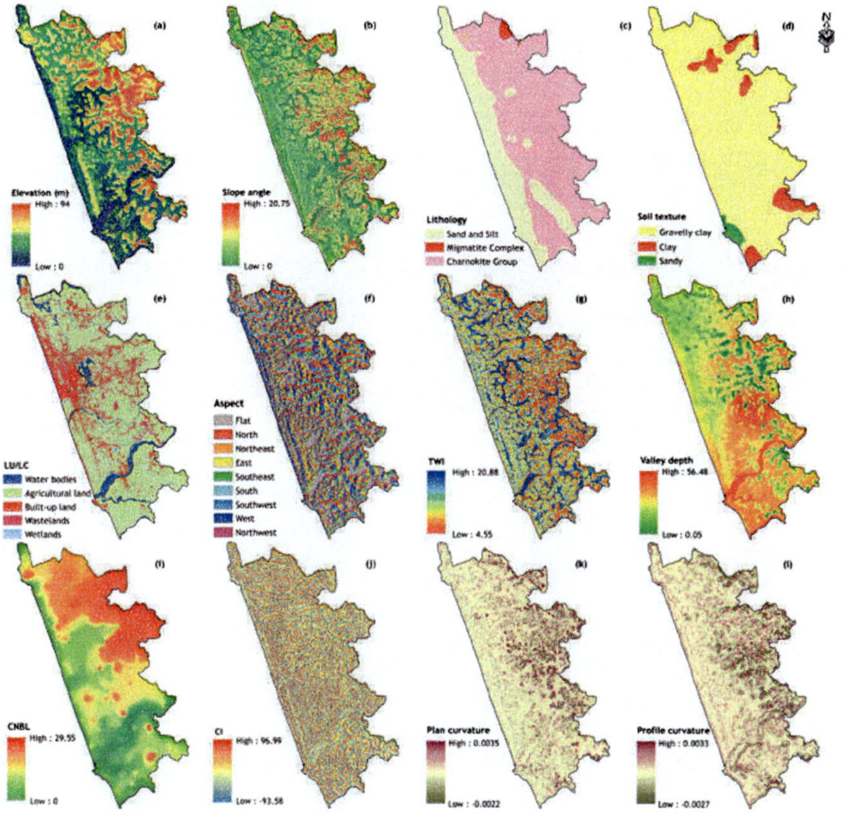

Fig. 2 Flood conditioning factors selected for the flood susceptibility modeling

of numerical or symbolic observations, and $B = \in Y$ where Y a set of class labels (here flood and non-flood). For classification problems, a classifier is a mapping $X \rightarrow Y$ [6]. The RF is working with two processes; the first is Breiman's "bagging" idea and the second is Ho's "random selection features. Bagging is an ensemble machine learning procedure to improve the prediction accuracy of a weak classifier by creating a set of classifiers.

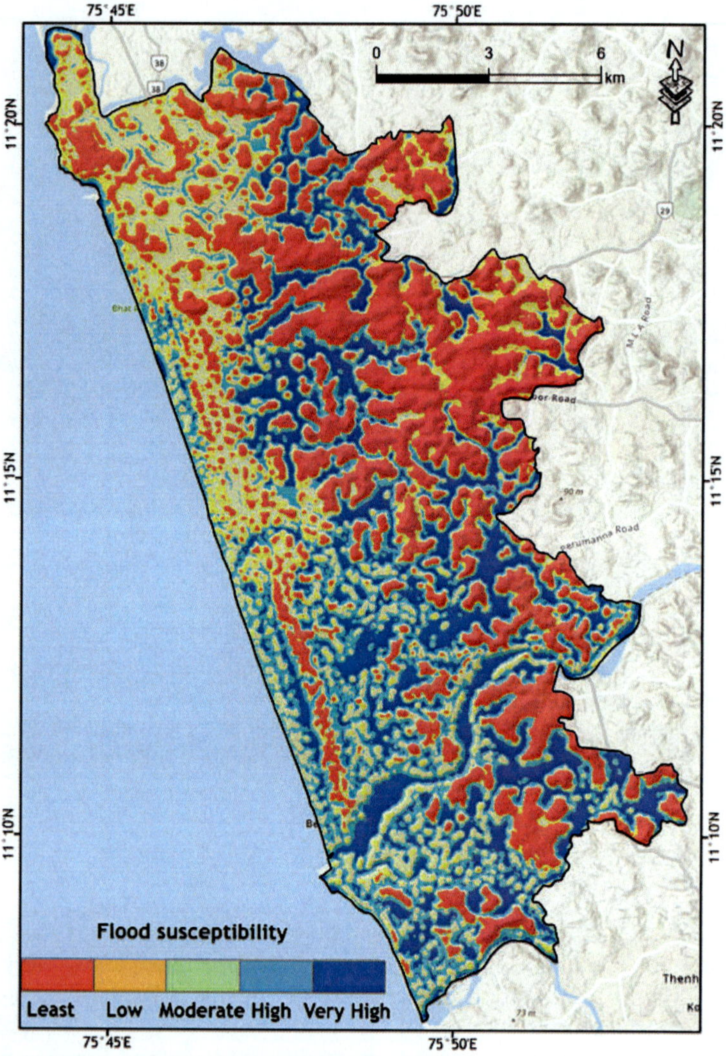

Fig. 3 Flood susceptibility map of the study area

3 Results and Discussions

3.1 Model Training and Validation

In the present study, the trained RF model is evaluated using different statistical methods before projecting to the geographical extend. The model performance evaluation in training and testing sections is summarized in Table 1. During training

Table 1 Model performance in training and testing sections

Models	TP	TN	FP	FN	N	Sensitivity	Specificity	Accuracy	K	AUC	RMSE
Training	186	159	29	56	430	0.769	0.846	0.802	0.605	0.852	0.380
Testing	88	59	4	33	184	0.727	0.937	0.799	0.598	0.785	0.407

Table 2 Model performance and error estimation in 10 fold cross-validations

Folds	Accuracy	AUC	RMSE
CV_1	0.916	0.937	0.372
CV_2	0.947	0.988	0.348
CV_3	0.682	0.943	0.352
CV_4	0.864	0.894	0.366
CV_5	0.894	0.818	0.398
CV_6	0.706	0.751	0.421
CV_7	1.000	1.000	0.283
CV_8	0.857	0.711	0.411
CV_9	0.733	0.768	0.413
CV_10	0.863	0.933	0.349

section, the ability of classifying flood locations or the sensitivity value is 0.769 whereas specificity higher value of 0.845. In the testing mode also, specificity value is higher than the sensitivity value (0.937 and 0.727, respectively) which shows that the model has better ability to classify the non-flood occurrences. It should be noted that overall accuracy in both training and testing sections is nearly same (i.e., 0.802 and 0.799, respectively) (Table 1). Kappa index also shows negligible differences in both training and testing phases. In the case of AUC values, which is generally considered as a robust measure of classification accuracy, model obtained a decent AUC value of 0.852 in training section and 0.785 in testing section.

In the case of RMSE, training phase shows lowest value (0.380) than validation phase (0.407). Besides to cross-check the efficiency of trained model, a ten-fold CV was implemented and summarized in Table 2, which shows overall good performance. Therefore, the model is finally projected for the entire study area.

3.2 Flood Susceptibility Modelling

The flood susceptibility map is prepared by transferring the probability of flood occurrences in the study area, which is further classified into five zones such as least susceptible area, low, moderate, high, and very high susceptibility areas (Fig. 3). About 34.31% of the study area is categorized under least flood susceptibility zone followed by 13.20% in low susceptibility, 19.93% in moderate susceptibility, 18.78% in high susceptibility, and 13.78% in very high susceptibility area.

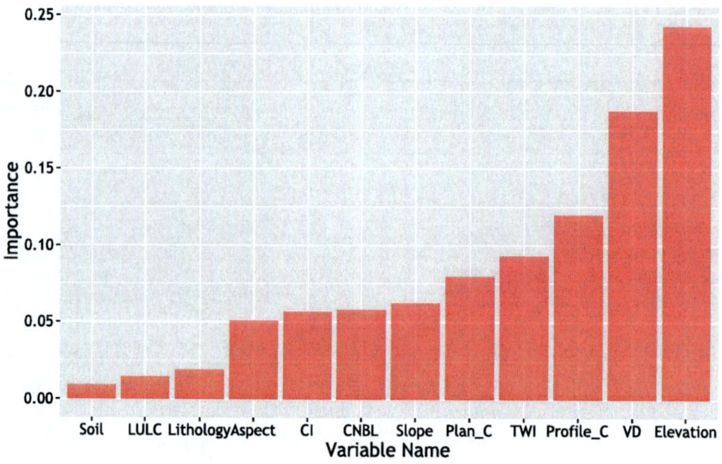

Fig. 4 Variable importance for the occurrence of flood in the study area

Variable importance analysis using random forest model is also carried out to identify the significant variables which influence the occurrence of flood in KUC. As shown in Fig. 4, elevation is the most important parameter which influences the flood occurrence followed by valley depth profile curvature and TWI. Other parameters such as plan curvature, slope, CNBL, CI, and slope aspect have moderate influence. Lithology, soil, and LULC are the least important parameters which affect the flood occurrences. In general, terrain parameters are the major flood influencing factors of KUC.

4 Conclusions

Urban flood susceptibility is estimated using RF method in a fast-growing urban agglomeration in southern India. Flood inundation locations are collected using field work, and thereafter, twelve independent variables such as land use/land cover, soil texture, lithology, elevation, slope angle, slope aspect, valley depth, topographical wetness index, profile curvature, plan curvature, convergence index, and channel network base level were analyzed and used for flood susceptibility mapping. The model obtained a decent AUC value of 0.852 in training section and 0.785 in testing section. Thereafter, the model is projected to geographical extend and classified into five zones such as least susceptible area, low, moderate, high, and very high. About 34.31% of the study area is categorized under least flood susceptibility zone followed by 13.20% in low susceptibility, 19.93% in moderate susceptibility, 18.78% in high susceptibility, and 13.78% in very high susceptibility area. The proposed flood susceptibility map is trust worthy for future infrastructure building and hazard mitigation.

References

1. Abedi R, Costache R, Shafizadeh-Moghadam H, Pham QB (2021) Flash-flood susceptibility mapping based on XGBoost, random forest and boosted regression trees. Geocarto Int:1–18
2. Achu AL, Thomas J, Aju CD, Gopinath G, Kumar S, Reghunath R (2021) Machine-learning modelling of fire susceptibility in a forest-agriculture mosaic landscape of southern India. Ecol Inform 64:101348. https://doi.org/10.1016/j.ecoinf.2021.101348
3. Ahmed N, Hoque MAA, Arabameri A, Pal SC, Chakrabortty R, Jui J (2021) Flood susceptibility mapping in Brahmaputra floodplain of Bangladesh using deep boost, deep learning neural network, and artificial neural network. Geocarto Int:1–22
4. Breiman L (2001) Random forests. Machine Learning 45(1):5–32
5. Census of India (2011) District Census Handbook, Kozhikode. Series-33, Part-XII-B
6. Chapi K, Singh VP, Shirzadi A, Shahabi H, Bui DT, Pham BT, Khosravi K (2017) A novel hybrid artificial intelligence approach for flood susceptibility assessment. Environ Model Softw 95:229–245
7. Fang Z, Wang Y, Peng L, Hong H (2021) Predicting flood susceptibility using LSTM neural networks. J Hydrol 594:125734
8. Islam ARMT, Talukdar S, Mahato S, Kundu S, Eibek KU, Pham QB, Linh NTT (2021) Flood susceptibility modelling using advanced ensemble machine learning models. Geosci Front 12(3):101075
9. Jesiya NP, Gopinath G (2019) A customized fuzzy AHP-GIS based DRASTIC-L model for intrinsic groundwater vulnerability assessment of urban and peri urban phreatic aquifer clusters. Groundw Sustain Dev 8:654–666
10. Kourgialas NN, Karatzas GP (2011) Flood management and a GIS modelling method to assess flood-hazard areas—a case study. Hydrol Sci J J des Sci Hydrol 56(2):212–225
11. Luu C, Pham BT, Van Phong T, Costache R, Nguyen HD, Amiri M, Trinh PT (2021) GIS-based ensemble computational models for flood susceptibility prediction in the Quang Binh Province, Vietnam. J Hydrol 599:126500
12. Panahi M, Jaafari A, Shirzadi A, Shahabi H, Rahmati O, Omidvar E, Bui DT (2021) Deep learning neural networks for spatially explicit prediction of flash flood probability. Geosci Front 12(3):101076
13. Pappenberger F, Matgen P, Beven KJ, Henry JB, Pfister L (2006) Influence of uncertain boundary conditions and model structure on flood inundation predictions. Adv Water Resour 29(10):1430–1449
14. Pham BT, Phong TV, Nguyen HD, Qi C, Al-Ansari N, Amini A, Tien Bui D (2020) A comparative study of kernel logistic regression, radial basis function classifier, multinomial naïve Bayes, and logistic model tree for flash flood susceptibility mapping. Water 12(1):239
15. Pham QB, Achour Y, Ali SA, Parvin F, Vojtek M, Vojteková J, Anh DT (2021) A comparison among fuzzy multi-criteria decision making, bivariate, multivariate and machine learning models in landslide susceptibility mapping. Geomatics, Nat Hazards Risk 12(1):1741–1777
16. Shankar MA, Bindu CA (2021) Appraising the need for disaster mitigation in existing planning documents of Municipal Corporations of Kerala in the event of past disasters. In IOP Conf Ser Mater Sci Eng 1114(1):012039). IOP Publishing
17. Tehrany MS, Pradhan B, Jebur MN (2014) Flood susceptibility mapping using a novel ensemble weights-of-evidence and support vector machine models in GIS. J Hydrol 512:332–343
18. Wang Y, Fang Z, Hong H, Costache R, Tang X (2021) Flood susceptibility mapping by integrating frequency ratio and index of entropy with multilayer perceptron and classification and regression tree. J Environ Manage 289:112449
19. Zzaman RU, Nowreen S, Billah M, Islam AS (2021) Flood hazard mapping of Sangu River basin in Bangladesh using multi-criteria analysis of hydro-geomorphological factors. J Flood Risk Manage 14(3):e12715

Comparative Assessment of Different Machine Learning Models to Estimate Daily Soil Moisture

G. E. Nagashree and M. K. Nema

Abstract Soil moisture is vital as it is the primary governing factor of agriculture production and natural vegetation growth. It plays an essential role in understanding the hydrological cycle and its effect on weather and climate, and its precise prediction helps to manage the water resources optimally. Prediction of soil moisture is dependent on surface meteorological variables and soil attributes. Existing soil moisture models/prediction methods are inaccurate, and developing an optimum mathematical model for it is difficult. This study evaluates the performance of four machine learning models (deep neural network (DNN) regression, support vector machine (SVM), multiple layer perceptron (MLP), and multi-linear regression (MLR) to estimate the soil moisture conditions. The models were tested for soil moisture at two depths (25 and 50 cm depth) using the meteorological data of two stations located in a Lesser Himalayan catchment. The model outputs were compared with the observed data, and intercomparison was also made. The model performance was evaluated based on MAPE, RMSE, Nash–Sutcliffe efficiency coefficient ($E_{\text{N–S}}$), and R^2. The study results indicated that the DNN model outperforms the other prediction models with the highest efficacy for both stations. Therefore, the DNN model can be endorsed to estimate soil moisture when primary meteorological data are available, and it can be promising for water-efficient agriculture applications and draught management.

Keywords Soil moisture · DNN · MLP · SVM · And LR

1 Introduction

Soil moisture is essential for the hydrological cycle and natural vegetation development. It is considered as one of the 50 essential climatic variables by the Global

G. E. Nagashree (✉)
Department of Water Resources and Ocean Engineering, NITK, Surathkal 575025, India
e-mail: genagashree@gmail.com

M. K. Nema
National Institute of Hydrology, Roorkee 247667, India

P. V. Timbadiya et al. (eds.), *Geospatial and Soft Computing Techniques*, Lecture Notes in Civil Engineering 339, https://doi.org/10.1007/978-981-99-1901-7_44

Climate Observing System. Due to the increase in global warming and rapid climate changes, there is a decline in every kind of natural resource, especially water [1]. As a result, groundwater levels and surface freshwater levels fall, affecting soil water content and disrupting irrigation practises in a specific location. It is essential to build a suitable irrigation system at the appropriate time, with optimum water resource optimization [2]. Soil moisture progression and regression have differing effects on water consumption and crop development. It helps in the determination of drought resistance [3], flood control [4, 5], and precision irrigation decisions in agriculture production [6]. To correctly manage agricultural water supplies and encourage higher crop yields, it is critical to precisely anticipate soil water regression patterns.

Many methods are proposed and implemented to predict soil moisture, like using empirical formulas, linear regression, and artificial intelligence [7]. Researchers [8] developed a formula for soil moisture, precipitation, and drought assessment prediction model by analysing the initial soil water content, daily rainfall, average air temperature, and daily average saturation difference, based on the multivariate linear relationship of soil moisture. Although the empirical formula is basic and straightforward, the model parameters exhibit significant regional independence. When transplanting to different places, they must be recalculated, which takes time. The advent of artificial intelligence in the field of hydrology has encouraged researchers to explore more areas with large amounts of data and obtain better accuracy. Researchers [9] have used machine learning models like Random Forest (RF) for root zone soil moisture estimation and found that ML models are best to adopt for data-poor regions compared to process-based models. One of the studies [10] have adopted recurrent neural network (RNN), multiple linear regression (MLR), and support vector machines (SVM) to predict soil moisture. As a result of this study, MLR has performed well by having minimum mean square error (MSE) compared to other models. Another study [11] used a shallow neural network, MLR, and SVM to forecast soil moisture one, two, and seven days in advance. The results demonstrate that a shallow neural network outperforms other models. Other studies [12] have adopted ANN, MLP, radial bias function, and ensemble Bayesian model for soil moisture computation, and it was found that Bayesian model outperformed with required output. Another study [13] employed soil moisture and weather data to develop SVM forecasts for the next 4 and 7 days. The prediction agrees well with actual soil moisture, and the findings outperformed the artificial neural network (ANN) model. Another study [14] found that deep neural network (DNN) performance was better than MLP and SVM.

Keeping these studies in mind, we have adopted four machine learning models, viz. deep neural network (DNN), support vector machines (SVM), multiple linear regression (MLR), and multi-layer perceptron (MLP) to predict soil moisture at the Lesser Himalayan region using meteorological data and soil moisture data. This model's performances were evaluated using four quantitative standard statistical measures like RMSE, R^2, MAPE, and $E_{N\text{-}S}$. This study aims to get the best model that could predict soil moisture with minor errors and more accuracy.

2 Materials and Methods

2.1 Model Descriptions

DNN, SVM, MLP, and LR models were among the ML approaches used in this work. The following is a basic outline of these techniques.

2.1.1 Multi-layer Perceptron (MLP)

A perceptron is the simplest neural network known, consisting of four major components: input values, weights and bias, net sum, and an activation function, as shown in Fig. 1a. MLP is a feed-forward neural network with one hidden layer sandwiched between the input and output layers. Data is fed forward from the input layer to the output layer. MLP is comprised of three layers: an input layer, a hidden layer, and an output layer [15], as shown in Fig. 1b. The neurons in the input layer receive the input data. The output values are passed to neurons in the hidden layer and then to the output layer after processing inside individual neurons in the input layer. The weights of neurons are related with their connections, and altering the weights in a certain way results in learning of the associated network. Weights may be computed using various activation functions, as it determines whether a neuron should be activated or not by computing a weighted total and then adding bias to it. There are multiple kinds of activation functions, including linear, sigmoid, Tanh, ReLU, and softmax functions. The learning or training algorithm refers to the process through which the weights in the network change. The back-propagation method is the most extensively used learning method for error reduction. It is used for gradient descent training; a step size, known as the learning rate, must be given. Gradient descent is a method for determining the minimum of a function. Gradient descent is classified into three types: batch gradient descent, mini-batch gradient descent, and stochastic gradient descent (SDG). The learning rate is a proportionality constant that governs the magnitude of weight changes. The weight change of a neuron is proportional to the effect of that neuron's weight on the error. Back-moment propagation's term defines how past weight changes effect present weight changes. As convergence is attained, many neural network applications will automatically reduce the learning rate and raise the momentum values.

2.1.2 Deep Neural Network (DNN)

Deep neural networks (DNNs) are densely linked neural networks with numerous hidden layers between the input and output layers [16]. DNN's algorithm is similar to MLP's, with the only difference being the number of hidden layers in the network. In DNN, the hidden layer can be added based on the number of features and sample data; that is, when the number of features rises, a separate hidden layer with a given number

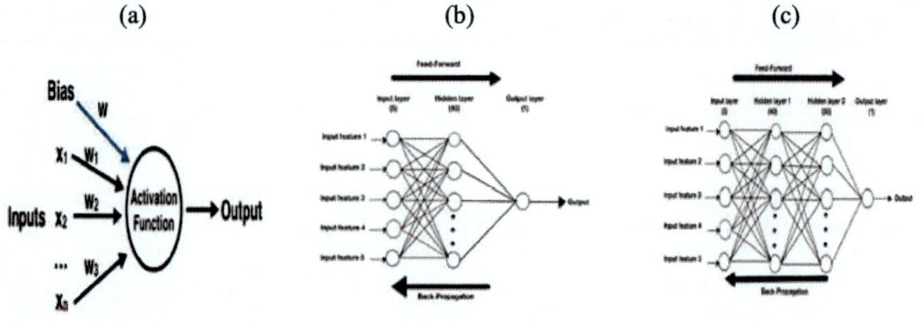

Fig. 1 Different ANN architecture of **a** single perceptron, **b** MLP, and **c** DNN

of neurons is introduced instead of increasing the neurons in the same hidden layer. This minimises the network's complexity. The number of nodes in the input layer is determined by the sample data's input feature. It is feasible to learn the relationships of a range of nonlinear data in the hidden layer. Complex nonlinear issues can also be solved, and information from high-dimensional data can be retrieved in a variety of ways. It is possible to combine with a value of an input variable, give a weight, extract a new value, and transport it to the output layer in the hidden layer connected to an input layer. The output layer, based on the estimated features in the hidden layer, allows for classification and prediction using the feed-forward network and back-propagation as described in the MLP model.

2.1.3 Support Vector Machine (SVM)

SVM is a prominent supervised learning technique that is used for both classification and regression issues [17]. The core idea behind SVM is to employ a linear model to construct nonlinear class boundaries via nonlinear mapping of input vectors into high-dimensional feature space. In the original space, the linear model built in the new space can reflect a nonlinear decision boundary. SVM creates an ideal separating hyperplane in the new space. If the data is linearly distributed, linear machines are trained to find an ideal hyperplane that distributes the data without error while minimising the distance between the hyperplane and the nearest training points. Support vectors are training sites that are close to the optimal separation hyperplane. Figure 2 exhibits the basic concept of SVM. There are an infinite number of decision functions, such as hyperplanes, that may successfully distinguish the negative and positive datasets with the greatest margin. This means that the distance between the nearest positive samples and a hyperplane will be maximised, while the distance between the closest negative samples will be minimised.

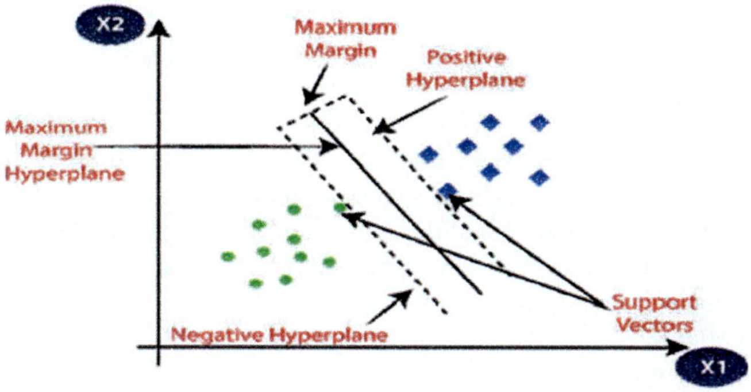

Fig. 2 Basis of SVM

2.1.4 Multiple Linear Regression (LR)

Multiple linear regression [18] is an extension of simple linear regression in which more than one independent variable is used to predict the result of a dependent variable. In both circumstances, we continue to use the term "linear" since we presume that the response variable is directly related to a linear combination of the explanatory factors. The equation for multiple linear regression is the same as for simple linear regression, but it contains extra terms:

$$y_i = \beta_0 + \beta_1 x_{1i} + \beta_2 x_{2i} + \cdots + \beta_p x_{pi} + e_i \tag{1}$$

where for $i = n$ observation: y_i = dependent variable, x_i = independent variable, β_p = coefficients of slope for each independent variable and e_i = the model's error. As for the simple case, β_0 is a constant, which will be the predicted value of y when all explanatory variables are zero. In a model with p explanatory variables, each explanatory variable has its own $\beta\_$ coefficients.

2.2 *Study Area and Data*

2.2.1 About the Study Area and Data

The meteorological data from two automatic weather stations (AWS) were considered for this study. The National Institute of Hydrology (NIH), Roorkee, installed these AWSs throughout a small Himalayan hilly watershed of the Henval River up to Devnagar in the upper Ganga basin in Uttarakhand state, India (Fig. 3). An AWS at Kumargaon (30° 21′ 00″ N, 78° 19′ 40.80″ E), which is located in mountain ridges at an elevation of 1798 m above mean sea level and another at Kanatal (30° 24′ 53.08″ N, 78° 19″ 17.98″ E) located 2590 m above mean sea level on the southerly sloping hill near an apple orchid. The research area has a mix of lesser Himalayan hilly temperate climatic conditions, with an annual rainfall range of 1200–1800 mm. The Himalayan subtropical woods give way to a belt of broad temperate leaf and mixed forest; the majority of which is pine forest. The total area under study is 102 km^2 approximately. This region's climate is mainly humid temperate, with valleys that are hot in summer and frigid in winter, with an average temperature ranging from 3 to 30° C. About 70 to 80% of rainfall occurs during June and September. The accurate measurement of several near-surface climatic, radiation, and soil variables from these two AWSs was used in this investigation, and data records for the year 2018 were utilised.

For this study, daily rainfall, average wind speed at 2 m above ground, maximum air temperature, and maximum soil temperature at depths of 2, 25, and 50 cm were all

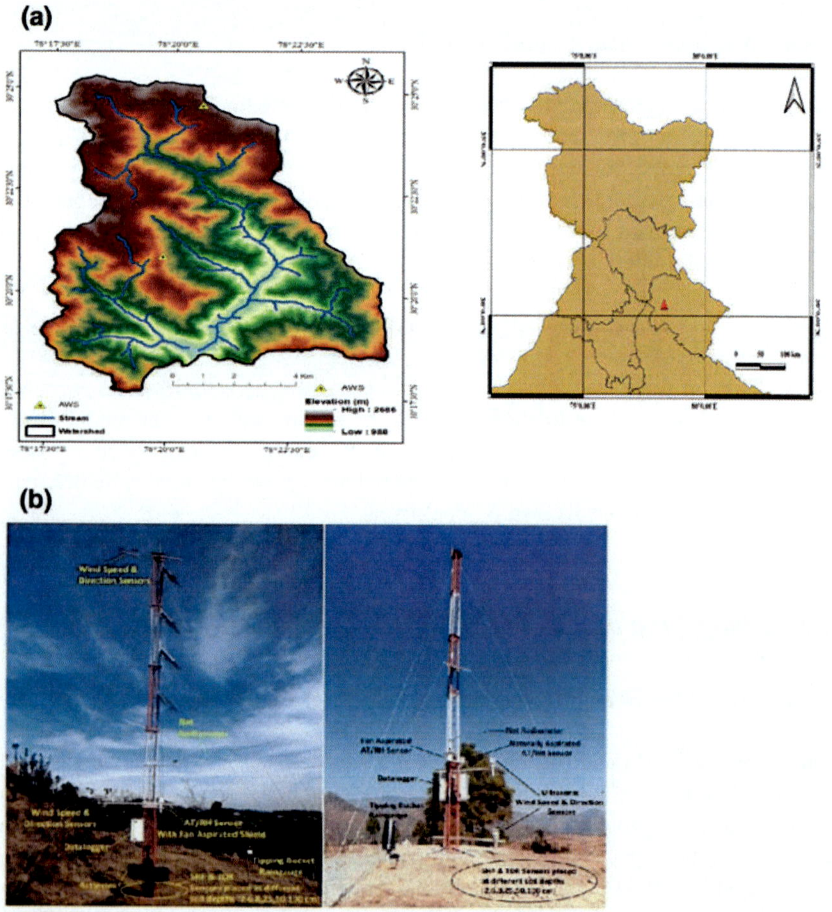

Fig. 3 Location of Henval watershed up to Devnagar (**a**) and AWS at Kumargaon and Kanatal (**b**)

Table 1 Correlation between various input features and soil moisture

Feature	Daily rainfall	Average wind speed at 2 m	Maximum air temperature	Maximum soil temperature at 2 cm	Maximum soil moisture at 2 cm	$(t-1)$ day soil moisture
Soil moisture correlation	0.346	−0.1342	0.015	0.892	−0.271	0.967

incorporated in the meteorological and soil moisture data [14]. Daily soil moisture measurements at depths of 2, 25, and 50 cm were used to produce soil moisture data. Rainfall, average wind speed at 2 cm, maximum air temperature, maximum soil temperature at 2 cm, soil moisture at 2 cm, and one-day prior soil moisture at 25 and 50 cm are input characteristics for the selected machine learning models. The current-day soil moisture levels at 25 and 50 cm depth were chosen as the model's output feature for both sites.

2.2.2 Data Processing and Analysis

Data formats and lengths vary depending on the source of meteorological and soil moisture data. Integration and matching of data are required. For training purposes, machine learning models necessitate a vast amount of data. The training and testing sets are chosen based on the number of soil moisture observations from 2018 to 2019. Missing values in the data should be handled carefully to avoid errors while training the model. In the dataset, there were few continuous missing data for 12 days, which were eliminated later. After processing the data, we had a total of 658 observations from 2/13/2018 to 12/14/2019 at Kumargaon AWS, which included 526 data from 2/13/2018 to 8/5/2019 as training set and 132 data from 8/6/2019 to 12/14/2019 as testing set. At Kanatal AWS, we obtained a total of 363 observations from 1/3/2018 to 12/31/2018, including 290 data from 1/3/2018 to 10/19/2018 as training set and 73 data from 10/20/2018 to 12/31/2018 as testing set. Large attribute values in modelling processes may generate numerical issues; thus, it is beneficial to normalise the features before applying MLP, DNN, SVM, and LR to prediction.

To know the correlation between six input features with the soil moisture, we used the Taylor diagram [19], because it is a necessary training feature to provide maximal weight for moisture prediction in order to increase regression accuracy. The data analysis is summarised in Table 1 and shown in Fig. 4.

2.2.3 Model Performance Evaluation

As shown by [20] to evaluate the overall performance of machine learning models, we use the Coefficient of Determination (R^2), Root Mean Square Error (RMSE),

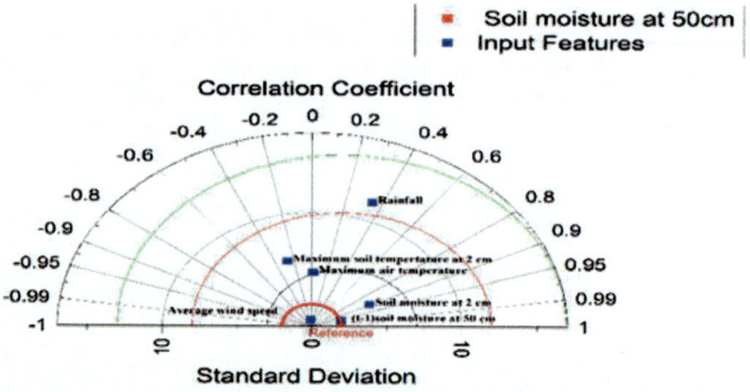

Fig. 4 Taylor diagram for soil moisture at Kumargaon AWS

Mean Absolute Percentage Error (MAPE), and Nsh-Sutcliffe efficiency coefficient ($E_{\text{N-S}}$) as the standard quantitative statistical performance evaluation measures.

$$R^2 = \left(\frac{\frac{1}{n}\sum_{i=1}^{n}(y_0(i) - \overline{y_0})\left(y_f(i) - \overline{y_f}\right)}{\sqrt{\frac{1}{n}\sum_{i=1}^{n}(y_0(i) - \overline{y_0})^2} * \sqrt{\frac{1}{n}\sum_{i=1}^{n}\left(y_f(i) - \overline{y_f}\right)^2}} \right)^2 \tag{2}$$

$$\text{RMSE} = \sqrt{\frac{1}{n}\sum_{i=n}^{n}\left(y_f(i) - y_0(i)\right)^2} \tag{3}$$

$$\text{MAPE} = \frac{1}{n}\sum_{i=1}^{n}\left|\frac{y_f(i) - y_0(i)}{y_0(i)}\right| \times 100 \tag{4}$$

$$E_{\text{N-S}} = 1 - \frac{\sum_{i=1}^{n}\left(y_f(i) - y_0(i)\right)}{\sum_{i=1}^{n}(y_0(i) - \overline{y_0})} \tag{5}$$

where $y_0(i)$ and $y_f(i)$ represent the actual and predicted soil moisture, respectively, and $\overline{y_0}$, $\overline{y_f}$ are their means, and n is the number of data points taken into account.

2.2.4 Development of Models

The experimental programme is coded using Keras API of TensorFlow framework [21] and scikit-learn for the selected models. As discussed about the data selection, we had total of 658 days of observation dataset (2/13/2018 to 12/14/2019) of the Kumargaon AWS site. We divided them into training and testing datasets with 526 and 132 observations, respectively. Similarly, for the Kanatal region, we had total of 363 days of observation dataset (1/3/2018 to 12/31/2018), where the training and testing dataset consists of 290 and 73 observations, respectively.

Figure 1b depicts the MLP network design, which comprises an input layer, one hidden layer, and an output layer. The input layer has six neurons because we chose six input characteristics, the number of neurons in the hidden layer was determined by the RMSE value, and the output layer has one neuron. ReLU function [22] used as an activation function, and we adopted Adam optimiser (type of SGD used for back-propagation) for training with mean square error as loss function. The DNN architecture (Fig. 1c) consists of one input layer, two hidden layers, and an output layer. The number of neurons in the input layer is six, while the number of neurons in the hidden layer is determined by the RMSE value. In each hidden layer, ReLU is utilised as an activation function, and the Adam optimiser is employed to train the model with a learning rate of 0.01. In SVM, radial kernel (RBF) was adopted with $C = 100$ and γ (gamma) $= 0.001$ which showed better output compared to other kernels. For multi-linear regression (LR) scikit-learn with linear regression was adopted.

3 Results and Discussion

Daily soil moisture data at depths of 25 and 50 cm from two AWS in the lesser Himalayan region were used to identify more acceptable models for predicting future daily soil moisture. All of the above models generated are compared using the same amount of training and testing datasets, respectively, while the quantitative standard statistical performance assessment measure is utilized to evaluate the performance of various models developed. The residual between actual and anticipated soil moisture is measured by RMSE, while the mean absolute percentage error of the forecast is measured by MAPE. R^2 measures the linear correlation between the actual and predicted soil moisture, while $E_{\text{N-S}}$ measures the capability of the model in predicting soil moisture level that differ from the mean.

Tables 2, 3, 4 and 5 illustrates numerous machine learning algorithms that perform well during training and testing. For Kumargaon region soil moisture at 25 cm, during training phase, the DNN model outperformed with R^2, RMSE, MAPE and $E_{\text{N-S}}$ of 0.99, 0.31, 1.1, and 0.99, respectively. In the training phase, the SVM model has the second-best performance, with 13% and 0.4% more RMSE and MAPE, respectively, compared to DNN. When the results for the testing period are examined, it is clear that the DNN model surpasses all other models. For Kanatal region soil moisture at 25 cm, in the training phase, both DNN and MLP shows similar R^2, MAPE and $E_{\text{N-S}}$ statistics, but the DNN model has RMSE reduced by 2% compared to MLP. But during the testing phase, we observe that SVM has R^2 0.97, which is best among other models, but its performance reduced after comparing other indices like, MAPE and $E_{\text{N-S}}$, yet DNN provide better statistics in the testing period. According to the results in Tables 2, 3, 4 and 5, the greatest performance of all machine learning models generated in this research differs from the various statistical measurements.

Table 2 Forecasting performance indices of models at a depth of 25 cm for Kumargaon AWS

Model	Training				Testing			
	R^2	RMSE	MAPE (%)	E_{N-S}	R^2	RMSE	MAPE (%)	E_{N-S}
DNN	**0.99**	**0.15**	**1.1**	**0.99**	**0.97**	**0.29**	**1.5**	**0.97**
SVM	0.98	0.28	1.5	0.98	0.96	032	**1.5**	0.96
MLP	0.98	0.27	2.1	0.98	0.95	0.38	2.2	0.95
LR	0.96	0.39	3	0.96	0.96	0.39	2.9	0.95

Bold represents the good estimation of different performace indices

Table 3 Forecasting performance indices of models at a depth of 50 cm for Kumargaon AWS

Model	Training				Testing			
	R^2	RMSE	MAPE (%)	E_{N-S}	R^2	RMSE	MAPE (%)	E_{N-S}
DNN	**0.99**	**0.31**	**1.2**	**0.99**	0.97	0.49	**1.4**	0.96
SVM	0.98	0.36	1.2	0.98	0.93	0.64	1.8	0.94
MLP	0.98	0.41	1.9	0.98	0.95	0.53	2.2	0.95
LR	0.97	0.54	2.6	0.97	**0.97**	**0.41**	2.9	**0.97**

Bold represents the good estimation of different performace indices

Table 4 Forecasting performance indices of models at a depth of 25 cm for Kanatal AWS

Model	Training				Testing			
	R^2	RMSE	MAPE (%)	E_{N-S}	R^2	RMSE	MAPE (%)	E_{N-S}
DNN	**0.99**	**0.49**	**0.9**	**0.99**	**0.96**	**0.23**	**0.5**	**0.96**
SVM	0.91	1.77	1.1	0.9	**0.97**	0.29	0.8	0.93
MLP	**0.99**	0.51	0.9	0.99	0.94	0.28	**0.6**	0.94
LR	0.94	1.39	2.4	0.94	0.39	1.52	4.1	−0.86

Bold represents the good estimation of different performace indices

Table 5 Forecasting performance indices of models at a depth of 50 cm for Kanatal AWS

Model	Training				Testing			
	R^2	RMSE	MAPE (%)	E_{N-S}	R^2	RMSE	MAPE (%)	E_{N-S}
DNN	**0.99**	**0.49**	0.8	**0.99**	**0.96**	0.15	0.4	0.96
SVM	0.89	1.68	1.0	0.88	0.97	0.12	0.3	0.97
MLP	**0.99**	**0.47**	**0.7**	0.99	**0.96**	0.2	0.6	0.92
LR	0.91	1.42	2.4	0.92	0.46	0.82	2.4	−0.26

Bold represents the good estimation of different performace indices

From all the observations mention in the tabulation, it can be seen that multi-linear regression (LR) shows good performance in the Kumargaon region, but as it comes to the Kanatal region, the performance has been deteriorated, having maximum error and minimum correlation between observed and predicted values.

When we compare the LR model with the DNN model of the Kanatal region, we find that RMSE and MAPE value has increased to 129% and 3.6%, respectively, and R^2 value has reduced to 57% in the testing period. Furthermore, in the testing phase, as given in Tables 2, 3, 4 and 5, the values with DNN, SVM, and MLP model prediction produced a well-near forecast when compared to the LR model. However, in the testing phase, as given in Tables 2, 3, 4 and 5, the values with DNN, SVM, and MLP model prediction produced a well-near forecast when compared to the LR model. Furthermore, the virtues or defects of prediction accuracy vary depending on the evaluation metrics used during the training and testing phases. Figures 5, 6, 7 and 8 depict the performance of all prediction models built in this paper during the training and testing periods in the two study sites at two depths.

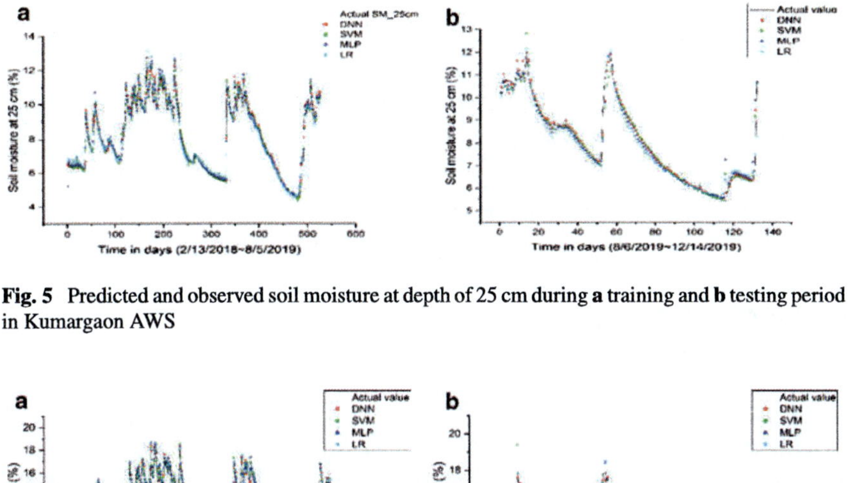

Fig. 5 Predicted and observed soil moisture at depth of 25 cm during **a** training and **b** testing period in Kumargaon AWS

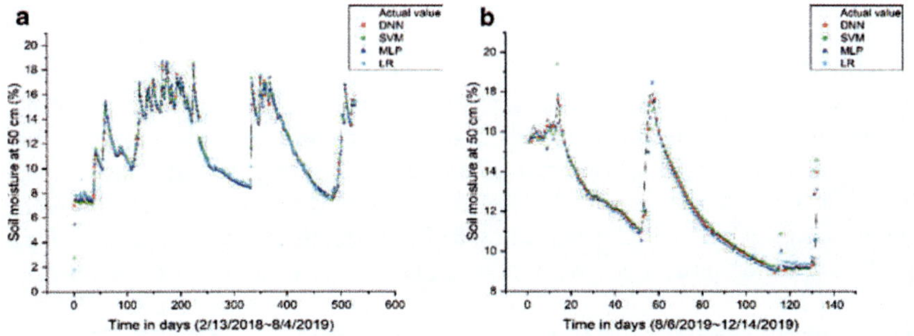

Fig. 6 Predicted and observed soil moisture at depth of 50 cm during **a** training and **b** testing period in Kumargaon AWS

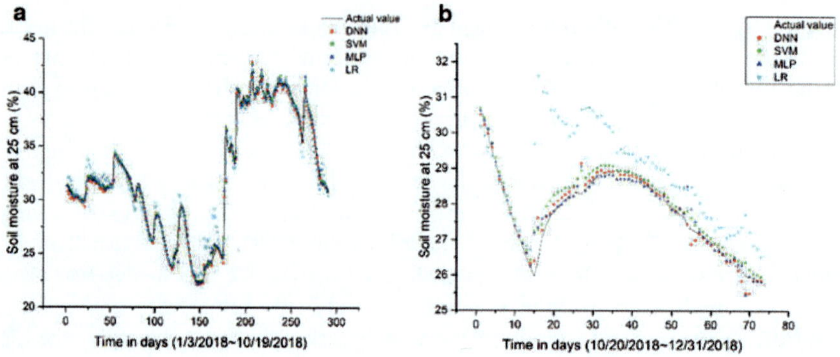

Fig. 7 Predicted and observed soil moisture at depth of 25 cm during **a** training and **b** testing period in Kanatal AWS

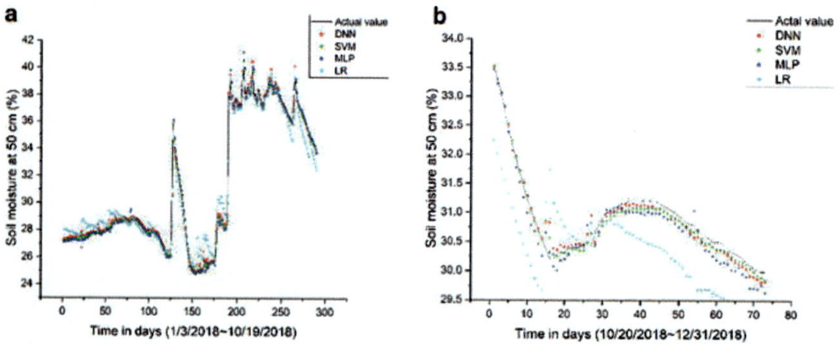

Fig. 8 Predicted and observed soil moisture at depth of 50 cm during **a** training and **b** testing period in Kanatal AWS

4 Conclusions

This study attempted to explore the efficacy of various machine learning approaches for predicting daily soil moisture. The predicting methods investigated include the DNN, SVM, MLP, and LR models. Meteorological data and daily soil moisture from actual field observations in Kumargaon and Kanatal were used to create various models investigated in this study at 25 and 50 cm depth. To assess the performance of many models produced, four conventional statistical performance evaluation methods are used. The findings of this study show that machine learning algorithms are excellent tools for modelling daily soil moisture and can provide good prediction performance. The results indicate that the DNN can obtain the best performance in different evaluation criteria during the training and testing phase. SVM and MLP models showed better performance in the training phase, but their performance was inferior in the testing phase compared to DNN. The LR model performance was at the least. Therefore, the study results are highly encouraging and suggest that the DNN approach is promising in modelling daily soil moisture. The work could be useful for researchers and engineers using machine learning approaches to forecast hydrological predictions.

References

1. Yeh TC, Wetherald RT, Manabe S (1984) The effect of soil moisture on the short-term climate and hydrology change—A numerical experiment. Mon Weather Rev 112:474–490
2. Li X, Huo Z, Xu B (2017) Optimal allocation method of irrigation water from river and lake by considering the field water cycle process. Water (Switzerland) 9. https://doi.org/10.3390/w9120911
3. Wang A, Lettenmaier DP, Sheffield J (2011) Soil moisture drought in China, 1950–2006. J Clim 24:3257–3271. https://doi.org/10.1175/2011JCLI3733.1
4. Mosavi A, Ozturk P, Chau KW (2018) Flood prediction using machine learning models: literature review. Water (Switzerland) 10:1–40. https://doi.org/10.3390/w10111536
5. Chifflard P, Kranl J, Zur SG, Zepp H (2018) The significance of soil moisture in forecasting characteristics of flood events. a statistical analysis in two nested catchments. J Hydrol Hydromechanics 66:1–11. https://doi.org/10.1515/johh-2017-0037
6. Pereira LS, Oweis T, Zairi A (2002) Irrigation management under water scarcity. Agric Water Manag 57:175–206. https://doi.org/10.1016/S0378-3774(02)00075-6
7. Raghavendra S, Deka PC (2014) Support vector machine applications in the field of hydrology: a review. Appl Soft Comput J 19:372–386. https://doi.org/10.1016/j.asoc.2014.02.002
8. Li Y, Yan S, Chen N, Gong J (2020) Performance evaluation of a neural network model and two empirical models for estimating soil moisture based on sentinel-1 sar data. Prog Electromagn Res C 105:85–99. https://doi.org/10.2528/pierc20071601
9. Carranza C, Nolet C, Pezij M, van der Ploeg M (2021) Root zone soil moisture estimation with Random Forest. J Hydrol 593:125840. https://doi.org/10.1016/j.jhydrol.2020.125840
10. Pandey A, Jha SK, Srivastava JK, Prasad R (2010) Artificial neural network for the estimation of soil moisture and surface roughness. Russ Agric Sci 36:428–432. https://doi.org/10.3103/s106836741006011x
11. Prakash S, Sekhar S (2020) Soil moisture prediction using shallow neural network. Int J Adv Res Eng Technol 11:426–435. https://doi.org/10.34218/IJARET.11.6.2020.038
12. Zounemat-Kermani M, Golestani Kermani S, Alizamir M, Fadaee M (2022) Soil moisture simulation using individual versus ensemble soft computing models. Int J Environ Sci Technol. https://doi.org/10.1007/s13762-022-04202-y
13. Gill MK, Asefa T, Kemblowski MW, McKee M (2006) Soil moisture prediction using support vector machines. J Am Water Resour Assoc 42:1033–1046. https://doi.org/10.1111/j.1752-1688.2006.tb04512.x
14. Cai Y, Zheng W, Zhang X et al (2019) Research on soil moisture prediction model based on deep learning. PLoS ONE 14:1–19. https://doi.org/10.1371/journal.pone.0214508
15. Victor Devadoss A, Antony Alphonnse Ligori T (2013) Forecasting of stock prices using multi layer perceptron. Int J Web Technol 002:52–58. https://doi.org/10.20894/ijwt.104.002.002.006
16. Baek JW, Chung K (2020) Context deep neural network model for predicting depression risk using multiple regression. IEEE Access 8:18171–18181. https://doi.org/10.1109/ACCESS.2020.2968393
17. Noble WS (2006) What is a support vector machine? Nat Biotechnol 24:1565–1567. https://doi.org/10.1038/nbt1206-1565
18. Tranmer M, Murphy J, Elliot M, Pampaka M (2020) Multiple linear regression, 2nd ed. Cathie Marsh Institute Work Paper 59

19. Taylor S (2005) Taylor diagram primer. Karl E. Taylor
20. Dawson CW, Abrahart RJ, See LM (2007) HydroTest: a web-based toolbox of evaluation metrics for the standardised assessment of hydrological forecasts. Environ Model Softw 22:1034–1052. https://doi.org/10.1016/j.envsoft.2006.06.008
21. Dillon JV, Langmore I, Tran D et al (2017) TensorFlow distributions
22. Sharma S, Sharma S, Anidhya A (2020) Understanding activation functions in neural networks. Int J Eng Appl Sci Technol 4:310–316

Artificial Intelligence-Based Reference Evapotranspiration Modelling with Minimum Climatic Parameters

K. Chandrasekhar Reddy

Abstract Artificial neural networks (ANN), an artificial intelligence-based technology, have yielded numerous favourable water resource and hydrology simulation results. The investigation goal is to find the efficient ANN architecture for estimating reference evapotranspiration with the fewest climate variables. The climatic parameters, namely wind speed (W), relative humidity (RH), air temperature (T), and sunshine hours (S) were used to estimate monthly reference evapotranspiration (ET_0). Partial correlation and multiple linear correlations between the climatic parameters and FAO-56 Penman–Monteith reference evapotranspiration ($PMET_0$) were carried out in order to determine the most influential factor by eliminating one factor at a time. The influence of parameters T, S, W, and RH was observed from highest to lowest, respectively. Therefore, the best ANN models to estimate artificial neural network reference evapotranspiration ($ANNET_0$) were developed using climatic parameters as inputs and eliminating one lowest influencing parameter each time. Training of model was done with a portion of data, and the remaining was used for testing the model. Performance indices have been used to assess the ability of the model by correlating the $PMET_0$ and $ANNET_0$. The viability of the generated models was verified by using the numerical indicators (i.e. efficiency coefficient, coefficient of determination, and root mean square error). ANN (1-5-1), ANN (2-5-1), ANN (3-4-1), ANN (4-3-1) with (T), (T, S), (T, S, W), and (T, S, W, RH) as inputs, respectively, were shown to have 89.58%, 94.36%, 95.20%, and 99.44% during the testing period, respectively. Thus, in the area of study and other locations with similar climatic conditions, these ANN models can assess monthly ET_0 with adequate accuracy.

Keywords Reference evapotranspiration · Performance indices · FAO-56 Penman–Monteith method · Multiple and partial correlation coefficients · Artificial neural networks

K. Chandrasekhar Reddy (✉)
Civil Engineering, Siddharth Institute of Engineering & Technology, Puttur, Andhra Pradesh 517583, India
e-mail: kamasani.kcr@gmail.com

P. V. Timbadiya et al. (eds.), *Geospatial and Soft Computing Techniques*, Lecture Notes in Civil Engineering 339, https://doi.org/10.1007/978-981-99-1901-7_45

1 Introduction

The optimization of irrigation water use is the need of the hour as the water demand is increasing rapidly. Determination of amount evapotranspiration is one of the feasible methods to decrease irrigation water consumption. The Lysimeter is perhaps the only accurate, direct field measuring, and appropriate instrument for evaluating evapotranspiration. The instrument, however, has a high initial cost and requires complicated maintenance. As a result, many semi-experimental and experimental models to predict reference evapotranspiration (ET_0) using climate data have been established by many investigators. ET_0 can be estimated by Hargreaves, Jensen-Haise, Blaney-Criddle, Makkink, Pan Evaporation, Priestley-Taylor, Radiation, Christiansen, FAO-56 Penman–Monteith (PM), Modified Penman, etc. However, in all weather conditions, the PM technique can produce results comparable to Lysimeter readings, therefore considered as the standard method to estimate ET_0 [1–5]. In FAO, Penman–Monteith method for estimating ET_0 requires data of climate parameters, namely temperature (highest and lowest), relative humidity (vapour pressure) (highest and lowest), wind speed, and sunshine (radiation) along with site location. Many methods of determining ET_0 don't adequately describe the nonlinear effects intrinsic in the ET_0 function. When the correlation among the independent and dependent parameters in literary modelling is nonlinear, artificial neural networks (ANNs) represent intricate nonlinear procedures relevant to ET_0 assessment for improved interpretation. Technology based on artificial intelligence (AI) is feasible [6] in contrast to others statistical interpretation approaches; ANN performed better even the data had errors and noise.

The present investigation is interfered by establishing efficient ANN architecture for estimating monthly ET_0 with the fewest climate variables.

2 Materials and Methodology

2.1 Research Area

The researcher acquired meteorological data for Tirupati, a well-known city of Chittoor, AP, India, with altitude 161.0 m and coordinate values of 13° 37' N and 79° 25' E (research region), from the India Meteorological Department. Climate data from 1992 to 2001 was utilized to find an efficient ANN model. Part of the data (1992–1998) was utilized to train ANN model and the rest to test it.

Table 1 Particulars of Penman–Monteith (PM) technique suggested by FAO-56

Primary reference	PM equation	Input data
Allen et al. [2]	$ET_0 = \frac{0.408\Delta(R_n - G) + \gamma \frac{900}{T_{mean}+273} u_2 (e_s - e_a)}{\Delta + \gamma(1 + 0.34u_2)}$	T_{max}, T_{min}, RH_{min}, RH_{max}, n, u_2

2.2 ET_0 Assessment Techniques

2.2.1 Penman–Monteith (PM) Technique Suggested by FAO-56

The Penman method, which combines an intern mass transfer and energy balance approach for the reference crop, inspired the PM method. An anticipated grass crop having 0.12 m height, 70 s/m surface resistivity, and 0.23 albedo is considered as a reference crop. It is equivalent to evaporation from the thickest green grass of equal height, which thrives and is adequately irrigated. ASCE and European studies resulted in relatively consistent and accurate performance in arid and humid climates [7]. As a result, the technique is regarded as the most accurate, and its necessary particulars are listed in Table 1. The PM technique used is suggested by FAO-56.

PM equation is a typical approach to estimate ET_0 as it represents physiological and physical characteristics that impact evapotranspiration. Climate variables like relative humidity, sunshine, extreme air temperatures, and wind velocity at 2 m elevations are considered inherent data for utilizing this technique.

2.2.2 Artificial Neural Networks (ANNs) Modelling

For the present investigation, a standard multilayer feed-forward ANN with the logistic sigmoid function was used, considering the momentum factor with a constant value of 0.9 and 0.1 for the learning rate. The input data was normalized between 0.1and 0.9 to avoid saturation. During calibration, error backpropagation, an iterative nonlinear optimization technique based on the gradient descent search method [8], was used. The standardization set to reduce error and the validation set to ensure appropriate neural network training was used not to overtrain the neural networks. The performance of the model was tracked throughout every repetition to avoid over-learning and thus improve it. The hit-end-error technique was used to create a network with the lowest mean squared error by eradicating too few or too many neurons, resulting in an ideal network. When modelled in MATLAB, it has augmented training time and discovered the ideal neural network.

2.3 Multiple Linear Correlation Analysis

The relationship among independent and dependent parameters in this investigation is calculated by the following coefficients.

2.3.1 Multiple Correlation Coefficient (R)

It is a metric for the linear relation among numerous parameters, consisting of one dependent parameter, y, (ET_0), and mutually independent variables, x_i (RH, T, S, and W). It is calculated as the ratio of the standard deviation of calculated value (s_{e1}) to the standard deviation of experimental measurements (s_1). Considering S_1 as the standard deviation of residuals, the coefficient may be represented as

$$R = \frac{s_{e1}}{s_1} = \sqrt{1 - \frac{S_1^2}{s_1^2}} \tag{1}$$

2.3.2 *Coefficient of Determination (D_1)*

It is defined as squared multiple correlation coefficient (R_1), i.e. $D_1 = R_1^2$. The coefficient of multiple non-determination $(1 - D_1) = 1 - R_1^2$.

The D_1 specifies the dependent variable's portion of the variance. Meanwhile, $(1 - D_1) = 1 - R_1^2$ denotes the proportion of variance not explained by the multiple linear correlation of the variable x_1 over $x_2, x_3, \ldots, x_m$.

2.3.3 *Partial Correlation Coefficient ($r_{1\text{-}i}$)*

It measures the relationship between independent variable x_i (T, RH, S, or W) and dependent variable x_1 (ET_0) after eliminating the linear influence of other parameters on them.

R_1 is the multiple correlation coefficient between x_1 and x_i. R_{1-i} is also a multiple correlation coefficient between x_1 and x_i, after omitting the chosen independent variable x_i. It is then calculated from

$$r_{1-i} = \sqrt{\frac{\left(1 - R_{1-i}^2\right) - \left(1 - R_1^2\right)}{\left(1 - R_{1-i}^2\right)}} = \sqrt{1 - \frac{\left(1 - R_1^2\right)}{\left(1 - R_{1-i}^2\right)}} \tag{2}$$

2.4 Performance Metrics

The following metrics were utilized to assess the capability of the generated models.

2.4.1 Coefficient of Determination (D)

It is equal to R^2, i.e. $D = R^2$, where R—correlation coefficient and it can be represented as

$$R = \frac{\sum_{i=1}^{n}(x_i - \overline{x})(y_i - \overline{y})}{\left[\sum_{i=1}^{n}(x_i - \overline{x})^2 \sum_{i=1}^{n}(y_i - \overline{y})\right]^{1/2}} \times 100 \quad (3)$$

Here, x_i = $PMET_0$ values and $\overline{x}$ = mean of x_i

y_i = $ANNET_0$ values and $\overline{y}$ = mean of y_i.

$i = 1$ to n.

n = number of data values.

The D value indicates the extent of relationship between $PMET_0$ and $ANNET_0$ values.

2.4.2 Root Mean Square Error (RMSE)

It is used to find the residual error between $PMET_0$ and $ANNET_0$ values. It is stated as [9].

$$\text{RMSE} = \sqrt{\frac{\sum_{i=1}^{n}(x_i - y_i)^2}{n}} \quad (4)$$

2.4.3 Efficiency Coefficient (EC)

Here, the efficiency coefficient (Nash and Sutcliffe 1970) is utilized to evaluate the skill of the developed ANN architectures. EC is more reliable alternative than the RMSE indicator when the training and testing data periods have different lengths [10]. EC is stated as

$$\text{EC} = \left(1 - \frac{\sum_{i=1}^{n}(x_i - y_i)^2}{\sum_{i=1}^{n}(x_i - \overline{x})^2}\right) \times 100 \quad (5)$$

Table 2 Coefficient of multiple and partial correlations

Multiple correlation coefficient					Partial correlation coefficient			
Independent variable omitted								
–	T	S	W	RH	T	S	W	RH
0.9967	0.9748	0.9835	0.9922	0.9958	0.9314	0.8937	0.7589	0.4625

If EC value more than 90% shows the model performing good, value 80–90% indicates satisfactory model, and 60–80% value intimates a not acceptable model.

3 Results and Discussions

Partial and multiple correlations coefficients are used to recognize the extremely influencing weather parameters. The results of the investigation among the climate parameters (RH, S, T, and W) and $PMET_0$ in the research region are given in Table 2. The ANN algorithm is functioned with the input factors in scope and the changing number of nodes in the hidden layer to attain optimal performance and efficiency, as given in Table 3. As a result, the scatter and comparison graphs obtained are illustrated in Figs. 1 and 2.

The coefficients of multiple and partial correlations given in Table 2 indicate that the influence in the order of lowest to highest is relative humidity, wind speed, sunshine hours, and temperature on ET_0 at the study area. This is due to the area located in the semi-arid zone, which is characterized mainly by radiation and high temperature. The high EC and low RMSE values are performing adequately for the area, and Table 3 is indicating the same.

Figures 1 and 2 show scatter and comparison graphs that show the same results. The graph drawn among ET_0 obtained from the PM method as the ordinate and ET_0 attained from ANN as the abscissa yields a straight line with a unit slope and a zero intercept, highlighting the fact that ANN outcomes are fairly significant compared to PM method results. The author suggests that the developed model for predicting monthly ET_0 may be implemented for the experimental area under consideration based on the results of the ANN model data because it yields better accuracy.

4 Conclusions and Recommendations

The ANN models were established to forecast monthly ET_0 using climate variables affecting the region chosen for the current investigation as inputs. During the testing period, ANN (1-5-1), ANN (2-5-1), ANN (3-4-1), ANN (4-3-1) with (T), (T, S), (T, S, W), and, (T, S, W, RH) as inputs, respectively, were shown to have

Table 3 Performance metrics of ANN models

Optimal ANN	Input parameters	Slope (m)		Intercept		R^2		RMSE		
		Training	Testing	Training	Testing	Training	Testing	Training	Testing	Training
4-3-1	T, S, W, RH	1.0001	0.9202	−0.0004	0.2184	0.9977	0.9944	0.07	0.10	99.7
3-4-1	T, S, W	0.9990	0.9077	0.0078	0.4433	0.9546	0.9520	0.29	0.29	95.4
2-5-1	T, S	1.0019	0.8921	−0.0097	0.5184	0.9664	0.9436	0.25	0.32	96.6
1-5-1	T	1.0000	0.9421	0.0001	0.3333	0.9099	0.8958	0.41	0.43	90.9

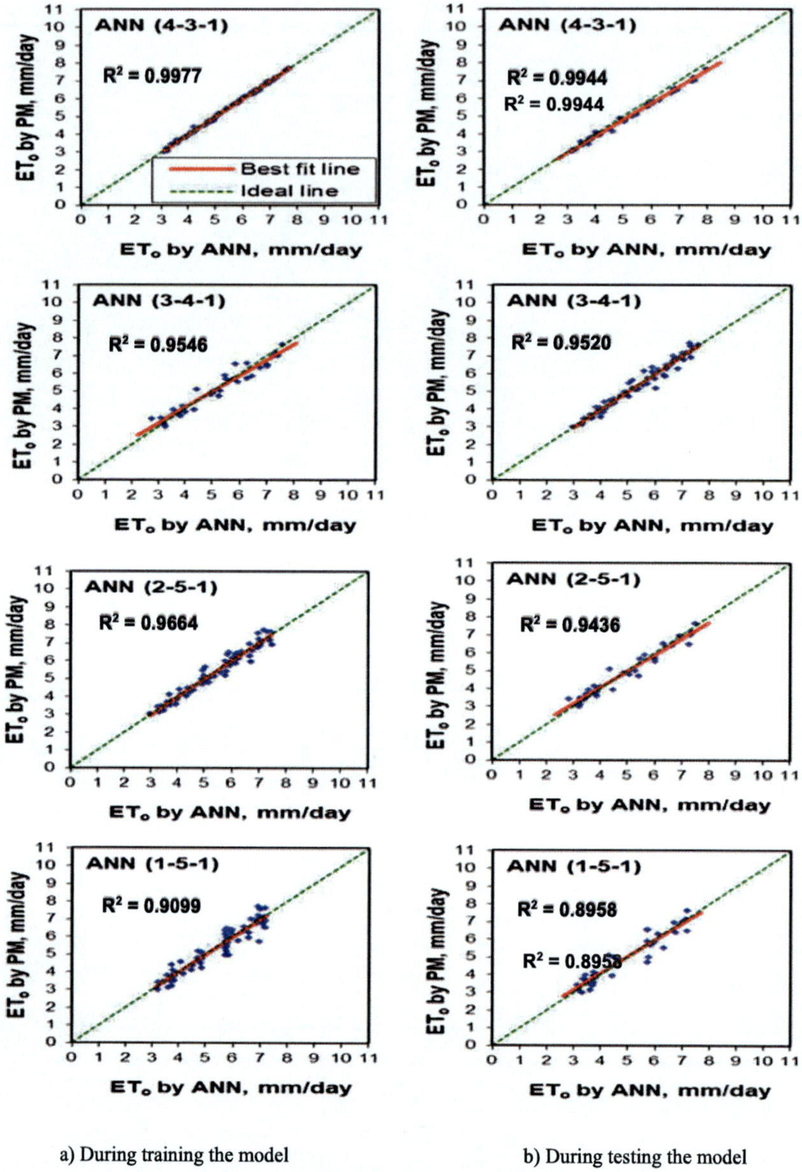

Fig. 1 Scatter plots of average monthly $PMET_0$ versus $ANNET_0$

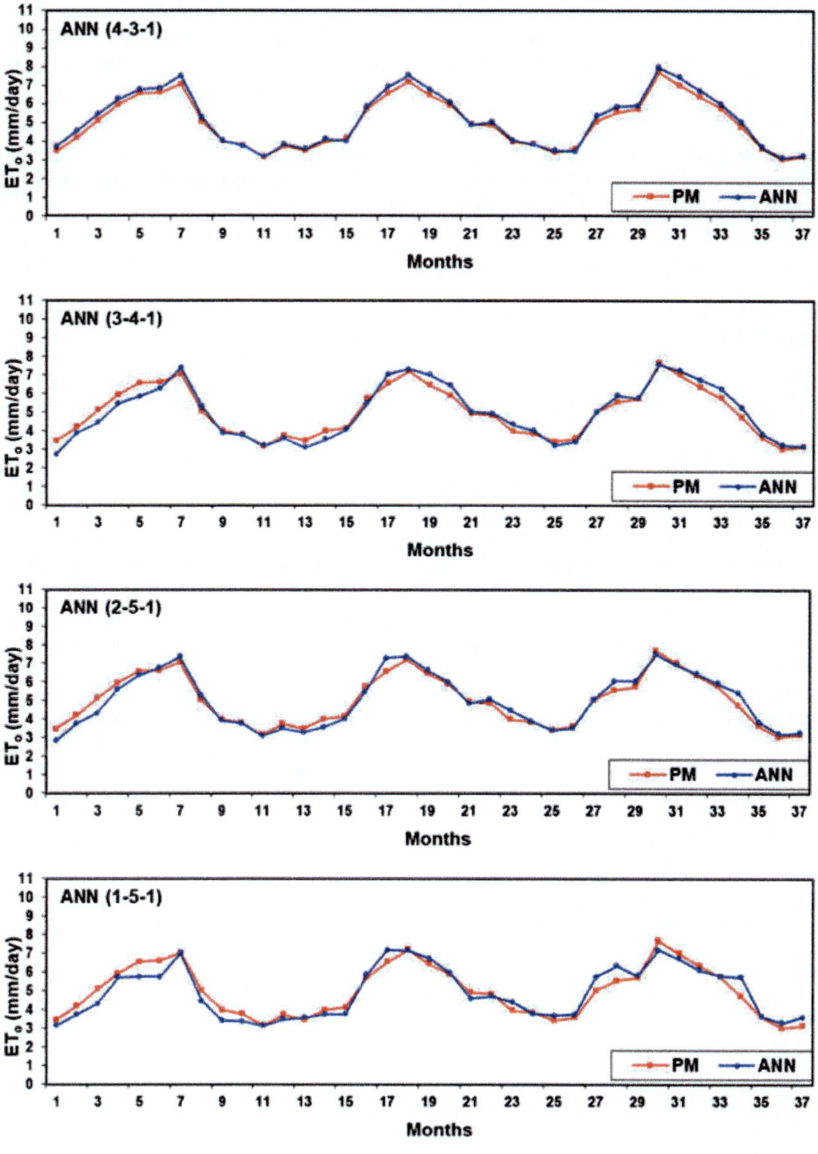

Fig. 2 Comparison graphs of mean monthly PMET$_0$ vs ANNET$_0$ during the testing period

89.58%, 94.36%, 95.20%, and 99.44%, respectively. However, the performance of ANN (4-3-1) model with inputs (T, S, W, RH) was better as compared to other.

These ANN models are suggested for forecasting monthly ET_0 with reasonable accuracy in the study location and other locations having alike climatic conditions.

Acknowledgements The corresponding author expresses gratitude to the Civil Engineering Department Staff and the college administration at SIETK, AP, India, for their suggestions and providing the necessary facilities throughout the research exploration.

References

1. Hargreaves GH, Samani ZA (1985) Reference crop evapotranspiration from temperature. Appl Eng Agric 1(2):96–99
2. Allen RG, Pereira LS, Raes D, Smith M (1998) Guidelines for computing crop water requirements-FAO irrigation and drainage paper 56, FAO—Food and Agriculture Organisation of the United Nations. Geophysics 156:178
3. Feng Y, Cui N, Zhao L, Hu X, Gong D (2016) Comparison of ELM, GANN, WNN and empirical models for estimating reference evapotranspiration in humid region of Southwest China. J Hydrol 536:376–383
4. Mallikarjuna P, Jyothy SA, Murthy DS, Reddy KC (2014) Performance of recalibrated equations for the estimation of daily reference evapotranspiration. Water Resour Manage 28(13):4513–4535
5. Nema MK, Khare D, Chandniha SK (2017) Application of artificial intelligence to estimate the reference evapotranspiration in sub-humid Doon valley. Appl Water Sci 7(7):3903–3910
6. Koch J, Berger H, Henriksen HJ, Sonnenborg TO (2019) Modelling of the shallow water table at high spatial resolution using random forests. Hydrol Earth Syst Sci 23(11):4603–4619
7. Govindaraju RS (2000) Artificial neural networks in hydrology. I: preliminary concepts. J Hydrol Eng 5(2):115–123
8. Pandey PK, Pandey V (2016) Evaluation of temperature-based Penman-Monteith (TPM) model under the humid environment. Model Earth Syst Environ 2(3):1
9. Laaboudi A (2020) Slama A (2000) Using Neuro-fuzzy and linear models to estimate reference Evapotranspiration in South region of Algeria (A comparative study). Italian J Agrometeorol 2:55–64
10. Liang GC, O'Connor KM, Kachroo RK (1994) A multiple-input single-output variable gain factor model. J Hydrol 155(1–2):185–198

Suitable Artificial Intelligence Techniques for Multispectral Image Classification

Ritica Thakur and V. L. Manekar

Abstract Hydrologic modelling is a complicated process depending on various factors. Since the evaluation of factors are exposed to high uncertainty due to high spatial variation. Hence, the precision of each factor becomes essential for modelling. Multispectral images are a highly informative set of data. It holds details of spectral characteristics and spatial structure of the objects in the image. Such data is beneficial to identify and classify the land covers on the bases of spectral signatures. Machine learning and AI techniques have made this work more efficient. This paper aims to understand the ability and suitability of AI techniques such as maximum likelihood classification (MLC), random trees (RT), and support vector machine to classify the image correctly. This paper discusses the basic principles of these AI techniques for land use classification. Accuracy assessment was considered as the criteria of comparison. Based on the obtained results, the performance of the SVM technique is found better than MLC and RT based on overall and Kappa coefficient (>0.80) for training and Kappa (0.64) for testing.

Keywords Artificial intelligence techniques · Support vector machine · Maximum likelihood classification · Random trees · LULC classification

1 Introduction

Quantification of Land Use and Land Cover (LULC) is always a tedious task. LULC is the basic information needed to plan, execute, and analyse the area. With the high-resolution satellites and multispectral data sensors, image processing becomes the most important part of information. This information can be used for many purposes. Nowadays, remotely sensed images are widely used for mapping and monitoring the

R. Thakur (✉) · V. L. Manekar
Civil Engineering Department, S.V. National Institute of Technology, Surat 395007, India
e-mail: riticathakur15@gmail.com

V. L. Manekar
e-mail: vlm@ced.svnit.ac.in

P. V. Timbadiya et al. (eds.), *Geospatial and Soft Computing Techniques*, Lecture Notes in Civil Engineering 339, https://doi.org/10.1007/978-981-99-1901-7_46

area. Image classification is the one of the data analyses used for extracting information to take decisions by many authorities. This study focusses on the LULC level 1 classification. Many LULC classification techniques are present in the literature. The classification can be broadly classified into supervised and unsupervised classification. However, supervised classification is tedious process, but more reliable than unsupervised classification. With new area of artificial intelligence techniques (AI), new classification and segregation techniques have become very efficient in classifying humongous complex pixel data sets. The most challenging nature of this job is the classification of the pixels in different classes produced from nearly similar spectral signatures. Different techniques have different working principles and techniques to refine the classified data. Many researchers have applied and studied techniques like support vector machine (SVM), K-means, and the maximum likelihood for the pixel-based classification of images [1, 2, 7, 9]. Al-doski et al. [2] studied the performance of SVM and k-mean method and recommended SVM over k-mean method. A classical approach of supervised image classification is maximum likelihood. It is a parametric approach of classification based on the training of model [9]. Though SVM is a nonparametric-based approach to segregating data sets in classes, another popular AI technique is Random Forest. It is a decision tree-based technique of generating rules from of nodes and leaves. This approach is also nonparametric and has high success rate of segregating data sets in the classes [6]. Thus, this study focuses on the applying these techniques to classify multispectral data sets into five classes. The main objective of this study is to find out the performance of the random tree (RT), maximum likelihood (ML), and support vector machine (SVM) for classification of Sentinel satellite image. Classification is restrained to level 1 classification using basic five classes: water, barren land, agriculture, forest and built up. The performance was evaluated using the classic approach of making a confusion matrix, commission and omission error, over all accuracy and kappa coefficient.

1.1 Materials and Methods

1.1.1 Random Forest Classifier

In Random Forest, each classifier is built using a random vector sampled independently from the input vector. It is made up of many tree classifiers. Each tree provides the most prevalent class to categories an input vector [3]. By integrating randomly selected elements or elements at each node, the random forest classifier creates a tree. To create a training data set for bagging, a method of random selection with alternative N examples—where N is the size of the first training set—was used for each feature/feature combination that was selected. The qualities used in decision tree induction can be chosen in a variety of ways, and most of these ways assign a quality measure directly to the attribute. Each time the maximum depth of a tree is reached using a mix of features and new training data. These old trees have not undergone pruning. This is one of the main advantages the random forest classifier has over other decision tree methods. According to the research, using the correct pruning techniques rather than the proper attribute selection techniques affects how effective tree-based classifiers are. Breiman [3] asserts that the Strong Law of Large Numbers avoids overfitting because the generalization error always converges as the number of trees increases, even if the tree is not trimmed (Feller). The quantity of features used at each node to generate a tree and the number of trees to be developed are two user-defined factors required to create a random forest classifier. Only a few attributes are considered at each node to determine the ideal split. The random forest classifier is composed of N trees since N is the number of trees to be created, which can be any number provided by the user. To categorize a new data set, the cases of each data set are passed down to each of the N trees. In that case, the class with the most N votes is selected by the forest.

1.2 Maximum Likelihood Classification (MLC)

MLC is popular parametric classification algorithm among researchers in many fields. This is a supervised classification technique. Basic principle of MLC is Bayes' classification. Initially, the algorithm is trained using supervised data sets. Then, further the image is classified on the basses of the likelihood of the pixels belonging to the trained group mean or covariance. For figuring the weighted distance or likelihood P of unclassified pixels measurement vector Y belong to one of the known classes N_c is based on the Bayesian equation [8, 9].

$$P = \ln(a_c) - [0.5 \ln(|\text{cov}_c|)] - [0.5(Y - N_c)\, T(\text{cov}_c - 1)\, (X - N_c)]$$

The Y vector is assigned to the class in which it has the maximum likelihood of belonging. The benefit of the MLC is a parametric classifier as it considers the variance–covariance within the class allocations and for normally distributed data [5].

1.3 SVM

A group of supervised algorithms used for regression and classification are called support vector machines (SVMs). SVMs are nonparametric classifiers as well. SVM was initially proposed by Vapnik and Chervonenkis (2015) and Vapnik (1999) [10, 11]. SVM performance is based on the training of the model. Most adopted linear separable classes are kernel density functions. This kernel density function is used to create hyperplanes. For the P number of data sets signified as $\{X_i, y_i\}$, $i = 1, \ldots, P$, where $X \in R^N$ is an N-dimensional space and $y \in \{-a, +a\}$ is no of class.

If a vector W exists perpendicular to the linear hyper-plane, these classes are regarded as linearly and hyperbolically separable. Two hyper-planes can be used to distinguish the data points in the two classes, i.e. class $+a$ represented as a and class 2 represented as $+a$. The design of these hyper-planes maximizes the separation between the two groups. When compared to traditional methods, SVMs yield more accurate results. However, the outcomes vary depending on the kernel used, the parameters selected for the chosen kernel and the SVM generation process [4].

2 Study Area and Data Source

2.1 Data Collection

For this study, a multispectral spectral image of Sentinel-2B has been used. Sentinel-2B is an optical imaging European satellite launched on 7 March 2017. The Sentinel-2B is the second satellite launched after Sentinel-2A as part of the European Space Agency's Copernicus Programme. The Sentinel-2B orbits will be placed phasing 180° opposite to Sentinel-2A. The Sentinel-2B has a wide, high-resolution multi-spectral imager with 13 spectral bands. Table 1 shows the details of the band properties.

Table 1 Shows the feature of each band of Sentinel-2B satellite

Sentinel-2B					
S. No	Sentinel 2 bands	Name of bands	Band width (nm)	Central wavelength (nm)	Spatial resolution (m)
1	1	Coastal aerosol	21	442.2	60
2	2	Blue	66	492.1	10
3	3	Green	36	559	10
4	4	Red	31	664.9	10
5	5	Vegetation red edge	16	703.8	20
6	6	Vegetation red edge	15	739.1	20
7	7	Vegetation red edge	20	779.7	20
8	8	NIR	106	832.9	10
9	8a	Narrow NIR	22	864	20
10	9	Water vapour	21	943.2	60
11	10	SWIR – Cirrus	30	1376.9	60
12	11	SWIR	94	1610.4	20
13	12	SWIR	185	2185.7	20

The data was downloaded by Earth Explorer (usgs.gov) https://earthexplorer.usgs.gov/ website

ArcMap 10.3 software is used for the present study for image processing. It provides the facility for pre-processing of the satellite image. Initially, the Sentinel 13 band data was read, and radiometric and atmospheric correction were given to each band before compositing the image. ArcMap 10.3 also provides the capability of performing AI-based image classification techniques. All the AI methods were performed using ArcMap 10.3 only. A random image of the location is selected for this study of 12,000 km^2.

2.2 Methodology

This study aims to understand the best supervised AI technique among parametric and nonparametric classifiers. The most popular maximum likelihood classifier, SVM, and random tree method were chosen for this analysis. Figure 1 shows the methodology adopted for the study. This classification was performed on the multispectral satellite data having 12 bands. A common training sample and testing samples were used to measure the classifier's capability.

Initially, a sets of signature files of around about 1800pixel per class using 330 shape files (sample count vary for each class to cover up the spatial variation in whole image) for water, built up, forest, agriculture and barren is used as training data. Figure 2 gives the detail of the training data. Three different classifiers were trained using the training data, and three images was generated using trained classifier of the random tree, maximum likelihood, and SVM. 499 Random points evenly distributed spatially across the image were generated, and ground truth data was generated using google earth image. Figure 3 represents all the training and testing data sets.

The testing and training point shape files were extracted from the image produced using random forest, maximum likelihood, and SVM. The data extracted were analysed against the ground truth data for the accuracy analysis. Commission error omission error, user accuracy, and producer accuracy were analysed for both training and testing points. Further, overall accuracy and Kappa coefficient were calculated to compare the classifier's performance.

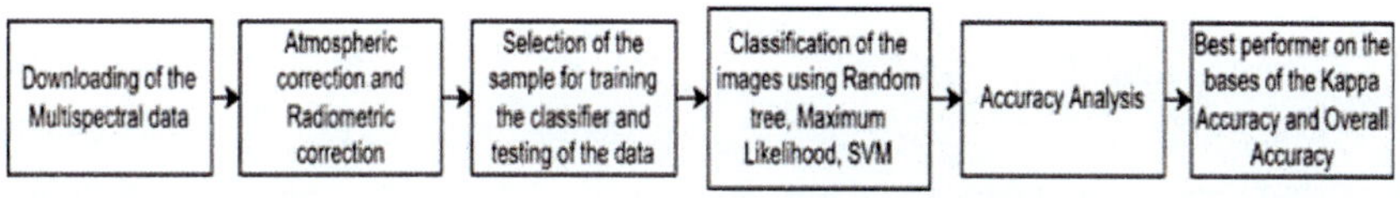

Fig. 1 Methodology adopted

Fig. 2 Training samples

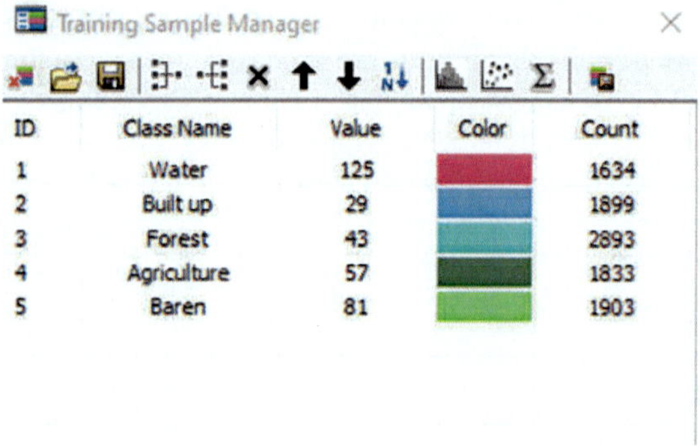

ID	Class Name	Value	Color	Count
1	Water	125		1634
2	Built up	29		1899
3	Forest	43		2893
4	Agriculture	57		1833
5	Baren	81		1903

Fig. 3 Training and testing points

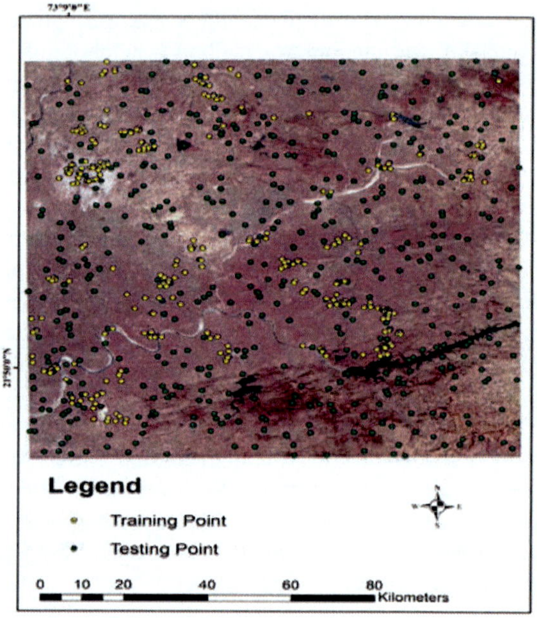

3 Results and Discussions

This study aimed to understand the best AI image classifier for the Sentinel multi-spectral satellite image. Among most popular supervised image classification technique, this study was constrained to random tree, maximum likelihood, and SVM techniques. Figure 4 shows the ground truth image from Google Earth, classified image from the random tree, maximum likelihood, and SVM. Training points and testing point data was extracted from the classified image, and confusion matrix was

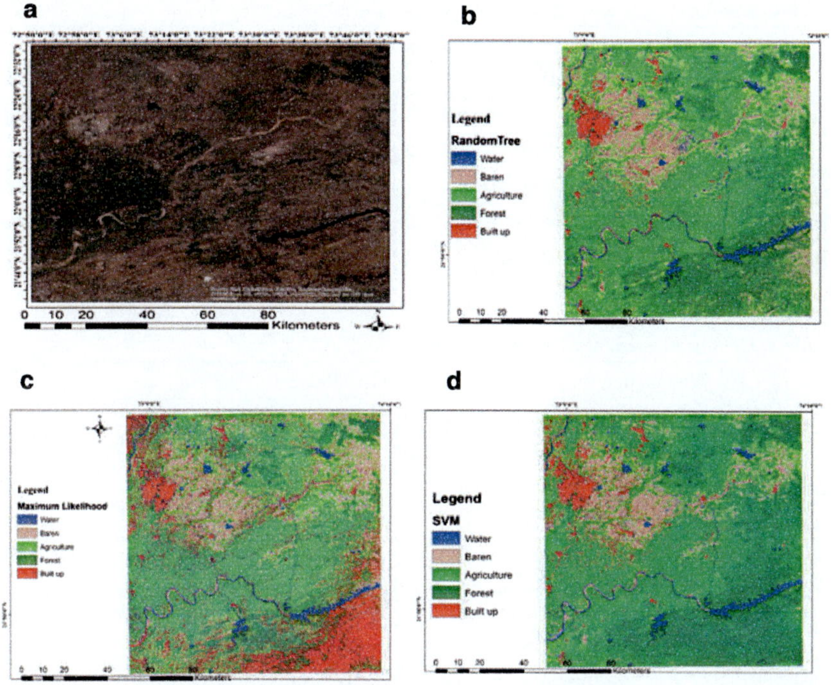

Fig. 4 **a** Google Earth image of the selected study area. **b** Classified image using random tree. **c** Classified image using maximum likelihood. **d** Classified image using maximum likelihood

prepared. Table 1 shows the confusion matrix for training point of a random tree, maximum likelihood, and SVM.

3.1 *Confusion Matrix for Training Data*

The confusion matrix is a 2D matrix with identity classes on both dimensions. It helps to understand the classification of the identity classes and accuracy related to them. Table 1 shows the confusion matrix for signature file given at the time of training of models. The confusion matrix based on the signature file (training data sets) represents how well the model is trained to classify the image. The confusion matrix (Table 2) of a random tree, maximum likelihood, and SVM representing well-trained models with training sample pixels is accurately classified in their classes. The Least error is present in SVM, and maximum error is present in maximum likelihood.

Table 3 shows the confusion matrix of the testing sample points. These are the randomly generated evenly spatially distributed point shape files with ground truth

Table 2 Confusion matrix training sample points

	Random tree						Maximum likelihood						SVM					
		A	B	C	D	E		A	B	C	D	E		A	B	C	D	E
Water	A	39	1	1	0	0	A	39	0	1	0	1	A	41	0	0	0	0
Barren	B	0	43	0	0	2	B	0	39	0	0	6	B	0	44	0	0	1
Agriculture	C	0	0	83	0	0	C	0	0	83	0	0	C	0	0	83	0	0
Forest	D	0	0	0	53	0	D	0	0	4	49	0	D	0	0	0	53	0
Built-up	E	0	1	0	0	107	E	2	10	0	0	96	E	0	1	0	0	107

Table 3 Confusion matrix testing sample points

	Random tree						Maximum likelihood						SVM					
		A	B	C	D	E		A	B	C	D	E		A	B	C	D	E
Water	A	11	0	1	2	0	A	10	0	1	0	3	A	12	1	1	0	0
Barren	B	3	95	19	11	3	B	2	66	17	1	45	B	3	75	27	17	9
Agriculture	C	5	22	187	37	5	C	16	9	193	7	31	C	8	5	208	28	7
Forest	D	1	6	8	66	0	D	0	5	15	51	10	D	3	0	5	73	0
Built-up	E	1	4	2	0	10	E	1	0	3	0	13	E	1	2	1	1	12

classes to test the predicted class. Hence, confusion matrix from testing sample point gives the real performance of the classifier. SVM seems to classify more accurately than random tree and maximum likelihood.

To account the errors in the image classification, commission and omission error is calculated.

3.2 *Commission and Omission Error*

Commission error is ratio of wrongly classified pixel to the total of classified pixel in that class. Similarly, the omission error is the ratio of omitted pixel from the class to the total number of assigned pixels in that class. The confusion matrix generated using training point shows least commission and omission errors. Figure 5 represents the commission and omission errors in the classified image using the training data set. It is evident from the graphs that the commission error (0–0.21) of the classifier is more than the omission error (0–0.13). ML classifier is shows the tendency of mostly wrong commissioned pixels in the classes, followed by RT. Forest class has no commissioned error by any of the classifiers. However, Fig. 5b shows that many pixels from the forest have been missing from ML-classified images. In omission error, the ML again having a high scorer of the errors following RT and SVM. It is concluded that SVM is best-trained classifier among RT and ML. Similarly, Fig. 6a shows the commission and omission error found in testing points.

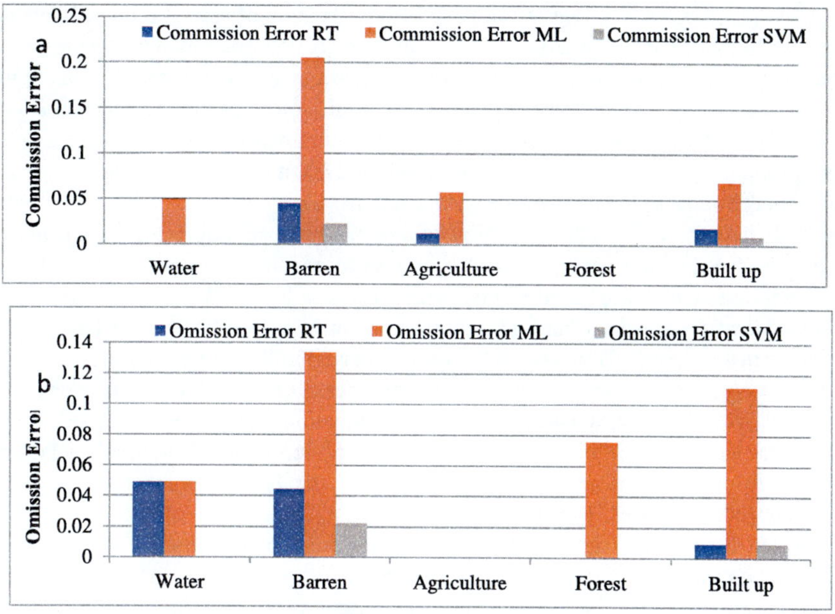

Fig. 5 **a** Commission error and **b** omission error of the classifier using training data set

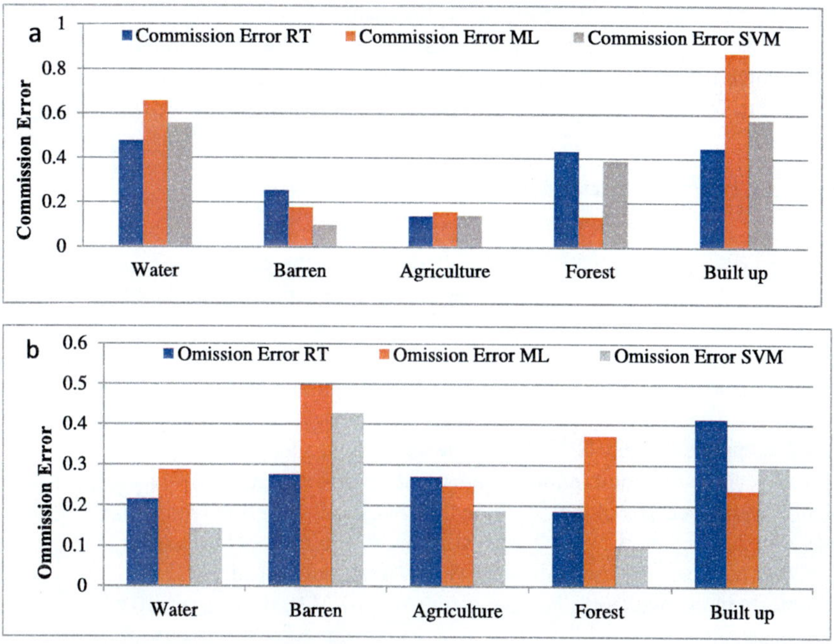

Fig. 6 **a** Commission error and **b** omission error of the classifier using testing data set

Commission error and omission error in classification of testing points show the real performance of the classifier. Number of wrongly commissioned pixels are more than the number of omitted pixels. Commission and omission error is significant in the ML classifier, though the commission and omission error in the water class are high for the ML classifier. For the barren land, commission error is high for RT, but omission error is high for the ML classifier. From the confusion matrix, it is evident that most of the pixel of the barren land has been classified under agriculture land. It happens due to the reflection of vegetation surrounding the barren land. From the confusion matrix, it is clear that SVM and RT have quite successfully commissioned the barren land and built-up pixel separately. RT has shown the least omission error for barren land. Maximum number of barren land pixels have been omitted in the classified image by RT classifier. For agriculture, commission error was low concerning to the omission error. Large number of the pixel are commissioned to other classes. SVM commissioned the maximum number of pixels within the class. With the least commissioned error and least omission error, SVM made a better classification than another classifier. The classified pixels of the forest of the testing sample, shows the least commissioned error and maximum omission error for ML classifier. It represents under prediction of the forest area by the ML. RT shows high commission error and higher omission error for the forest. SVM shows a high commission error and low omission error making it over predicting the forest area. For the built-up area, the ML classifier shows high commission and low omission errors, making it over predicting the built-up area. RT has nearly same number of omission and commission errors under this scenario though the pixel placement to the classes is inaccurate. For SVM, omission error is low, and commission error is high, which represents again over prediction of built-up area.

3.3 Producer Accuracy and User Accuracy

Producer and user accuracy are complementary to the commission and omission errors. It is essential to consider user accuracy to understand the classifier's performance. Figure 7 shows the producer and user accuracy of the training sample, and Fig. 8 shows the producer and user accuracy of the testing samples. From the figures, the producer and user accuracy of the training samples are very high and acceptable range. The user accuracy of the ML classifier is very low, being less accurately classified pixels. RT and SVM have a better performance.

The accuracy of the randomly generated point with ground truth gives a better picture of the accuracy of the classifier, though the training shows more than 0.9 accuracy. The testing shows all together a different accuracy. ML classifier is showing high user accuracy for the built-up only. SVM has performed better than RT in both accuracies in many classes. For the water class, SVM has a high rate (0.85) of accurately classified pixels with average producer accuracy. Water pixels are accurately classified with high producer accuracy using the RT classifier. Similarly, RT is classifying the barren pixel with 0.7 user accuracy, but the producer accuracy is low.

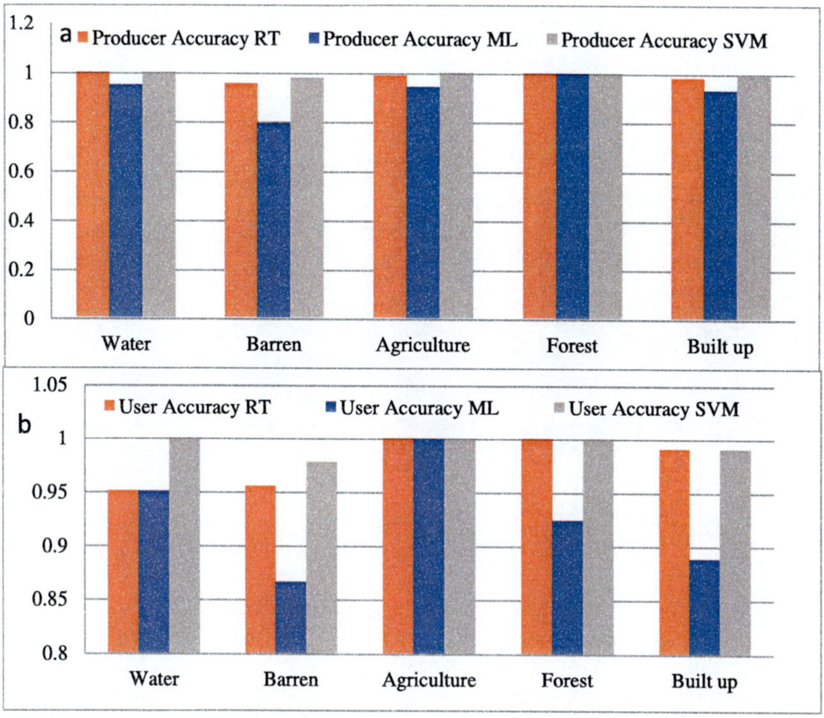

Fig. 7 **a** Producer accuracy of training samples. **b** User accuracy of the training samples

SVM gives user accuracy of 0.57 with high producer accuracy for barren land. All three classifiers have accurately classified agriculture class. Forest was classified most accurately by SVM then by ML and RT. Built up is accurately classified by the ML classifier, but the producer accuracy of ML classifier is very low.

3.4 Overall Accuracy and Kappa Coefficient

Overall accuracy and Kappa coefficient give the efficiency of the classifier. Table 3 shows the overall accuracy and Kappa coefficient of RT, ML, and SVM on trained and testing samples. From the training data sets of the classifier, SVM has performed best among RT and ML in overall accuracy and Kappa coefficient. It signifies that the SVM is the best trained among the three classifiers. SVM performs best on the testing data set among all the classifiers, followed by RT and ML. Table 4 shows the overall accuracy of the training and testing of samples.

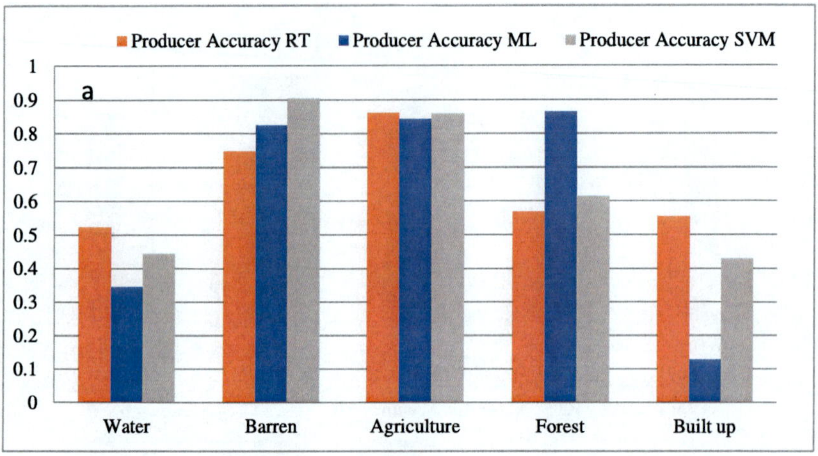

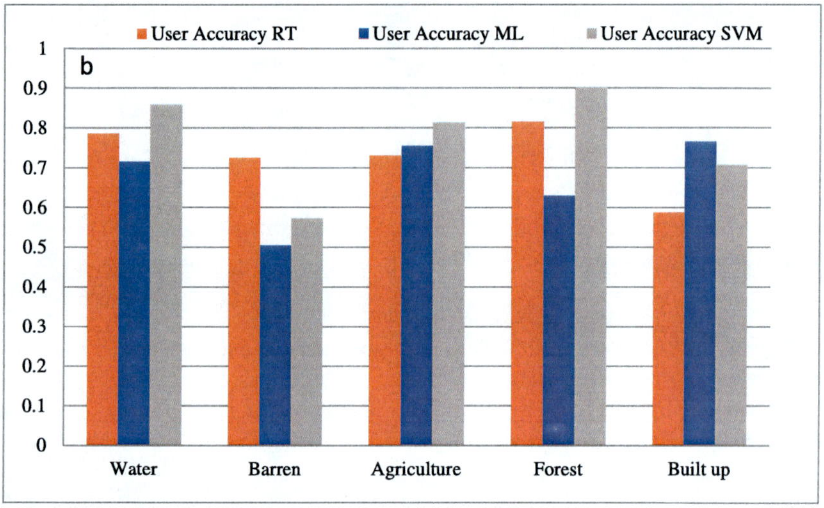

Fig. 8 **a** Producer accuracy of testing samples. **b** User accuracy of the testing samples

Table 4 Results obtain at the time of training and testing

Method	Training		Testing	
	Overall accuracy	Kappa coefficient	Overall accuracy	Kappa accuracy
Random tree	0.98	0.98	0.74	0.61
Maximum likelihood	0.93	0.91	0.67	0.52
SVM	0.99	0.99	0.76	0.64

4 Conclusions

This study was conducted to understand the performance of the supervised AI classifier to classify the multispectral Sentinel satellite data set. The Image was classified into four elementary class water, barren, agriculture, forest, and built-up. All the tree technique shows acceptable classified image with an overall accuracy more than 90 per cent in the training phase and more than 67 per cent in the testing phase. SVM and RT being a nonparametric classifiers, performed well than a parametric classifier. Among all, the SVM performed best as a classifier. Sentinel image is a high resolution to the resolution of 10 m giving a clear image and better LULC classification of the area.

References

1. Abburu S, Babu Golla S (2015) Satellite image classification methods and techniques: a review. Int J Comput Appl 119(8):20–25
2. Al-doski J, Mansor SB, Zulhaidi H, Shafri M (2013) Image classification in remote sensing. 3(10):141–148.
3. Breiman L (1999) Random forests. UC Berkeley TR567
4. Huang C, Davis LS, Townshend JRG (2002) An assessment of support vector machines for land cover classification. Int J Remote Sens 23(4):725–749
5. Inc E (1999) Erdas field guide. Erdas Inc. Atlanta
6. Jagannathan G, Pillaipakkamnatt K, Wright RN (2009) A practical differentially private random decision tree classifier. In 2009 IEEE international conference on data mining workshops. IEEE, pp 114–121
7. Jog S, Dixit M (2016) Supervised classification of satellite images. In: Conference on advances in signal processing, CASP 2016, X, pp 93–98
8. Otukei JR, Blaschke T (2010) Land cover change assessment using decision trees, support vector machines and maximum likelihood classification algorithms. Int J Appl Earth Obs Geoinf 12:S27–S31
9. Sisodia PS, Tiwari V, Kumar A (2014) Analysis of supervised maximum likelihood classification for remote sensing image. In: International conference on recent advances and innovations in engineering (ICRAIE-2014). IEEE, pp 1–4
10. Vapnik VN (1999) An overview of statistical learning theory. IEEE Trans Neural Netw 10(5):988–999
11. Vapnik VN, Chervonenkis AY (2015) On the uniform convergence of relative frequencies of events to their probabilities. In: Measures of complexity. Springer, Cham, pp 11–30

Data-Driven Approaches for Estimation of Particle Froude Number in a Sewer System

Deepti Shakya, Mayank Agarwal, Vishal Deshpande, and Bimlesh Kumar

Abstract The deposition of sediments has a major impact on the hydraulic capacity of a channel. The deposition of sediment over time may result in a diminished ability of the sewers to carry waste and other materials. For designing sewers and drainage systems in urban areas, the self-cleansing mechanism is often deployed. In this study, the accurate prediction of the particle Froude number (F_r) plays a significant role. This study investigates the insights of the performance of the data-driven approaches to determine the particle F_r with regard to non-deposition with a deposited bed. The dataset obtained comprises variety of studies published in literature having a range of values for volumetric sediment concentration (C_v), the dimensionless grain size of particles (D_{gr}), the median size of sediment (d), hydraulic radius (R), and the friction factor of the pipe (λ). Three data-driven approaches, namely Random Forest (RF), M5Prime (M5P), and Reduced Error Pruning Tree (REPT), have been utilized in this study for modeling purposes. Results show that the RF approach is superior in comparison with M5P and REPT approaches. The best performing method in this study is RF (CC = 0.966, NSE = 0.932, RMSE = 0.64, and R^2 = 0.933) followed by M5P, REPT.

Keywords Data-driven · Sedimentation · Froude number · Machine learning · Sewer

D. Shakya (✉) · M. Agarwal
Department of Computer Science & Engineering, Indian Institute of Technology, Patna 801106, India
e-mail: deepti_1921cs14@iitp.ac.in

M. Agarwal
e-mail: mayank265@iitp.ac.in

V. Deshpande
Department of Civil & Environmental Engineering, Indian Institute of Technology, Patna 801106, India
e-mail: deshpande@iitp.ac.in

B. Kumar
Department of Civil Engineering, Indian Institute of Technology, Guwahati 781039, India
e-mail: bimk@iitg.ac.in

P. V. Timbadiya et al. (eds.), *Geospatial and Soft Computing Techniques*, Lecture Notes in Civil Engineering 339, https://doi.org/10.1007/978-981-99-1901-7_47

1 Introduction

The irregular flow, industrial discharge, and infiltration cause frequent solids deposition in the sewer system. The sediment properties may change if the deposit is left in the sewer system for an extended period of time, leading to further deposition and sewer blockage. The sewer system gradient must produce self-cleansing velocities at various discharges in order to prevent this. Drainage and sewer pipes are rigid boundary channels. A rigid boundary channel is defined as the channel with an unmovable bed and sides. The flow must carry sediment from upstream and not exceed the channel's sediment transport capacity in order to avoid sediment deposition in a rigid boundary channel. Changes in shear stress and velocity result from sediments deposition to the channel bottom. Based on self-cleansing criteria, drainage and sewer systems are developed. Sediment particles are continually transported without being deposited under self-cleansing criteria [11]. Self-cleaning criteria have the following properties [2, 3, 11]: A sewer system is considered to be self-cleaning if its capacity to carry material is sufficient to preserve the balance between deposition and erosion with respect to the depth of time-averaged sediment deposit. One of the following requirements must be satisfied for a channel to be self-cleaning: (1) Flow must remove depositions from the bottom of the channel and (2) flow must remove deposits from the channel's bottom.

To attain the self-cleansing goal, several empirical equations use minimum velocity and shear stress as mentioned in the literature. In order to estimate sediment movement in circular cross-sectional channels with various flatbed thicknesses, El-Zaemey [6] investigated how the deposited bed impacts channel velocity and shear stress distributions in the channel by developing empirical equations.

Perrusquía [14] proposed a model for the prediction of bed load transport and flow resistance when water depth exceeds half-full and for isolated dunes with permanent deposited beds [15]. Furthermore, May [10] presented a model that predicts flow resistance, sediment transport rate, and flow conditions with deposited sediment, where the particle F_r is correlated with the flow conditions.

In terms of pipe size, Ab Ghani [1] and May [10] proposed empirical equations for large pipe channels, while El-Zaemey [6], Perrusquia [14, 15] proposed empirical equations for narrow pipe channels. They discovered that for non-deposition with deposited bed, all of these studies performed well. When used on the more recent datasets, the existing empirical equations overfit, which causes a problem [4].

Recent studies have used machine learning (ML) techniques to solve problems in sediment transport. Ebtehaj et al. [5] established a method which combines a feed-forward neural network (FFNN) with extreme learning machine (ELM). Safari and Mehr [17] suggested a design tool using MGGP to predict particle F_r for larger sewer pipes. Safari [16] compared decision tree (DT), generalized regression neural network (GR), and multivariate adaptive regression splines (MARS) for predicting particle F_r. Artificial neural network (ANN) methods of radial basis function (RBF) and FFNN predicted the critical velocity for a shallow and (slightly) thicker sediment layer (Wan Mohtar et al. 2018). Both FFNN and RBF methods had correlation coefficients of 0.884 and 0.863. ANN methods can predict and forecast sediment critical velocity. More data and thickness variation require mathematical optimization to improve ML method performance. Wan Mohtar et al. (2018) provide convincing V_c prediction using RBF and FFNN, but no universal equation was possible.

In general, there are various neuron-based models such as ANN, adaptive neuro-fuzzy inference system (ANFIS), and support vector machine (SVM) which have some flaws like the models requires larger dataset, overfitting problem, and hyper-parameters tuning.

To address some of the shortcomings of conventional machine learning methods, new machine learning methods have recently been developed, such as rule-based and tree-based techniques. For instance, Hussain and Khan [8] used RF and discovered that RF offers greater accuracy for hydrological modeling than SVM and ANN.

The paper contributes as follows: (1) We predict particle F_r in the sewer system using data-driven methods and compared with the state-of-the-art methods. (2) We propose data-driven approaches (RF, M5P, and REPT) to predict particle F_r in a sewer system. This study uses five datasets for non-deposition with a deposited bed [1, 6, 10, 14, 15]. We use volumetric sediment concentration (C_v), dimensionless grain size (D_{gr}), hydraulic radius to sediment median size (d/R), and channel friction factor (λ) variables to predict particle F_r. (3) CC, NSE, RMSE, and R^2 are used to measure methods effectiveness.

2 Materials and Methods

2.1 Dimensional Analysis and Functional Formula

Pipe diameter (D), hydraulic radius (R), flow depth (Y), deposited bed thickness (t_s), and deposited bed size (W_b), volumetric sediment concentration (C_v), sediment median size (d), non-deposition flow mean velocity (V_n), sediment relative density (s), channel friction factor (λ), and channel bed slope ($S\%$) are some of the variables that have been shown to have an effect on the self-cleaning phenomenon in the research that has been done previously [1, 6, 10, 14, 15].

According to Safari et al. [18], the notion of self-cleaning is examined under two conditions of non-deposition: with deposited bed and without deposited bed. A functional relationship between the variables can be represented as follows, as illustrated in Eq. 1:

$$V_n = f(D,\ R, Y,\ t_s,\ W_b,\ C_v,\ \rho_s, \rho,\ d,\ g,\ v,\ \lambda) \tag{1}$$

Since D, R, and Y describe the flow depth characteristics, the hydraulic radius (R) has been maintained in the functional formulation due to its applicability in reflecting the condition of movement of sediments. Accordingly, ρ_s and ρ can be characterized by $\frac{\rho_s}{\rho} = s$.

$$V_{\mathrm{n}} = f(R, t_{\mathrm{s}}, W_{\mathrm{b}}, d, C_{\mathrm{v}}, s, d, g, v, \lambda) \tag{2}$$

As can be seen in Eq. 3, Function 2 can be transformed into the appropriate dimensionless form that is as follows:

$$\frac{V_{\mathrm{n}}}{\sqrt{gd(s-1)}} = f\left(C_{\mathrm{v}}, D_{\mathrm{gr}}, d/R, W_{\mathrm{b}}/t_{\mathrm{s}}, \lambda\right) \tag{3}$$

According to the criteria for the design of self-cleaning systems, it is safe to assume that the influence of the parameters $W_{\mathrm{b}}/t_{\mathrm{s}}$ will remain the same. In light of this, Eq. 3 can be rewritten as follows:

$$\frac{V_{\mathrm{n}}}{\sqrt{gd(s-1)}} = f\left(C_{\mathrm{v}}, D_{\mathrm{gr}}, d/R, \lambda\right) \tag{4}$$

where the right side of Eq. 4 is regarded to be the independent variable, while the left side of Eq. 4 represents a dependent variable that is F_{r}.

D_{gr} can be calculated as follows:

$$D_{\mathrm{gr}} = \left(\frac{(s-1)gd^3}{v^2}\right)^{1/3}$$

where C_{v} represents the volumetric concentration of sediment particles, D_{gr} represents the dimensionless grain size, and d/R represents the ratio of the hydraulic radius to the median size of the sediment; λ represents pipe friction factor; v represents kinematic viscosity of fluid; and g represents acceleration due to gravity.

3 Data Source

This study uses five non-depositions with deposited bed data sets [1, 6, 10, 14, 15] for the prediction of particle F_{r} in a sewer system, and their ranges of data are given in Table 1. El-Zaemey [6] evaluated the hydraulic flow resistance and investigated the influence that continual deposits had on the sediment carrying capacity of pipe channels. Experiments are performed in a canal with a diameter of 305 mm and beds that are either level or rough and smooth. The flow conditions that characterize the stream traction in pipe channels were investigated and studied by Perrusquía [14], along with their correlations to sediment transport rate and flow resistance. Experiments were carried out inside a concrete conduit that measured 225 mm in

Table 1 Data ranges

	C_V (ppm)	V_n (m/s)	d (mm)	s	R (mm)	λ
El-Zaemey [6]	7–917	0.385–0.962	0.53–8.4	2.56–2.61	24–86	0.01–0.06
Perrusquía [14]	96–252	0.476–0.668	0.90–2.50	2.65	44–60	0.0354–0.053
Perrusquia [15]	44–186	0.509–0.537	0.9	2.65	49–50	0.0263–0.0427
Ab Ghani [1]	21–1269	0.492–1.332	0.72	2.63	75–128	0.0031–0.0644
May [10]	3.5–1280	0.375–1.317	0.47–0.73	2.63–2.64	64–117	0.0037–0.1456

diameter, 23 m in length, and different sizes of pipe diameter were also tested such as 2.5 and 0.9 mm. The primary concentration of Perrusquia [15] was on the calculation of flow resistance on a permanent deposit for the scenario of isolated dunes and also carried out their tests on a pipe that was 2 m in length.

May [10] provided equations that could be used to predict the hydraulic resistance of flow, the pace at which transport of sediments occurred with a deposited bed, as well as the flow conditions that were necessary to prevent deposition from taking place.

Experiments are carried out on a concrete pipe that is 21 m long and has a diameter of 450 mm. The average depth of the deposit can range anywhere from 13 to 27% of the pipe diameter at flow rates ranging from 0.4 to 1.3 m/s. There are four distinct sand gradings that are utilized, and their sizes are 0.73, 0.61, 0.58, and 0.47 mm. The two intermediate sizes are generated by mixing the sands that measure 0.47 mm and 0.73 in varying quantities.

4 Workflow

The following is an explanation of the workflow (Fig. 1):

- Dataset: A dataset that includes non-deposition without and with deposited bed data has been provided by Safari et al. [18]. We take into consideration the investigations that were carried out for non-deposition using deposited bed data by El-Zaemey [6], Perrusquia [14], Perrusquia [15], Ab Ghani [1], and May [10].

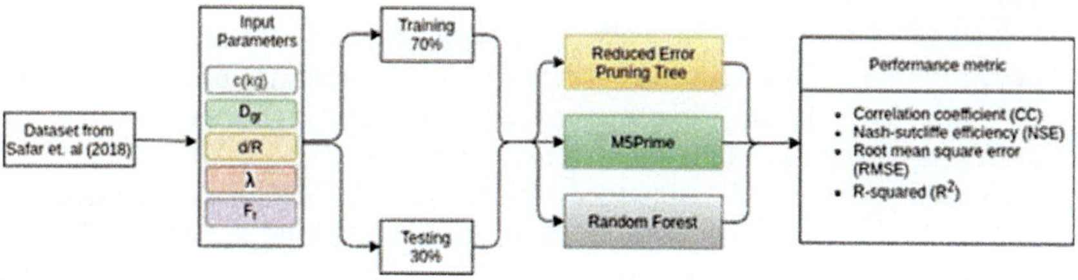

Fig. 1 Workflow of the study

- Data split: The dataset is split into a train and a test set, in the ratio of 7:1 where 277 data values of the total 397 data values allocated for training set and the remaining 120 data values allocated for testing set.
- Methods: In this study, a novel method based on data-driven techniques is presented to predict particle F_r in a sewer system (RF, M5P, and REPT). We tested the effectiveness of data-driven methods for the prediction of particle F_r by comparing them with each other and the state-of-the-art methods.
- Performance metric: A number of different performance metrics, including correlation coefficient (CC), Nash–Sutcliffe efficiency (NSE), root mean square error (RMSE), and R^2, are used to assess the efficacy of the methods.

5 Methods Description

1. Random Forest (RF): RF is an ensemble learning method whose base learner is a decision tree which means many decision trees are formulated in a definite random way to produce a random forest. Each tree is constructed from a distinct row and for splitting at each node of the tree, select a different sample of features. Then, each of the trees makes its own individual prediction and averages these predictions to generate an output. Interestingly, this averaging makes an RF better than a single decision tree.
   In RF, various parameters are used including the number of iterations (number of trees), i.e., 100 used in this study, number of samples per leaf is 1, the maximum depth of the tree is set to default value, i.e., 0.
2. M5Prime (M5P): M5P is an evolutionary learner that uses genetic algorithms to solve regression problems. This method begins by fitting a linear regression to the terminal node and then applies a multivariate linear regression to each subspace. To do this, it divides the dataset into several separate subspaces using the different sets of data. This technique, which is driven by the data, addresses continuous class problems rather than discrete class problems with very high dimensionality.
   Error estimation is provided, along with information on the M5P method's division criteria, on each node of the tree. The variation in the default value of the class that is entering the node is used to determine the error. By determining each characteristic of the node, the function ensures that the expected error reduction is minimized to the greatest extent possible. The standard deviation of the continuous class values at any node is used to determine the tree error. Overfitting occurs because of this split since it creates a large tree structure. Now, the enormous tree-like structure is trimmed down, and the linear regression functions are used to take the place of the subtrees that were removed.
3. Reduced Error Pruning Tree (REPT): REPT is like a decision tree that reduces the risk of overfitting by verifying the predictive utility of all nodes of a tree. Firstly, this approach uses the regression tree to generate various trees in multiple iterations.

Table 2 Summary of performance metrics

Metrics	Formulae	Range	Significance
Correlation coefficient (CC)	$CC = \frac{\sum_{t=1}^{n}(a_t-\overline{a})(p_t-\overline{p})}{\sqrt{\sum_{t=1}^{n}(a_t-\overline{a})^2}\sqrt{\sum_{t=1}^{n}(p_t-\overline{p})^2}}$	−1 to +1	Higher the better
Nash–Sutcliffe efficiency (NSE)	$NSE = 1 - \frac{\sum_{t=1}^{n}(a_t-p_t)^2}{\sum_{t=1}^{n}(a_t-\overline{a})^2}$	1 to $-\infty$	Higher the better
Root mean square error (RMSE)	$RMSE = \sqrt{\frac{1}{n}\sum_{t=1}^{n}(a_t - p_t)^2}$	0 to ∞	Lower the better
R^2	$R^2 = \left(\frac{\sum_{t=1}^{n}(a_t-\overline{a})(p_t-\overline{p})}{\sqrt{\sum_{t=1}^{n}(a_t-\overline{a})^2}\sqrt{\sum_{t=1}^{n}(p_t-\overline{p})^2}}\right)^2$	0–1	Higher the better

where a_t is actual values, p_t is predicted values, $\overline{a}$ is average of actual values, $\overline{p}$ is average of predicted values, and n is number of observations

Secondly, it selects the best tree method which has the minimum error from various trees. Then, the REPT is utilized to prevent overfitting. The last step of this approach is to handle the missing values and arrange the values with the help of the embedded method. The main aim of this approach is to eliminate some branches of the tree to attain the accurate sub-tree through the post-pruning method.

6 Performance Metrics

See Table 2.

7 Results

This section presents the performance of the data-driven approaches of machine learning for the testing set. Table 3 presents the comprehensive findings about the three data-driven methods' performance metrics after taking into account all of the possible combinations of the input variables. The default hyper-parameters are used for modeling. Table 3 presents the correlation coefficient (CC = 0.966) of RF in predicting the particle F_r is higher than those from M5P (CC = 0.943) and REPT (CC = 0.924) methods. Similarly, the NSE values show that the RF (0.932) method has the highest prediction accuracy followed by M5P (0.887) and REPT (0.851). Compared with the REPT (RMSE = 0.953) and M5P (RMSE = 0.829) methods, the particle F_r prediction error of the RF method (RMSE = 0.646) is smaller. The R^2 value shows that the RF method (0.933) can predict particle F_r better than the M5P (0.890) and REPT (0.854) methods. The evaluation of performance metrics of

Table 3 Comparison of methods for the prediction of F_r

Methods	Metrics			
	CC	NSE	RMSE	R^2
RF	**0.966**	**0.932**	**0.646**	**0.933**
M5P	0.943	0.887	0.829	0.890
REPT	0.924	0.851	0.953	0.854

Bold values indicate the best performance

the data-driven approaches shows that the RF method performs better than the M5P and REPT.

Regression plots and line plots are shown in Fig. 2 to illustrate a comparison of the predictive power of different data-driven approaches for the testing dataset that is utilized in this study. The figures present the values of particle F_r that are derived from actual values as well as the values predicted by all of the methods that are utilized in this study. Figure 2b, d, and f demonstrate the actual values versus predicted scatter plots to predict particle F_r in a sewer system. These plots compare the actual value to the predicted values of different methods. The fact that the data points are getting closer to the regression line, which shows how accurate the RF is in predicting thee particle F_r in scatter plot (Fig. 2f). In addition, both the M5P and the REPT show a significant amount of dispersion around the regression line. The comparison of the RF method's output to those of the M5P and REPT shown in Fig. 2 reveals that the RF method produces a superior outcome.

The violin plot illustrates the actual values of particle F_r and the predicted particle F_r of three data-driven approaches. From Fig. 3, we can see that there is a significant disparity in the patterns of distribution between the methods used in this study. The plot that corresponds to the RF has a shape that is most similar to that of the actual values when compared to all of the data-driven techniques.

8 Conclusions

The range of data and modeling techniques used in this study affects experimental methods performance. By using four experimental data sets from the literature, data-driven machine learning approaches are used to predict particle F_r having a deposited bed. The four (C_v, D_{gr}, d/R, λ) dimensionless variables are used as an input and one (F_r) as an output. Among all input variables, C_v is the most viable variable followed by λ, d/R, and D_{gr} to predict particle F_r. All methods perform better when all variables are combined, according to the results. RF came out on top, followed by M5P and REPT in the performance rankings. Data-driven approaches outperform soft computing techniques like MGGP, GEP, and MLP methods as cited in the literature. Therefore, in order to predict particle F_r, the methods used in this study can be utilized as reliable design tools.

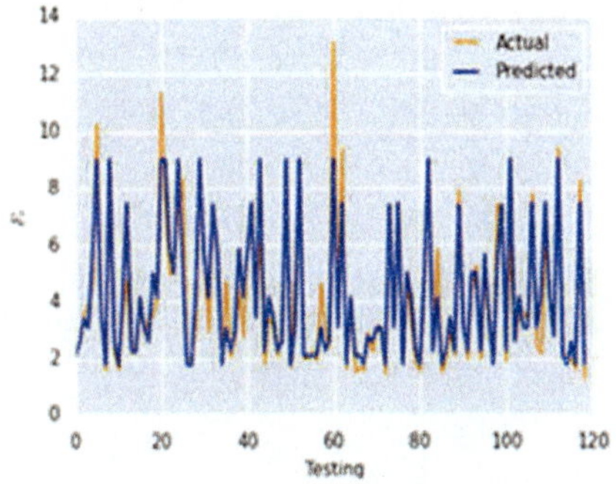

(a) Line plot for REPT

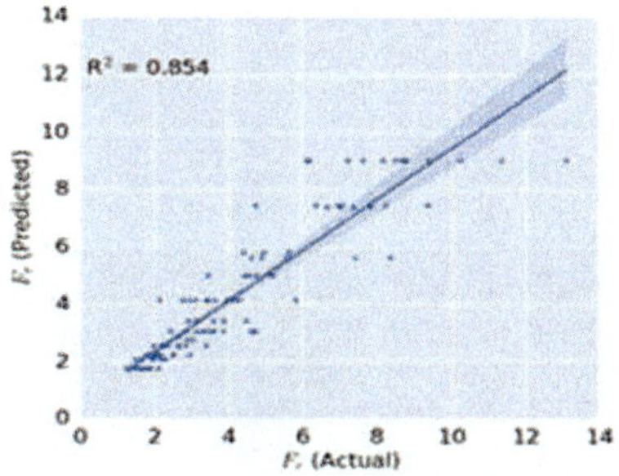

(b) Regression plot for REPT

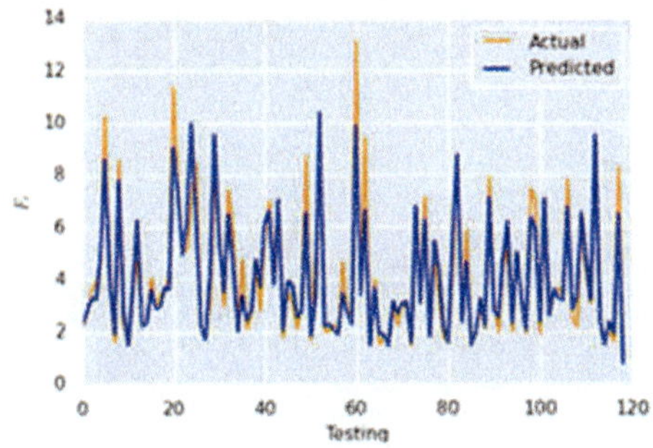

(c) Line plot for M5P

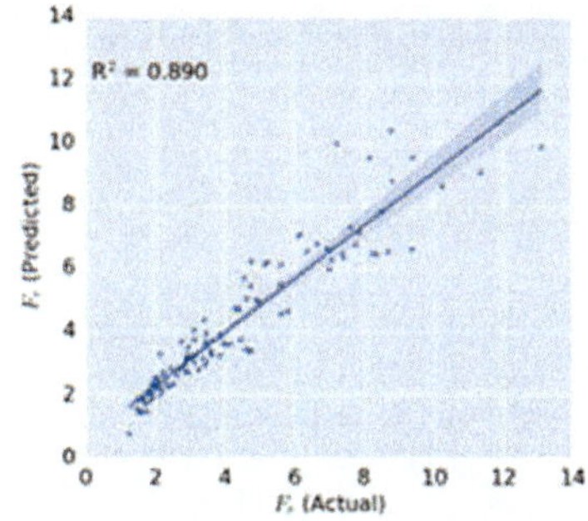

(d) Regression plot for M5P

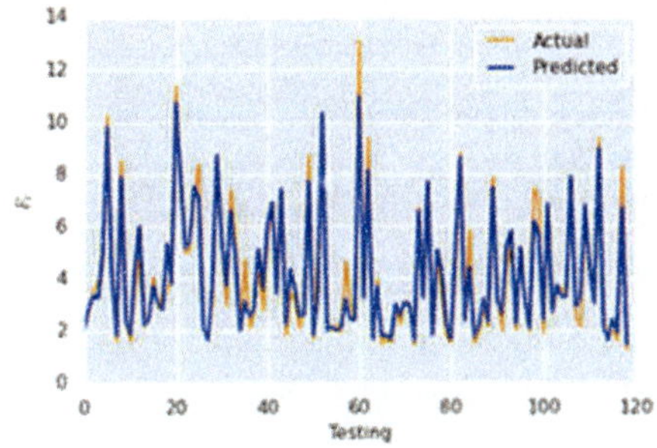

(e) Line plot for RF

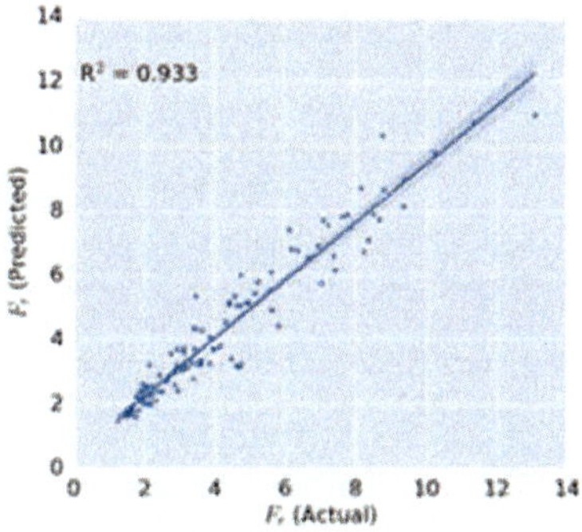

(f) Regression plot for RF

Fig. 2 Line plots and regression plots of actual and predicted particle F_r in a sewer system for methods

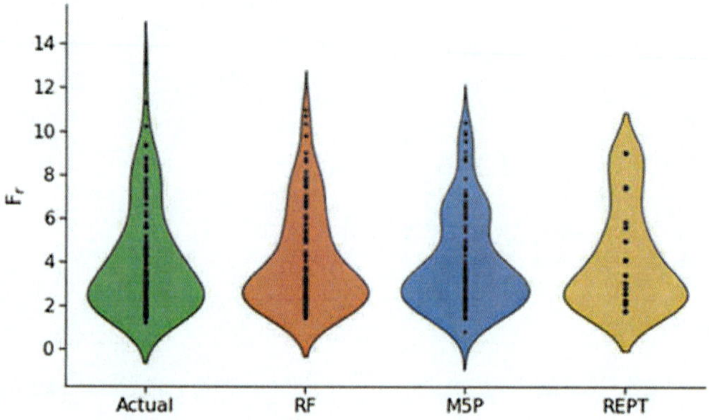

Fig. 3 Violin plot of the actual data values and predicted data values for each method to predict particle F_r

References

1. Ab Ghani AA (1993) Sediment transport in sewers. Ph.D. thesis. Newcastle University
2. Ackers J, Butler D, May R (1996) Design of sewers to control sediment problems. Construction Industry Research and Information Association, London
3. Butler D, May R, Ackers J (2003) Self-cleansing sewer design based on sediment transport principles. J Hydraul Eng 129(4):276–282
4. Ebtehaj I, Bonakdari H, Safari MJS, Gharabaghi B, Zaji AH, Madavar HR, Khozani ZS, Eshaghi MS, Shishegaran A, Mehr AD (2020) Combination of sensitivity and uncertainty analyses for sediment transport modeling in sewer pipes. Int J Sediment Res 35(2):157–170
5. Ebtehaj I, Bonakdari H, Shamshirband S (2016) Extreme learning machine assessment for estimating sediment transport in open channels. Eng Comput 32(4):691–704
6. El-Zaemey AKS (1991) Sediment transport over deposited beds in sewers. Ph.D. thesis. Newcastle University
7. Freund Y, Schapire RE (1997) A decision-theoretic generalization of on-line learning and an application to boosting. J Comput Syst Sci 55(1):119–139
8. Hussain D, Khan AA (2020) Machine learning techniques for monthly river flow forecasting of Hunza river, Pakistan. Earth Sci Inf 13(3)
9. Kargar K, Safari MJS, Mohammadi M, Samadianfard S (2019) Sediment transport modeling in open channels using neuro-fuzzy and gene expression programming techniques. Water Sci Technol 79(12):2318–2327
10. May R (1993) Sediment transport in pipes, sewers and deposited beds. Report no
11. May RW, Ackers JC, Butler D, Siân J (1996) Development of design methodology for self-cleansing sewers. Water Sci Technol 33(9):195
12. Montes C, Vanegas S, Kapelan Z, Berardi L, Saldarriaga J (2020) Non-deposition self-cleansing models for large sewer pipes. Water Sci Technol 81(3):606–621
13. Nalluri C, El-Zaemey A, Chan H (1997) Sediment transport over fixed deposited beds in sewers—An appraisal of existing models. Water Sci Technol 36(8–9):123–128
14. Perrusquía G (1992) An experimental study on the transport of sediment in sewer pipes with a permanent deposit. Water Sci Technol 25(8):115–122
15. Perrusquia G (1993) An experimental study from flume to stream traction in pipe channels. Report no. Chalmers University of Technology

16. Safari MJS (2019) Decision tree (DT), generalized regression neural network (GR) and multivariate adaptive regression splines (MARS) models for sediment transport in sewer pipes. Water Sci Technol 79(6):1113–1122
17. Safari MJS, Mehr AD (2018) Multigene genetic programming for sediment transport modeling in sewers for conditions of non-deposition with a bed deposit. Int J Sediment Res 33(3):262–270
18. Safari MJS, Mohammadi M, Ab Ghani A (2018) Experimental studies of self-cleansing drainage system design: a review. J Pipeline Syst Eng Practice 9(4):04018017

Estimation of Time-Dependent Pier Scour Depth Using Ensemble and Boosting-Based Data-Driven Approaches

Sanjit Kumar, Mayank Agarwal, Vishal Deshpande, and Manish Kumar Goyal

Abstract The scour phenomenon around the vertical piles in rivers and oceans can have a significant impact on the stability of the structures. As a result, accurate prediction of the scour depth forms an important challenge in the design of piles. Various empirical approaches proposed in the literature are often confined to specific environmental and bed conditions. So, when such empirical approaches are applied to a new environment, they either underestimate or overestimate the scour depth, which may lead to improper design of the piles. This study aims to develop two data-driven approaches: extra trees regressor (ETR) and extreme gradient boosting regressor (XGBR), which are ensemble and boosting-based machine learning-based approaches, respectively, to estimate the temporal variation of pier scour depth with non-uniform sediments under clear water conditions. The motivation behind using a boosting and an ensemble-based approach is that they provide superior results as compared to standard machine learning-based approaches. The dataset is compiled using various sources from existing literature. For each of the data-driven approaches, nine different combinations of features (shallowness of the flow, sediment coarseness, densimetric Froude number, sediment particle size distribution, pier Froude number, and three different dimensionless time scales) are tried in order to determine the best combination that can be used for prediction of scour depth. Both extra trees regressor and XGBR excel at prediction of the scour depths, but extra trees regressor performs better in most of the models as compared to XGBR. The highest r^2 and NSE across

S. Kumar (✉) · M. Agarwal (✉)
Department of Computer Science and Engineering, Indian Institute of Technology Patna, Patna 801106, India
e-mail: sanjit_2021cs07@iitp.ac.in

M. Agarwal
e-mail: mayank265@iitp.ac.in

V. Deshpande (✉)
Department of Civil and Environmental Engineering, Indian Institute of Technology Patna, Patna 801106, India
e-mail: deshpande@iitp.ac.in

M. K. Goyal (✉)
Department of Civil Engineering, Indian Institute of Technology Indore, Indore 453552, India
e-mail: mkgoyal@iiti.ac.in

P. V. Timbadiya et al. (eds.), *Geospatial and Soft Computing Techniques*, Lecture Notes in Civil Engineering 339, https://doi.org/10.1007/978-981-99-1901-7_48

nine models for extra trees regressor are 0.956 and 0.9544, respectively, while in the case of XGBR, the highest r^2 and NSE across nine models are reported as 0.9474 and 0.9461, respectively.

Keywords Data driven · Scour depth · Froude number · Machine learning · Empirical equation

1 Introduction

Bridge failure is mostly due to scouring around the pier during floods. Many researchers Khosravi et al. [1], Kothyari et al. [2], and Ahmadianfar et al. [3] have included this problem in their studies. A schematic diagram of pier of bridge, vortex, and scour depth with different times is shown in Fig. 1. The velocity of clear water in that scour depth occurs is less than the threshold velocity to sediment transport. So no sediment moves from upstream to hole of scour after floods.

Study of time-dependent scour depth and time is divided into three stages according to changes in scour depth. The first stage, initial to time t_1, is where scour depth increases faster. The second stage, from time (t_1) to time (t_2), is where scour rate increases fast. The last stage is an equilibrium stage where depth rate approximately stops; equilibrium scour depth and time are represented as $d_{s,e}$ and t_e, respectively.

Pandey et al. [4] introduced to prediction of maximum scour depth near spur dikes in uniform bed sediment. Boosting regression tree (BRT) and bagging regression tree (BGT) novel multilayer stacked generalization frameworks are developed to model. Tao et al. [5] proposed to XGB and genetic algorithm (GA) for modeling scour depths

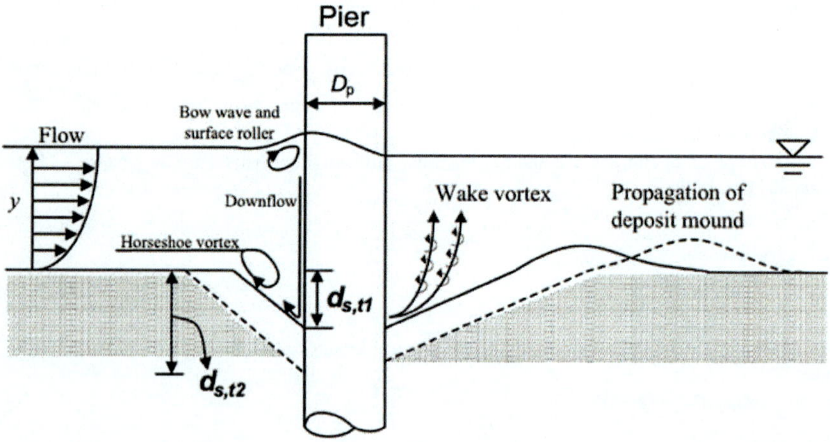

Fig. 1 Familiar with basic term of time-dependent scour depth

under submerged weir. Lim and Cheng [6] proposed a semi-empirical equation for the estimation of time-averaged equilibrium scour depth at 45° wing wall abutment under live bed conditions. Ballio et al. [7] experimentally investigated temporal development of local scour at a vertical wall abutment. Machine learning is used in scouring by many authors (Ebtehaj et al. [8], Hong et al. [9], Goyal and Ojha [10], Kumar et al. [11], Dang et al. [12], Qaderi et al. [13], and Sattar et al. [14]).

With correct prediction of scour depth, we prevent the damage of bridge that stand on piers and abutments. In this study, we proposed two methods that help in evaluation of sour depth. We describe basic definition of method. Convert basic dataset into dimensionless that used in this study. Split dataset into training and testing dataset that used in training and testing stage, respectively. We compare the accuracy of each method with all models and finally conclude the best model with methods for better prediction of time-dependent scour depth, which can be help in saving money and time of construction of bridge and human life with other things damage caused by bridge collapse.

2 Materials and Methods

2.1 *Study Area and Data Source*

Scour depth (d_{st}) varies with time around bridge pier with mixture of a non-cohesive sediment under conditions of steady flow which can be influenced by multi-groups, viz. properties of bed of pier (d_{50}, σ_g, ρ_s, U_c), properties water flow around pier (ρ, U, μ, g, y), , bridge pier geometry (Al, Sh, D_p), and time (t, t_R), where the mean of sediment diameter is represented as d_{50}; standard deviation of distribution of the sediment size as σ_g; the density of the sediment as ρ_s; the critical average velocity of approach flow as U_c for movement of bed sediment; the fluid density as ρ; the average velocity of water at the depth over a flow measuring weir is recorded represented as U; the viscosity as μ; the gravitational acceleration as g; the undisturbed depth of approach flow as y; the diameter of the pier as D_p; the alignment as Al and Sh as shape of the pier; the time as t; moreover, the timescale as t_R for scour to develop. From these fundamental parameters (see Table 1) of this study, we calculate the dimensionless parameters (see Table 2).

This study uses 555 data points, with 71 points from Oliveto and Hager [15], 46 points from Kothyari [16], and 438 points from Chang et al. [17]. The importance of this study is to investigate the effectiveness of ensemble algorithm ETR and XGBR on time-dependent scour depth. Overall, approximately 58% of the data used for training of model (23, 33, and 232 Kothyari [16], Oliveto and Hager [15], and Chang et al. [17], respectively) and the rest 42% of whole dataset (23, 38, and 206 for Kothyari [16], Oliveto and Hager [15], and Chang et al. [17], respectively) used for testing. Given the uniformity of the sediments, the whole dataset to divide into

Table 1 Basic characteristics of the fundamental dataset collected from the sources

Parameters	Min.	Max.	Mean	STD
D_p	0.065	0.17	0.1027	0.0133
U	0.0093	0.5863	0.2845	0.0821
y	0.001	0.301	0.171	0.068
d_{50}	0	0.003	0.001	0.001
σ_g	1.2	3	2.047	0.678
T	1	32,170.97	9162.469	7907.171

Table 2 Basic characteristics of calculated the dimensionless dataset from fundamental dataset

	Parameters	Min.	Max.	Mean	STD
Input	y/D_p	0.006	3.014	1.7	0.687
	D_p/d_{50}	35.484	414.634	134.634	53.935
	F_d	0.042	4.426	2.486	0.65
	$F_d - F_{d\beta}$	−1.637	2.421	0.51	0.579
	$U/\sqrt{gD_p}$	0.009	0.564	0.285	0.081
	σ_g	1.2	3	2.047	0.678
	$\log(T_1)$	−0.002	4.768	3.689	0.826
	$\log(T_2)$	−0.145	4.942	3.492	0.825
	$\log(T_3)$	0.03	4.725	3.645	0.847
Output	d_{st}/D_p	0	1.42	0.485	0.37

288 data points for training and 267 data points for verification with sediment non-uniformity parameter σ_g. The σ_g-values for homogeneous sediment are less than and equal to 1.4 (Chiew [18] and Chiew and Melville [19]). In this study, it range from 1.2 to 3, which means datasets contains both homogeneous and heterogeneous sediments.

2.2 Selection of Input Parameters

Examination of the correlation coefficient value on the prediction of time-dependent scour depth based on dimensionless parameters shows that a higher correlation coefficient value has a more significant influence in predicting scour depth. Input combination was created based on the correlation coefficient, and the highest correlation coefficient value was found to select the best input combination from five different input combinations. Figure 3 shows correlation coefficient value of all dimensionless parameters. Out of these, $\log(T_n)$ is a higher correlation coefficient value and first input parameter of the model. For the second parameter, select the second highest correlation coefficient value. This will continue until the parameter has the least correlation coefficient value. In all steps, get five input combinations. The selection of input no. 5 (see Table 3), an effective input combination, has a more significant effect in predicting scour depth. Here $\log(T_n)$ shows that it can be one of $\log(T_1)$, $\log(T_2)$, and $\log(T_3)$. And Froude is one of F_d, $F_d - F_{d\beta}$, and $U/\sqrt{gD_p}$.

All parameters have individual significance for the results. Figure 3 shows the Pearson correlation coefficient (r) value, which is used to calculate the importance of the parameters to the results. Furthermore, it was observed that $\log(T_3)$ has an r-value of 0.605, higher than the $\log(T_1)$ ($r = 0.538$), σ_g ($r = -0.469$), $\log(T_2)$ ($r = 0.464$), F_d ($r = 0.452$), y/D_p ($r = 0.429$), $F_d - F_{d\beta}$ ($r = 0.373$), $U/\sqrt{gD_p}$ ($r = 0.273$), and D_p/d_{50} ($r = 0.176$), which means that it has a greater impact on the results. Finally, we get five input combinations (see Table 3). With the help of the best input combination, we obtain nine models (see Table 4). This is done to allow for more comprehensive analysis about the parameters affecting the scour depth.

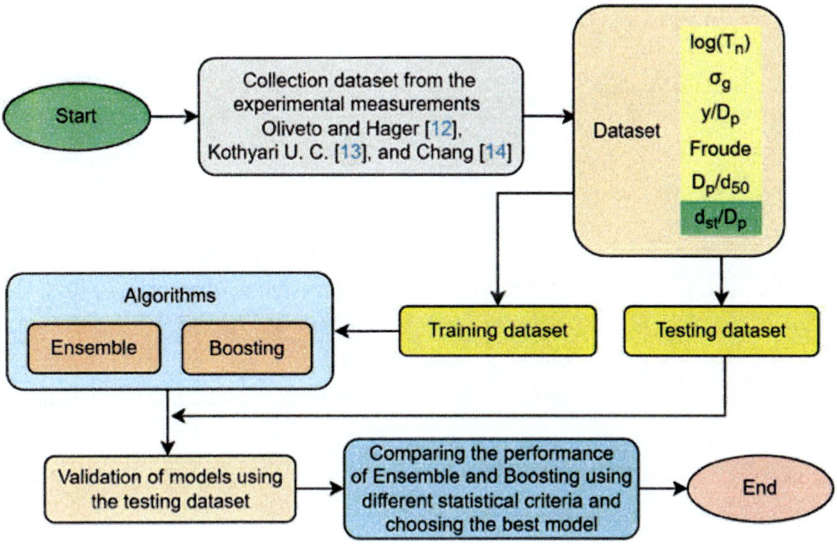

Fig. 2 Work flow chart of this study

Table 3 Input combination basis of r-value

Input no.	Input parameters
1	$\log(T_n)$
2	$\log(T_n)$, σ_g
3	$\log(T_n)$, σ_g, y/D_p
4	$\log(T_n)$, σ_g, y/D_p, Froude
5	$\log(T_n)$, σ_g, y/D_p, Froude, D_p/d_{50}

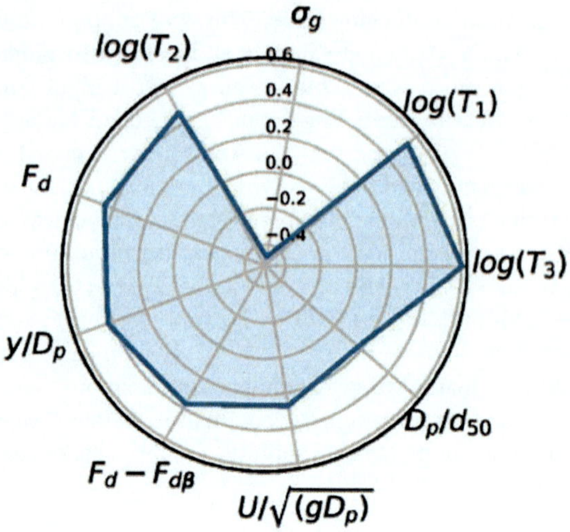

Fig. 3 Radar chart of correlation coefficient of input parameters

Table 4 Model list by input combination

Model	Inputs used
M1	$d_{st}/D_p = f(\log(T_1), \sigma_g, y/D_p, F_d, D_p/d_{50})$
M2	$d_{st}/D_p = f(\log(T_2), \sigma_g, y/D_p, F_d, D_p/d_{50})$
M3	$d_{st}/D_p = f(\log(T_3), \sigma_g, y/D_p, F_d, D_p/d_{50})$
M4	$d_{st}/D_p = f(\log(T_1), \sigma_g, y/D_p, (F_d - F_{d\beta}), D_p/d_{50})$
M5	$d_{st}/D_p = f(\log(T_2), \sigma_g, y/D_p, (F_d - F_{d\beta}), D_p/d_{50})$
M6	$d_{st}/D_p = f(\log(T_3), \sigma_g, y/D_p, (F_d - F_{d\beta}), D_p/d_{50})$
M7	$d_{st}/D_p = f(\log(T_1), \sigma_g, y/D_p, U/\sqrt{gD_p}, D_p/d_{50})$
M8	$d_{st}/D_p = f(\log(T_2), \sigma_g, y/D_p, U/\sqrt{gD_p}, D_p/d_{50})$
M9	$d_{st}/D_p = f(\log(T_3), \sigma_g, y/D_p, U/\sqrt{gD_p}, D_p/d_{50})$

3 Model Description

3.1 Extra Trees Regressor (ETR)

Extra trees is an ensemble machine learning algorithm that combines predictions from multiple decision trees. It is analogous to the popularly used random forest algorithm. It can often perform as well or better than the random forest algorithm, although it uses a more simplistic algorithm to construct decision trees used as members of the ensemble machine algorithm. It is also easy to use because it has several vital hyperparameters and analytical reasoning techniques to compose those hyperparameters.

Extremely random trees, or extra trees for short, is an ensemble machine learning algorithm. In particular, it is a set of decision trees associated with other sets of decision tree algorithms such as bootstrap aggregation (bagging) and random forest. The extra trees algorithm works by generating many decision trees without removing them from the training dataset. The predictions are made by averaging the predictions of decision trees in the case of regression or majority vote in classification.

3.2 Extreme Gradient Boosting Regressor (XGB Regressor)

XGBR stands for extreme gradient boosting regressor, which was introduced by Chen and Guestrin [20]. Since it introduces, it has become one of the most popular assembly algorithms. This model's first iteration starts with a combination of weak regressor like other ensemble models and tries to fit complete data the loss functionaries to reduce these error residuals by adding weaker regressors. New weak learners are added to focus on areas where existing learners are not doing well. After several iterations, we see that the model fits the data better. This process is done iteratively until the residuals are zero. In this model, the tree can have a differing number of leaf nodes.

Moreover, they eliminate the weight of the tree calculated with less evidence which decreases further. The Newton boosting function uses an approximation of Newton–Raphson that gives a direct path to the minimum value of the gradient. XGBR can use additional random parameters to reduce the correlation between trees.

Figure 2 displays a flowchart for the data collection, training and testing, proposed methods, and selection of best models with respect to methods to the evaluation of scour depth around piles.

3.3 Model Evaluation Criteria

All approaches are finally compared in terms of their power of prediction. Here a portion of 288 rows of the original dataset was used for training purposes, and 267 rows of the original dataset were used for testing purposes. We used both graphical methods (line graphs, scatter plots, and box plots) and quantitative metrics to evaluate the performance of each approach. A ranking of performance was achieved using quantitative metrics including r-square (r^2), Pearson correlation coefficient (r), Nash–Sutcliffe efficiency (NSE), mean absolute error (MAE), mean square error (MSE), and percent bias (Pbias) that was computed as follows:

Pearson correlation coefficient (r):

$$r = \frac{\sum_{i=1}^{n}(y_i - \overline{y})\left(\hat{y}_i - \overline{\hat{y}}\right)}{\sqrt{\sum_{i=1}^{n}(y_i - \overline{y})^2}\sqrt{\sum_{i=1}^{n}\left(\hat{y}_i - \overline{\hat{y}}\right)^2}}$$

r-square (r^2):

$$r^2 = \left(\frac{\sum_{i=1}^{n}(y_i - \overline{y})\left(\hat{y}_i - \overline{\hat{y}}\right)}{\sqrt{\sum_{i=1}^{n}(y_i - \overline{y})^2}\sqrt{\sum_{i=1}^{n}\left(\hat{y}_i - \overline{\hat{y}}\right)^2}}\right)^2$$

Nash–Sutcliffe efficiency (NSE):

$$\text{NSE} = 1 - \frac{\sum_{i=1}^{N}(y_i - \hat{y}_i)^2}{\sum_{i=1}^{N}(y_i - \overline{y})^2}$$

Mean absolute error (MAE):

$$\text{MAE} = \frac{1}{N}\sum_{i=1}^{N}\left|y_i - \hat{y}_i\right|$$

Mean square error (MSE):

$$\text{MSE} = \frac{1}{N}\sum_{i=1}^{N}(y_i - \hat{y}_i)^2$$

Percent bias (Pbias):

$$\text{Pbias} = 100 \times \frac{\sum_{i=1}^{N}(y_i - \hat{y}_i)}{\sum_{i=1}^{N} y_i}$$

4 Results and Discussions

In this study, ETR and XGBR both these data-driven methods are used to the prediction of time-dependent scour depth around bridge pier under clear water steady flow.

ETR method uses ensemble technique for training and updating the weight for testing. As we know, ensemble learning technique used weak learner to make strong learner in parallel form. In ETR, extra tree is used as weak learner. Number of estimator is 90 and max feature selection is auto are tune to training purpose. In case of XGBR, it uses boosting technique. In this weak learner used sequential to make strong leaner. The parameter set as learning rate is 0.2 with max depth is 5.

Here six model evaluation criteria are used to getting a better performance of method with different models. These are r^2, r, NSE, MAE, MSE, and Pbias. ETR has best r^2 value (0.9560) with model 6 and other performance evaluation values ($r = 9.777$), (NSE = 0.9544), (MAE = 0.0397), (MSE = 0.0060), and (Pbias = 2.4038) followed by r^2 value 0.9546, 0.9517, 0.9511, 0.9486, 0.9419, 0.9410, and 0.9384 of models 9, 5, 8, 2, 7, 4, and 1, respectively. XGBR has best r^2 value (0.9474) with model 4 and other performance evaluation values ($r = 9.734$), (NSE = 0.9461), (MAE = 0.0420), (MSE = 0.0071), and (Pbias = 2.9243) followed by r^2 value 0.9456, 0.9402, 0.9399, 0.9397, 0.9326, 0.9208, 0.9174, and 0.9137 of models 7, 2, 1, 8, 5, 9, 3, and 6, respectively. All these values are listed in Table 5. With these values, we can get that ETR outperforms than XGBR method. XGBR with model 4 and model 7 better perform compared to ETR with model 7, 4, and 1.

Table 5 Model evaluation criteria values of proposed method in testing phase

Algorithm	Models	r^2	r	NSE	MAE	MSE	Pbias
ETR	1	0.9384	0.9687	0.9377	0.0437	0.0082	1.9617
	2	0.9486	0.9740	0.9465	0.0406	0.0070	3.2139
	3	0.9528	0.9761	0.9510	0.0426	0.0064	2.4240
	4	0.9410	0.9701	0.9406	0.0448	0.0078	1.6457
	5	0.9517	0.9756	0.9496	0.0376	0.0066	3.3444
	6	**0.9560**	0.9777	0.9544	0.0397	0.0060	2.4038
	7	0.9419	0.9705	0.9411	0.0438	0.0077	1.5134
	8	0.9511	0.9753	0.9494	0.0398	0.0066	2.6965
	9	0.9546	0.9770	0.9529	0.0418	0.0062	2.8440
XGBR	1	0.9399	0.9695	0.9389	0.0460	0.0080	2.4616
	2	0.9402	0.9697	0.9374	0.0451	0.0082	4.2372
	3	0.9174	0.9578	0.9165	0.0460	0.0110	1.3409
	4	**0.9474**	0.9734	0.9461	0.0420	0.0071	2.9243
	5	0.9326	0.9657	0.9304	0.0475	0.0091	3.7868
	6	0.9137	0.9559	0.9123	0.0478	0.0115	2.5995
	7	0.9456	0.9724	0.9444	0.0422	0.0073	2.8287
	8	0.9397	0.9694	0.9358	0.0450	0.0084	4.9877
	9	0.9208	0.9596	0.9197	0.0453	0.0106	2.0470

Bold indicates the best performance model

Figure 4 presents line plot of actual and predicted value of d_{st}/D_p with scatter plot of actual versus predicted values with proposed method and best model. From line plot, we can visualize predicted values near to actual values. In scatter plot, predicted and actual values less deviate to regressor line plot as $y = x$ line. Figure 5 helps to find best model among all model with proposed methods. All values present with star symbol and the best values circle with hollow circle that is the closest to 1 in case of r^2. Figure 5a shows circle on model 6. So ETR better performs with model 6, and Fig. 5b shows circle on model 4. Hence, XGBR better performs with model 4.

In Fig. 6, relative errors of training and testing dataset with ETR method model 4 and XGBR method model 6 are evaluated with $RE = 100 * (\hat{y}_i - y_i)/y_i$, where y_i is time-dependent scour depth (d_{st}/D_p) around bridge piers under clear water. The

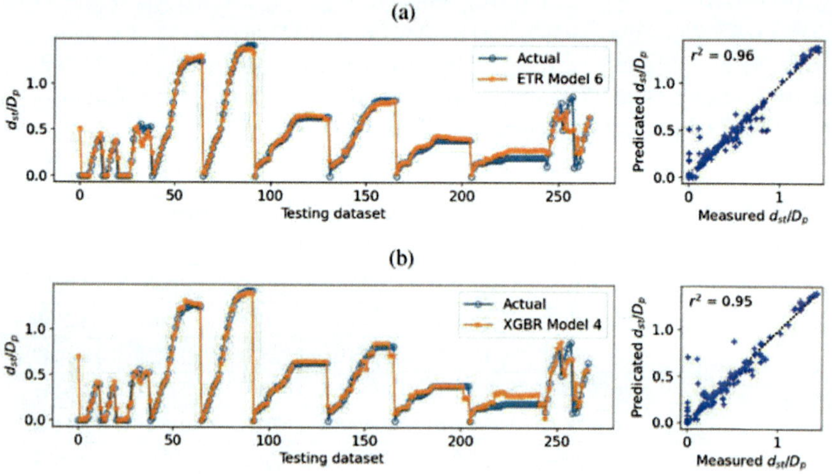

Fig. 4 Line plot of actual and predicted values with scatter plot of actual versus predicted values with method **a** ETR model 6 and **b** XBGR model 4

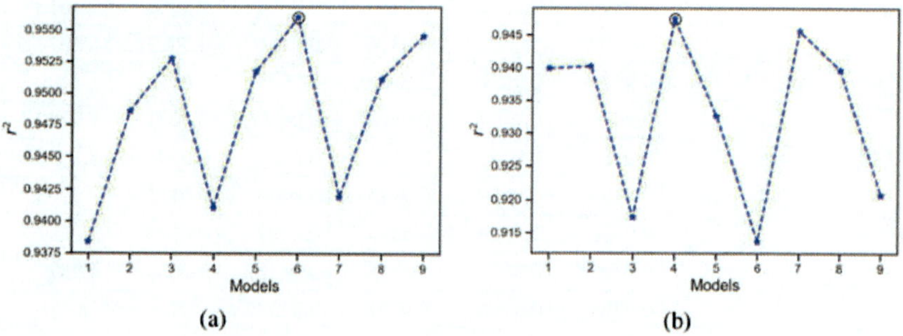

Fig. 5 Compare all models of the proposed methods **a** ETR and **b** XGBR with r^2

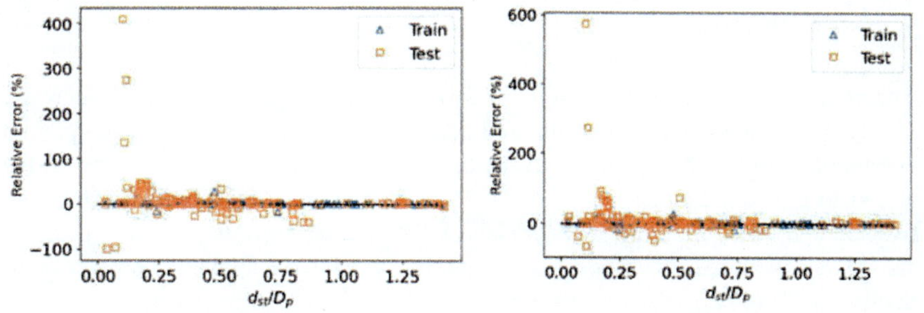

(a) (b)

Fig. 6 Relative error of time-dependent scour depth for **a** ETR method model 4 and **b** XGBR method model 6 with testing dataset

straight line at 0 separates out the positive and negative relative error values. Figure 6a presents ETR method. In this, we can visualize that more points are around the zero value of relative error compared to Fig. 6b, which presents XGBR method. So we can say that ETR method with model 6 outperforms XGBR method with model 4.

Figure 7 shows standard deviation with correlation value of actual value of scour depth and predicted value of scour depth by ETR method with model 6 and XGBR method with model 4. From this figure, we visualize that standard deviation of ETR is less than XGBR, and correlation value of ETR is greater than XGBR method. So from this figure, we can conclude that ETR has better performance than XGBR.

Fig. 7 Taylor diagram of ETR and XGBR methods with testing dataset

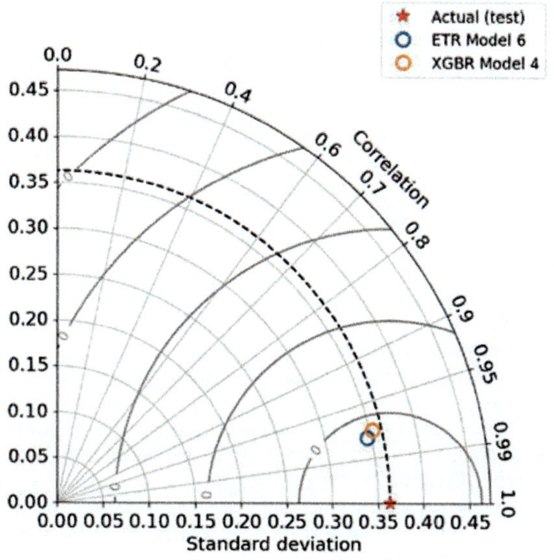

5 Conclusions

The prediction of time-dependent scour depth around bridge piers under clean water steady flow is complex. Due to its nonlinear structure, so basic empirical equation not able to good prediction of scour depth, that can be reason of damage the bridge and loss of time and money spend on bridge construction. In this study, we proposed two methods with nine different models, ETR method used as ensemble learning and XGBR method used as boosting technique, which were applied to predict scour depth. The following conclusions are derived from the foregoing study:

1. The $log(T_n)$ has the higher r-value followed by σ_g, y/D_p, Froude, and D_p/d_{50}.
2. The overall result concludes that ETR method gives the best prediction at equilibrium time and $F_d - F_{d\beta}$.
3. XGBR method gives a better result at sharp increasing initial time of scouring with $F_d - F_{d\beta}$. That is good for ETR with sharp increasing initial time T_1.
4. Comparison of all proposed methods that ETR predict outperforms XGBR.

References

1. Khosravi K, Khozani ZS, Mao L (2021) A comparison between advanced hybrid machine learning algorithms and empirical equations applied to abutment scour depth prediction. J Hydrol 126100
2. Kothyari UC, Garde RCJ, Ranga Raju KG (1992) Temporal variation of scour around circular bridge piers. J Hydraul Eng 118:1091–1106
3. Ahmadianfar I, Jamei M, Karbasi M, et al (2021) A novel boosting ensemble committee-based model for local scour depth around non-uniformly spaced pile groups. Eng Comput 1–23
4. Pandey M, Azamathulla HM, Chaudhuri S et al (2020) Reduction of time-dependent scour around piers using collars. Ocean Eng 213:107692
5. Tao H, Habib M, Aljarah I et al (2021) An intelligent evolutionary extreme gradient boosting algorithm development for modeling scour depths under submerged weir. Inf Sci (Ny) 570:172–184
6. Lim S-Y, Cheng N-S (1998) Prediction of live-bed scour at bridge abutments. J Hydraul Eng 124:635–638
7. Ballio F, Radice A, Dey S (2010) Temporal scales for live-bed scour at abutments. J Hydraul Eng 136:395–402
8. Ebtehaj I, Sattar AMA, Bonakdari H, Zaji AH (2017) Prediction of scour depth around bridge piers using self-adaptive extreme learning machine. J Hydroinf 19:207–224
9. Hong JH, Goyal MK, Chiew YM, Chua LH (2012) Predicting time-dependent pier scour depth with support vector regression. J Hydrol 468–469:241–248. ISSN 0022-1694. https://doi.org/10.1016/j.jhydrol.2012.08.038
10. Goyal MK, Ojha CSP (2011) Estimation of scour downstream of a ski-jump bucket using support vector and M5 model tree. Water Resour Manage 25:2177–2195. https://doi.org/10.1007/s11269-011-9801-6
11. Kumar S, Goyal MK, Deshpande V, Agarwal M (2023) Estimation of time dependent scour depth around circular bridge piers: application of ensemble machine learning methods. Ocean Eng 270:113611. ISSN 0029-8018. https://doi.org/10.1016/j.oceaneng.2022.113611
12. Dang NM, Tran Anh D, Dang TD (2021) ANN optimized by PSO and Firefly algorithms for predicting scour depths around bridge piers. Eng Comput 37:293–303

13. Qaderi K, Javadi F, Madadi MR, Ahmadi MM (2021) A comparative study of solo and hybrid data driven models for predicting bridge pier scour depth. Mar Georesour Geotechnol 39:589–599
14. Sattar AMA, Plesiński K, Radecki-Pawlik A, Gharabaghi B (2018) Scour depth model for grade-control structures. J Hydroinf 20:117–133
15. Oliveto G, Hager WH (2005) Further results to time-dependent local scour at bridge elements. J Hydraul Eng 131:97–105
16. Kothyari UC (1989) Scour arouand bridge piers. Ph.D. thesis. University of Roorkee, Roorkee
17. Chang W-Y, Lai J-S, Yen C-L (2004) Evolution of scour depth at circular bridge piers. J Hydraul Eng 130:905–913
18. Chiew YM (1984) Local scour at bridge piers. Publ Auckl University of New Zeal
19. Chiew YM, Melville BW (1989) Local scour at bridge piers with non-uniform sediments. Proc Inst Civ Eng 87:215–224
20. Chen T, Guestrin C (2016) Xgboost: A scalable tree boosting system. In: Proceedings of the 22nd ACM SIGKDD international conference on knowledge discovery and data mining. pp 785–794